Tools for Conservation

Environmental Management for Collections

Alternative Conservation Strategies for Hot and Humid Climates

Shin Maekawa

Vincent L. Beltran

Michael C. Henry

The Getty Conservation Institute Los Angeles

The Getty Conservation Institute works to advance conservation practice in the visual arts, broadly interpreted to include objects, collections, architecture, and sites. It serves the conservation community through scientific research, education and training, model field projects, and the broad dissemination of the results of both its own work and the work of others in the field. And in all its endeavors, it focuses on the creation and dissemination of knowledge that will benefit professionals and organizations responsible for the conservation of the world's cultural heritage.

The GCI's Tools for Conservation series provides practical procedures and methodologies for the practice of conservation. The series is specifically intended for conservators, conservation scientists, and technical experts in related fields. Previous titles in the series are *Monitoring for Gaseous Pollutants in Museum Environments*, by Cecily M. Grzywacz (2006); *Use of Oxygen-Free Environments in the Control of Museum Insect Pests*, by Shin Maekawa and Kerstin Elert (2002); *Effects of Light on Materials in Collections*, by Terry T. Schaeffer; *Color Science in the Examination of Museum Objects: Nondestructive Procedures*, by Ruth Johnston-Feller (2001); *Infrared Spectroscopy in Conservation Science*, by Michele R. Derrick, Dusan Stulik, and James M. Landry (1999); and *Thin-Layer Chromatography for Binding Media Analysis*, by Mary F. Striegel and Jo Hill (1996).

Published by the Getty Conservation Institute
Getty Publications
1200 Getty Center Drive, Suite 500
Los Angeles, California 90049-1682
www.getty.edu/publications/

Tevvy Ball, *Project Editor*
Hespenheide Design, *Designer*
Suzanne Watson, *Production Coordinator*

Distributed in the United States and Canada by The University of Chicago Press, Chicago.
Distributed outside the United States and Canada by Yale University Press, London.

Printed in China

Library of Congress Cataloging-in-Publication Data

Maekawa, Shin, 1952– author.
Environmental management for collections : alternative conservation strategies for hot and humid climates / Shin Maekawa, Vincent L. Beltran, Michael C. Henry.
pages cm — (Tools for conservation)
Includes bibliographical references and index.
ISBN 978-1-60606-434-4 (pbk.)
1. Museums—Climatic factors. 2. Museums—Collection management. 3. Museum conservation methods. 4. Humidity—Control. 5. Temperature control. 6. Cultural property—Protection. I. Beltran, Vincent L., 1974– author. II. Henry, Michael C., 1949– author. III. Getty Conservation Institute, issuing body. IV. Title. V. Series: Tools for conservation.
TH6057.M87M34 2014
069'.53—dc23

2014046477

Contents

PART 2: CASE STUDIES

Foreword

Since its earliest days, the Getty Conservation Institute (GCI) has focused on the management of museum environments as a core research activity. In 1997 the GCI initiated the Collections in Hot & Humid Environments project to investigate strategies for the preservation of the many important cultural heritage collections found in tropical and near-tropical regions. The initiative sought to define methodologies for assessing threats to collections and the buildings housing them, and to develop economically viable strategies using locally sustainable technologies to address these threats. The Alternative Climate Controls for Historic Buildings project, undertaken between 2003 and 2010, built on this work, researching specific alternatives to conventional air-conditioning systems by studying the control of relative humidity through ventilation, dehumidification, and heating, thereby allowing greater variations in temperature. This volume brings together the research from these projects.

Historically, the standards for museum environments were developed largely in Europe and North America, responding to the climatic conditions in these regions with strategies and resources that were widely available in those places at the time. However, collections in hot and humid climates do not necessarily require elaborate and expensive heating, ventilation, and air-conditioning (HVAC) systems. Research conducted at the GCI has shown that simple, reliable, sustainable, low-energy systems can provide appropriate solutions to issues of museum environmental management. Additionally, the introduction of HVAC systems can have a negative impact on historic buildings—and when these buildings contain important museum collections, this impact must be carefully weighed against any advantages climate management systems may afford.

Drawing lessons from GCI project case studies, *Environmental Management for Collections* addresses the challenges faced by managers of collections in hot and humid climates. This book provides practical guidance for architects, engineers, facilities managers, and practitioners involved in the design, installation, commissioning, and maintenance of environmental management strategies for both historic and new museum buildings. It represents many years of research by GCI scientists and our partners, led by GCI senior scientist Shin Maekawa, whom I particularly would like to thank. I would also like to acknowledge Shin's coauthors,

Vincent Beltran, assistant scientist at the GCI, and Michael C. Henry, principal of Watson & Henry Associates. All three have endeavored to synthesize years of research into this practical guide. I am also grateful to Susan Macdonald, head of Field Projects for the GCI, for helping to steer this process, and to Cynthia Godlewski, the GCI senior project manager who worked to bring this book into print.

Timothy P. Whalen
Director
The Getty Conservation Institute

Acknowledgments

Many people contributed to the development and the publication of this book.

After completing her doctoral research at the Institute of Archaeology, University of London, Franciza L. Toledo joined the Getty Conservation Institute (GCI) in 2000 to participate in projects at Hollybourne Cottage (Jekyll Island, United States) and the Historic Archive of San Cristóbal de La Laguna (Tenerife, Spain), continuing her research on the use of natural ventilation for museum collections in hot and humid climates. After returning to her native Brazil in 2002, she continued to work with the GCI as a consultant on projects at the Museu Paraense Emílio Goeldi in Belém and at the Museu Casa de Rui Barbosa in Rio de Janeiro, as well as by her participation in the GCI's 2007 Experts' Roundtable on Sustainable Climate Management Strategies. Toledo also participated in the initial preparations of this manuscript until her death in 2010. The authors acknowledge her major contribution to those projects and this publication.

In 1997, following a ten-year research project on the use of inert gases to control the biodeterioration of cultural objects, Shin Maekawa met with Nieves Valentín, a research biologist at the Instituto del Patrimonio Cultural de España, Madrid, to discuss a new collaboration to examine nontoxic methods for controlling microbial activity in historic buildings in hot and humid climates. This meeting led to a series of GCI-sponsored laboratory experiments conducted at Valentín's laboratory in Madrid, to initial field testing at the Archivo Histórico Nacional in Madrid in 1999, and to projects in Tenerife, Spain. The major contributions of Valentín and her Spanish institution are acknowledged.

The case studies presented in part 2 of this publication were the result of collaborations with local organizations and professionals. While their participation and contributions are noted at the end of each case study, the following individuals must be acknowledged here for their exceptional efforts toward making these case studies possible:

Maria Garcia Morales, then a conservator for the Organismo Autónomo de Museos y Centros of Tenerife, and Rafael Martín Cantos, then chief paper conservator for the Municipality of San Cristóbal de La Laguna, for their coordination and participation during field trials in San Cristóbal de La Laguna and Valle Guerra on Tenerife Island, as well as for the 2007 Experts' Roundtable on Sustainable Climate Management Strategies, also organized on Tenerife.

Warren Murphey, then senior director of operations and museum for the Jekyll Island Authority, and John Hunter, current director of the same organization, for their collaboration throughout nearly ten years of testing and trials of alternative environmental management strategies at Hollybourne Cottage.

Henry Ng, vice president of the World Monuments Fund (WMF), and T. K. McClintock, contracting project manager and president of Studio TKM, for their management and coordination of the Juanqinzhai project in Beijing, China. This project was part of the Qianlong Garden restoration project, which is collaboratively funded and operated by the WMF and the Palace Museum in Beijing.

Ana Pessoa, director of the Memories and Information Department at the Fundação Casa de Rui Barbosa, and Claudia Carvalho, its architect, for their collaboration and local project operation at the Museu Casa de Rui Barbosa in Rio de Janeiro.

Peter Toledo, former director of the Museu Paraense Emílio Goeldi, and Lucia van Velthem, former curator of the Goeldi's Amazonian ethnographic collection, for their collaboration and local coordination at the museum's storage facility.

The above two Brazilian projects were generously funded by Fundação Vitae (São Paulo, Brazil), which supplied both equipment and personnel. Gina Machado, former project manager for the foundation, closely collaborated with the GCI project team throughout the course of the projects. Without this funding and collaboration, these projects would not have been possible.

At the GCI, the initial project proposal was developed with the encouragement of Alberto de Tagle, former director of Science. Following his departure from the GCI, Jeanne Marie Teutonico, the associate director, continued to support the project and strongly encouraged this publication. Preparation of the manuscript was made possible by the enthusiastic and patient backing of Susan Macdonald, head of Field Projects. Publication was diligently coordinated at the GCI by Cynthia Godlewski, senior project manager, and at Getty Publications by Tevvy Ball, editor, and Suzanne Watson, senior production coordinator.

Editing of the manuscript was conducted by Elizabeth Maggio, our highly skilled technical editor, and Gary Hespenheide was the book's designer. Additionally, we would like to acknowledge our extremely thorough and helpful external reviewers.

Finally, but not least, all three coauthors would like to acknowledge the support and encouragement of each of our families.

Shin Maekawa
Vincent L. Beltran
Michael C. Henry

Introduction

Hot and humid climates pose unique challenges in environmental management for the preservation of museum collections and heritage buildings. As shown in figure 1, these climate conditions prevail over large expanses of the world where many museums are located.

Hot and humid climates are abundant in thermal energy and moisture, which are the enabling factors for increased biological activity and accelerated chemical reactions that damage collections materials. In seasonally hot and humid climates, humidity variation can also cause large dimensional changes in hygroscopic materials. Many museums in these geographical areas contain objects that were created and used locally and were historically exposed to environmental conditions that differ significantly from the conventionally accepted museum environmental specifications of 21°C and 50% RH—recommendations that were primarily developed for major European, British, and North American museums in cooler and less humid climates (Brown and Rose 1996). In hot and humid climates, environmental conditions appropriate for the object and region should be established for conserving collections and historic interiors.

In hot and humid climates, the use of conventional air-conditioning systems to control interior environments for collections, especially in heritage buildings, can introduce unexpected risks to both the buildings and the collections. These risks include increased microbial damage to collections and building interiors; the mobilization of salts to buildings' surfaces, damaging building materials and architectural finishes; and dimensional changes in hygroscopic materials. The challenges of providing appropriate collections environments in hot and humid climates are compounded by the use of air-conditioning for the thermal comfort of a building's occupants, especially when cooling without adequate dehumidification. When conserving cultural heritage in hot and humid climates, stewards of collections and heritage buildings must understand the potential risks so they can select and implement appropriate environmental management strategies for conservation.

For conservation environments in hot and humid climates, a number of simple, reliable, low-energy alternatives to conventional air-conditioning systems have been identified, developed, and tested for collections and heritage buildings. These strategies include combinations of dehumidifiers, fans, and heaters. While this publication is specific to hot

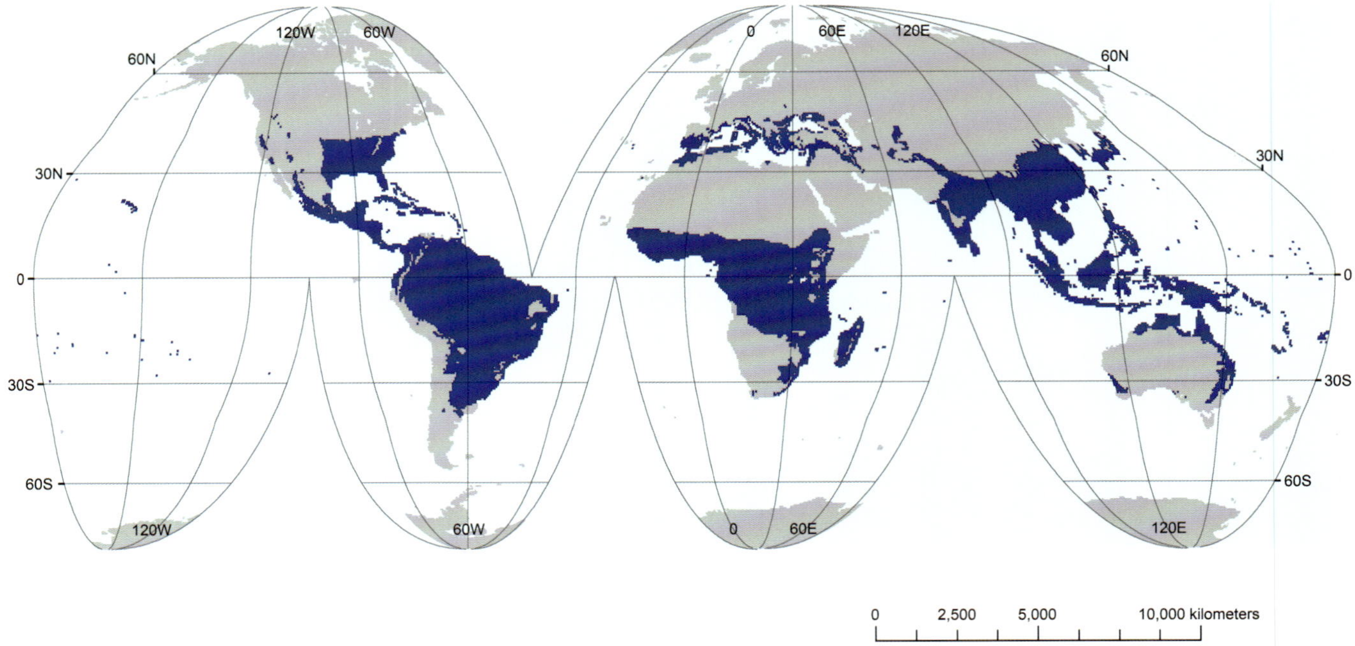

Figure 1

An approximation of hot and humid climates (in blue) based on a world map of Köppen-Geiger climate classifications. Adapted from Kottek et al. 2006.

and humid climates, the underlying principles, decision-making methodologies, and design processes presented here can also inform development of alternative environmental management systems for collections in other challenging climates.

Part 1—chapters 1 through 8—of this publication presents background, principles, and guidance on the application of alternative strategies; part 2 presents case studies that document the use of these strategies.

Chapter 1, "Hot and Humid Climates," presents the defining characteristics of hot and humid climates, their classification, and their geographic locations.

Chapter 2, "Risks of Hot and Humid Climates for Collections and Buildings," explores the climatic and man-made risks that hot and humid climates present to movable and immovable cultural property. The risks are prioritized from the standpoint of preventive conservation, with an emphasis on priorities for environmental management.

Chapter 3, "Balancing Risks and Establishing Conservation Priorities," reviews the underlying principles of current risk-based environmental management for museums in temperate climates and concludes by describing a conservation environment classification protocol, developed by the authors, that specifically targets risks to mixed cultural heritage collections in hot and humid climates.

Chapter 4, "Occupant Health and Comfort in Hot and Humid Climates," presents the implications of temperature and humidity on human health and occupant comfort. Understanding the physiological aspects of comfort is essential if one is to successfully resolve the conflicting requirements of conditions for comfort and conditions for collections conservation.

Chapter 5, "Psychrometric Strategies for Environmental Management," describes psychrometric objectives for collections conservation and thermal comfort and illustrates the basic psychrometric strategies for managing thermal energy and moisture, such as dehumidification, heating, cooling, and humidification.

Chapter 6, "Nonmechanical Strategies for Environmental Management," discusses the thermal and moisture performance of building envelopes commonly encountered in hot and humid climates and addresses their implications for interior environmental management, particularly for collections conservation. This chapter also discusses essential passive interior and exterior strategies, such as source control of moisture.

Chapter 7, "Mechanical Strategies for Environmental Management," describes the mechanical equipment used by dehumidification, ventilation, heating, cooling, and humidification systems to manage the interior environment. This chapter also discusses nonpsychrometric devices—such as filters, controllers, and sensors—and presents risks associated with the installation and operation of mechanical air-conditioning systems.

Chapter 8, "Design and Implementation of Environmental Management Strategies," offers guidance on how to apply the information contained in chapters 1 through 7 to create suitable environments for collections conservation in hot and humid climates. Specific examples of these approaches, in the form of case studies, are provided in part 2.

Part 2 presents seven case studies in which readers will find the background, data, and results of field testing of selected alternative approaches to the environmental management of museums and storage facilities; the case studies are all located in hot and humid climates, with two exceptions. The first is an archive in a marine climate, which presents conservation challenges related to cool damp winters. The second is a Chinese emperor's retirement studio in Beijing, a mixed-humid climate with hot and humid summers that can dessicate and embrittle objects and building fabric. These examples were research studies undertaken by the Getty Conservation Institute and project partners. They provide insights into real-world issues encountered during system implementation and operation. The case studies are:

- Hollybourne Cottage, a heritage house museum on Jekyll Island, Georgia, United States (two environmental management approaches are presented)
- Historic Archive at San Cristóbal de La Laguna, an archive in a heritage building in Tenerife, Canary Islands, Spain
- Valle Guerra museum storage for Organismo Autónomo de Museos y Centros, a mixed-collection storage in a contemporary building in Tenerife, Canary Islands, Spain
- Storage for the Museu Paraense Emílio Goeldi, an ethnographic storage room in a contemporary building in Belém, Pará, Brazil
- Library of the Museu Casa de Rui Barbosa, a suite of rooms in a heritage house museum in Rio de Janeiro, Brazil

- Juanqinzhai in Qianlong Garden, a restored heritage house museum containing an elaborate interior and an object collection in the Forbidden City, Beijing, China

The four appendices discuss environmental monitoring for alternative climate management, the Köppen-Geiger climate classification system, climate calculations for Bangkok and Istanbul, and structured decision-making strategies. A glossary is also included for consistency of terminology and ease of comprehension. Citations of books, publications, and reports on subjects related to environmental management for museums are provided in the bibliography.

This publication is specifically intended for several primary readerships, including architects, engineers, and practitioners involved in the design or rehabilitation of buildings and systems containing cultural heritage collections; collections stewards (managers and conservators) responsible for establishing environmental criteria for cultural heritage collections; and facilities managers who maintain buildings that house cultural heritage collections and operate building systems in order to maintain collections environments within the criteria.

A tacit assumption underlying this book is that the development and application of target conditions and strategies for environmental management of any project is the result of the collaborative engagement of all of the professionals listed above. It is hoped that providing information on various aspects of environmental management will engender informed discussion within multidisciplinary project teams.

References

Brown, J. P., and W. B. Rose
1996 Humidity and moisture in historic buildings: The origins of building and object conservation. *APT Bulletin* 27 (3): 12–24. A longer version of the article can be found at http://cool.conservation-us.org/byauth/brownjp/humidity1997.html.

Kottek, M., J. Grieser, C. Beck, B. Rudolf, and F. Rubel
2006 World map of the Köppen-Geiger climate classification updated. *Meteorologische Zeitschrift* 15 (3): 259–63.

PART 1

FUNDAMENTALS AND APPROACHES

Chapter 1

Hot and Humid Climates

Climate imposes most of the thermal energy and moisture loads upon a building and its interior environment, acting through the building envelope and by way of the outside air brought in for ventilation. Hot and humid climates present particularly high thermal and moisture loads that create challenges for effectively managing the interior environment for the preservation of collections and the historic building fabric, as well as for occupant comfort.

This chapter discusses the special characteristics of hot and humid climates, including:

- characterization of the thermal and moisture loads
- climate zone classification
- identification of hot and humid climate zones

In this publication, atmospheric moisture content is expressed as dew point temperature rather than as absolute moisture content.

Characterization of Thermal and Moisture Loads in Hot and Humid Climates

Climate results from the interaction of multiple large-scale environmental factors, including solar radiation, wind, moisture availability, and nighttime radiative cooling. The amount of solar radiation reaching the Earth's surface is greatest (>6 kWh/m^2/day) in the region bounded geographically by the Tropic of Cancer (latitude 23°26'16" N) and the Tropic of Capricorn (latitude 23°26'16" S).

Between the Tropic of Cancer and the Tropic of Capricorn, which delineate the tropics, the prevailing surface winds, called trade winds, blow toward the equator from the northeast in the Northern Hemisphere and from the southeast in the Southern Hemisphere (fig. 1.1). These winds move over a large expanse of ocean, which supplies water vapor to the atmosphere. As this humid air mass reaches the hot equatorial region, it is heated. The heated air rises and then cools, a process that results in the condensation of water in the form of rain. As a result, the tropics typically receive more than 1 m of precipitation annually (fig. 1.2). Because a large portion of the available solar radiation energy is expended in the evaporation of surface water, precipitation occurs in a daily cycle.

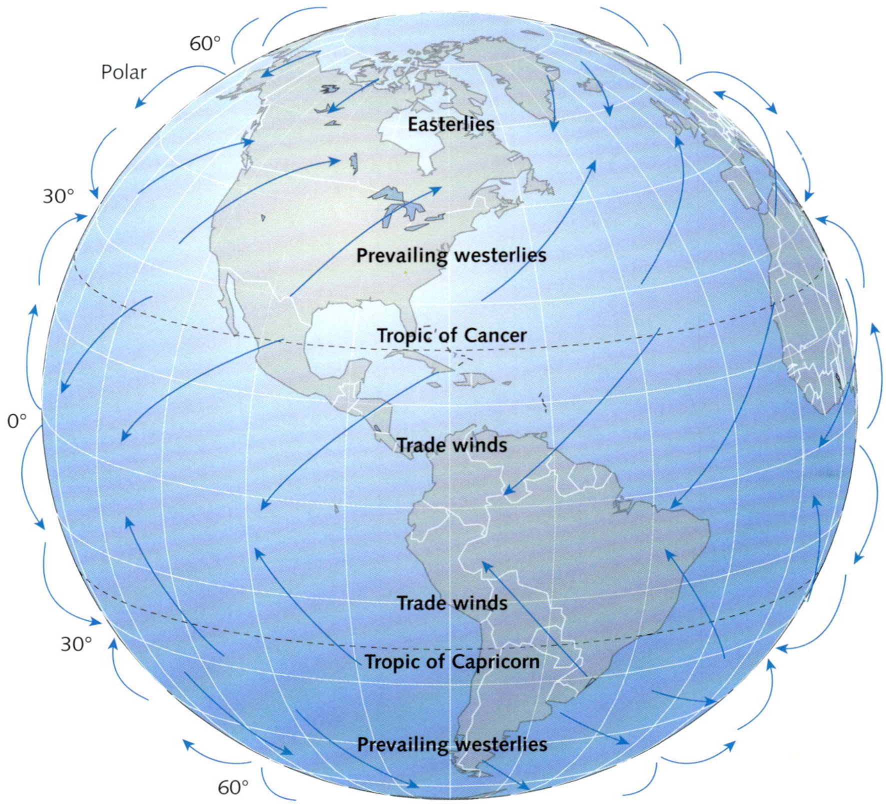

Figure 1.1
Prevailing surface winds, which are representative of winds around the world. Graphic: aerospaceweb.org (http://www.aerospaceweb.org/question/atmosphere/q0117.shtml).

Figure 1.2
Map of the world showing average annual precipitation. Source: planetolog.com/map-world-detail.php?type=RES+id=2.

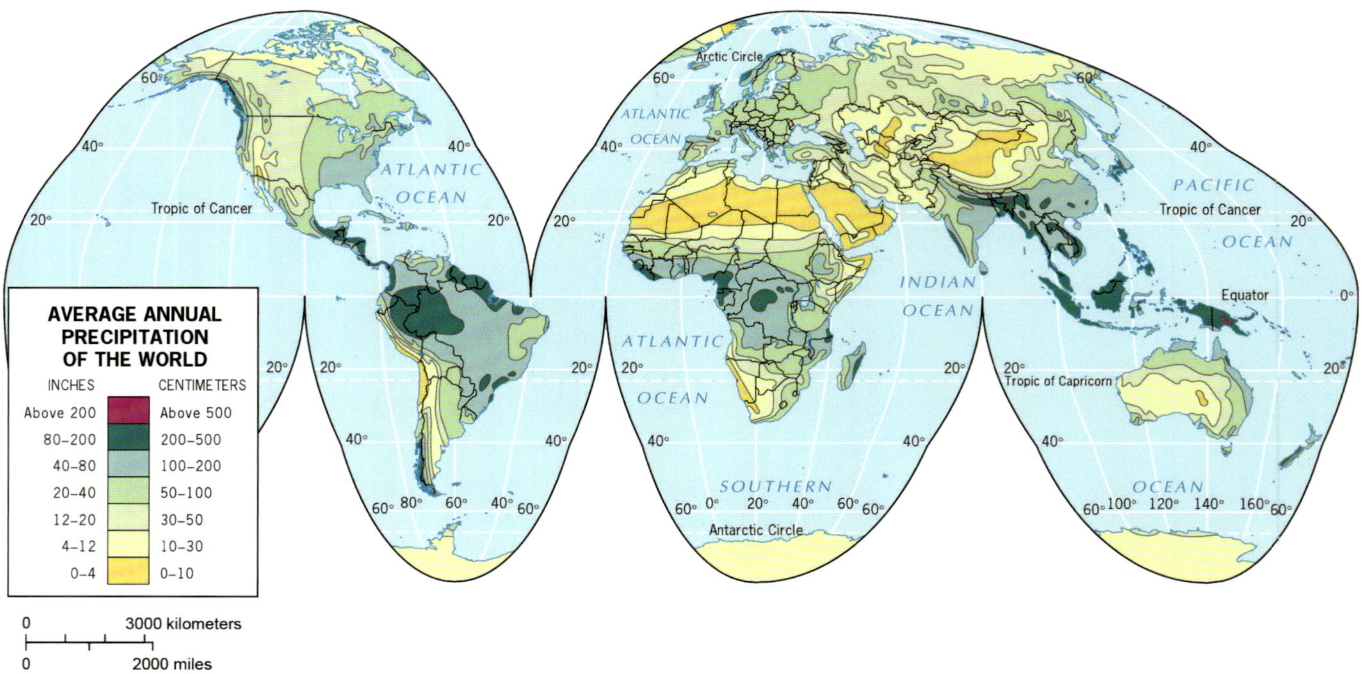

In the tropics, the air generally remains humid, and the temperature over a year typically ranges from degree Celsius values in the high tens to the mid-thirties. Owing to the suppression of radiative cooling by nighttime cloud cover and high atmospheric moisture, the diurnal temperature fluctuation is also small, typically less than 10°C.

In the tropics, localized climates can vary due to geographical influences. For example, at Belém, Brazil, which is near the Equator, high temperature and precipitation are consistent throughout the year. In contrast, at Rio de Janeiro, Brazil, which is located near the Tropic of Capricorn, temperatures are consistently high, though precipitation is seasonal. As a result, humidity is consistently high (70%–100% RH) during the wet season, but during the dry season, the humidity range is much wider (40%–100% RH). Both locations are the sites of case studies discussed in part 2.

In regions located between the latitudes of 30° and 60°, prevailing westerly winds blow from the southwest in the Northern Hemisphere and from the northwest in the Southern Hemisphere (see fig. 1.1). Weather systems, transported by the winds, produce seasonal precipitation in the regions. North and south of the tropics, the subtropic regions extend to roughly 35°N and 35°S, and the climate factors encountered in the tropics are limited to summer months in the subtropics, when solar radiation and precipitation are high. Jekyll Island, Georgia, USA, and San Cristóbal de La Laguna and Valle Guerra, both in Tenerife, Spain, are located in the subtropics. These sites are subjects of case studies described in part 2.

In the temperate regions, which extend poleward from the subtropics to the Arctic Circle (66.5°N) and Antarctic Circle (66.5°S), seasonal variability is typically more moderate; however, localized climates can show extreme differences, particularly if they are not near the ocean. For example, Beijing, located at an approximate latitude of 40°N and situated 150 km inland from the nearest large body of water (Bohai Sea), exhibits a wide variety in climate, which ranges from a monsoon-influenced hot and humid summer to a cold and dry continental winter. Beijing is the site of another case study discussed in part 2.

Frigid regions—regions north of the Arctic Circle and south of the Antarctic Circle—receive the least amount of solar energy. Since they have sufficient precipitation, the regions remain under ice or snow, which reflects the majority of solar radiation throughout the year. Therefore, the regions remain cold at all times.

Also addressed in this publication is the oceanic or maritime climate, because it too challenges conservation efforts. Although it does not have the hot summers of the tropics and subtropics, it does have cool damp winters. The oceanic climate, in opposition to the continental climate, is found in areas adjacent to or surrounded by ocean but outside the tropical latitudes. Cool temperatures and high atmospheric moisture result in high humidity in winter, which is manifested as fog or as overcast skies. In spite of being at a subtropical latitude, San Cristóbal de La Laguna in Tenerife is an example of the oceanic or maritime climate because of its 550 m elevation. San Cristóbal is discussed as a case study in part 2.

Overview of Climate Zone Classifications by Empirical Methods

Early methods of classification divided climates into five geographical zones according to latitude. This division gave rise to the standard terms of frigid, temperate, and tropic zones, as well as the subtropic, subarctic, and subantarctic zones at the edges of the temperate zones (Oliver 1991). Modern climate classification is based on empirical methods and focuses on the effects or indications of different climate types.

The most widely known empirical method is the Köppen-Geiger system of world climate classification that is used by various fields of science and engineering. The Köppen-Geiger system, developed initially in the late 1800s, is based on the distribution of native vegetation that results from climatic factors, such as precipitation amounts and patterns and monthly and annual average temperatures. A detailed description of this classification system is found in appendix 2.

In the 1980s, the US Department of Energy's Building Energy Codes Program developed an alternate empirical method of climate classification for use in selecting and designing energy-efficient residential and commercial buildings. This climate classification method utilized the cluster analyses of a large volume of climate data from cities and counties in the United States and the thermal energy criteria for the heating and cooling needs of buildings. However, the resulting number of climate zones was unwieldy, and zone classifications were not linked to those of the Köppen-Geiger system, limiting the possibility for the international application of the method.

In 2002 a new climate classification system was published by scientists at the Pacific Northwest National Laboratory (PNNL) of the US Department of Energy (Briggs, Lucas, and Taylor 2002). This method was based on the combination of cluster analysis of climate data and traditional systems of climate classification used in many disciplines, including the previously developed building-energy-demand-based classification and the Köppen-Geiger classification system. PNNL's method addressed the dual criteria of thermal energy and moisture availability and provided the ability to relate this information to the Köppen-Geiger system. Linkage to the Köppen-Geiger system allows and encourages the international use of the PNNL system, even if local climate data are limited or missing.

Building scientists and engineers must be able to classify climate zones when analyzing the energy performance of buildings to guide the selection of building envelopes and climate management systems. To fulfill this need, the American National Standards Institute (ANSI), ASHRAE (formerly the American Society of Heating, Refrigerating and Air-Conditioning Engineers), and the Illuminating Engineering Society (IES) adopted the PNNL climate classification system, referencing it in ANSI/ASHRAE/IES Standard 90.1-2013, *Energy Standard for Buildings except Low-Rise Residential Buildings* (ANSI, ASHRAE, and IES 2013), and ANSI/ASHRAE Standard 90.2-2007, *Energy-Efficient Design of Low-Rise Residential Buildings* (ANSI and ASHRAE 2007). These standards established energy efficiency codes for buildings in various climate zones.

Given that climate management strategies discussed in this publication deal with buildings and their interior environments, and that two of the factors influencing environmental risks to collections and occupant comfort are thermal energy and available moisture, the climate classification method of ANSI/ASHRAE/IES Standard 90.1-2013 (referred to hereafter as Standard 90.1-2013) is particularly relevant and useful. This method is described below.

Standard 90.1-2013 Climate Classification System

Climate classifications help users to categorize various climate types and develop general considerations for establishing realistic environmental goals as well as approaches for achieving the goals. Therefore, it is essential for those who are considering environmental improvements for collections to be able to determine the climate type where the project site is located. The case studies listed in part 2 can be compared to the reader's specific project site and building by determining the climate zone as well as the building type.

In the development of the Standard 90.1-2013 climate classification system, cluster analysis was performed on thermal and moisture data for 237 US cities available from the National Climatic Data Center (NCDC) of the National Oceanic and Atmospheric Administration (NOAA). Thermal energy criteria for climate classifications in Standard 90.1-2013 were defined by the cooling and heating needs of occupants in buildings. Indices were derived from daily temperature measurements and from the heating requirements of a given structure at a specific location. The energy use was considered to be directly proportional to the value of heating degree days (HDD) or cooling degree days (CDD) at that location. HDD and CDD are quantitative indices designed to reflect the demand for energy to heat or cool a residence or business property; HDD and CDD are defined relative to base temperatures—the outside temperature below which a building needs heating or above which a building needs cooling.

Historically, revisions of Standard 90.1-2013 have used 18°C and 10°C as the base temperatures for the calculation of HDD and CDD values, respectively (see sidebar "Calculation of HDD and CDD Values"). The 2013 revision separates the HDD18°C and CDD10°C thermal criteria into eight thermal zones. Table 1.1 shows the eight thermal climate zones and their thermal criteria. The application of cooling criteria to zones 1 through 4 indicates regions that range from very hot to seasonally warm, pointing to the need for cooling if the thermal comfort of occupants is considered. Application to the United States resulted in thermal zones arrayed as nearly horizontal bands, with the hottest zone occupying the southern tip of Florida and the coldest zone covering the northern half of Alaska.

The thermal criteria only address one aspect—heating and cooling loads—of energy consumption in buildings. In order to address the humidification or dehumidification loads, it is necessary to consider

Table 1.1

Thermal climate zone definitions according to ANSI/ASHRAE/IES Standard 90.1-2013 (ANSI, ASHRAE, and IES 2013).

Zone Number	Standard 90.1-2013 Climate Zone Classification	Cooling Criteria (cooling degree days, CDD)	Heating Criteria (heating degree days, HDD)
1	Very Hot-Humid and Very Hot-Dry	5000 < CDD10°C	
2	Hot-Humid and Hot-Dry	3500 < CDD10°C ≦ 5000	
3	Warm-Humid and Warm-Dry	2500 < CDD10°C ≦ 3500	
	Warm-Marine	CDD10°C ≦ 2500	HDD18°C ≦ 2000
4	Mixed-Humid and Mixed-Dry	CDD10°C ≦ 2500	2000 < HDD18°C ≦ 3000
	Mixed-Marine		2000 < HDD18°C ≦ 3000
5	Cool-Humid, Cool-Dry, and Cool-Marine		3000 < HDD18°C ≦ 4000
6	Cold-Humid and Cold-Dry		4000 < HDD18°C ≦ 5000
7	Very Cold		5000 < HDD18°C ≦ 7000
8	Subarctic		7000 < HDD18°C

Calculation of HDD and CDD Values

A daily CDD10°C value is determined by the difference between the daily average temperature and the CDD base temperature, with a zero value given if the base temperature exceeds the daily average. Similarly, a daily HDD18°C value is the difference between the HDD base temperature and the daily average temperature, with a zero value given if the daily average exceeds the base temperature. Annual CDD10°C and HDD18°C values are calculated by summing daily temperature differences between the daily average and the base temperature over a year.

CDD and HDD values are calculated from actual temperature data, and websites, such as http://www.degreedays.net/#, are available to assist energy professionals in evaluating CDD and HDD for specific base values for specific geographical locations.

moisture-based criteria. However, moisture is mostly a function of precipitation and temperature and their daily and seasonal patterns and variations. Thus, the moisture criteria for climate classification are more complex than the thermal criteria. Climate classification with respect to moisture in Standard 90.1-2013 is determined by a process of elimination, resulting in one of three possible classifications: Humid (A), Dry (B), and Marine (C). Figure 1.3 shows the standard's decision tree—starting at Marine (C) and ending at Humid (A)—and includes the criteria and steps for classifying climate by moisture.

In the Marine-type climate, the dry season coincides with summer, and the rainy season occurs during mild winter months, resulting in a cool damp winter. The Dry- or Humid-type climates will not meet either the annual precipitation pattern or the summer and winter temperatures

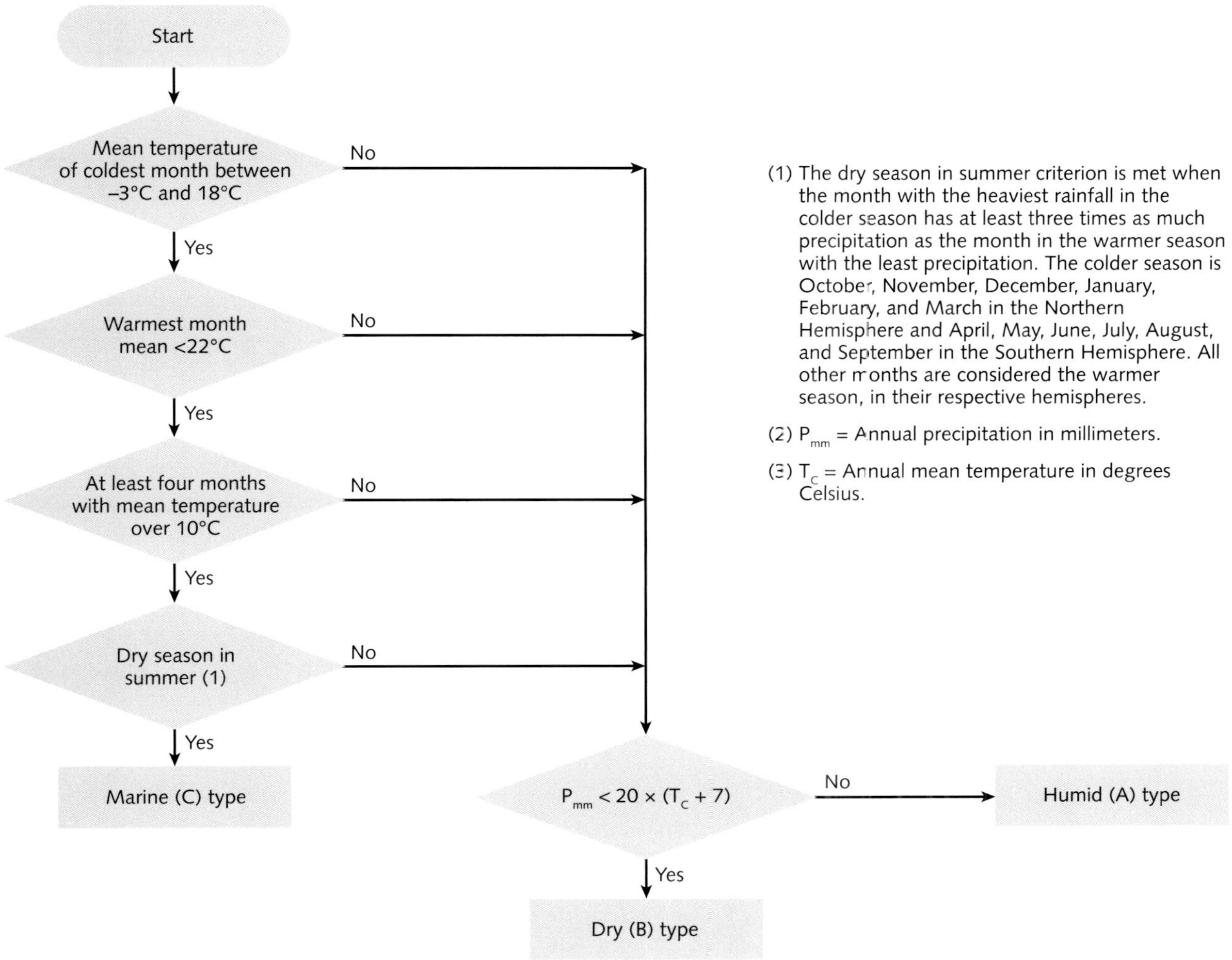

Figure 1.3

Decision tree for classifying climate with respect to moisture; including climate zone definitions and criteria. From Standard 90.1-2013 (ANSI, ASHRAE, and IES 2013).

of the Marine-type climate. The distinction between Dry- and Humid-type climates is based on the annual rainfall amount with respect to its annual mean temperature (referred to here as the Dry-Humid Index). As a consequence, Humid-type climates will be characterized by a hot and humid summer, while Dry-type climates will have a hot and dry summer. For the continental United States, application of the moisture criteria resulted in three moisture-based climate subzones: the Marine type along the west coast, the Dry type in the central mountain region, and the Humid type in the eastern half (Briggs, Lucas, and Taylor 2002).

The combination of moisture- and thermal-based criteria in Standard 90.1-2013 produced a climate classification method that can be distinguished by both temperature and humidity and is indicative of the heating/cooling and dehumidification/humidification portions of climate loads on the building. In the United States, the resulting climate classifications can be thought of as three vertical moisture zones (A, B, and C) overlaying eight horizontal thermal bands (1–8). Climate zones are designated with a number and letter notation corresponding to their thermal and moisture criteria—for example, 1A is Very Hot-Humid. Since the climate-moderating effect by a large body of water is limited, the Marine type (C) climate can be introduced only to the intermediate thermal zones (3, 4, and 5). Cooling or dehumidification is not needed in the Very Cold and Subarctic thermal zones, 7 and 8, since the cold air can only retain a small amount of water vapor; therefore, the dry-humid classification is not applied to the thermal zones. These considerations result in a total of seventeen possible climate zones.

While hot and humid conditions, the primary focus of this publication, are consistent throughout the year in climate zone 1A (Very Hot-Humid), these conditions may occur only seasonally in climate zones 2A (Hot-Humid), 3A (Warm-Humid), and 4A (Mixed-Humid). To avoid repetition in this publication, these four thermal- and moisture-based climate zones—1A, 2A, 3A, and 4A—will be collectively referred to as "hot and humid climates" or "seasonally hot and humid climates" unless a specific climate zone is of interest. The designation "Marine climates," also mentioned in this publication because of their cool and damp winter conditions, refers to climate zones 3C (Warm-Marine) and 4C (Mixed-Marine).

Earlier in this chapter, the sites of case studies to be discussed in part 2 were identified by their geographical regions. Table 1.2 shows the sites' corresponding climate zones based on the criteria from Standard 90.1-2013.

Table 1.2

Comparison of geographical regions and ANSI/ASHRAE/IES Standard 90.1-2013 climate zone classifications for case study sites discussed in part 2.

Case Study Sites	Geographical Region	Standard 90.1-2013 Climate Zone Classification
Belém, Brazil	Tropical	Very Hot-Humid (1A)
Rio de Janeiro, Brazil	Tropical	Very Hot-Humid (1A)
Jekyll Island, United States	Subtropical	Hot-Humid (2A)
San Cristóbal de La Laguna (Tenerife), Spain	Oceanic	Mixed-Marine (4C)
Valle Guerra (Tenerife), Spain	Subtropical	Hot-Humid (2A)
Beijing, China	Temperate	Mixed-Humid (4A)

World Climate Zone Classification

Application of Standard 90.1-2013 to regions for which suitable climate data are not available requires linking their climate zone classifications to existing Köppen-Geiger climate classes, since the Standard 90.1-2013 climate zones have not been mapped outside the United States. The moisture criteria in Standard 90.1-2013 are based in large part on annual totals and patterns of precipitation, a variable that is also used in the Köppen-Geiger system. Because precipitation—quantitative and widely recorded—forms a basis common to the climate classifications derived from the two methods, any location that has been mapped by the Köppen-Geiger system can be assigned a climate zone based on criteria from Standard 90.1-2013.

Table 1.3 shows a comparison of climate zones determined by Standard 90.1-2013 and the Köppen-Geiger system for cities around the world.

It should be noted that the link between climate classifications for the two methods may not translate one-to-one, as cities with the same Köppen-Geiger climate classification may be separately classified with Standard 90.1-2013. For example, while Brisbane and Sydney, Australia, are both classified by Köppen-Geiger as having *Cfa* climates (Humid Subtropical), the criteria for Standard 90.1-2013 classify the Brisbane climate as 2A (Hot-Humid) and the Sydney climate as 3A (Warm-Humid). In these cases, the use of CDD10°C and/or HDD18°C data specific to each location is what differentiates their classifications derived from the method of Standard 90.1-2013. Sample calculations of the Standard 90.1-2013 method are shown for Bangkok and Istanbul in appendix 3.

Table 1.3

Comparison of ANSI/ASHRAE/IES Standard 90.1-2013 climate zone classifications and corresponding Köppen-Geiger classifications for cities around the world (modified from Briggs, Lucas, and Taylor 2002, table 2B).

	Standard 90.1-2013 Climate Classification		Köppen-Geiger Climate Classification	
City	Zone Number	Type	Key	Description
Chennai (India), Darwin (Australia)	1A	Very Hot-Humid	*Af/Am/Aw*	Tropical Rainforest/Tropical Monsoon/Tropical Wet and Dry
Riyadh (Saudi Arabia)	1B	Very Hot-Dry	*BWh*	Tropical Desert
Saint George's (Grenada), Brisbane (Australia)	2A	Hot-Humid	*Cfa*	Humid Subtropical (Warm Summer)
Cairo (Egypt)	2B	Hot-Dry	*BWh/BSh*	Arid Subtropical
Nairobi (Kenya), Sydney (Australia)	3A	Warm-Humid	*Cfa*	Humid Subtropical (Warm Summer)
Crete (Greece), Kalgoorlie (Australia)	3B	Warm-Dry	*BSk/BWh*	Semiarid Middle Latitude/Arid Subtropical/Highlands
Naples (Italy), Bilbao (Spain)	3C	Warm-Marine	*Csa/Cfb*	Dry Summer Subtropical (Mediterranean)
Seoul (Korea), Melbourne (Australia)	4A	Mixed-Humid	*Cfa/Dfa*	Humid Subtropical/Humid Continental (Warm Summer)
Kabul (Afghanistan), Adelaide (Australia)	4B	Mixed-Dry	*BSk/BWh*	Semiarid Middle Latitude/Arid Subtropical/Highlands
Brussels (Belgium), Santiago (Chile)	4C	Mixed-Marine	*Csb/Cfb*	Marine (Cool Summer)
Sapporo (Japan), La Paz (Bolivia)	5A	Cool-Humid	*Dfa*	Humid Continental (Warm Summer)
Ulan Bator (Mongolia), Rio Gallegos (Argentina)	5B	Cool-Dry	*BSk/BWk*	Cold Semiarid/Cold Desert
Kraków (Poland)	5C	Cool-Marine	*Cfb*	Marine (Cool Summer)
Oslo (Norway), Punta Arenas (Chile)	6A	Cold-Humid	*Dfa/Dfb*	Humid Continental (Warm Summer/Cool Summer)
Volgograd (Russia)	6B	Cold-Dry	*BSk*	Semiarid Middle Latitude/Highlands
Reykjavik (Iceland)	7	Very Cold	*Dfb*	Humid Continental (Cool Summer)
Fort Smith (Canada)	8	Subarctic	*Dfc*	Subarctic

Summary

This chapter presented several methods for classifying climate, including the widely used Köppen-Geiger system, which is based on the distribution of native vegetation resulting from climatic factors such as temperature and precipitation. A recent classification method described in Standard 90.1-2013 bases its method on the climatic factors that drive thermal and moisture loads on buildings and their interior environments. Links between climate classifications of the two methods are generally consistent, and where there are inconsistencies, the classes determined by Standard 90.1-2013 are better correlated to thermal energy and moisture factors, which affect collections, buildings, and occupants.

The use of thermal energy and moisture availability as the basis for the Standard 90.1-2013 classification system is also relevant to conservation efforts for museums in hot and humid climates, as these climatic variables are considered enabling factors for increased biological and microbiological activities, dimensional changes, and chemical reaction rates in collections. Thus, Standard 90.1-2013 is well suited for indicating the relative impact of the climate on a building and its interior environment. The Standard 90.1-2013 classification system also serves as a preliminary indicator of the potential risks to collections and occupant comfort posed by different climate zones and types.

Using the Standard 90.1-2013 classification method, the remainder of this book will primarily focus on four hot and humid climate zones, ranging from 1A (Very Hot-Humid) to 4A (Mixed-Humid). Two Marine-type climate zones, 3C (Warm-Marine) and 4C (Mixed-Marine), are also discussed as marine climates. The four hot and humid climate zones are characterized by consistently or seasonally hot and humid conditions, while the two Marine climate zones are typified by cool and damp winters. These six climate zones encompass approximately one-quarter of the surface area of the continental United States and large portions of Southeast Asia, Central and South America, and the sub-Saharan region of Africa.

Chapter 2 will consider the specific risks to cultural heritage, including collections and historic buildings, posed by hot and humid climates.

References

ANSI [American National Standards Institute] and ASHRAE

2007 *Standard 90.2-2007—Energy-Efficient Design of Low-Rise Residential Buildings*. Atlanta: ASHRAE.

ANSI, ASHRAE, and IES [Illuminating Engineering Society]

2013 *Standard 90.1-2013—Energy Standard for Buildings except Low-Rise Residential Buildings* (SI edition). Atlanta: ASHRAE.

Briggs, R. S., R. G. Lucas, and T. Taylor
2002 Climate classification for building energy codes and standards. In *Technical and Symposium Papers, ASHRAE Winter Meeting, Chicago, IL, January 2003*. Atlanta: ASHRAE.

Oliver, J.
1991 The history, status and future of climatic classification. *Physical Geography* 12 (3): 231–51.

Chapter 2

Risks of Hot and Humid Climates for Collections and Buildings

Stewards of cultural heritage collections in hot and humid climates must provide for the collections' long-term conservation as well as for access to the collections by the public and scholars. The challenge is twofold: Environmental factors such as humidity, temperature, pollution, and natural light can cause deterioration of both movable and immovable cultural heritage. Yet the use of air-conditioning systems to reduce these environmental risks can unintentionally introduce additional environmental risks to collections in hot and humid climates.

This chapter reviews the environmental risks to collections in hot and humid climates as well as the risks posed by interior environmental management to heritage buildings. By presenting these topics together, we believe that museum managers, conservators, and curators will better understand the interrelationship of the hot and humid climate, the collection environment, and the building envelope, and the potential impacts of each of these factors on the others.

This chapter is directed to all readers of this book. For more advanced information on conserving collections and buildings, refer to the list of references at the end of this chapter.

Environmental Risks for Collections

In 1994 the Canadian Conservation Institute (CCI), in its Framework for Preservation of Museum Collections (CCI 2013), introduced the following ten agents of deterioration that pose a risk of damage to collections:

1. direct physical forces (sudden or gradual)
2. thieves and vandals
3. dissociation (institutional neglect)
4. fire
5. water
6. pests
7. pollutants
8. light (ultraviolet and infrared)
9. incorrect temperature
10. incorrect humidity

Half of these ten agents, from pests (number 6) to incorrect humidity (number 10), are particular risks for collections and heritage buildings in hot and humid climates. These agents fall into three mechanisms of deterioration and damage—biological, mechanical, and chemical—which are described below.

Biological Risk and Damage

Biological risk to cultural heritage collections is largely due to the activity of organisms that directly cause deterioration or damage to objects and materials. Ranging in size from nanometers to meters, these organisms include, in descending order of the risk they pose to collections: microorganisms, insects, and vermin.

Microorganisms

Microorganisms are pervasive in the environment. They include fungi and bacteria, both of which play a major role in the breakdown of organic material. Fungi include molds, yeast, and mushrooms. Molds grow by the extension of threadlike structures, 2–10 μm in diameter, called hyphae, which form a mass of interwoven filaments called mycelium. The mycelium is the vegetative portion of fungi and is commonly visible on damp organic surfaces; it can damage collections by staining or through digestion (fig. 2.1). Bacteria, another group of microorganisms, may inflict damage similar to that caused by fungi.

In hot and humid climates, the abundance of water and the elevated temperatures for substantial periods of the year make microorganism activity the dominant biological risk to collections. In particular,

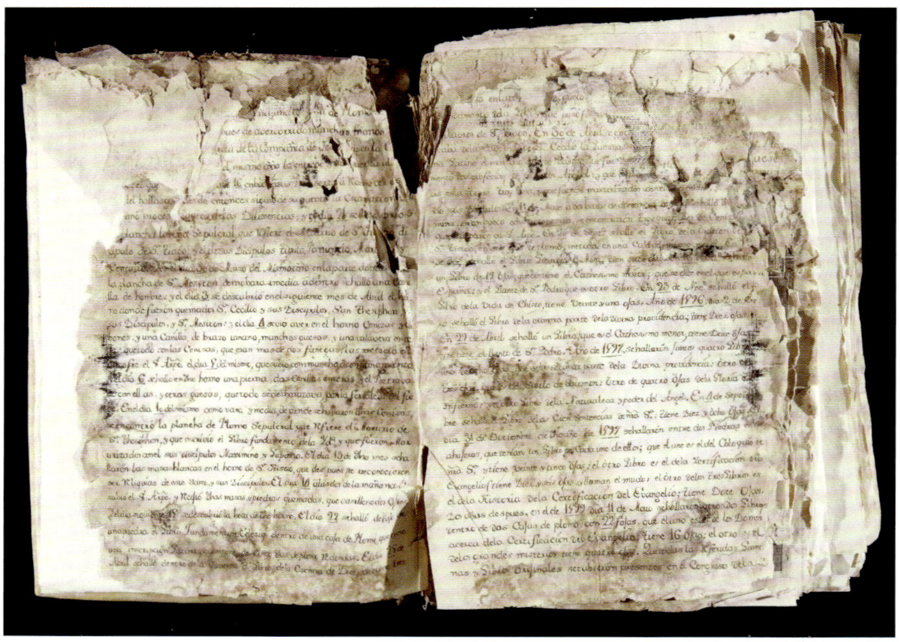

Figure 2.1
A book, found in a humid area of a monastery in Granada, Spain, exhibiting deterioration due to fungal growth. Fungi of the following genera were isolated from the book: *Stemphylium, Alternaria, Chaetomium, Aspergillus, Rhizopus,* and *Trichoderma*. Photo: Fernando Suarez (Servicio de Libros y Documentos, Instituto del Patrimonio Cultural de España).

fungal activity poses a greater risk than bacterial activity because fungal activity can occur at a lower level of humidity (~75% versus ~90% RH).

The potential for collection damage from biological activity can be addressed by effectively implementing an Integrated Pest Management (IPM) program (see sidebar "Integrated Pest Management"). However, prevention of fungal activity requires management of the environment, because fungal activity is dependent on the presence of optimal temperature and humidity conditions, on critical thresholds for temperature and humidity, and on sufficient time available for germination.

The nutrients needed to support fungi are plentiful in organic collections—particularly, materials such as paper; parchment; leather; skin; and starched, sized, or soiled textiles that may contain soluble proteins, starches, or sugars. Optimal temperature and humidity conditions for fungal spore germination in the presence of these nutrients are specific to species. For example, *Penicillium chrysogenum* can germinate on book covers in less than 25 uninterrupted hours at 30°C and 100% RH (fig. 2.2). Generally, higher humidity will shorten germination time, and departures from optimal temperature will increase germination time.

Integrated Pest Management

First introduced in the 1970s as a response to increasing concern over the safety of synthetic pesticides, Integrated Pest Management (IPM) is a multidisciplinary approach that seeks to integrate a range of pest control and prevention methods that minimize risks to the environment and public health.

The basic principles of IPM are:

- Set action thresholds—Since the goal is control of a pest population rather than eradication, acceptable pest levels need to be determined, above which appropriate control measures can be taken.
- Monitor and identify pests—Consistent pest monitoring provides information on pests' distribution and seasonal trends, while accurate identification will provide information on life cycles, preferred food sources and environmental conditions, and potential treatment options.
- Prevention—It is always preferable to prevent rather than to treat infestations, so pest prevention policies examine entry points into the building (building envelope) and conditions that might sustain (collection vulnerability, environment, housekeeping) or introduce pests (incoming materials).
- Control—If an infestation is deemed active, the pest must be identified and an appropriate treatment (including, but not limited to, the judicious use of pesticides) chosen based on collection type, extent of infestation, institutional capabilities, and budget.

For more information, visit the website of the Integrated Pest Management Working Group (IPM-WG) at http://www.museumpests.net/.

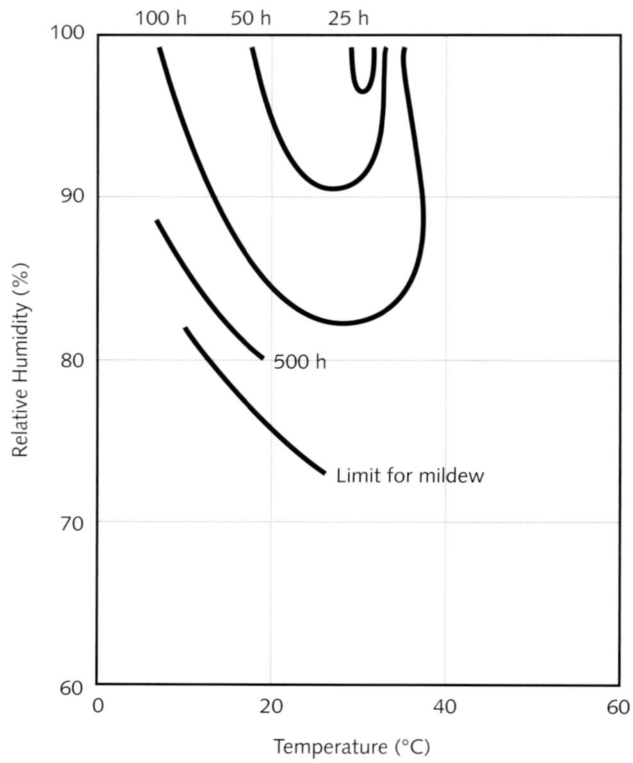

Figure 2.2

Germination time of *Penicillium chrysogenum* on book covers. From Brundrett 1990, fig. 4.3, which references Groom and Panisset 1933.

In addition to having specific optimal conditions required for its rapid germination, each fungal species has specific temperature and humidity threshold levels, below which spore germination will not take place. The threshold humidity values may be referred to in the literature as a fractional "water activity." Assuming that moisture vapor is evenly distributed in a space, the humidity and water activity values are numerically equivalent (75% RH = 0.75 A_w). For *Penicillium chrysogenum*, the temperature and humidity thresholds are approximately 5°C and 73% RH. If temperature or humidity conditions fall below these species-specific threshold values, the germination process will cease and then restart when the threshold values are satisfied. When conditions are below the threshold values, many species may enter into a dormant phase, from which they can be reactivated when threshold values are satisfied.

Following germination, the fungal growth rate typically remains dependent on temperature and humidity conditions similar to those favorable for germination, though some molds may continue to grow in less-than-optimal conditions (Block 1953). Once a mycelium structure has developed, a fungus can retain moisture within its structure, allowing it to maintain growth during periods of reduced humidity (Brundrett 1990). Some fungal species are also able to support growth in unfavorable humidity conditions by transporting water through the hyphae from a remote source (Strang and Kigawa 2013).

The exposure of absorbent materials to extremely high humidity can also extend their period of vulnerability to biological

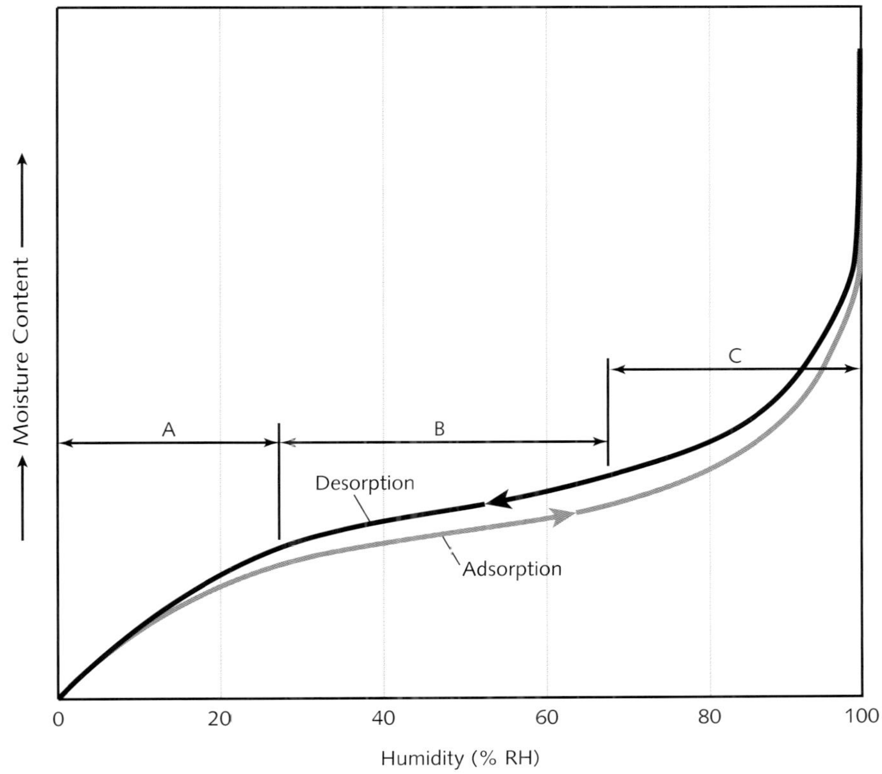

Figure 2.3
Adsorption and desorption isotherms for a generic hygroscopic material.

attack. Textiles and paper can rapidly adsorb moisture vapor and absorb liquid, but it takes longer to remove the same amount of moisture by desorption or evaporation. The equilibrium moisture content of materials at specific humidity levels is greater during desorption than during adsorption (fig. 2.3). As a result, the ambient humidity for drying must be lower than the ambient humidity for adsorption. Materials with a large capacity for moisture storage can remain active sites of fungal growth even after the surrounding atmospheric environment has returned to "safe" humidity levels.

In hot and humid climates, the temperature requirements for fungal germination are satisfied more often than are the humidity requirements, and both are satisfied more often than in other climate zones. For example, in Rio de Janeiro, which is classified as climate zone 1A (Very Hot-Humid), monthly temperature typically ranges from 20°C to 30°C, and monthly humidity ranges from 60% to 90% RH (weatherbase.com). For *Penicillium chrysogenum*, the temperature in Rio de Janeiro is sufficient (>5°C) for both germination and growth, but diurnal humidity falls below the required threshold of humidity (~73% RH), and fungal activity is interrupted. Thus, in hot and humid climates, management of humidity conditions rather than of temperature may be a more effective means of limiting mold activity in the collection space.

While threshold activity levels are species specific, and some fungi remain viable at 65% RH, maintaining humidity levels below 75% RH can prevent germination and growth for a wide range of fungal species (Florian 1997; Valentín et al. 1998). This threshold value represents a maximum upper limit for interior humidity for operation of an environmental management system. Lower values for this upper limit, sometimes as low as 60%–65% RH, may be beneficial in the event that operation of the environmental management system is interrupted. The difference between the lower operational value (60%–65% RH) and the maximum limit (75% RH) may provide extra time before the onset of a high risk of germination; however, for buildings and collections that can effectively buffer interior humidity and have low vapor infiltration rates, this extra margin may be unnecessary. It should be noted that a greater margin of safety with respect to biological risk may come at the cost of potential mechanical risk if humidity control is lost—this point will be elaborated upon in the section below on mechanical risk to collections. Even if the 75% RH limit is exceeded, the duration of the event remains an important parameter; excursions lasting less than the germination time will have very low risk, but excursions lasting longer than the germination time will likely result in fungal outbreak.

When humidity in a space is near the threshold value for germination, temperature gradients become an important consideration. If locally depressed air temperature occurs near a surface that is cooler than room air, local humidity will be elevated. When coinciding with a nutrient-rich surface or material, extended periods of high localized humidity above the mold threshold value can lead to germination. Gradients in temperature and humidity in a space can be minimized by fostering air movement, effectively reducing the risk of fungal activity. Strategies for promoting air movement are presented in chapters 6 and 7.

Insects

Insects constitute over half of the known living organisms (Chapman 2009) and are often transported into a collection space on infested objects, perhaps loans or new acquisitions. Certain species of moths, beetles, and termites (see sidebar "Insects") are of particular concern to cultural heritage collections because they can damage organic objects, such as textiles, paper-based material, wooden items, and natural history specimens. Damage to objects typically occurs when insects digest materials, bore into materials to lay their eggs, or transition to the pupal stage (fig. 2.4).

Vermin

The term *vermin* refers to animals with disproportionately high populations due to favorable conditions and a lack of natural predators. Rodents such as rats and mice are typically categorized as vermin, but the term may also include snakes, pigeons, and small predators. Rapid breeding and population growth by rats and mice must be sustained by food sources, the search for which may lead to a collections space. Rodent damage to collections can occur from the gathering of food and nesting materials and from gnawing, which limits the length of their constantly growing incisors.

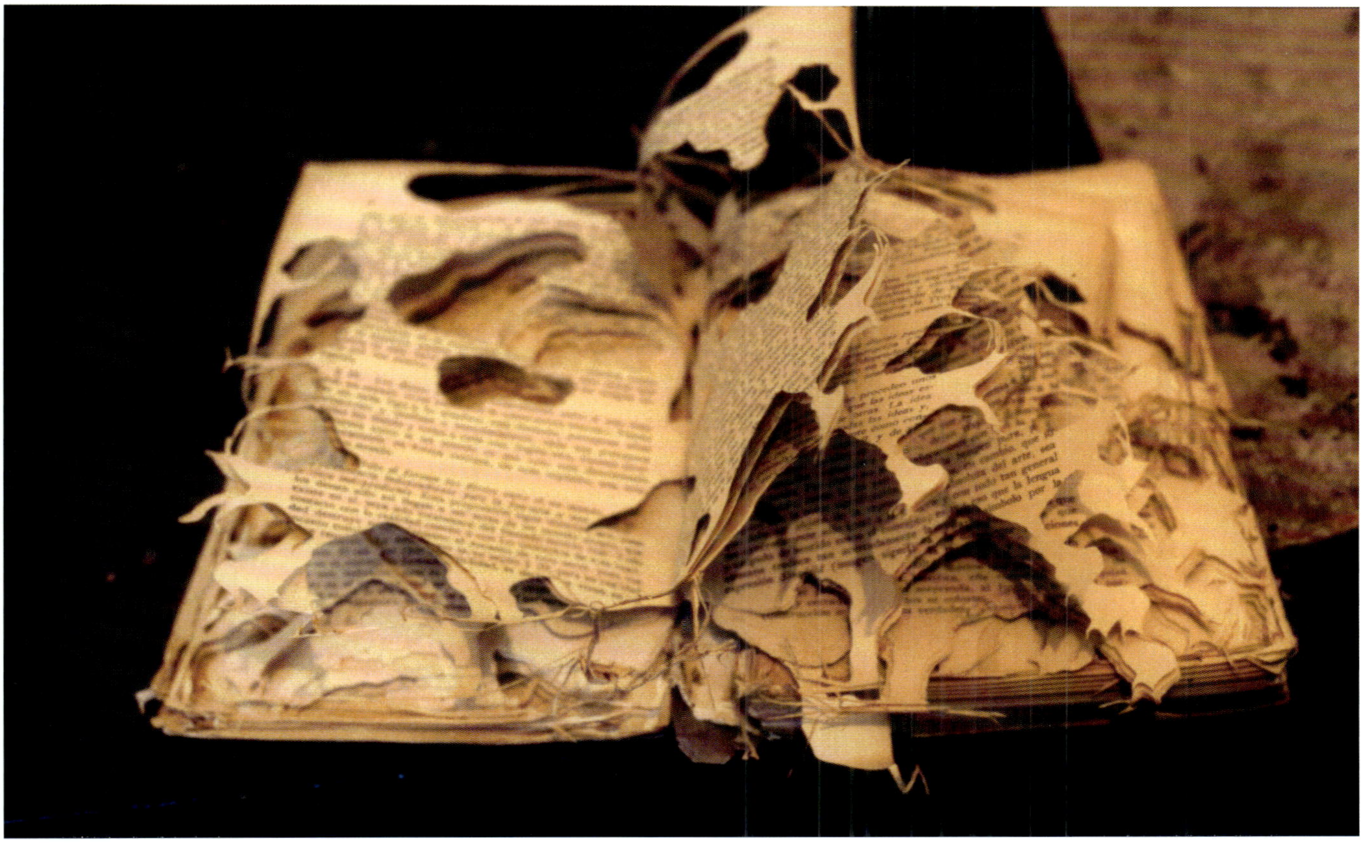

Figure 2.4
Book damaged by termites at the Museu de Arte Sacra in Salvador, Bahia, Brazil. Photo: James Druzik. © J. Paul Getty Trust.

Insects

Insects known to cause damage to collections are listed below. Although some insects not listed may not directly damage collections, their presence may be an indicator of insects that are of concern. Source: Science Museum of Minnesota, http://www.smm.org/conservation/agents/pests.

Tineola bisselliella (clothes moth)
Lasioderma serricorne (cigarette beetle)
Stegobium paniceum (drugstore beetle)
Subfamily Lyctinae (powderpost beetle)
Family Dermestidae (carpet beetle)
Family Dermestidae (larder beetle)
Family Dermestidae (hide beetle)
Lepisma saccharina (silverfish)
Plodia interpunctella (Indianmeal moth)
Genera Tribolium (flour beetle)
Oryzaephilus surinamensis (sawtooth grain beetle)
Necrobia rufipes (ham beetle)
Order Blattaria or Blattodea (cockroach)
Epifamily Termitoidae (termite)

Mechanical Risk and Damage

Mechanical damage to objects results from environmental factors or mechanical forces such as abrasion. Environmentally induced mechanical damage can include:

- dimensional change due to shifts and/or fluctuations in temperature and humidity,
- embrittlement or deformation of materials due to extreme temperature or humidity conditions,
- material fatigue or embrittlement due to repeated cycling of temperature and humidity,
- material rupture when moisture content cycling results in crystal formation of soluble salts.

The risk of mechanical damage is affected by many factors, such as the range of temperature and humidity, the dimensional constraint in an object, and the coefficients of thermal and moisture (hygroscopic) expansion of the materials involved. In most organic materials, such as wood, the influence of humidity on dimensional change is greater than the influence of temperature.

In hot and humid climates, the annual ranges of ambient temperature and humidity are small compared to the wider ranges of these variables in temperate climates. As a result, the risk of mechanical damage may be lower in hot and humid climates than in other climate zones. However, there remains a risk for mechanical damage to collections when they are relocated from unconditioned ambient environments to interior environments that are substantially different with respect to average temperature and humidity. Once an object is acclimated in an artificially controlled environment, the risk due to change in temperature or humidity is not eliminated, since failure of the mechanical equipment used to maintain the environment may cause rapid changes or extremes in temperature and humidity as the interior condition rebounds to the unconditioned state.

Dimensional change

Many organic and inorganic materials respond to changes in temperature and/or humidity by dimensional change, the extent of which is material specific. As described previously, humidity typically exerts a much greater influence than does temperature on the mechanical response of many organic materials. For example, the humidity-driven dimensional change of wood in the longitudinal direction (along the grain) dominates over dimensional change due to temperature, even though longitudinal mechanical response is generally small (Simpson and TenWolde 1999). Shifts in either temperature or humidity may result in dimensional changes that can lead to warping, nonuniform deformation causing internal stress, or fracture of the material or its assembly. The interdependence of temperature and humidity is also important—cooling of the air will cause humidity to increase, and warming of the air will cause a decrease in humidity. This relationship is discussed in greater detail in chapter 5.

The dimensional change of a material is a direct function of its change in moisture content, and the moisture content of a material is a function of the humidity of the air surrounding the material. The relationship of dimensional change to moisture content change is typically linear at constant temperature for most materials. However, the isothermal relationship of moisture content to humidity is nonlinear, as seen in the equilibrium moisture content curve (also known as sorption isotherm) in figure 2.3. As shown in the figure, the equilibrium moisture content of the material changes very little with humidity in area B (moderate conditions) but changes more rapidly in regions A (dry conditions) and C (humid conditions). Therefore, for the same change in humidity, the risk of mechanical damage is greater in the humidity ranges corresponding to areas A and C. In contrast, equilibrium moisture content varies much less within the humidity range corresponding to area B, reflecting a lower risk of mechanical damage.

In many instances, an object's vulnerability to mechanical damage from dimensional change is influenced more by its construction—how it is assembled and constrained—than by the environmental response of an individual material. Objects made from composite materials and/or containing delicate marquetry, veneers, glue joints, and restrained panels are particularly vulnerable to differential dimensional change between adjacent or joined and constrained materials within the object. See figure 2.5 for an example of warped and cracked door panels.

The response time of an object to humidity changes depends on its surface area and mass. For example, furniture constructed of dense wood members may require days or even months to fully respond to changes in the surrounding environment; as a consequence, shorter-term fluctuations in humidity and temperature will have little effect on

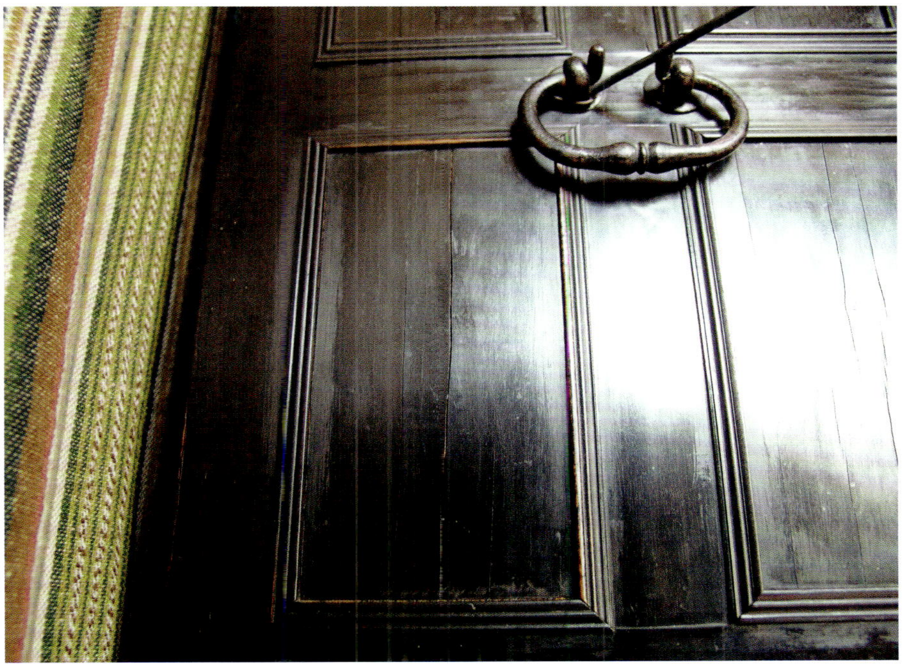

Figure 2.5

Warped and cracked door panels of a chest in Our Lord in the Attic Museum, Amsterdam. The door is subjected to humid summers and cold winters. Operation of a comfort heating system during winter caused extended period of low humidity in the church and may have resulted in splitting of the wood panels. Photo: Shin Maekawa. © J. Paul Getty Trust.

moisture content—and therefore little effect on dimensional change—in these objects. However, for objects with large surface area relative to mass, such as a painting on bark or a thin wood panel painting, rapid changes in environmental conditions will result in rapid moisture content changes and dimensional change. Other factors that influence the response time to humidity changes include surface coatings, finishes, and treatments. For wood objects, the orientation of the surface grain will potentially be a strong influence.

Embrittlement

The vulnerability of materials to mechanical damage can increase under extreme environmental conditions. A variety of materials, including acrylic and oil paints, become brittle at temperatures below 5°C and thus more vulnerable to shock, vibration, and handling. At temperatures above 60°C, many common plastics may start to deform. Low-humidity conditions increase brittleness in paper and paintings, and high-humidity conditions can lead to delamination, exfoliation, and deformation of materials such as veneers, paper, and photographs.

Material fatigue

Over extended periods of time, repeated small fluctuations in temperature and humidity that result in cycles of reversible, nondamaging dimensional change can weaken a material, increasing the risk of fatigue failure. Material fatigue is often quantified as a fractional stress applied over a specified number of cycles—the fractional stress being a portion of the ultimate stress value that would result in failure in a single application. For example, for brittle materials, such as glass, ceramic, and aged oil paint, fatigue failure can occur with applications of 0.25 times the ultimate stress over 1 million cycles. For tougher materials, such as wood, paper, and leather, fatigue failure can occur after 1 million cycles of 0.5 times the ultimate stress value (Strang and Kigawa 2013). In reality, the million-cycle mark may represent an underestimation, because each cycle may not be of sufficient duration for the material to fully respond. Just as objects may be proofed to singular environmental events (proofed single fluctuation), exposure to cycled environmental fluctuations similar to or less than what an object has previously experienced (proofed repetitive fluctuations) are not likely to result in mechanical damage.

Material rupture from crystal formation of soluble salts

Objects containing salts, such as those in archaeological collections or building materials collections, are at risk of mechanical damage due to phase change in the salt from fluctuations in humidity. Deliquescent salts, such as sodium chloride and sodium nitrate, absorb and desorb moisture vapor from the air. Such salts have a critical humidity above which the salt will absorb moisture, forming a liquid solution that can migrate, and below which the salt will crystallize. The critical humidity for sodium chloride (75% RH) varies little with temperature, but the critical humidity for many other salts is also dependent on temperature. When ambient humidity rises above its critical value, the solubilized salt can migrate in

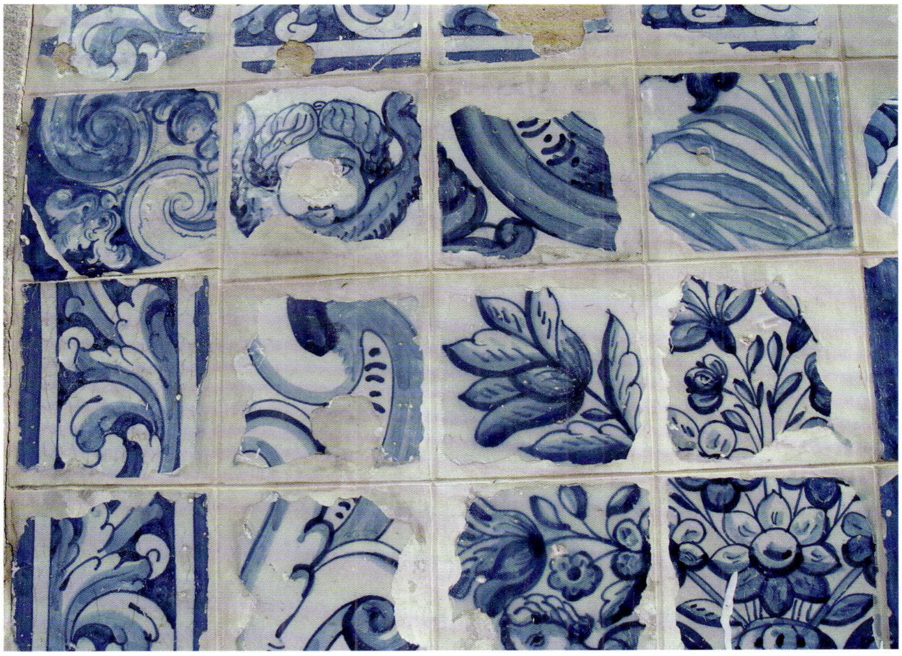

Figure 2.6
Salt-induced damage of tiles at the Museu de Arte Sacra in Salvador, Bahia, Brazil. Photo: Shin Maekawa. © J. Paul Getty Trust.

a permeable material, and when ambient humidity falls below the critical humidity value, the expansive recrystallization of the salt within the pores of a material or between materials, such as a coating and a substrate, can result in material rupture or delamination, blistering, and the flaking of finishes (fig. 2.6). In hot and humid climates, the typical ranges of ambient humidity generally span the critical humidity values for common salt compounds.

Additional Issues Related to Mechanical Damage

Particulates damage

Abrasive mechanical damage can be caused by coarse particulates (diameter ≧ 10 μm) on material surfaces, often as a result of frequent cleaning. Particulates can enter the collection space because of a mechanical environmental management system that lacks effective filtration or because of reliance on natural ventilation through open windows. In museums, people are the dominant source of particulates, owing to the transport of material via shoes and clothing and the flaking of skin (Yoon and Brimblecombe 2001). Building construction, renovation, and maintenance activities can also generate large amounts of particulates.

In addition to causing mechanical damage, coarse and fine particulates (diameter ≦ 2.5 μm) in the forms of dust and soot can contribute to the degradation of an object's surface finish and appearance.

Prior environmental exposure

An object's history of environmental exposure can also affect its tolerance of environmental change. Previous exposure to extreme environments may have "proofed" an object to those conditions and to further damage, such as warping and fractures, that may already have taken place. Thus, only environmental conditions beyond those that the "proofed" object has experienced will likely result in additional mechanical damage; from a mechanical damage standpoint, an object that has experienced a wide range of past conditions will tolerate a similar range of conditions in the future (Michalski 1993). Estimates of the former environmental conditions to which an object has been exposed should be realistic and not underestimated. However, if an object has undergone conservation treatment, such as the repair of cracks or damaged joints, the former environment to which that object had been exposed may not be relevant: in such a case, the treatment has effectively reset the starting point of the history of environmental exposure to the time of treatment.

Risk levels for mechanical damage

The ranges of temperature and humidity that pose low mechanical risk are object specific and dependent on the distribution of stress within an object and its past use, environmental exposure. and conservation treatment histories. For mixed historic collections (excluding objects with very high sensitivity), humidity fluctuations of ±10% RH and temperature fluctuations of ±10 K from the historic average conditions present low risk of mechanical damage (ASHRAE 2011). Widening of the humidity fluctuation to ±20% RH and the temperature fluctuation to ±20 K will only increase the risk of severe mechanical damage for objects with high sensitivity (e.g., wooden furniture with focused stresses) and very high sensitivity (e.g., large prints adhered at corners). Gradual environmental fluctuations over an extended period, as in seasonal shifts, will reduce the mechanical stress from dimensional change experienced by an object because the slow rate of change allows for stress relaxation within an object (Mecklenburg and Tumosa 1991; Erlebacher et al. 1992; Michalski 1991, 1993; Erhardt and Mecklenburg 1994; Oreszczyn, Cassar, and Fernandez 1994; Erhardt et al. 1996; Mecklenburg, Tumosa, and Erhardt 1998, 2005). An excellent summary of past and present environmental museum guidelines can be found on the wiki page of the American Institute for Conservation of Historic and Artistic Works (AIC Conservation Wiki).

Chemical Damage

The various materials of an object continuously interact chemically with the surrounding environment. These chemical reactions, which can result in deterioration or damage, are strongly influenced by:

- temperature,
- humidity,
- reactive pollutants,
- light.

In hot and humid climates, the abundance of thermal energy and moisture vapor increases the risk of chemical deterioration and damage for organic and inorganic materials. This risk is further increased by recent industrialization and the use of windows for natural ventilation and natural light in collections spaces.

Temperature-induced reactions

Many chemical reactions are dependent on temperature, and increasing temperature results in faster chemical reaction rates, since more molecules are colliding with sufficient activation energy for a reaction to take place. The accelerating effect of temperature on chemical reactions can be highlighted by the chemical deterioration of aging cellulose acetate film through acid hydrolysis (fig. 2.7). Although this process cannot be stopped or reversed, storage at reduced temperatures will slow the rate of chemical change and increase the lifetime of the material (Reilly 1993). (A reduction in moisture, which also serves as a reactant in acid hydrolysis, can further limit the reaction rate.) Other materials susceptible to accelerated chemical damage due to high temperature include: cellulose nitrate film, magnetic media, unstable photographic materials such as color prints, and acidic paper. Even for objects with low sensitivity to high temperature, a general rule of thumb regarding the benefits of reduced temperature suggests that an object's lifetime will double for every 5°C decrease in ambient temperature (Sebor 1995).

Humidity-induced reactions

Humidity also plays an important role in the promotion of chemical deterioration processes, such as the electrochemical reaction of corrosion (Brundrett 1990). As humidity rises, the rate of corrosion increases

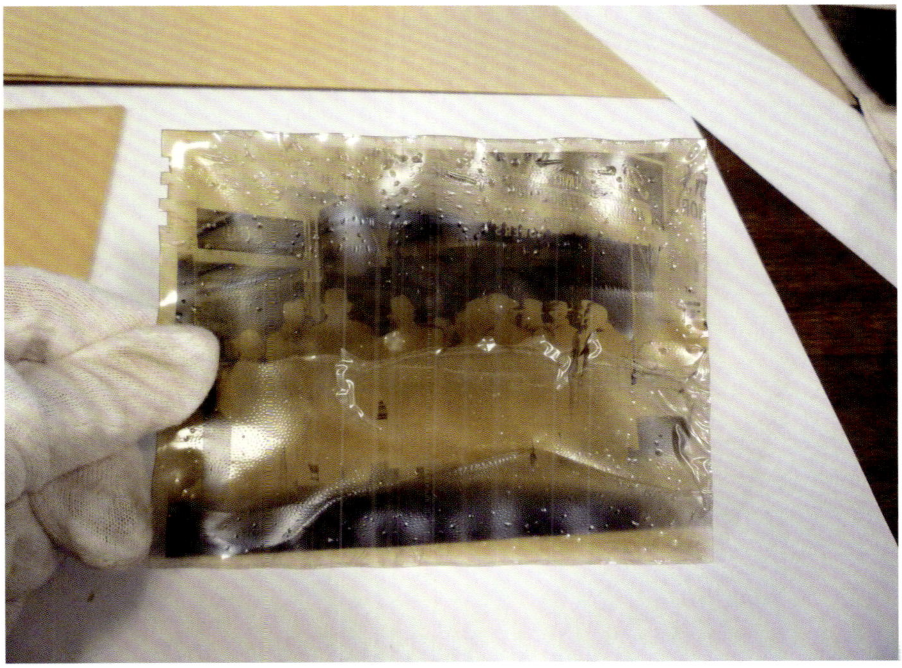

Figure 2.7
Deteriorated cellulose acetate negative from the study collection at the Harry Ransom Center in Austin, Texas. Photo: Barbara Brown.

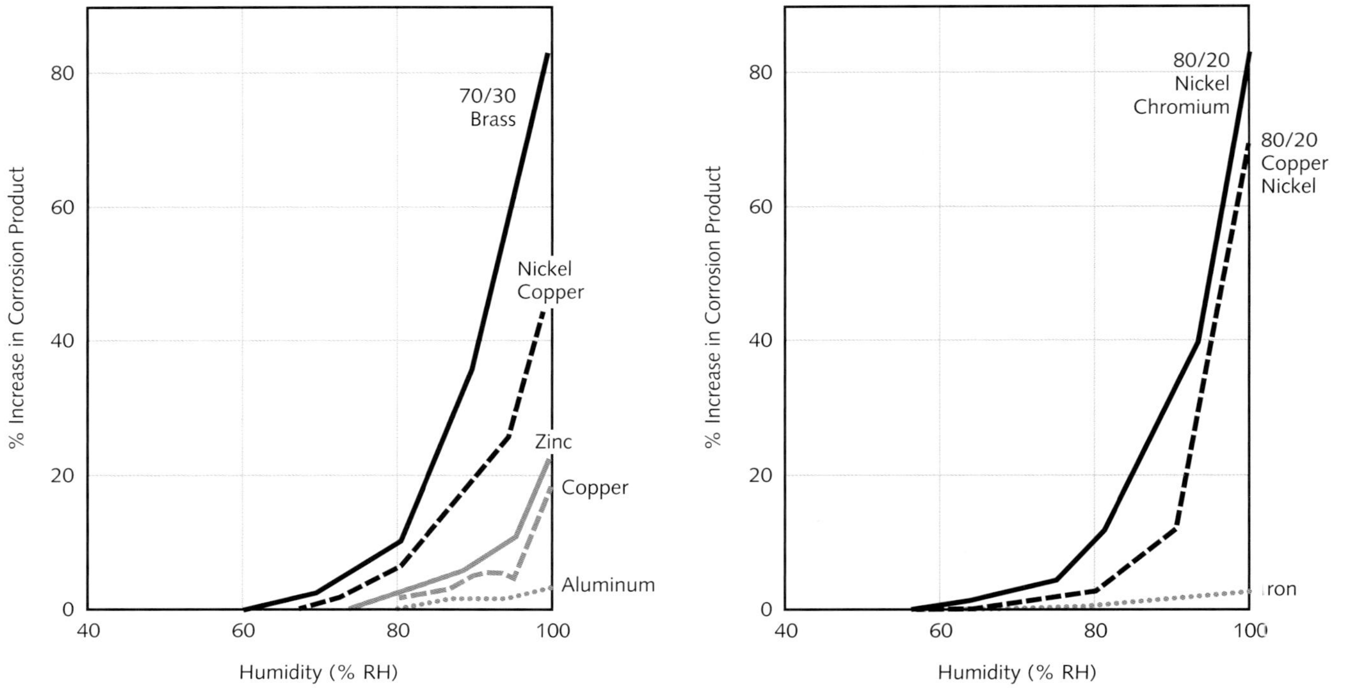

Figure 2.8

Percentage increase in corrosion product following exposure of various metals to successive 24 hour periods of increasing humidity at 20°C (Hudson 1929).

(fig. 2.8). Each metal has a critical humidity above which corrosion accelerates. This critical humidity is dependent on the state of the metal surface (e.g., surface texture or prior surface corrosion) and the presence of reactive pollutants on the surface or in the air. In the case of archaeological metals, corrosion reactions are dependent on base metal composition and on either the soil environment during burial or the ambient environment above ground. Corrosion products from post-excavation exposure to low-humidity, oxygen-rich air differ from corrosion products due to in situ exposure to moist soil conditions. Metal collections will generally benefit from storage and display in low-humidity conditions, and limiting exposure to cool temperatures will also reduce the potential for condensation and corrosion.

The environmental factors of temperature and humidity are key parameters in determining the theoretical lifetime of an object due to chemical deterioration. Using a comparative lifetime analysis, this relationship has been quantified for the chemical deterioration of book papers by acid hydrolysis (Sebor 1995). Using ambient conditions of 50% RH and 20°C as a baseline for an object's theoretical lifetime, a higher humidity of 80% RH at 20°C or a higher temperature of 25°C at 50% RH will reduce the object's theoretical lifetime by one-half (table 2.1). Conversely, humidity of 30% RH at 20°C, or temperature of 15°C at 50% RH will increase theoretical lifetime by a factor of two.

In hot and humid climates, typical humidity and temperature conditions are 70% RH and 25°C. Reductions in humidity will extend an object's theoretical lifetime to a greater extent than reduced temperature. While the specific effect will vary with each object type, the overall trend of drier and cooler conditions providing an improved conservation environment holds true for many chemically unstable objects, such as cel-

Lifetime Multiplier	Temperature (°C)	Humidity (% RH)
0.31	25	70
0.38	20	70
0.5	25	50
0.5	20	80
1	20	50
2	15	50
2	20	30

Table 2.1

Lifetime multipliers based on the deterioration of book papers by acid hydrolysis when exposed to specific temperature and humidity conditions (adapted from Sebor 1995).

lulose acetate and nitrate films, magnetic media, and silver gelatin photographic prints.

Pollutant-induced reactions

Chemical deterioration of objects can also result from airborne pollutants such as sulfur dioxide, nitrogen oxides, and ozone. Short- and long-term exposure to elevated pollutant concentrations can damage a variety of collections materials by acidification (paper), corrosion (metals), discoloration (pigments and dyestuffs), or loss of strength (textiles). Sources of these pollutants include industrial processes, urban activities, transportation, and agriculture.

Pollutants may also be transferred to objects by contact (e.g., plasticizer in PVC) or may be intrinsic to an object's material (e.g., self-generated acid from degradation of cellulose acetate film).

Where countries industrialize and populations concentrate in cities, pollutant concentrations may increase, and chemical degradation of cultural heritage will become a more pressing issue. A summary of major gaseous pollutants that are risks to cultural heritage and their sources can be found in ASHRAE "Museums, Galleries, Archives and Libraries" (ASHRAE 2011).

Light-induced reactions

Exposure of an object to natural or artificial light can induce photochemical reactions that may result in damage. The general reaction mechanism involves an object's absorption of light energy, which, if not dissipated as heat, has the potential to trigger irreversible reactions that may change the object's appearance. In hot and humid climates, the high levels of solar radiation and the use of open windows for natural ventilation can expose objects to damaging light levels. Light-induced damage to objects can be reduced by limiting light exposure levels (ranging from 50 to 200 lux depending on the object), limiting exhibition durations, and filtering ultraviolet or infrared wavelengths.

Rate of chemical damage

Though surviving collections in hot and humid climates remain subject to chemical damage, the rate of chemical decay may be dependent on the availability of reaction sites. While chemical deterioration may initially be

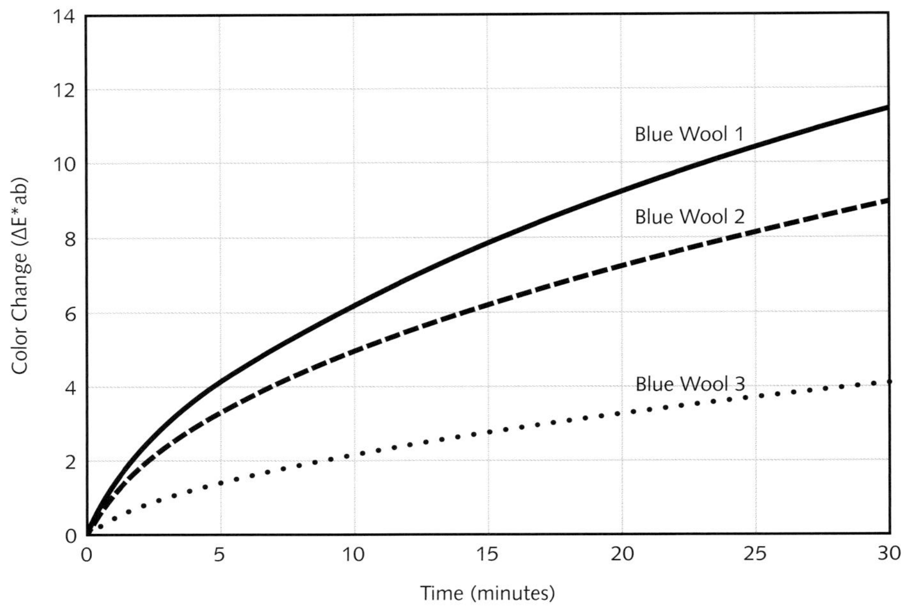

Figure 2.9
Decline in rate of light-induced color change of ISO Blue Wool Standards over time. Samples were exposed to a high-intensity xenon lamp for 30 minutes, and color change was determined from the difference between initial color and color at subsequent time intervals.

rapid for a relatively new object, the eventual loss of reaction sites could slow the rate of chemical damage over time. This shift in chemical reactivity is shown graphically in figure 2.9, where the rate of light-induced chemical color change declines following an initial period of rapid color change.

Risks to Heritage Buildings Posed by Interior Environmental Management

Climate factors and their risks to collections cannot be viewed in isolation. Attempts to mitigate the influence of climate on collections through interior environmental management can also present risks to the building, especially if it is a heritage building.

Cultural institutions often exhibit and store their collections in structures that are architecturally and historically significant; these heritage buildings may have originally been constructed for a different purpose, such as residential, religious, civic, agricultural, or industrial uses. These buildings incorporate the environmental management methods and building envelope features available at the time of their construction and suited to their original use. Most heritage buildings predate mechanical environmental control systems, including mechanical air-conditioning. Expectations for both human thermal comfort and the science of collections conservation have substantially evolved since the construction of most heritage buildings.

Heritage buildings, built prior to the advent of mechanical climate control, include many passive climate management features for moving air through the building. These features, some of which may no longer be used, can result in temperature and humidity gradients within and between spaces in the building. For example, cooling and moisture loads may vary greatly from room to room owing to large windows and exterior doors that can allow solar heat gain and the infiltration of humid outside air. Open stairways and hallways with high ceilings, originally intended to promote nonmechanical ventilation by buoyancy, may aggravate temperature gradients between floor levels. Gradients between spaces and floors are further influenced by room-specific factors, such as shading by trees and surrounding buildings or thermal cooling from floor or wall contact with relatively cool below-grade soil.

Since many heritage buildings lack the space necessary to properly distribute conditioned air, achieving a balanced distribution of conditioned air throughout the individual spaces and rooms of the building may be difficult. Installation of ducts and registers to transfer conditioned air within the building can be restricted by wall, floor, or ceiling assemblies lacking cavities for ductwork or by interior decorative finishes that, because they cannot be disturbed, preclude access to cavities behind them. Large spaces, such as ballrooms, churches, and halls, often cannot be adequately conditioned because of difficulties in delivering and circulating sufficient conditioned air to achieve uniform temperature and humidity and because the volume of the space far exceeds the concealed space available for ducts and registers of adequate size (fig. 2.10).

The adaptation of heritage buildings for the storage and exhibition of collections often requires installation of mechanical equipment and systems for heating, ventilation, and air-conditioning (HVAC) in order to satisfy conventional criteria for collections conservation or to address the thermal comfort of building occupants.

Common environmental risks to heritage buildings related to the introduction of mechanical systems for environmental management include:

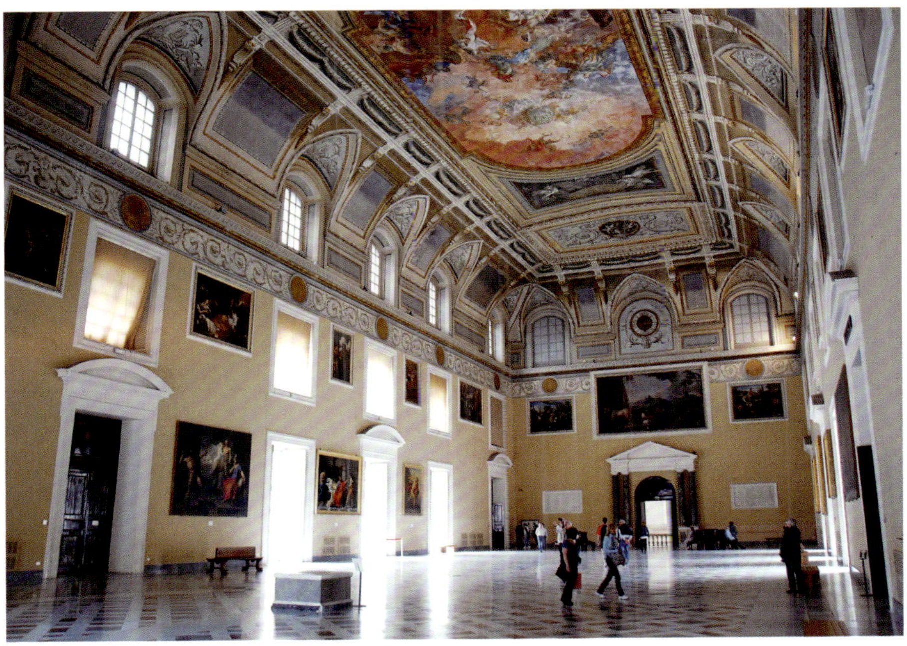

Figure 2.10
Ballroom at the Naples National Archaeological Museum in Italy, with a large space volume and limited areas for ducts and diffuser installation. Photo: Shin Maekawa. © J. Paul Getty Trust.

- temperature and humidity gradients within building envelope assemblies,
- mobilization of salts in heritage building envelope and fabric,
- risk of condensation and high moisture content in building envelopes.

The damage resulting from these risks often occurs out of sight, within the envelope and structural assemblies; symptomatic damage, such as efflorescence on observable surfaces, may be only part of the problem.

Spatial issues, visual intrusion, and loss of historic fabric related to the introduction of mechanical systems in heritage buildings are addressed in chapter 7.

Temperature and humidity gradients within building envelope assemblies

In hot and humid climates, the introduction of mechanical systems to heritage buildings results in interior conditions that are cooler and drier than the exterior conditions. The differences between the interior and exterior conditions occur across the various assemblies that make up the building envelope, such as the roof, walls above or below the ground, and operable elements such as windows and doors.

In the case of moisture differences, moisture in either its liquid or vapor state will move from the exterior (higher concentration) to the interior (lower concentration). Because the total exterior surface areas of buildings can be quite large, the amount of moisture transported through the envelope may be substantial. As discussed below, moisture transport can mobilize soluble salts.

Thermal gradients across envelope assemblies are not necessarily damaging on their own, but they can result in differential expansion/contraction within the envelope assembly, causing cracks or fissures that can permit water entry. When thermal gradients approach the dew point temperature, condensation can form on the assembly surface; this problem is discussed in greater detail below.

When thermal and moisture gradients coincide at or near dew point temperature, moisture-driven deterioration mechanisms, such as fungal activity, corrosion, and salt mobilization, can take place.

Mobilization of salts in heritage building envelope and fabric

Air-conditioning, and in particular dehumidification, can create moisture gradients in the building envelope between the interior and exterior environments, resulting in the transport of moisture and deliquescent or soluble salts through permeable materials in the building envelope. Often encountered in building materials such as masonry and plasters, soluble salts will migrate by moisture flow from areas of high concentration to areas of low concentration. When the salt solution approaches an area where the humidity is below the salt's critical humidity, the salt will recrystallize. The exact location of recrystallization, on or below the material surface, is dependent on humidity and on the material moisture content of the material substrate. While the formation of salt on the material surface (efflorescence) is relatively harmless (but aesthetically intrusive), the recrystallization of salt within the material or below its surface finishes (subflorescence) is likely to be destructive. Subflorescence is especially problematic for masonry materials, which commonly contain sodium chloride (critical humidity of 75% RH at 25°C) and sodium sulphate (critical humidity of 83% RH at 25°C).

An example of salt-related damage due to interior environmental management can be seen in the Museu do Círio in Bélem, Brazil, whose masonry walls contain sodium chloride. An air-conditioning system was installed in the restored nineteenth-century two-story building in 2002. Prior to installation, the building was naturally ventilated and, owing to its location in climate zone 1A (Very Hot-Humid) and its proximity to a large river, it is likely that humidity consistently exceeded 75% RH, allowing the sodium chloride within the masonry to remain in solution. Following installation in 2002, the air-conditioning system was operated intermittently, and the museum building was subjected to daily cycles of humidity between 60% RH and 80% RH (fig. 2.11). While efflorescence was observed on wall surfaces, it was the recrystallization of salt within the material that resulted in the delamination of surface layers and deterioration of the masonry.

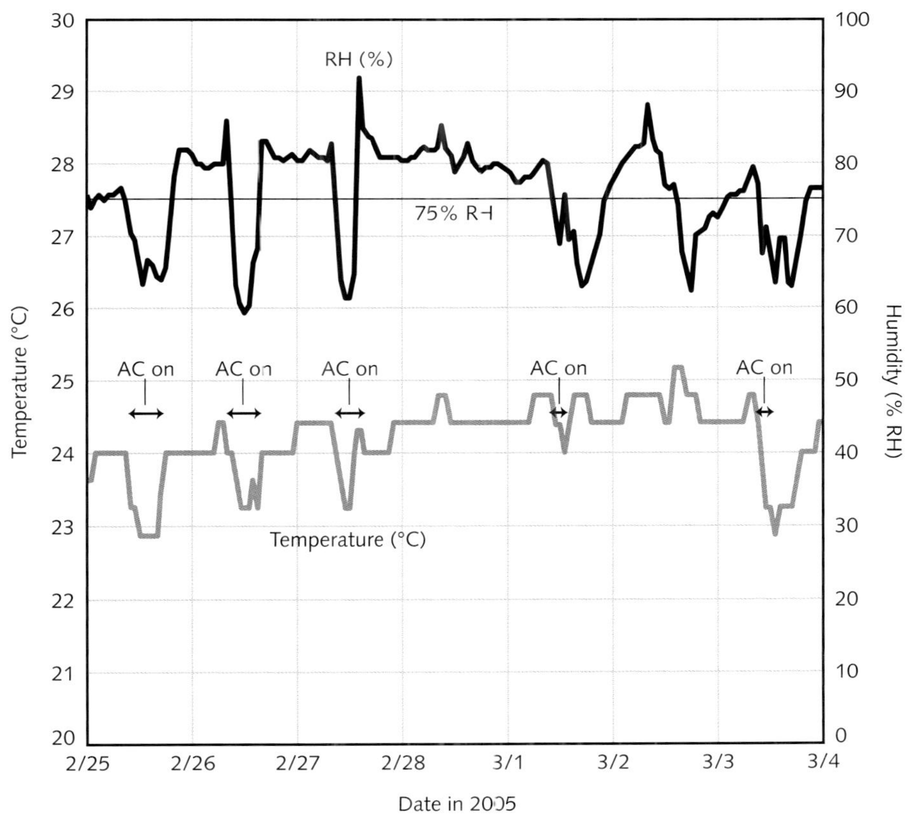

Figure 2.11
Plot of temperature and humidity in the Museu do Círio in 2005. Operation of the air-conditioning system only during visitor hours (to maintain a temperature of 23°C) resulted in the incidental daily cycling of humidity above and below 75% RH, which is the critical humidity of the sodium chloride contained in the masonry walls. Data courtesy of Franciza Toledo.

Risk of condensation and high moisture content in building envelopes

Condensation can form on building surfaces that are cooled to near or below the dew point temperature of the surrounding air. Interior surface condensation can occur on supply air registers of the air-conditioning systems or on wall or ceiling surfaces when the temperature of the supply air is below the dew point temperature of the mixing room air, resulting in paint flaking or microorganism activity (fig. 2.12a). Condensation may also occur on exterior surfaces of thermally conductive materials—such as glass, metals, and masonry—when cooled below the dew point temperature of the outside air (fig 2.12b).

Because of the inverse relationship between temperature and humidity, reduced interior temperatures can also produce high moisture contents within porous materials of exterior walls, which can support microorganism activity within exterior walls or on their external faces. For example, the air-conditioning of the interior space used for entomological storage at the research campus of the Museu Paraense Emílio Goeldi in Belém, Brazil, resulted in the growth of black mold on its porous exterior wall surfaces (fig. 2.13). Similar microorganism growth can be expected at cracks in exterior walls where cooled air may leak to the outside, lowering the surface temperature of the exterior wall to near or below the dew point temperature of the outside air.

The installation and operation of mechanical environmental management systems in heritage buildings in hot and humid climates can introduce new and unintended risks to collections and the building. The design, installation, and operation of such systems in these buildings requires an understanding of numerous factors including: the conservation needs of the collection, the historic significance of the building, the climate, the building envelope and materials, the surrounding landscape, the spatial distribution and use of the building, and the thermal and moisture characteristics of the building.

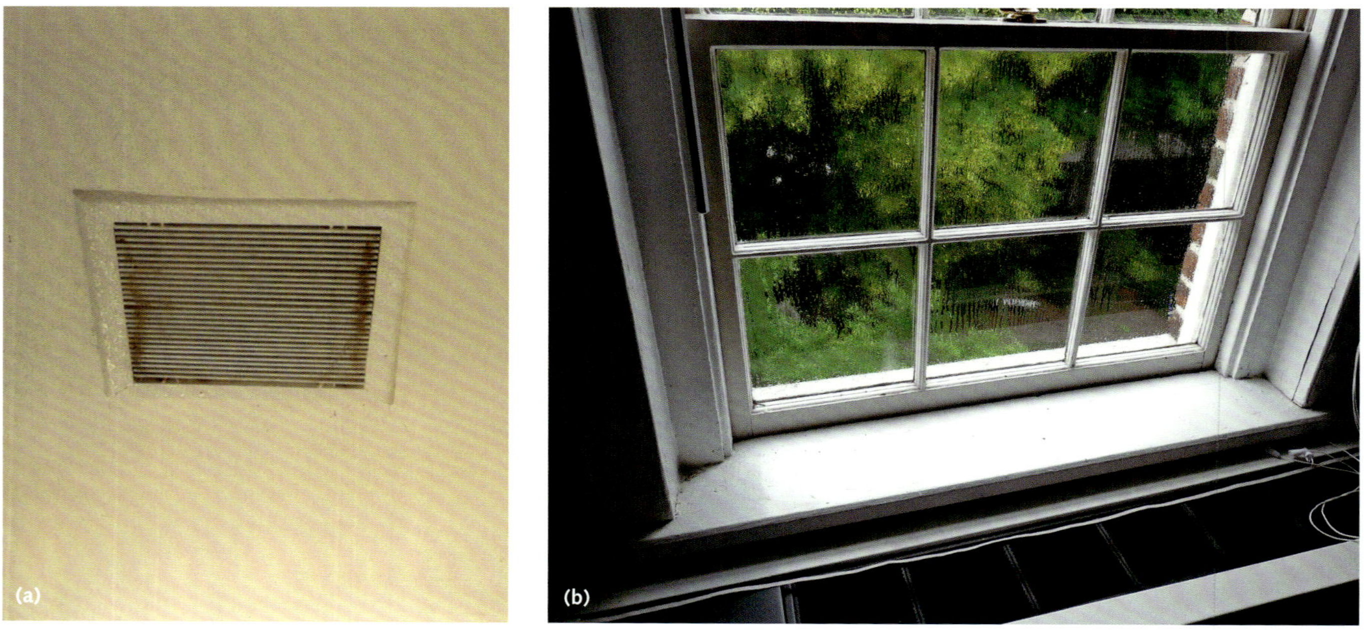

Figure 2.12
Microorganism activity on a supply air grille (a), resulting from condensation, and exterior condensation on a window that is chilled by supply air on the interior (b). Photos: Michael C. Henry (a); Shin Maekawa (b). © J. Paul Getty Trust.

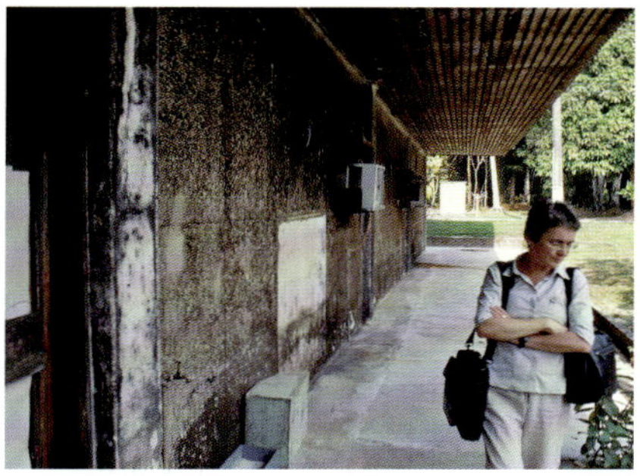

Figure 2.13
Black mold on the exterior surfaces of the building envelope of an entomological storage at the Emílio Goeldi Museum in Belém, Brazil. Photo: Shin Maekawa. © J. Paul Getty Trust.

Summary

This chapter reviewed the environmental risks to movable and immovable cultural heritage in hot and humid climates. Also reviewed were the potential risks environmental management systems pose to heritage buildings. The following chapter describes an environmental classification system based on the risk of damage to collections and buildings.

References

ASHRAE
2011 Museums, galleries, archives and libraries. Chapter 23 in *ASHRAE Handbook—HVAC Applications*. Atlanta: ASHRAE.

Block, S. S.
1953 Humidity experiments for mould growth. *Journal of Applied Microbiology* 1, 287–93.

Brundrett, G. W.
1990 *Criteria for Moisture Control*. Butterworth & Co.
http://www.cci-icc.gc.ca/caringfor-prendresoindes/articles/10agents/chap06-eng.aspx.

CCI [Canadian Conservation Institute]
2013 Framework for preservation of museum collections.
http://www.cci-icc.gc.ca/discovercci-decouvriricc/preventive/15-eng.aspx.

Chapman, A. D.
2009 Australian biodiversity information services, Toowoomba, Australia. A report for the Australian Biological Resources Study. ISBN (online) 978 0 642 56861 8.

Erhardt, D., and M. F. Mecklenburg
1994 Relative humidity re-examined. In *Preventive Conservation: Practice, Theory, and Research. Preprints of the Contributions to the Ottawa Congress, 12–16 September, 1994*, ed. A. Roy and P. Smith, 32–38. London: International Institute for Conservation of Historic and Artistic Works.

Erhardt, D., M. F. Mecklenburg, C. S. Tumosa, and T. M. Olstad
1996 New versus old wood: Differences and similarities in physical, mechanical, and chemical properties. In *ICOM Committee for Conservation Preprints, 11th Triennial Meeting, Edinburgh, Scotland*, 903–10. Paris: ICOM.

Erlebacher, J., E. D. Brown, M. F. Mecklenburg, and C. S. Tumosa
1992 The effects of temperature and relative humidity on the mechanical properties of modern painting materials. In *Materials Issues in Art and Archaeology*, vol. 3, ed. P. B. Vandiver et al., 359–70 San Francisco: Materials Research Society.

Florian, M.-L.
1997 *Heritage Eaters: Insects and Fungi in Heritage Collections*. London: James and James.

Groom, P., and T. Panisset
1933 Studies on *Penicillium chrysogenum* in relation to temperature and relative humidity of the air. *Annals of Applied Biology* 20:633–60.

Hudson, J. C.
1929 Atmospheric corrosion of metals: Third report to the Atmospheric Corrosion Research Committee of the British Non Ferrous Metals Research Association. *Transactions of the Faraday Society* 25:177–252.

Integrated Pest Management Working Group (IPM-WG)
http://www.museumpests.net.

Mecklenburg, M. F., and C. S. Tumosa
1991 Mechanical behavior of paintings subjected to changes in temperature and relative humidity. In *Art in Transit: Studies in the Transport of Paintings*, ed. M. F. Mecklenburg, 173–216. Washington, DC: National Gallery of Art.

Mecklenburg, M. F., C. S. Tumosa, and D. Erhardt
1998 Structural response of painted wood surfaces to changes in ambient relative humidity. In *Painted Wood: History and Conservation*, ed. V. Dorge and F. C. Howlett, 464–83. Los Angeles: Getty Conservation Institute.

———.
2005 The changing mechanical properties of aging oil paints. In *Materials Issues in Art and Archaeology VII: Symposium Held November 30–December 3, 2004, Boston, Massachusetts, U.S.A.*, ed. P. B. Vandiver, J. L. Mass, and A. Murray, 13–24. Warrendale, PA: Materials Research Society.

Michalski, S.
1991 Paintings: Their response to temperature, relative humidity, shock, and vibration. In *Art in Transit: Studies in the Transport of Paintings*, ed. M. F. Mecklenburg, 223–48. Washington, DC: National Gallery of Art.

———.
1993 Relative humidity: A discussion of correct/incorrect values. In *ICOM Committee for Conservation 10th Triennial Meeting, Washington, DC, 22–27 August 1993: Preprints*, ed. Janet Bridgland, 624–29. Paris: ICOM Committee for Conservation.

Oreszczyn, T., M. Cassar, and K. Fernandez
1994 Comparative study of air-conditioned and non air-conditioned museums. In *Preventive Conservation: Practice, Theory and Research: Preprints of the Contributions to the Ottawa Congress, 12–16 September 1994,* 144–48. London: IIC. http://www.conservation-wiki.com/wiki/Environmental_Guidelines.

Reilly, J. M.
1993 *IPI Storage Guide for Acetate Film; Instructions for Using the Wheel, Graphs, and Table; Basic Strategy for Film Preservation.* Rochester: Image Permanence Institute. http://www.cci-icc.gc.ca/caringfor-prendresoindes/articles/10agents/chap09-eng.aspx.

Science Museum of Minnesota
http://www.smm.org/conservation/agents/pests.

Sebor, A. J.
1995 Heating, ventilation and air-conditioning systems. In *Storage of Natural History Collections: Ideas and Practical Solutions,* ed. C. L. Rose, C. A. Hawks, and H. H. Genoways. Washington, DC: Society for the Preservation of Natural History Collections.

Simpson, W., and A. TenWolde
1999 Physical properties and moisture relations of wood. In *Wood Handbook: Wood as an Engineering Material,* 3.1–3.24. General Technical Report FPL; GTR-113. Madison, WI: USDA Forest Service, Forest Products Laboratory.

Valentín, Níeves, R. Garcia, O. De Luis, and S. Maekawa
1998 Microbial control in archives, libraries and museums by ventilation systems. *Restaurator* 19 (2): 85–107.

Yoon, Y. H., and P. Brimblecombe
2001 The distribution of soiling by coarse particulate matter in the museum environment. *Indoor Air* 11 (4): 232–40.

Chapter 3

Balancing Risks and Establishing Conservation Priorities

The previous chapter presented the extraordinary environmental risks to cultural heritage collections in hot and humid climates and identified the potential unintended consequences of using conventional heating, ventilating, and air-conditioning (HVAC) systems to control interior environments for collections, especially in heritage buildings. In order to minimize the climate-driven risks to collections without introducing unacceptable consequential risks to the collections and building, a nonprescriptive risk management approach is needed to balance risks and establish conservation priorities for environmental management specific to hot and humid climates. This risk management approach should prioritize the risks to a collection and/or a heritage building, distinguishing the greatest risks from those that are less significant. This process provides the collections steward and environmental management professional with important criteria for evaluating the most appropriate environmental management strategies for the conservation of the collection and the building.

This chapter addresses the following topics:

- risk-based environmental management versus prescriptive environmental control
- existing risk-based guidance for managing the collection environment
- underlying principles of risk-based environmental management
- prioritizing risks of collections damage in hot and humid climates
- environmental considerations unique to hot and humid climates

The chapter concludes by describing a conservation environment classification protocol, developed by the authors, that specifically targets risks to mixed cultural heritage collections in hot and humid climates.

Risk-Based Environmental Management versus Prescriptive Environmental Control

In the last half of the twentieth century, the widespread availability of mechanical HVAC systems and the publication of Garry Thomson's *The*

Museum Environment in 1978 (2nd ed., Thomson 1994) resulted in the general acceptance of prescriptive guidelines for tight control of temperature and humidity for collections conservation in museums and archives. These guidelines centered on nominal values for average temperature and humidity (21°C ± 2°C, 50% ± 5% RH). Although Thomson recommended a range of values for interior environmental conditions specific to climate zone and collection type, his median target values were generally adopted by the museum community.

Since the 1990s, the conventional wisdom of prescriptive requirements for conservation environments has been reassessed. The necessity of prescriptive environmental management for collections conservation has been challenged by the realities of high energy consumption and the financial cost of tight mechanical control of humidity and temperature, as well as an increasing global awareness of the impact of human activity on climate change. These large-scale societal, economic and environmental issues will continue to pressure the cultural heritage community to seek and implement energy-efficient strategies for collections conservation.

As conservation science expands our understanding of the environmental risks to cultural heritage collections, specifications for the collections environment are moving away from tight control that is centered on universally applied values for temperature and humidity and its fluctuations. It is now seen as desirable to manage a collections environment within the context of a set of risk factors, rather than controlling it according to a prescriptive set of generic criteria. The risk-based environmental management approach is particularly appropriate to collections located outside the temperate climate zones or to places where infrastructure and available technological and economic resources may limit strategic options. Therefore, this approach is a fundamental underpinning of the current publication.

Existing Risk-Based Guidance for Managing the Collections Environment

A risk-based approach to defining conditions for conservation environments is not novel. Two publications, ASHRAE's "Museums, Galleries, Archives and Libraries" (2011), in the United States, and PAS 198 (British Standards Institute 2012), in the United Kingdom, apply a risk-based approach to conservation environments for collections in temperate climates.

"Museums, Galleries, Archives and Libraries" was first published as a chapter in the ASHRAE *Applications Handbook* in the 1990s and was the work product of a subcommittee of conservation scientists and mechanical engineers organized by ASHRAE to address conservation environments for collections (Michalski 2007). The document emphasizes the importance of balancing environmental risks to collections with realistically achievable interior temperature and humidity targets. "Museums, Galleries, Archives and Libraries" provides a set of classes of environmental control for general collections, and each class is matched to certain collections vulnerabilities as well as to the performance capabilities and limitations of the building envelope. There are five "classes of control" for general collections, as well as environmental specifications for archival cold storage and dry storage for metals. The classes of control for general collections range from Class AA (precision control without seasonal variation) to Class D (prevent dampness). For each class of control, collections risks and benefits are identified (table 3.1).

Each environmental control class in table 3.1 is defined by a set of allowable ranges for temperature and humidity within the collections space and for seasonal and short-term intervals; these ranges center

Table 3.1

Temperature and relative humidity specifications for collections (from ASHRAE 2011, table 3). The shaded part of the table delineates specifications for general museums, art galleries, libraries, and archives.

		Maximum Fluctuations and Gradients in Controlled Spaces			
Type	**Set Point or Annual Average**	**Class of Control**	**Short Fluctuations Plus Space Gradients**	**Seasonal Adjustments in System Set Point**	**Collections Risks and Benefits**
General museums, art galleries, libraries, and archives All reading and retrieval rooms, rooms for storing chemically stable collections, especially if mechanically medium to high vulnerability.	50% RH (or historic annual average for permanent collections) Temperatures set between 15°C and 25°C *Note:* Rooms intended for loan exhibitions must handle set point specified in loan agreement, typically 50% RH, 21°C, but sometimes 55% or 60% RH.	**AA** Precision control, no seasonal RH adjustment, very limited seasonal temperature adjustment possible	±5% RH, ±2 K	RH: no change Up 5 K; down 5 K	No risk of mechanical damage to most artifacts and paintings. Some metals and minerals may degrade if 50% RH exceeds a critical RH. Chemically unstable objects unusable within decades.
		A Precision control, some gradients or seasonal changes, not both	±5% RH, ±2 K	Up 10% RH, down 10% RH Up 5 K; down 10 K	Small risk of mechanical damage to high-vulnerability artifacts; no mechanical risk to most artifacts, paintings, photographs, and books. Chemically unstable objects unusable within decades.
			±10% RH, ±2 K	RH: no change Up 5 K; down 10 K	
		B Precision control, some gradients plus winter temperature setback	±10% RH, ±5 K	Up 10%, down 10% RH Up 10 K, but not above 30°C	Moderate risk of mechanical damage to high-vulnerability artifacts; tiny risk to most paintings, most photographs, some artifacts, some books; no risk to many artifacts and most books. Chemically unstable objects unusable within decades, less if routinely at 30°C, but cold winter periods double life.
		C Prevent all high-risk extremes	Within 25% to 75% RH year-round Temperatures rarely over 30°C, usually below 25°C		High risk of mechanical damage to high-vulnerability artifacts; moderate risk to most paintings, most photographs, some artifacts, some books; tiny risk to many artifacts and most books. Chemically unstable objects unusable within decades, less if routinely at 30°C, but cold winter periods double life.
		D Prevent dampness	Reliably below 75% RH		High risk of sudden or cumulative mechanical damage to most artifacts and paintings because of low-humidity fracture; but avoids high-humidity delamination and deformations, especially in veneers, paintings, paper, and photographs. Mold growth and rapid corrosion avoided. Chemically unstable objects unusable within decades, less if routinely at 30°C, but cold winter periods double life.
Special archives, libraries Storing chemically unstable collections	Cold store: −20°C, 40% RH	±10% RH, ±2 K			Chemically unstable objects usable for millennia. RH fluctuations under one month do not affect most properly packaged records at these temperatures (time out of storage becomes lifetime determinant).
	Cool store: 10°C 30% to 50% RH	(Even if achieved only during winter setback, this is a net advantage to such collections, as long as damp is not incurred)			Chemically unstable objects usable for a century or more. Such books and papers tend to have low mechanical vulnerability to fluctuations.
Special metal collections	Dry room: 0% to 30% RH	RH not to exceed some critical value, typically 30% RH			

Note: Short fluctuations means any fluctuation less than the seasonal adjustment.

on a target annual average temperature and humidity, as shown under the column heading "Set point or annual average."

The methodology of ASHRAE "Museums, Galleries, Archives and Libraries" hinges on selection of the appropriate class of control guided by the specific vulnerabilities of the collections, shown in the right-hand column of table 3.1, rather than on the notion "more control is better." "Museums, Galleries, Archives and Libraries" also links the class of control achievable in practice for a collections conservation environment in a building to the building's construction and envelope type and past use (table 3.2).

Other guides and standards, including those used by individual countries, utilize risk-based approaches to managing environmental conditions for cultural heritage collections. These include:

- PAS 198:2012—*Specification for Managing Environmental Conditions for Cultural Collections* (British Standards Institute): Develops an understanding of the sensitivity of collections to temperature, humidity, light, and pollution and specifies the relative risk of collections damage from each of these deterioration agents.
- PD 5454:2012—*Guide for the Storage and Exhibition of Archival Materials* (British Standards Institute): Focuses largely on storage facilities and presents temperature and humidity recommendations for various collection types; additional topics include HVAC systems and environmental monitoring.
- BS EN 15757:2010—*Conservation of Cultural Property. Specifications for Temperature and Relative Humidity to Limit Climate-Induced Mechanical Damage in Organic Hygroscopic Materials* (British and European Standard): Describes methodology based on the historical climate and condition surveys for examining the risk of mechanical damage from fluctuating temperature and humidity.
- ANSI/NISO Z39.79-2001—*Environmental Conditions for Exhibiting Library and Archival Materials* (American National Standard developed by National Information Standards Organization): Recommends values or ranges for light, humidity, temperature, and pollutants for temporary exhibitions and delineates acceptable practice for construction and use of exhibition cases and supports.

Underlying Principles of Risk-Based Environmental Management

No matter which guidance is followed, target criteria for interior collections environments for conservation can only be effective if they are realistic and can be maintained within the context of the climate and building envelope performance, as well as institutional capacity. This principle can be applied to the decision-making process for establishing conservation

Table 3.2

Classification of climate control potential in buildings (from ASHRAE 2011, table 4; adapted from Conrad 1995).

Category of Control	Building Class	Typical Building Construction	Typical Type of Building	Typical Building Use	System Used	Practical Limit of Climate Control	Class of Control Possible
Uncontrolled	I	Open structure	Privy, stocks, bridge, saw-mill, well	No occupancy, open to viewers all year	No system.	None	D (if benign climate)
	II	Sheathed post and beam	Cabins, barns, sheds, silos, ice-house	No occupancy. Special event access.	Exhaust fans, open windows, supply fans, attic venting. No heat.	Ventilation	C (if benign climate) D (unless damp limate)
Partial control	III	Uninsulated masonry, framed and sided walls, single-glazed windows	Boat, train, lighthouse, rough frame house, forge	Summer tour use. Closed to public in winter. No occupancy.	Low-level heat summer exhaust ventilation, humdistatic heating for winter control.	Heating, ventilating	C (if benign climate) D (unless hot, damp climate)
	IV	Heavy masonry or composite walls with plaster. Tight construction, storm windows	Finished house, church, meeting house, store, inn, some office buildings	Staff in isolated rooms, gift shop. Walk-through visitors only. Limited occupancy. No winter use.	Ducted low-level heat. Summer cooling, on/off control, DX cooling, some humidification. Reheat capability.	Basic HVAC	B (if benign climate) C (if mild winter) D
Climate controlled	V	Insulated structures, double glazing, vapor retardant, double doors	Purpose-built museums, research libraries, galleries, exhibits, storage rooms	Education groups. Good open public facility. Unlimited occupancy.	Ducted heat, cooling, reheat, and humidification with control deadband.	Climate control, often with seasonal drift	AA (if mild winters) A B
	VI	Metal wall constructions, interior rooms with sealed walls and controlled occupancy	Vaults, storage rooms, cases	No occupancy Access by appointment.	Special heating, cooling, and humidity control with precision constant stability control.	Special constant environments	AA A Cool Cold Dry

environments in any climate zone, and it can result in the balanced protection of the building fabric, historic and nonhistoric, from unintended consequences and damage caused by attempts to maintain inappropriate interior conditions. The New Orleans Charter for the Joint Preservation of Historic Structures and Artifacts (see sidebar) provides a framework for an integrated conservation approach; it is aimed at managers and administrators.

New Orleans Charter for Joint Preservation of Historic Structures and Artifacts

The New Orleans Charter, which evolved from two symposia called "Museums in Historic Buildings," held in 1990 (Montreal) and in 1991 (New Orleans), recognizes the need to place equal importance on the care of a building and of the collection housed within. Among the ten principles adopted were the identification of conservation needs only after adequate study of the building and collection, the need for interdisciplinary collaboration, and the need for appropriate documentation of all project stages. The New Orleans Charter has been adopted by the American Institute of Conservation (AIC), the Association for Preservation Technology International (APTI), the National Conference of State Historic Preservation Officers (NCSHPO), the American Institute of Architects (AIA), and the American Association of Museums (AAM). Text of the New Orleans Charter can be found at: http://cool.conservation-us.org/bytopic/ethics/neworlea.html.

Although ASHRAE "Museums, Galleries, Archives and Libraries" does not specifically address the special considerations of hot and humid climates, the underlying principles on which the ASHRAE chapter is founded are still relevant to the challenges of these climates. The principles include:

- consideration of the implications of historical climate exposure on the collections and their present state;
- determination of the historical average humidity that an object has experienced, which must be taken into account when setting target averages for humidity in a managed environment—a task that requires consideration of the material implications of a target humidity that is substantively different from the historical average annual humidity;
- determination of the specific environmental factors presented by the climate and climatic context of the location and of the ways in which these factors might present risks to the collections;
- identification of the environmental vulnerabilities of the collections materials, and of the ways in which these vulnerabilities, combined with specific climatic factors, result in risks;
- consideration of the environmental risks to the collections in the context of their relative significance to all risks, including fire, flooding, and theft or vandalism;
- recognition of the thermal and moisture performance limitations of the building envelope, which sets the constraints on what interior conditions can be maintained in a given climate, and, in the case of a historic building containing collections, recognition that the building may have its own

set of material vulnerabilities and climate risks, particularly if the target interior conditions are substantially different from exterior conditions;

- strategies for mitigating the risks presented by the exterior climate, given the specific collections vulnerabilities and the performance capacity of the building envelope, including setting of target specifications that are realistic and achievable and that can be met through the use of simple, robust, reliable, energy-efficient strategies and systems.

Prioritizing Risks of Collections Damage in Hot and Humid Climates

Here we summarize chapter 2, which described the three general mechanisms—biological, mechanical, and chemical—through which collections deterioration and damage occur in hot and humid climates.

Biological damage—the greatest risk

Chapter 2 argued that the greatest risk to collections in hot and humid climates is biological damage due to prolonged periods of high humidity that are conducive to the growth of microorganisms such as fungi and bacteria.

Mechanical damage—a risk where climate varies widely

Chapter 2 concluded that mechanical deterioration in hot and humid climates constitutes a serious risk to collections in regions where dry seasons and humid or wet seasons result in wide variations of humidity and possibly temperature. However, in regions with narrow humidity variations, mechanical damage is not as rapidly destructive as biological damage.

Chemical damage—a lesser risk

Chapter 2 showed that, from a conservation standpoint, chemical damage in hot and humid climates, enhanced by elevated temperature and possibly humidity, represents a lesser risk to collections relative to biological or mechanical damage. The chapter postulates that the majority of surviving collections in hot and humid climates have already undergone initially rapid rates of chemical deterioration by processes such as hydrolysis or corrosion, and subsequent reaction rates have slowed with the depletion of reaction sites on objects.

Environmental Considerations Unique to Hot and Humid Climates

In hot and humid climates, the average environmental conditions of temperature and humidity to which cultural heritage has been historically exposed differ significantly from the conventionally accepted average museum environment of 21°C ± 2°C, 50% ± 5% RH. For example, collections and heritage buildings in Belém, Brazil, which is classified as climate zone 1A (Very Hot-Humid) and which is the location for a case study (see chap. 13), have been exposed to an average annual temperature of 27°C and daily humidity ranging from roughly 73% to 96% RH (weatherbase.com). Subjecting collections and heritage building envelopes in this hot and humid location to conventional museum environmental conditions could activate dimensional changes and mechanical stresses and possibly result in new damage. Therefore, for hot and humid climates, a different approach to specifying average humidity criteria is required.

Consider the choice of a target average humidity. Since 75% RH is the nominal risk threshold for fungal activity, the humidity in the collections space must be maintained below this threshold—but it must not be so low that other risks, such as rebound on loss of system or mechanical stress, are introduced as negative consequences.

Temperature targets in hot and humid climates require similar evaluation. However, in hot and humid climates, historical average annual and seasonal temperatures are typically very high, and lowering temperature in the collections space without adding dehumidification will exacerbate the high-humidity problem.

A Conservation Environment Classification Protocol for Hot and Humid Climates

Conventional standards and guides for environmental collections management in temperate climates are problematic when applied in hot and humid climates.[1] In response, the authors have developed a classification protocol for target environmental conditions that is specific for risks to mixed cultural heritage collections in hot and humid climates. This protocol, referred to here as the Conservation Environment Classification–Hot and Humid (HH) protocol, is based on the principles of risk-based guidance for managing the collections environment of ASHRAE "Museums, Galleries, Archives and Libraries" and is intended for Very Hot-Humid (1A), Hot-Humid (2A), Warm-Humid (3A), and Mixed-Humid (4A) climate zones, as well as Marine (C) climate zones, which are subject to high moisture loads, though they lack high thermal energy.

The Conservation Environment Classification–HH protocol addresses the following conservation objectives and criteria:

- minimize the primary risk of biological damage, especially microorganism damage, to a collection due to elevated humidity levels, common in hot and humid climates;
- minimize dimensional change and resultant mechanical stresses in objects due to short-term and seasonal variations in humidity and temperature;
- keep collections at temperature and humidity conditions close to the values of historical averages previously experienced by the collection, avoiding the risk of mechanical stress that would be induced by a shift away from an object's native climate (see sidebar "Historical Data");

Historical Data

While an object's or collection's history of humidity and temperature exposure provides helpful information for use in determining target conditions for an environmental management system, choosing what "historical" data to analyze can pose difficulties for a project. For example, an object may have been exposed to a variety of ambient (unsheltered or sheltered) and interior (unconditioned or conditioned) environments for differing lengths of times. An object originally created in an unsheltered ambient environment may have been repaired and housed in a conditioned museum storage or gallery for decades. Though analysis of historical interior conditions may seem reasonable in some cases, it may be prudent to examine historical ambient conditions from the locality in which the object was created, as this would provide a more conservative estimate of mechanical risk due to deviation from the historical mean. It should also be noted that in the absence of a mechanical conditioning system, mean interior temperature and humidity will be similar to ambient conditions although they will display a smaller range of variation. In the case of a collection, determining the appropriate historical dataset to analyze may be further complicated by a collection that contains individual objects with different environmental histories.

After deciding upon a historical dataset to analyze for an object or collection, one is often confronted with a lack of information specific to one's needs. For example, historical climate data from the nearest weather station may not be representative of a site because of geographical and contextual or situational differences (including natural and artificial features). While one may establish a weather station onsite to record local conditions, the limited time frame of this particular data would restrict its use as a historical benchmark for the objects.

Knowing these caveats, however, one may still be justified in the use of a historical dataset that is less than ideal. For example, use of a less-than-representative historical dataset from a nearby weather station may be the best option as a general site descriptor, particularly if the differences between the two sites are reasonably small. If the appropriate dataset is truly site specific and the differences in climate between the nearest weather station and the site are large, a limited, recently assembled dataset might be suitable as a proxy for historical data.

- identify the degree of risk for accelerated chemical damage from elevated temperature.

The Conservation Environment Classification–HH protocol (table 3.3) is based on three prioritized risk factors—microbial, mechanical, and chemical—and presents a series of humidity and temperature environmental targets and classification criteria for conservation environments for mixed collections in hot and humid climates.

How to Read the Conservation Environment Classification–HH Protocol Table

In table 3.3, the three risk factors—microbial, mechanical, and chemical—are organized vertically at left as risk categories. The next column to the right identifies the specific risk in each category. Similar in methodology to ASHRAE "Museums, Galleries, Archives and Libraries," the Conservation Environment Classification–HH protocol uses historical annual averages as reference values, since mechanical risk will increase if the environment shifts away from an object's native climate. The next six columns—humidity (first three columns) and air temperature (second three columns)—provide the statistical parameter (see sidebar "Percentiles"), criteria, and class for the environmental conditions of concern.

Table 3.3

Conservation Environment Classification–Hot and Humid (HH) protocol showing humidity and temperature criteria for mixed collections in hot and humid climates.

Risk		Humidity (% RH)			Temperature (°C)		
Category	**Specific Risk (statistic)**	**Parameter**	**Criteria**	**Class**	**Parameter**	**Criteria**	**Class**
Microbial risk[1] (dominant risk in hot and humid climates)	Germination threshold	97.5 percentile (P97.5, x)	$x \leqq 65$	A[6]	*Microbial activity will typically remain dormant at temperatures above 40°C.*		
			$65 < x \leqq 70$	B			
			$70 < x \leqq 75$	C			
			$x > 75$	F			
Mechanical risk[2,3] (overall class determined by lowest class of any specific mechanical risk)	Short-term variation	Rolling 24 hr variation, 95th percentile (P95, x)	$\Delta 0 \leqq x \leqq \Delta 10$	a	Rolling 24 hr variation 95th percentile (P95, x)	$\Delta 0 \leqq x \leqq \Delta 4$	a
			$\Delta 10 < x \leqq \Delta 20$	b		$\Delta 4 \leqq x \leqq \Delta 10$	b
			$x > \Delta 20$	f		$x > \Delta 10$	f
	Seasonal variation	Absolute difference in seasonal means (means ≦ 70% RH, x)[5]	$\Delta 0 \leqq x \leqq \Delta 10$	a	Absolute difference in seasonal means (x)	$\Delta 0 \leqq x \leqq \Delta 10$	a
			$\Delta 10 < x \leqq \Delta 20$	b		$\Delta 10 < x \leqq \Delta 20$	b
			$x > \Delta 20$	f		$x > \Delta 20$	f
	Deviation from historical mean	Absolute difference between annual and historical mean (x)	$\Delta 0 \leqq x \leqq \Delta 10$	a	Absolute difference between annual and historical mean (x)	$\Delta 0 \leqq x \leqq \Delta 10$	a
			$\Delta 10 < x \leqq \Delta 20$	b		$\Delta 10 < x \leqq \Delta 20$	b
			$x > \Delta 20$	f		$x > \Delta 20$	f
Chemical risk[4]	Deviation from historical mean	*Consider condensation risk at high humidity and depressed surface temperatures.*			Difference between annual and historical mean (x)	$x > \Delta 5$	++
						$\Delta 3 < x \leqq \Delta 5$	+
						$\Delta{-3} \leqq x \leqq \Delta 3$	0
						$\Delta{-5} \leqq x \leqq \Delta{-3}$	–
						$x < \Delta{-5}$	– –

1 While adherence to Class C criteria will limit most microbial activity, meeting criteria for Class A or Class B will further extend the time period until mold germination and will restrict growth.
2 Due to variations in material response times to humidity and temperature fluctuations and construction techniques, the risk of mechanical damage is object specific.
3 If deliquescent salts are present, one must maintain humidity below the deliquescence point to limit the risk of salt-related mechanical damage.
4 Chemically unstable collections should be stored in cold (–20°C) or cool (10°C) conditions, while only stable metal collections treated against corrosion or having natural patina may be stored at conditions above 60% RH.
5 If seasonal mean humidity exceeds 70% RH, mechanical risk due to seasonal humidity variation is assigned to Class f. Above 70% RH, small shifts in humidity can result in large changes in a material's equilibrium moisture content and may lead to irreversible dimensional change.
6 Though a lower humidity limit is not given for Class A microbial risk, the maintenance of an excessively low humidity condition may introduce mechanical risk by deviating far from the historical humidity mean.

Class Designations and Risk Levels

Microbial Class
A: no risk B: low risk C: moderate risk F: high risk

Mechanical Class
a: no risk b: moderate risk f: high risk

Chemical Class
++: significantly increased risk
+: moderately increased risk
0: similar risk
–: moderately reduced risk
– –: significantly reduced risk

Percentiles

While datasets that exhibit normal or Gaussian distributions can be adequately described statistically using only mean values and standard deviation (a measure of the dispersion of data around its mean), non-Gaussian datasets, commonly found in environmental monitoring, require a more detailed analysis. Percentiles, which are thresholds below which a percentage of observations fall, can be used to provide a more nuanced statistical view of a dataset with a non-normal distribution.

The Conservation Environment Classification–HH protocol uses a percentile of the sample data, which is a common technique to strengthen reliability of the dataset. In the case of microbial risk, the protocol specifies use of the P97.5 value, which represents the top value within the middle 95% of the dataset (extending from P2.5 to P97.5). Short-term mechanical risk examines the P95 value of variations within rolling 24 hour periods.

Microbial risk

Microbial risk constitutes the primary risk to collections in hot and humid climates, and its importance is reflected in the use of capital letters in the "Class" column of table 3.3. The specific risk is the germination threshold for fungal and bacterial spores. For humidity, the parameter is the 97.5 percentile (shown as P97.5) rather than its maximum value; thus, short and uncommon excursions of humidity are excluded. Such brief periods of high humidity, if randomly distributed as expected in hot and humid climates, will not harm the object. The allowable ranges of acceptable humidity are less than or equal to 75% RH and are segmented into Class A (P97.5 $\leqq$ 65% RH) for no risk, Class B (P97.5 $>$ 65% RH $\leqq$ 70% RH) for low risk, and Class C (P97.5 $>$ 70% RH $\leqq$ 75% RH) for moderate risk. While adherence to Class C criteria will significantly restrict most microbial activity, collections environments meeting the criteria for Class A or Class B will extend the time period necessary for mold germination and will restrict growth. No threshold for low humidity is given for Class A, but humidity conditions that are substantially lower than the historical average will result in mechanical risk. Interior environments with a P97.5 exceeding 75% RH are designated Class F, high risk, and pose significant risk of biological damage to collections.

Mechanical risk

Mechanical risk is subdivided into three specific environmental risks in humidity and temperature—short-term variation, seasonal variation, and deviation from historical mean. In the Class column in table 3.3, mechanical risk is represented by lowercase letters that reflect its lesser importance with respect to microbial risk. For each specific factor for mechanical risk, the criteria for Class a (no risk) and Class b (moderate risk) roughly correspond to the short-term and seasonal humidity and temperature ranges given by ASHRAE "Museums, Galleries, Archives and Libraries" for precision environmental control classes. Interior environments meeting Class f (high risk) criteria exceed these ranges and pose a risk of mechanical damage to collections.

Short-term variation

The Conservation Environment Classification–HH protocol defines short-term variations as P95 percentile values of fluctuations within a rolling 24 hour period, rather than as their maximum values, thereby excluding short and uncommon excursions of humidity—corresponding to the environmental response times of objects such as paper, textiles, bone, ivory, and marquetry—and short-term variations so defined, therefore, represent a conservative assessment; short-term variations over longer durations may be more relevant for objects such as furniture with extended equilibration times.

Seasonal variation

The protocol for mechanical risk due to seasonal variation specifies that seasonal humidity means must be less than or equal to 70% RH; above this threshold, small shifts in humidity may have a disproportionate effect on an object's equilibrium moisture content and, as a consequence, on its mechanical stress. Thus, if a seasonal humidity mean exceeds 70% RH, the mechanical risk from seasonal humidity variation is Class f. If objects or building fabrics contain deliquescent salts, mechanical risk may result from in situ salt crystallization when humidity cycles about the deliquescence point. In this case, the salt compound must be identified, and the deliquescent humidity for the salt becomes the upper threshold for humidity (if <70% RH).

Deviation from historical mean

For mechanical risk due to a deviation from the historical mean, absolute differences between the annual means and the historical means are evaluated, as movement away from an object's native environmental conditions can result in mechanical stress in the form of dimensional change or permanent deformation.

Overall class of mechanical risk

While mechanical risk is evaluated in the six categories of specific risk—short-term variations, seasonal variations, and deviations from historical means of both temperature and humidity—the overall class of mechanical risk is determined by the lowest class (the highest risk) of any specific risk. For example, if five of the six specific mechanical risks are assigned to Class a and the sixth specific mechanical risk is assigned to Class b, the overall mechanical risk of the interior environment would be Class b.

Chemical risk

Following microbial and mechanical risk in table 3.3, chemical risk is indicated by the deviation from historical mean temperature and is classified as Class ++ (significantly above historical mean), Class + (above historical mean), Class 0 (similar to historical mean), Class – (below historical mean), or Class – – (significantly below historical mean). The maximum (above +5°C) and minimum (below −5°C) thresholds were based on the principle that a 5°C increase or decrease in temperature will result approximately in a doubling or halving, respectively, of the chemical reaction rate (Michalski 2002). Although increased temperature will accelerate chemical deterioration processes, the Conservation Environment Classification–HH protocol criteria allow moderate shifts in interior temperature so that conservation heating strategies can be used to manage interior humidity and reduce microbial risk.

It should be reemphasized that the Conservation Environment Classification–HH protocol applies only to mixed cultural heritage collections. Chemically unstable collections and special metal collections will require particular storage conditions beyond the specifications listed above.

Using the Protocol: Sample Overall Classification for a Collections Environment

The following hypothetical example demonstrates how to apply the Conservation Environment Classification–HH protocol to the environment of a gallery collection.

Museum staff gathered the temperature and humidity data from environmental monitoring conducted in the gallery, as well as from the local weather station, and statistically processed the data to obtain the following information:

1. P97.5 of humidity: 68% RH
2. P95 of rolling 24 hour humidity and temperature variations: Δ7% RH and Δ3.1°C
3. seasonal mean humidity and temperature: wet season, 70% RH and 22°C; dry season, 58% RH and 27.4°C
4. annual mean humidity and temperature: 64% RH and 24.7°C
5. historical mean humidity and temperature at the weather station: 72% RH and 21.3°C

Microbial risk

Based on the information in list item 1, the microbial risk is Class B (low risk), as P97.5 of interior humidity (68% RH) is between 65% and 70% RH.

Mechanical risk

The mechanical risk is evaluated for six specific criteria: short-term humidity and temperature variation, seasonal humidity and temperature variation, and deviation from historical mean humidity and temperature.

Calculation of short-term variation

Short-term variation is calculated from the information in list item 2. The P95 value of the rolling 24 hour variation for humidity is Δ7% RH (criteria range: Δ0% ≦ x ≦ Δ10% RH), and P95 for temperature is Δ3.1°C (criteria range: $\Delta 0^{\circ}\mathrm{C} \leqq x \leqq \Delta 4^{\circ}\mathrm{C}$). Thus, the mechanical risk due to short-term variation is Class a (no risk) for both humidity and temperature.

Calculation of seasonal variation

Seasonal variation is calculated from the information in list item 3. The mechanical risk due to seasonal humidity variation is Class b (moderate risk), owing to an absolute difference in seasonal humidity means of Δ12% RH (criteria range: $\Delta 10\% < x \leqq \Delta 20\%$ RH). Note that both seasonal humidity means are less than or equal to 70% RH, satisfying one of the criteria for this specific mechanical risk. The mechanical risk due to seasonal temperature variation is Class a, resulting from an absolute difference in seasonal temperature means of Δ5.4°C (criteria range: $\Delta 0^{\circ}\mathrm{C} \leqq x \leqq \Delta 10^{\circ}\mathrm{C}$).

Calculation of deviation from historical mean

Deviation from historical mean is calculated from information in list items 4 and 5. Mechanical risk due to deviation from the historical mean—determined by the absolute difference in annual and historical means—is Class a for both humidity (Δ8% RH; criteria range: $0\% \leqq x \leqq 10\%$ RH) and temperature (Δ3.4°C; criteria range: $\Delta 0^{\circ}\mathrm{C} \leqq x \leqq \Delta 10^{\circ}\mathrm{C}$).

Determining overall class of mechanical risk

Given the above, the overall mechanical risk for this collection is Class b. This determination is based on the highest risk classification—in this case, Class b, due to seasonal humidity variation—for any of the six specific mechanical risks.

Chemical risk

Chemical risk is based on the difference between annual and historical temperatures. In this case, the risk is Class + (moderately increased risk) owing to a difference of Δ3.4°C (criteria range: $\Delta 3^{\circ}\mathrm{C} < x \leqq \Delta 5^{\circ}\mathrm{C}$).

Overall risk classification results

The overall risk classification for this hypothetical interior environment is described as Bb+: low microbial risk, moderate mechanical risk, and moderately increased chemical risk.

In part 2, the Conservation Environment Classification–HH protocol will be applied to environmental datasets collected in seven environmental management case studies located in Very Hot-Humid (1A), Hot-Humid (2A), Mixed-Humid (4A), and Mixed-Marine (4C) climate zones.

Summary

This chapter emphasized the importance of historical average exposures to temperature and humidity for collections in nontemperate climates, and it addressed the considerations that must be applied when setting target conditions for collections environments in hot and humid climates. It set out the necessity of balancing the environmental vulnerabilities and needs of the collections with the performance capabilities and vulnerabilities of the building envelope. Building upon the risk-based logic used to develop ASHRAE's "Museums, Galleries, Archives and Libraries," the authors developed a Conservation Environment Classification–HH protocol specific to hot and humid climates.

This classification protocol can be used as a basis for the analysis of existing interior environments and can help identify opportunities for improving conservation environments in relation to the specific vulnerabilities of a collection, as will be discussed in chapter 8.

The next chapter will take up another important consideration of interior environmental management in hot and humid climates: occupant health and safety and thermal comfort.

References

ASHRAE
2011 Museums, galleries, archives and libraries. Chapter 23 in *ASHRAE Handbook—HVAC Applications*. Atlanta: ASHRAE.

Conrad, E.
1995 A table for classification of climatic control potential in buildings. Norwalk, CT: Landmark Facilities Group.

Michalski, S.
2002 Double the life for five degrees drop, double the life for half the RH. *Thirteenth Triennial meeting, ICOM-CC, Rio de Janeiro, 22–27 September 2002*. London: James and James.

———.
2007 The ideal climate, risk management, the ASHRAE chapter, proofed fluctuations, and toward a full risk analysis model. Contribution to the Experts' Roundtable on Sustainable Climate Management Strategies, held in April 2007 in Tenerife, Spain. http://getty.museum/conservation/our_projects/science/climate/paper_michalski.pdf.

Thomson, G.
1994 *The Museum Environment*. 2nd ed. Butterworth-Heinemann Series in Conservation and Museology. Oxford: Butterworth-Heinemann.

Note

1 The exception would be museum buildings and systems specifically designed and constructed to maintain conventionally accepted conservation conditions (21°C ± 2°C, 50% ± 5% RH) in hot and humid climates.

Chapter 4

Occupant Health and Comfort in Hot and Humid Climates

From earliest times, the most basic function of shelter has been to protect occupants and contents from weather. As shelter evolved into sophisticated buildings with systems for interior heating and cooling, occupant expectations also evolved as to what constitutes an acceptable level of thermal comfort, beyond essential protection from weather. At present, people spend roughly 90 percent of their time inside buildings, and potential exposure to indoor pollutants is a concern (EPA 2001). As a result, building occupants expect that a building and its systems will provide an interior environment that is safe and healthy and has a high degree of thermal comfort.

This chapter considers these issues of occupant health and thermal comfort in the context of hot and humid climates. These issues are important, because museum and archives buildings must address both the thermal comfort and the safety of occupants and the conservation needs for the collections. An understanding of thermal comfort in hot and humid climates is also necessary if alternatives to conventional air-conditioning are to be successfully applied. An understanding of thermal comfort is also needed for Marine-type climates, described in chapter 1. Although they are not subject to elevated high summer temperatures, they may occasionally have temperatures in the cool range and high humidity in winter.

The environmental health of building occupants is more important than their thermal comfort, although the needs may coincide in extreme climatic conditions. However, regulations concerning occupant health vary greatly between governmental jurisdictions, and it is impractical to address all health requirements in this publication. Therefore, this chapter will cover only the essential matters related to occupant health, especially with respect to ventilation.

This chapter presents information on:

- occupant health and indoor air quality
- occupational noise and acoustic comfort
- thermal comfort for building occupants
- priorities for occupant health and comfort

Since architecture and engineering professionals are likely already aware of the issues posed by occupant health and thermal comfort, this chapter is directed to the museum professional in order to provide basic information to enable discussion with other team members.

Occupant Health and Indoor Air Quality

An important aspect of occupant health in buildings is the quality of the indoor air. As a result, acceptable indoor air quality (IAQ) is a principal objective of interior environmental management and includes control of airborne chemical and biological hazards. Chemical hazards are typically gaseous or particulate, and biological hazards include fungi, bacteria, viruses, and other organisms. In both cases, occupant exposure to the air contaminant occurs by inhalation, ingestion, or dermal contact/absorption. Occupant risk is dependent on the toxicity of the specific contaminant, the degree of exposure, and the physiological vulnerability of the occupant.

Indoor air quality has become an important consideration for a number of reasons:

- People spend more time inside buildings.
- Unintentional ventilation due to air infiltration through building envelopes has been reduced in order to reduce energy consumption by mechanical systems.
- Intentional ventilation through mechanical systems has been reduced for the same reason.
- Unintentional ventilation of building spaces containing collections must be reduced to limit penetration of light and infiltration of pests, pollutants, and moisture vapor.
- The increased use of synthetic and manufactured building materials, finishes, and furnishings can generate additional contaminants.

The net effect of these factors has been the reduction of outside air available for dilution of contaminants. The transformation of a naturally ventilated building into a building with a mechanically controlled interior environment may reveal interior air contaminant sources that were previously diluted by high air infiltration rates.

This section provides an overview of topics relating to occupant health, including:

- airborne contaminants,
- hazardous collections,
- determining target conditions for occupant health,
- control strategies for indoor air quality.

Airborne contaminants

In general, airborne contaminants that present occupant health concerns in cultural institutions fall into four classifications:

- particulates
- bioaerosols
- volatile organic compounds
- inorganic gases

Particulates

Particulate contaminants are small solid materials suspended in the air. Outdoor particulates originate from natural sources (e.g., dust storms and fires) and anthropogenic activities (e.g., diesel engine exhaust, power plant operation, and industrial and agricultural pollution). Indoor particulates originate from activities such as cleaning, construction and renovation, the operation of office equipment (e.g., photocopiers, printers, and paper shredders), and combustion (e.g., tobacco smoking and fuels). Additional interior particulate sources include insulation, carpets, and other surfaces.

Particulates can be categorized by size: coarse (diameter 2.5–10 μm), fine (diameter <2.5 μm), and ultrafine (diameter <0.1 μm). From a health perspective, ultrafine particles are of heightened interest, as this size of particulates has been correlated with respiratory illness due to deeper penetration into the lungs (Mühlfeld, Gehr, and Rothen-Rutishauser 2008).

Bioaerosols

A bioaerosol is an airborne material that contains or was released by a living organism; bioaerosols include fungi, bacteria, viruses, protozoa, algae, pollen, and animal matter. Living organisms require adequate nutrients for growth, and the interior environment can provide an abundant source (e.g., dust and wood). Most living organisms survive in a relatively narrow range of thermal conditions, which also fall within the temperature range for human comfort.

Living organisms also require adequate moisture for growth, and in hot and humid climates, moisture is naturally abundant in liquid and vapor states, both of which can infiltrate to the interior environment. Additional interior moisture sources include water supply and waste systems, water-based mechanical systems, envelope leaks, and envelope drying. Warm, pooled water in mechanical equipment, drip pans, fountains, or basements provides an environment conducive to biological growth.

Building occupants can introduce bioaerosols and biological contaminants through sneezing, coughing, and talking. Building materials can store bioaerosols that are later resuspended when the material is disturbed by activity.

Occupant health effects from bioaerosols may include respiratory impairment, infectious diseases such as Legionnaires' disease, and cancer from mycotoxins such as *Aspergillus flavus* and *Stachybotrys chartarum* (Douwes et al. 2003).

Volatile organic compounds

Volatile organic compounds (VOCs) encompass a wide range of naturally produced or man-made hydrocarbon-based chemicals that readily evaporate or off-gas at room temperature; hot and humid climates may accelerate the rate of evaporation or off-gassing. Indoor sources of VOCs include building materials, furnishings and finishes, office equipment such as copiers and printers, graphics and craft materials, pesticides, and cleaning supplies. As with bioaerosols, building surfaces such as carpet or ceiling tiles can store VOCs and release them later (Berglund, Johansson, and Lindvall 1988). In some buildings, indoor VOC concentrations can exceed outdoor concentrations by a factor of ten (EPA 2014b).

Symptoms of VOC exposure in building occupants include: irritation of the eyes, nose, and throat; headaches; and nausea. Some commonly occurring VOCs, such as benzene, have been identified as carcinogenic (EPA 2014b).

Inorganic gases

Inorganic gases can also pose a hazard to occupant health. Combustion is the major nonindustrial source of inorganic gases such as carbon monoxide, carbon dioxide, nitrogen oxides and sulfur oxides (EPA 2014a). Indoor sources of inorganic gases include human respiration (carbon dioxide), office equipment (ozone and nitrogen oxides from photocopiers), and cleaning supplies (ammonia).

Potential occupant health effects of inorganic gases include pulmonary and respiratory symptoms, such as perceived shortness of breath (carbon dioxide), mucous irritation (nitrogen oxides and sulfur dioxide), inflammation (ozone), and asphyxiation (carbon monoxide).

Hazardous collections

Certain collections present health risks for building occupants interacting with the collection or its environment. In the past, many collections, especially those in hot and humid climates, have undergone treatments involving toxic chemicals, such as arsenic, formaldehyde, and naphthalene to reduce biological risks. Other collections may be intrinsically unsafe because of the content of their materials; examples include radioactive or chemically reactive mineralogical collections and industrial collections containing solvents or mercury. Chemically unstable collections such as those containing cellulose nitrate or acetate, polyvinyl chloride, or deteriorating polymers are also potential occupant health risks.

Collections with the above characteristics will require safety measures such as environmental segregation or ventilation and room exhausts (see sidebar "Risks Posed by Conservation Laboratories").

Determining target conditions for occupant health

Recognizing the significant impact of indoor air quality on population health, international and national organizations have established guidelines for indoor air quality, typically focusing on "classical" pollutants such as carbon monoxide, nitrogen oxides, and VOCs (Index project 2005). There are fewer regulations for bioaerosols because of the wide variety of types of microbial particulate matter and their effects on individual occupants (Morey 1990). Regulations for occupant health are specific to the location and jurisdiction of the project; therefore, environ-

Risks Posed by Conservation Laboratories

The laboratory setting, in which many artifacts are studied and treated, poses a special set of risks to occupants because of the potential for exposure to harmful physical, chemical, radioactive, or biological agents. Local and state regulations address the health and safety concerns in the laboratory environment. Typical laboratory requirements are the following:

- fume hood with regulated escape velocity
- exhaust with activated carbon to reduce emissions
- 100% outdoor air makeup
- 10 air changes per hour (ACH) or higher

mental management project teams should obtain appropriate professional guidance for identifying minimum targets for indoor air quality.

Strategies for indoor air quality

An industrial hygienist may suggest several strategies to limit chemical or biological hazards in the indoor air.

The most direct strategy is source elimination or reduction of the air contaminant. Examples of source control include: storing solvents and pesticides in special closed containers in well-ventilated areas, prohibiting indoor tobacco smoking, and effective filtration of outside air for ventilation. For bioaerosols, source control is applied to moisture in both liquid and vapor states in order to limit or arrest the activities of microorganisms. While improved housecleaning can be an effective strategy for mitigating bioaerosols, the potential effects of biocides or disinfectants on cultural heritage collections and historic interiors must be investigated before implementation.

Ventilation is a common strategy for improving indoor air quality (see sidebar "Implications of Ventilation"), and ventilation strategies include dilution ventilation and local exhaust ventilation. Dilution ventilation introduces filtered outside air into a space to reduce indoor contaminant concentration and is effective for low-toxicity contaminants. Local exhaust ventilation is necessary for the control of highly toxic contaminants and removes indoor contaminants at their source to prevent dispersal into the surrounding air. A fume extractor at a worktable is a common example of local exhaust ventilation. If exhaust flow rates are high relative to room size, a supply of conditioned or unconditioned outside air will be needed to replace the air removed by exhaust. The balance of supply and exhaust air affects room pressure, and positive room pressure can be used to limit the infiltration of uncontrolled outside air, while negative room pressure will prevent contaminants from exiting a room. Contaminant concentrations can also be controlled by recirculating room air through particulate filters, chemical filters, or air washers, as necessary.

Implications of Ventilation

While the introduction of outside air into a building by ventilation can improve indoor air quality, caution should be exercised so that potentially negative aspects of ventilation do not deteriorate the interior environment. Ventilation during periods of high outdoor humidity and/or temperature can negatively impact the interior environmental comfort or conservation conditions. The stability of the interior environment will be affected by the introduction of outside air at different temperature and humidity conditions than the interior air. Of particular concern in highly urban environments is the potential for introducing corrosive outdoor pollutants, such as nitrogen oxides or sulfur dioxide, through ventilation.

Typical HVAC systems are able to prevent problems associated with ventilation by conditioning outside air before it is introduced to the interior space. In doing so, one can apply psychrometric strategies to affect the humidity and temperature of the incoming air. The use of filtration and activated carbon systems can also reduce the pollutant load from the exterior.

The use of ventilation in the case studies described in part 2 relies on its conditional use when specific environmental conditions are met. Through programmed logic in a programmable logic controller (PLC) or other control device, the humidity and temperature conditions of the inside and outside air are continuously monitored to determine if ventilation is both needed and safe.

ANSI/ASHRAE Standard 62.1-2013, *Ventilation for Acceptable Indoor Air Quality* (ANSI and ASHRAE 2013b), is the basis for ventilation codes in the United States and specifies minimum ventilation rates for commercial buildings for occupant health. This standard provides an approach—the ventilation rate procedure (VRP)—that separates ventilation into occupancy- and area-based components and accounts for contaminants generated by both people and the building, while considering the efficiency of outside air transfer for various ventilation systems. The standard also provides a design-based indoor air quality procedure that uses mass balance analysis to maintain specific contaminant levels and uses a prescriptive natural ventilation procedure that provides design conditions for natural ventilation systems.

Although generated in the United States, ANSI/ASHRAE Standard 62.1-2013 has been referenced by international and regional indoor air quality standards related to ventilation.

Occupational Noise and Acoustic Comfort

An environmental management system will generate operational noise that can affect occupant health. Operational noise may result from rotating or moving equipment located inside and outside the building. System operation may also produce aerodynamic or fluid dynamic noise such as high-frequency sounds from air diffusers and piping or low-frequency rumble from ductwork. Noise may be transmitted as vibrations through building components and can be audible at locations distant from the source.

Noise

Acting as the fundamental unit of sound pressure level (SPL), the decibel (dB) is expressed as a logarithm of the comparative ratio of sound pressure with reference to the lowest threshold of hearing, which is 20 μPa. Healthy hearing typically extends three orders of magnitude, from 20 Hz to 20 kHz, but deterioration of hearing with age will narrow this range, particularly at higher frequencies. The human ear also responds better to higher frequencies. Thus, SPL is often calculated using standard frequency weighting curve "A," in which case, the unit is abbreviated as dBA.

High levels of background noise in the workplace can reduce worker productivity and lead to a high absentee rate. Subjective occupant assessments of typical noise sources range from "loud" for a window air conditioner at a distance of 1 m (~70 dBA) to "perceptible" for a flying insect at 1 m (~20 dBA) (see sidebar "Noise"). Design guidelines for HVAC-related background noise suggest maximum noise levels ranging from 25 dBA (theaters and concert halls) to 45 dBA (open-plan offices), depending on room use (ASHRAE 2011, 2013).

Thermal Comfort for Building Occupants

Human thermal comfort is a result of the body's interactions with its surrounding environment. These interactions can take the form of physical responses such as perspiring or shivering, psychological cues such as recent thermal history or control (or lack thereof) over the conditions in the space, or behavioral factors such as clothing choice or the extent of physical activity. Quantifiable definitions of thermal comfort conditions vary among individuals, but statistical analysis of empirical data leads to a definition of environmental parameters that result in acceptable thermal comfort for a majority of occupants.

This section focuses on:

- thermal comfort models
- physiological responses to the environment
- parameters for thermal comfort
- thermal comfort guidelines

Thermal comfort models

Current research proposes two approaches by which human thermal comfort can be assessed: the heat balance model and the adaptive thermal comfort model.

Heat balance model

The heat balance model is based on experiments in a well-controlled laboratory setting and focuses on the physics of heat transfer and the body's physiological response to its environment. Used widely since the 1980s to predict thermal comfort, this approach assumes a steady state heat balance to predict the impact of any combination of four environmental variables (air temperature, relative humidity,[1] radiant temperature, and air velocity) and two personal variables (metabolic activity and clothing) on the thermal comfort of individuals in a large population of test subjects. As a result, this model emphasizes static, consistent interior environmental conditions.

In the heat balance model, human thermal comfort is quantified by the predicted mean vote (PMV) and predicted percentage dissatisfied (PPD) method. The PMV is the average comfort vote of occupants based on a seven-point index of cold (–3) to hot (+3), with neutral (0) as ideal and the limits of variability for neutrality ranging from –0.5 to +0.5. The PPD indicates the percentage of a population that would consider the environment to be unacceptable (Fanger 1970). Acknowledging the variability in individual thermal comfort, this model typically seeks 80% approval for a group of sedentary occupants, allowing for 10% dissatisfaction in both general and local thermal discomfort.

Adaptive thermal comfort model

The adaptive thermal comfort model is based on the principle that "if a change occurs such as to produce discomfort, people react in ways which tend to restore their comfort" (Dear and Brager 1998; Nicol and Humphreys 2002). A person's response to environmental stress can take the form of physiological, behavioral, and psychological adaptations. While the heat balance model assumes that the human body is a passive receptor of thermal stimuli, the adaptive approach treats an occupant as an active participant in determining personal thermal comfort. The adaptive model validates its approach through results obtained from field studies on dynamic thermal comfort, as opposed to the static laboratory studies relied upon by the heat balance model.

Physiological response to the environment

The metabolic reactions that drive human life produce heat that is exchanged with the environment in order to maintain a core body temperature of approximately 37°C. In the adaptive model of thermal comfort, physiological adaptations to thermal comfort are achieved through genetic adaption (generation to generation) and through acclimatization (within a generation). The physiological processes by which the human body exchanges heat with the environment are controlled by the hypothalamus in the brain.

In hot environments, body heat is dissipated by increased blood flow to the skin (vasodilation) and by production of perspiration. Vasodilation transfers internal heat to the skin surface, where cooling takes place by radiant or convective heat transfer from the exposed skin to the cooler environment. However, heat transfer by cooling alone is inadequate when metabolic heat generation is large or when the surrounding environment is near, or above, core body temperature. Under these conditions, heat transfer by the evaporation of perspiration must supplement cooling by radiation and convection.

Parameters for thermal comfort

As previously stated, individual thermal comfort is a function of environmental, behavioral, and psychological factors; specifics of each category will be discussed in the following sections.

Environmental parameters for thermal comfort

Heat dissipation from the human body to the Hot-Humid environment is affected by four environmental variables:

- air temperature
- humidity
- mean radiant temperature
- air velocity

The influence of these parameters on the sensation of thermal comfort is complex, synergistic, and subject to time-dependent factors, such as conditions that are not steady state, and spatial factors, such as asymmetries in the surrounding air and surface temperatures. The resultant sensation of thermal comfort may apply to all, or only part of, a person's body.

Physiological cooling mechanisms are temperature dependent. The rate of cooling decreases with increasing temperature, and increasing humidity decreases cooling by limiting the evaporation of perspiration. As evaporation decreases, the percentage of skin covered with perspiration increases; wet skin correlates with sensations of thermal discomfort and unpleasantness (ASHRAE 2009). Where wet skin is covered by clothing, a clinging or sticky feeling exacerbates the sensation of thermal discomfort; this problem may be resolved by loose-fitting clothing or moisture-wicking fabrics.

Mean radiant temperature—the average temperature of surfaces surrounding the human body—affects the body's ability to exchange heat via radiative heat transfer. Lowering the mean radiant temperature will increase radiant heat removal from the body, and vice versa.

Air velocity affects the rate of evaporative and convective heat removal from a body. Increased air velocity results in a larger volume of air passing over the skin surface, allowing for more heat transfer by cooling and by evaporation of perspiration. Moving air at temperatures below that of the skin increases the rate of convective heat transfer from the skin. Even at high ambient humidity, increased air velocity will improve the rate of heat transfer by evaporation of perspiration. At elevated air speeds, however, moving air on bare skin may be considered unpleasant, despite more effective cooling.

Behavioral factors for thermal comfort

Behavioral factors include a range of individual adaptations to the environment with the goal of improving thermal comfort. Occupant actions can include personal choices such as changes in activity level or clothing. Behavioral responses have typically taken the form of cultural conventions such as midday breaks or siestas that shift work to the cooler, late afternoon hours. Recently, these cultural traditions have been altered by technology, as can be seen in the effect of HVAC systems on occupant clothing choice and the shortening or elimination of the midday break.

Physical activity affects the body's metabolic rate and, therefore, the amount of body heat that must be dissipated to the environment. Metabolic rates for various activities are defined in units of *met*; one met equals the amount of heat generated by a typical sedentary adult (58 W/m^2). Walking on a level surface at 0.9 m/s is 2 mets (115 W/m^2), and vigorous housecleaning ranges from 2.0 to 3.4 met (115–200 W/m^2).

While physical activity generates the body heat that must be removed (or retained) to maintain comfort, clothing acts as an insulator and affects the rate of convective heat transfer between the body and the environment. The thermal resistance provided by clothing is typically defined in units of *clo*, where 1.0 clo (0.155 $m^2 \cdot K/W$) is "comfortable" for a person at rest when the air temperature is 21.1°C. A short-sleeve shirt and trousers have a combined clothing value of 0.5 clo, and a business suit has a clothing value of 1.0 clo. The total clothing value is the sum of the clothing values for each article of clothing worn.

In hot and humid climates, with temperatures typically exceeding 21.1°C, the use of clothing with an overall rating of 1.0 clo can result in a sensation of being overly warm. While the clothing value refers specifically to convective heat exchange, appropriate clothing choices can address evaporative cooling through the use of natural fabrics like cotton and silk, which are of particular use at elevated humidity. As mechanical air-conditioning has become prevalent in buildings and transportation, many people no longer choose to wear traditional clothing that may have evolved in response to the ambient climate.

Psychological factors for thermal comfort

Psychological responses to ambient environmental conditions play a major role in an occupant's perception of thermal comfort in an environment. Though difficult to quantify, psychological factors affecting thermal comfort include expectations, thermal memory or history, and adaptive opportunity.

A person's expectations for a comfortable thermal environment will affect perception of thermal comfort. Occupants of naturally ventilated buildings are accustomed to, and therefore expect, larger variations in temperature and will tend to tolerate a wide temperature range. This tolerance may be increased with the availability of adaptive opportunities, as will be described below. In contrast, occupants in mechanically conditioned spaces will expect smaller variations in temperature; exposure to large temperature variations will likely result in complaints.

An occupant's short- and long-term thermal histories shape expectations of what constitutes a comfortable thermal environment. Short-term thermal history refers to recent experiences that influence thermal perception. For example, increased physical activity even for a limited period will typically shift thermal preference to cooler interior temperatures. Changes in the outdoor environment can also affect expectations of indoor thermal comfort, particularly for occupants of naturally ventilated buildings (Humphreys 1978). As exterior temperatures rise, building occupants may better tolerate a warmer interior environment. Long-term thermal history refers to an extended exposure to a set of environmental conditions. For example, the long-term thermal history of persons in hot climates allows them to work more effectively at warmer temperatures compared to persons acclimated to cold climates (ASHRAE 2009).

The opportunity for building occupants to exert individual control on their local environment can affect tolerance of interior environmental variability. It has been shown that a wider range of thermal conditions remain satisfactory when occupants are given some degree of control over their personal environment (e.g., Paciuk 1990; Williams 1995). In naturally ventilated buildings, these adaptive opportunities may include actions such as opening windows, closing shutters, drawing blinds, or operating fans. When adaptive opportunities for occupant control of the interior environment are limited, as in large open-plan, air-conditioned buildings with fixed windows, occupants have a narrower range of thermal tolerance.

Guidelines for thermal comfort

The physiological, environmental, behavioral, and psychological factors described above interact as a complex set of variables that result in occupant perceptions as to whether or not an environment is thermally "comfortable." While the physiological principles of metabolic heat generation and heat transfer between the body and the environment are well established and measurable, individual definitions of thermal comfort can vary greatly between building occupants. Because of the subjective nature of thermal comfort, guidelines have been developed that seek to provide acceptable thermal comfort for a significant percentage of typical occupants. These guidelines, based on field and laboratory data, detail combinations of thermal and personal variables. Even so, national regulations may sometimes deviate from the thermal range recommended by a standard. For example, a European survey of national thermal regulations found that roughly half of the countries had thermal limits that did not match the recommended values of European Standard EN 15251:2007–08 (Wargocki and Wyon 2007).

Several of the major thermal comfort standards are:

- ANSI/ASHRAE Standard 55-2013—*Thermal Environmental Conditions for Human Occupancy* (ANSI and ASHRAE 2013a);
- ISO 7730-2005—*Ergonomics of the Thermal Environment; Analytical Determination and Interpretation of Thermal Comfort Using Calculation of the PMV and PPD Indices and Local Thermal Comfort Criteria* (International Organization for Standardization). Based on the heat balance model, this standard is supplemented by a number of technical standards, including ISO 7726 (*Measurement*), ISO 8996 (*Metabolic Heat Production*), ISO 9920 (*Clothing Properties*), and ISO 10551 (*Subjective Assessment Methods*);

- EN 15251:2007–08—*Indoor Environmental Input Parameters for Design and Assessment of Energy Performance of Buildings Addressing Indoor Air Quality, Thermal Environment, Lighting and Acoustics* (European Standard). Presented earlier as *Indoor Air Quality and Acoustic Guidance*, this wide-ranging standard designates thermal design values for the indoor environment.

Aligned closely with ISO 7730:2005 and EN 15251:2007–08, ANSI/ASHRAE Standard 55-2013 informs the thermal comfort goals for interior spaces. Based largely on the heat balance model described above, this standard acknowledges the variability in individual thermal comfort and seeks acceptable conditions for a majority of occupants, permitting considerable latitude in establishing the thermal comfort target at the start of an environmental management project. The predicted mean vote (PMV)–predicted percentage dissatisfied (PPD) method also allows engineers to define boundaries of the thermal comfort zone for a specific set of conditions, including air temperature, mean radiant temperature, relative humidity, metabolic activity, clothing insulation, and air speed. The PMV index provides a mean characterization of occupant perception of the comfort of a set of conditions in a building environment, using a thermal sensation scale that extends from +3 (hot) to –3 (cold), with 0 being neutral. The PPD index expresses the thermal dissatisfaction of the same test group for the same set of conditions. The relationship between PMV and PPD is defined in figure 4.1 (see also sidebar "Three Classes of Comfort Control").

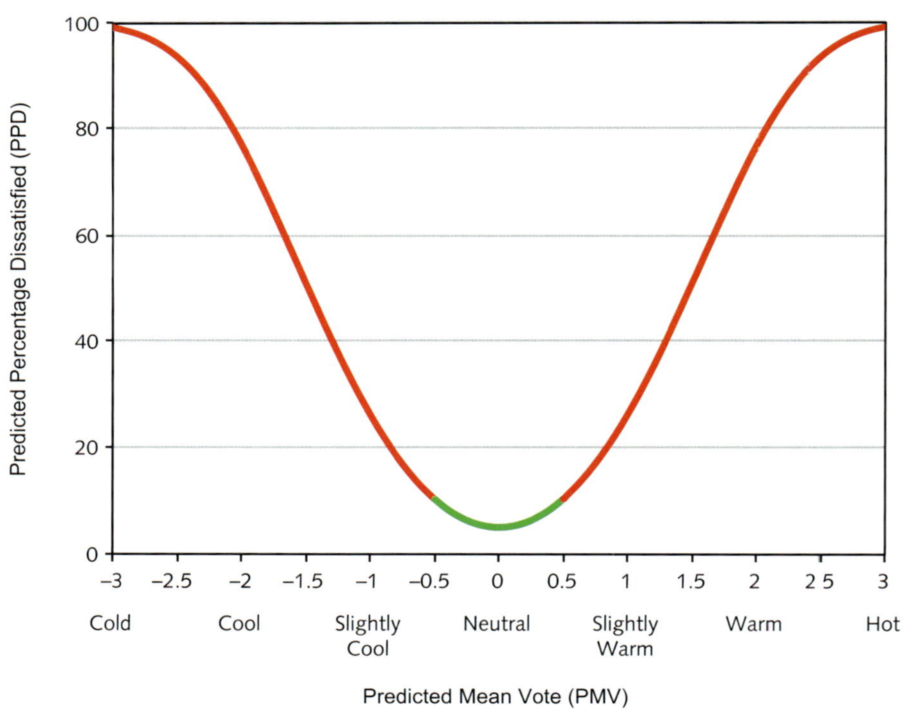

Figure 4.1

Relation between predicted percentage dissatisfied (PPD) and predicted mean vote (PMV). Adapted from ANSI/ASHRAE Standard 55-2013, fig. F3 (ANSI and ASHRAE 2013a).

Three Classes of Comfort Control

Paralleling the conservation classes determined for environmental control, the 2003 version of ASHRAE Standard 55 defined three classes of comfort based on varying levels of PMV and PPD. Class B (PPD < 10; $-0.5 <$ PMV < 0.5) was considered the typical level of comfort control for most applications, while Class A (PPD < 6; $-0.2 <$ PMV < 0.2) and Class C (PPD < 15; $-0.7 <$ PMV < 0.7) represented tighter and more relaxed levels of comfort, respectively. While the graphical method applied in the 2010 version of the ASHRAE Standard 55 assumes Class B control, use of a less-stringent level of thermal comfort may be more applicable to the transient occupancy of museum settings.

ANSI/ASHRAE Standard 55-2013 builds upon the PMV-PPD relation and presents graphic comfort zone and analytical comfort zone methods for determining the thermal comfort zone. The graphic comfort zone method delineates acceptable thermal comfort zones for occupants with activity levels between 1.0 and 1.3 met and clothing insulation between 0.5 and 1.0 clo, and for dew point temperatures below 16.8°C. The analytical comfort zone method is applicable up to metabolic rates of 2.0, clothing insulation above 1.0 clo, and dew point temperatures above 16.8°C. These methods delineate the thermal comfort zone on a psychrometric chart, which will be discussed in chapter 5. The use of elevated air speeds can also be incorporated into the analytical comfort zone method, resulting in an increase in the acceptable maximum operative temperature (defined as ½ dry bulb temperature + ½ mean radiant temperature). For museums and collection storage spaces in hot and humid climates, these methods for determining acceptable thermal conditions for occupied spaces may be overly conservative, since the standard assumes situations where comfort is related to the productivity of full-time occupants, rather than to the short-term occupation by visitors.

Recent iterations of ANSI/ASHRAE Standard 55 have also begun incorporating the principles of adaptive thermal comfort into their guidelines. ANSI/ASHRAE Standard 55-2013 presents a method based on the adaptive model for determining acceptable thermal conditions in occupant-controlled naturally conditioned spaces. (EN 15251:2007–08 also proposes an adaptive approach to thermal comfort for free-ventilation buildings.) This method assumes that the occupant is responsible for the regulation of personal thermal comfort within the space through actions such as the opening of exterior windows, the operation of local fans, and the adaptation of clothing insulation. A climatic criteria of this method, relevant to hot and humid regions, also requires that the mean exterior temperature be less than 33.5°C. While occupant control would typically not be permitted in the conservation environment of a museum, application of this method may be valid because of the transient nature of occupants in museum galleries.

The effect of shifting environmental conditions on thermal comfort can be demonstrated through the use of thermal comfort calculators. Based on empirical research and on the thermal comfort criteria discussed above, thermal comfort calculators provide an estimate of occupant satisfaction regarding thermal comfort for a given set of conditions. One example of a thermal comfort calculator is the ASHRAE Thermal Comfort Tool (see sidebar "Application of ANSI/ASHRAE Standard 55-2013: Using the ASHRAE Thermal Comfort Tool").

Application of ANSI/ASHRAE Standard 55-2013: Using the ASHRAE Thermal Comfort Tool

The ASHRAE Thermal Comfort Tool allows one to define an interior environment and classify occupation in order to estimate occupant response with respect to thermal comfort. The following examples of how to use the tool refer to points A, B, and C in figure 4.2.

Using the PMV model with elevated air speed, a hot and humid interior environment is defined as air and mean radiant temperature of 29°C and humidity of 65% RH, with an interior air speed of 0.1 m/s. Occupant activity is defined as 1 met (sedentary) and occupant clothing as 1 clo (light business suit). This initial set of environmental and personal conditions results in an expected perception of a warm environment by the occupant; the calculated PMV is 1.75, and the PPD is 64% (point A). These results significantly exceed the acceptable PMV and PPD thresholds of 0.5% and 10%, respectively.

If the same set of environmental and personal conditions is used except for changing personal clothing from 1.0 to 0.5 clo (trousers and shirt), the result is an occupant perception of a slightly warm environment, with a reduction in PMV (from 1.75 to 1.26) and PPD (from 64% to 38%) (point B). If interior air speed is then increased to 0.5 m/s, the occupant perceives a neutral environment with a PMV of 0.25 and a PPD of 6%, which are within the thermal comfort zone (point C). Alternatively, reductions in both the interior air and mean radiant temperatures from 29°C to 24.5°C in the initial example would also result in acceptable thermal comfort conditions (PMV, 0.45; PPD, 9%) (not depicted in fig. 4.2).

Priorities for Occupant Health and Comfort

When determining the appropriate interior environment for building occupants, occupant health is of utmost importance. Historically, the use of intentional ventilation (natural or mechanical) was prompted by health concerns. Issues of cooling and comfort have emerged more recently, as the use of centralized HVAC systems has become more common.

Therefore, the environmental issues associated with occupant health must be addressed first, through indoor air quality guidance such as ANSI/ASHRAE Standard 62.1-2013. Once occupant health and safety

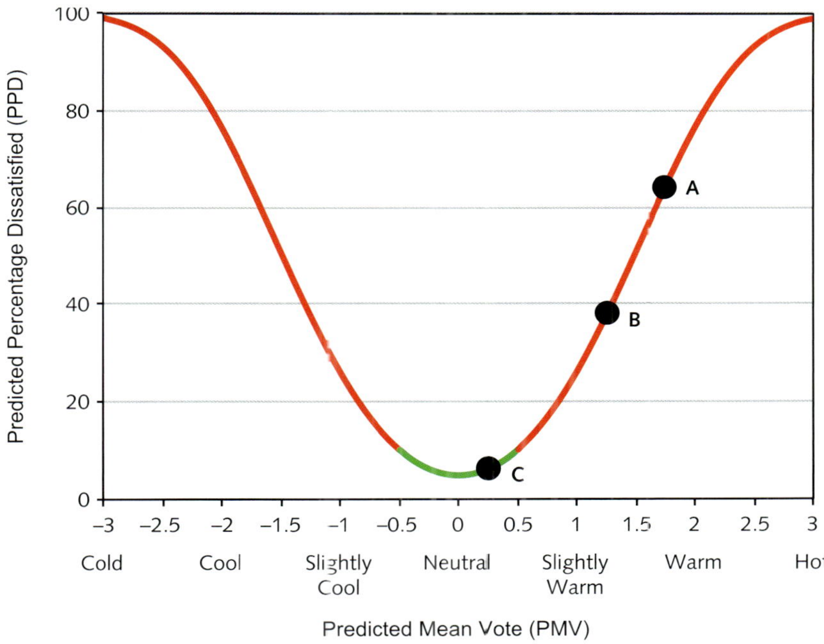

Figure 4.2

Calculated PMV-PPD relationship for various thermal comfort scenarios. Thermal comfort parameters that are common to three points are 29°C (air and mean radiant temperature), 65% RH, and 1 met.

have been adequately addressed, modifications to the interior environment can be carried out to address the thermal comfort of occupants using guidance such as ANSI/ASHRAE Standard 55-2013.

Conflict between Collections Conservation and Thermal Comfort

A fundamental environmental management consideration for many cultural institutions is the balance between human comfort, discussed in this chapter, and collections conservation, discussed in chapter 2. For the majority of HVAC systems, the main objective is to create an environment that is safe with respect to occupant health and that is thermally comfortable, thus maximizing worker efficiency and the visitor experience. While humidity control plays a role in thermal comfort, particularly through the elimination of extreme conditions, a narrow temperature range is the major focus of environmental control for thermal comfort.

Differences in the objectives for health, thermal comfort, collections conservation, and energy efficiency can lead to conflicts in determining the appropriate environmental conditions. This challenge will be discussed in greater detail in chapter 8.

Summary

This chapter discussed indoor air quality (IAQ) for building occupants, including exposure to airborne contaminants and noise. Several environmental strategies or processes employed to address IAQ were also presented. When modifying environments in existing buildings, one should be alert to the potential for exacerbating unhealthy conditions that were previously masked by excessive infiltration and natural ventilation. The transition from natural ventilation has the potential to reduce IAQ, particularly with respect to air contaminants generated inside the building.

This chapter also provided an overview of the factors influencing the thermal comfort of building occupants and the methods for analyzing a set of interior conditions and assessing thermal comfort. Of particular note are the distinction between controlled and adaptive environments and the discussion of ways in which these influence occupant perceptions of comfort. In hot and humid climates, the adaptive model for thermal comfort may be appropriate for museums and cultural heritage buildings, where the transient nature of visitor occupancy offsets the lack of direct temperature control.

Chapter 5 will examine the effects of psychrometric strategies for environmental management—such as heating, ventilation, dehumidification, and cooling—on the physical and thermodynamic properties of the interior environment.

References

ANSI [American National Standards Institute] and ASHRAE
2013a *Standard 55-2013—Thermal Environmental Conditions for Human Occupancy*. Atlanta: ASHRAE.

———.
2013b *Standard 62.1-2013—Ventilation for Acceptable Indoor Air Quality*. Atlanta: ASHRAE.

ASHRAE
2009 Thermal comfort. Chapter 9 in *ASHRAE Handbook—Fundamentals*, sect. 9.1.

———.
2011 Noise and vibration control. Chapter 48 in *ASHRAE Handbook—HVAC Applications* (SI edition). Atlanta: ASHRAE.

———.
2013 Sound and vibration. Chapter 8 in *ASHRAE Handbook—Fundamentals* (SI edition). Atlanta: ASHRAE.

Berglund, B., I. Johansson, and T. Lindvall
1988 Adsorption and desorption of organic compounds in indoor materials. *Proceedings of Healthy Buildings* 3:299–309.

Dear, R. J. de, and G. S. Brager
1998 Developing an adaptive model of thermal comfort and preference. *ASHRAE Transactions* 104, pt. 1, 1–19.

Douwes, J., P. Thorne, N. Pearce, and D. Heederik
2003 Bioaerosol health effects and exposure assessment: Progress and prospects. *Annals of Occupational Hygiene* 47 (3): 187–200. http://annhyg.oxfordjournals.org/content/47/3/187.full.

EPA [Environmental Protection Agency]
2001 Healthy buildings, healthy people: A vision for the 21st century. http://www.epa.gov/iaq/pubs/hbhp.html.

———.
2014a An introduction to indoor air quality (IAQ): Sources of combustion products. http://www.epa.gov/iaq/combust.html.

———.

2014b An introduction to indoor air quality (IAQ): Volatile organic compounds (VOCs). http://www.epa.gov/iaq/voc.html.

Fanger, P. O.

1970 *Thermal Comfort.* Copenhagen: Danish Technical Press.

Humphreys, M. A.

1978 Outdoor temperatures and comfort indoors. *Building Research and Practice* 6 (2): 92–105.

Index project

2005 Critical appraisal of the setting and implementation of indoor exposure limits in the EU. EUR 21590 EN. European Commission, Directorate General, Joint Research Centre. http://ec.europa.eu/health/ph_projects/2002/pollution/fp_pollution_2002_frep_02.pdf.

Morey, P. R.

1990 The practitioner's approach to indoor air quality investigations. In *Proceedings of the Indoor Air Quality International Symposium.* Akron, OH: American Industrial Hygiene Association.

Mühlfeld, C., P. Gehr, and B. Rothen-Rutishauser

2008 Translocation and cellular entering mechanisms of nanoparticles in the respiratory tract. *Swiss Medical Weekly* 138 (27–28): 387–91.

Nicol, J. F., and M. A. Humphreys

2002 Adaptive thermal comfort and sustainable thermal standards for buildings. *Energy and Buildings* 34:563–72.

Paciuk, M.

1990 The role of personal control of the environment in thermal comfort and satisfaction at the workplace. *Journal of the Environmental Design Research Association* 21:303–12.

Wargocki, P., and D. Wyon

2007 Ventilation and health in nonindustrial indoor environments. Report from a European Multidisciplinary Scientific Consensus Meeting. *Indoor Air* 12 (2): 113–28.

Williams, R. N.

1995 A field investigation of thermal comfort, environmental satisfaction and perceived control levels in UK office buildings. *Proceedings of Healthy Buildings* 3:1181–86.

Note

1 The term *air temperature* is used instead of *temperature* to distinguish it from other temperature types, such as surface, radiant, and dew point. Similarly, the term *relative humidity* is used rather than *humidity*, as it is this specific form of humidity that is relied upon by the heat balance model.

Chapter 5

Psychrometric Strategies for Environmental Management

Chapter 1 set out the characteristic conditions of hot and humid climates. Subsequent chapters described the interior environmental conditions necessary for conservation of cultural collections (chaps. 2 and 3) and for human health and thermal comfort (chap. 4).

In this chapter we describe basic and combined psychrometric processes—such as heating, cooling, dehumidification, and humidification—that may be used as part of an overall interior environmental management strategy. It will show how a psychrometric chart can be used to depict temperature and humidity conditions of a given space, target ranges of conditions for conservation and thermal comfort, and psychrometric processes needed to achieve the target ranges of the conditions.

The Psychrometric Chart

A psychrometric chart is a graphical representation of the interrelationships among the thermodynamic properties of moist air in an enclosed space, such as a room in which collections are displayed in a museum. Such a chart is produced for a specific barometric pressure or elevation (accounting for changes in air density). The format of psychrometric charts varies. A simplified version, shown in figure 5.1, relates three common psychrometric properties:

- *Temperature*—the temperature of the air recorded by a typical thermometer; it is shown in the figure on the x-axis in units of degrees Celsius.
- *Dew point temperature*—the temperature at which an air mass will reach saturation, indicating the moisture content of the air. It is the temperature at which water vapor starts to condense out of air that is cooling; it is shown in the figure on the right y-axis in units of degrees Celsius.
- *Humidity*—the percent ratio of water vapor pressure to the theoretical water vapor pressure of saturated air for a given temperature. If the dew point temperature and the temperature of an air mass are close, humidity will be near 100% RH. Humidity is shown in the figure by the chart lines curving upward from left to right; the unit is % RH.

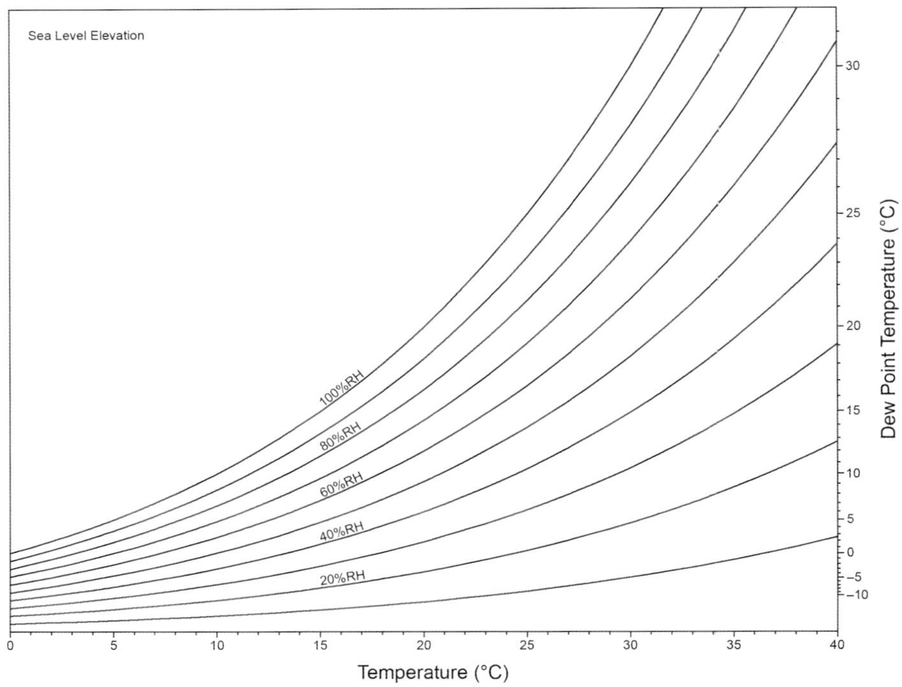

Figure 5.1

Psychrometric chart for sea level elevation.

Though not appearing in the simplified version presented in figure 5.1, other psychrometric properties typically shown on the psychrometric chart include humidity ratio, wet-bulb temperature, enthalpy, specific volume, and sensible heat ratio.

The utility of the psychrometric chart is its ability to determine graphically the properties of an air mass through the knowledge of only two variables—temperature and humidity are the most common measurements performed—for a fixed barometric pressure or elevation. The following sections will demonstrate how a psychrometric chart can be used to depict a given environment and target environment as well as the outcome of applying psychrometric processes to alter an environment by tracking change from the initial state point (defined by the initial set of environmental conditions) to its new state points.

Environmental Conditions for a Given Space

Paired measurements of temperature and humidity can be individually plotted on a psychrometric chart. This dataset can produce a cluster of data points that signifies the environmental range for a given space. Statistical values obtained from a dataset—such as the maximum, percentile 97.5 (P97.5); average or mean, percentile 2.5 (P2.5); and minimum—can describe the dispersion of data or define representative environmental conditions on the psychrometric chart without presenting all data points. (Note that the maximum rolling 24 hour deviations of temperature and humidity, two of six mechanical risk criteria from the Conservation Environment Classification–HH protocol, cannot be depicted on the psychrometric chart because of its time-dependent nature.) The application of various psychrometric processes can also be shown on the psychrometric chart, whereby existing environmental conditions can be shifted toward more desirable target environmental conditions.

Target Ranges of Environmental Conditions for Conservation and Thermal Comfort

The range of interior environmental conditions that are acceptable for collections conservation and/or thermal comfort can be illustrated on the psychrometric chart as environmental targets. The target and its boundaries can be defined by a set point or average value for temperature and humidity, bounded by acceptable limits in short-term variations from the set point or average value. Alternatively, the environmental target can be defined by establishing maximum and minimum range limits.

Environmental target ranges for collections conservation

In chapter 3 (table 3.3), the Conservation Environment Classification–HH protocol defines the level of risk posed by microbial, mechanical, and chemical factors for mixed collections. Although the best theoretical conservation environment is Class A (no risk) for microbial risk, Class a (no risk) for mechanical risk, and Class 0 (same risk as historical condition) for chemical risk, the temperature and humidity conditions necessary for these classifications may not be advisable if the target conditions differ significantly from the historical means of temperature and humidity to which the object or collection was native. Using the psychrometric chart, figure 5.2 shows the historical mean temperature (24.2°C) and humidity (77% RH) of Rio de Janeiro, Brazil, the location of one of our case studies (chap. 14), and two examples of target ranges of interior environmental conditions for collections conservation produced by the application of an environmental management strategy.

Figure 5.2a shows a target interior environment (annual means of 25°C and 60% RH) that focuses on reducing microbial risk. Minimizing the microbial risk to Class A sets the percentile 97.5 of humidity value to 65% RH or less, as indicated by the top border of the blue region. Assuming the use of a mechanical environmental management system programmed to maintain the indoor environment between 55% RH (bottom of blue region) and 65% RH, thus limiting the risk of mechanical damage due to short-term and seasonal variations, the resulting annual mean humidity in the interior space would be approximately 60% RH. Comparisons between the historical annual mean humidity of Rio de Janeiro (77% RH) and the target annual mean interior humidity (60% RH) results in a differential of Δ17% RH and a Class b designation for mechanical risk. (In this example, the remaining mechanical risks—short-term humidity and temperature variation, seasonal humidity and temperature variation, and deviation from historical annual mean temperature—are classified as Class a.) Similarly, the difference between

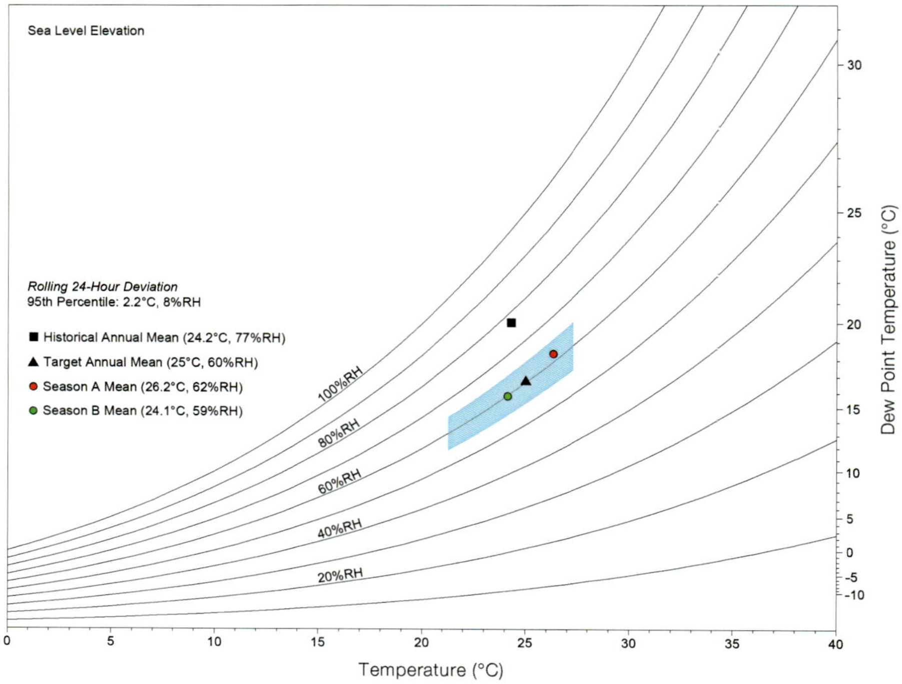

Figure 5.2a

Psychrometric depiction of a target interior environmental condition focused on minimizing microbial risk in Rio de Janeiro, Brazil.

the historical annual mean temperature of Rio de Janeiro (24.2°C) and the target annual mean interior temperature (25°C) results in a Class 0 designation for chemical risk. (The temperature range shown by the blue region indicates the Class 0 chemical risk criteria of Δ3°C around the historical annual mean temperature.)

In summary, application of the Conservation Environment Classification–HH protocol to the example in figure 5.2a results in a Class A (no risk) designation for microbial risk, a Class b (moderate risk) designation for overall mechanical risk, and a Class 0 (same risk as historical condition) designation for chemical risk.

Figure 5.2b shows a target interior environment (annual means of 25°C and 68% RH) that focuses on minimizing mechanical risk. A target annual mean interior humidity of 68% RH is selected to limit the deviation from the historical mean humidity (77% RH) to less than Δ10% RH, thus achieving a Class a designation for mechanical risk owing to deviation from historical mean humidity. To achieve this mean interior humidity condition, a mechanical environmental management system may be programmed to maintain the environment between 73% and 63% RH (top and bottom of blue region), again limiting the risk of mechanical damage due to short-term and seasonal humidity variations. (In this example, mechanical risks due to short-term/seasonal humidity and temperature variations and deviation from historical annual mean temperature are also Class a.) As a result of the humidity range set by the mechanical system, the 97.5 percentile of interior humidity will also be approximately 73% RH, resulting in a Class C designation for microbial risk. Comparing the historical annual mean temperature (24.2°C) and the target annual mean temperature (25°C) results in a difference of Δ0.8°C

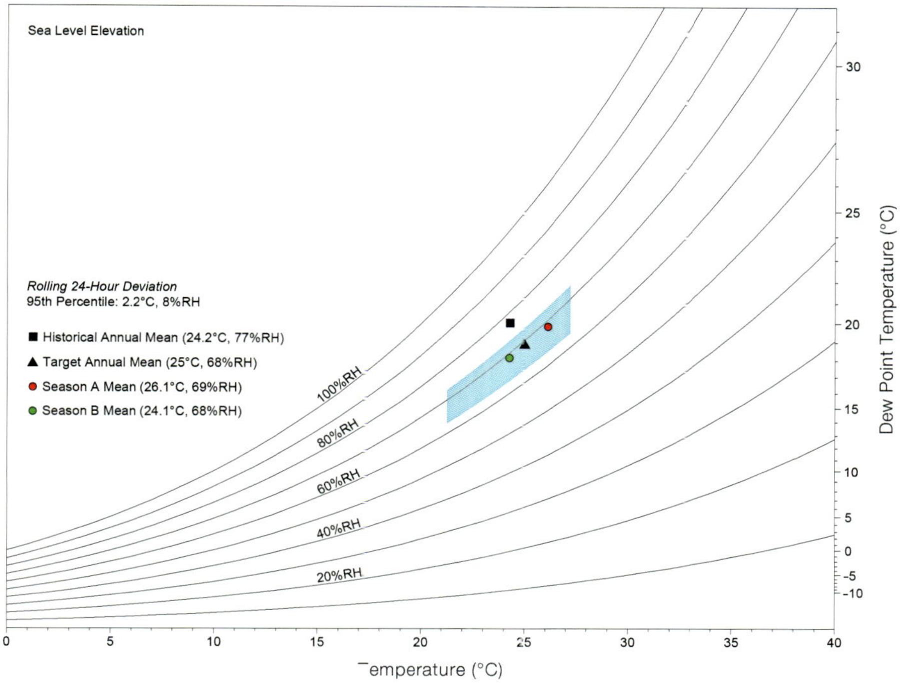

Figure 5.2b

Psychrometric depiction of a target interior environmental condition focused on minimizing mechanical risk in Rio de Janeiro, Brazil.

and a Class 0 designation for chemical risk. (The temperature range of the blue region again indicates Class 0 criteria, Δ3°C, for chemical risk.)

Thus, application of the Conservation Classification Environment–HH protocol to the example shown in figure 5.2b results in a Class C (moderate risk) designation for microbial risk; a Class a (no risk) designation for overall mechanical risk, and a Class 0 (same risk as historical condition) designation for chemical risk.

As these two examples demonstrate, it is essential to balance microbial, mechanical, and chemical risks when selecting an appropriate environmental target range for collections.

While figure 5.2 graphically illustrates the classification and target range of an interior environment for collections conservation, in practice the selection of an appropriate environmental classification is also influenced by collection needs, building envelope performance, opportunities for passive strategies, and complexity of mechanical strategy.

Environmental target ranges for thermal comfort

Chapter 4 presented the environmental conditions that are considered acceptable for thermal comfort, recognizing that the perception of comfort will vary among individuals.

ANSI/ASHRAE Standard 55-2013—*Thermal Environmental Conditions for Human Occupancy*—introduced in chapter 4, defines psychrometric zones of thermal comfort based on laboratory and field surveys of occupant perceptions of the environment (ANSI and ASHRAE 2013). On the psychrometric chart shown in figure 5.3, the graphical comfort zone method is used to delineate acceptable thermal comfort zones for dew point temperatures below 16.8°C; use of the analytical comfort zone method is required to delineate comfort zones for

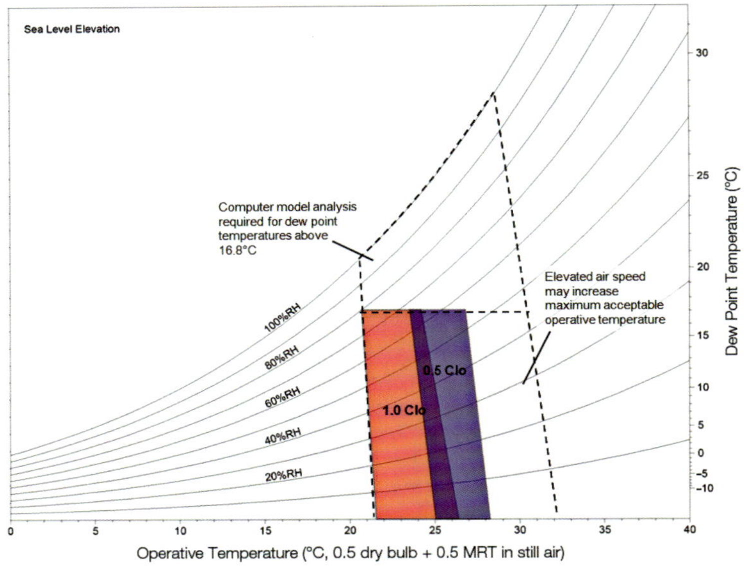

Figure 5.3

Acceptable occupant thermal comfort zones generated by the graphical comfort zone method from ANSI/ASHRAE Standard 55-2013 (ANSI and ASHRAE 2013). Note that the x-axis represents operative temperature, which is defined as ½ temperature plus ½ mean radiant temperature (MRT) in still air.

environments with dew point temperatures above 16.8°C, as is common in hot and humid climates. As discussed in chapter 4, the use of elevated air speeds can also provide a cooling effect for occupants and will expand the thermal comfort zone by allowing for higher acceptable maximum operative temperatures (defined as: ½ dry bulb temperature + ½ mean radiant temperature).

Figure 5.3 illustrates that unlike the target environmental zones for collections conservation, the target environmental zones for thermal comfort are dependent on a variety of factors in addition to temperature and humidity, including clothing, activity level, mean radiant temperature, and air speed. By comparing figure 5.3, which focuses on thermal comfort zones, and figure 5.2, which focuses on collections conservation, we see a graphical demonstration of the differences between thermal comfort environments and conservation environments.

Displaying Psychrometric Processes from Initial to Final State Points

Psychrometric processes—for example, heating and cooling—are depicted on the psychrometric chart by indicating an initial state point, or initial set of environmental conditions, and the final state point following application of the process. As shown in figure 5.4, the net direction of change from the initial state point defines the overall psychrometric process. Horizontal and vertical changes of the state point indicate changes in temperature and dew point temperature, respectively. A process to the right (higher temperature) from the initial condition indicates heating, while cooling is denoted by a process to the left (lower temperature). Similarly, an upward process shows humidification by the addition of moisture, while a downward process indicates dehumidification. A diagonal process represents a combination of two adjacent processes. For example, a shift toward the lower left would indicate both cooling and dehumidification. It should be noted, however, that the actual psychrometric path taken by a

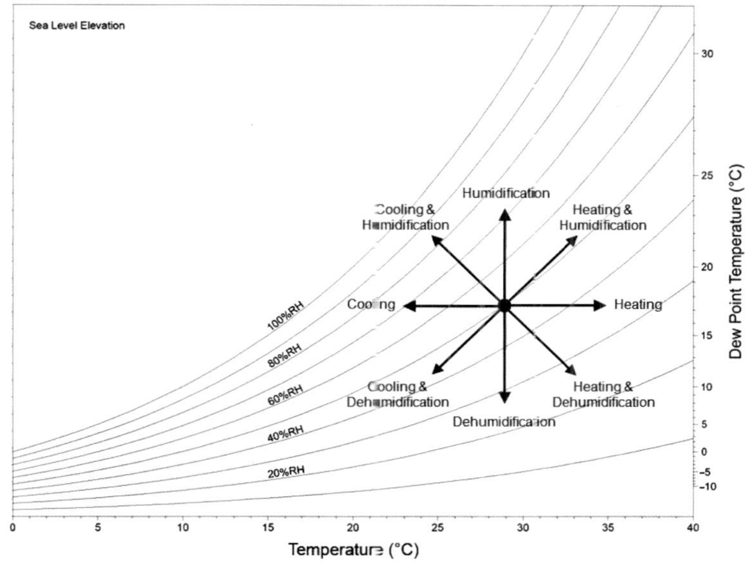

Figure 5.4
Directions of basic psychrometric processes depicted on a psychrometric chart.

mechanical or nonmechanical process may not be the straight line shown in the illustration. For the purposes of this publication, these processes are illustrated simply as initial and final state points on the psychrometric charts, rather than as the actual *process paths* that must occur between the two state points.

When an initial environmental state is viewed on a psychrometric chart, the location of its initial state point can indicate limitations on the extent to which certain processes can be utilized. Figure 5.4 shows that while the heating process (shift to the right on the psychrometric chart) can be used to reduce humidity with no change in moisture content, its practical use is limited by a threshold temperature above which the benefits of lower humidity are outweighed by the disadvantages that elevated temperature can have on conservation (e.g., increased chemical deterioration) or on thermal comfort (e.g., occupant dissatisfaction). Similarly, continued application of the cooling process (shift to the left) can lead to issues of condensation as conditions approach the saturation line (100% RH).

Psychrometric Strategies

There are two strategic approaches—nonmechanical and mechanical—to effecting change in the psychrometric state of an environment toward a target range of end conditions. This section describes how both strategic approaches can be presented as changes on psychrometric charts. Chapter 6 then presents examples of practical methods for implementing nonmechanical strategies, and chapter 7 presents examples of methods for implementing mechanical strategies.

Nonmechanical strategies

Nonmechanical strategies are not psychrometric processes in the conventional engineering sense, but are methods that can be employed to reset the initial state point to more favorable conditions closer to, or within, the target zones. Nonmechanical strategies reduce the moisture and/ or thermal energy gains to a space and can aid in limiting variations of humidity and temperature. These strategies can reduce the energy that must be expended by a mechanical environmental management system to arrive at a final state point.

Source control of moisture

Source moisture control reduces the dew point temperature of a given space by reducing or eliminating the infiltration of humid air or the moisture available for evaporation. On the psychrometric chart, this action results in resetting the initial state point downward, decreasing moisture content and humidity without changing temperature. For example, in figure 5.5, a hot and humid air mass at an initial state point A is reset to a final state point A' following implementation of source moisture control measures. The final state point maintains the same temperature (30°C) but has lower values for dew point temperature (from 27°C to 25°C) and humidity (from 85% to 75% RH). By resetting the initial state point of the air mass, this nonmechanical strategy shifts the environment closer to the target range of conditions.

Source control of thermal energy

Strategies for source control of thermal energy reduce the thermal energy inputs to a given space. This reduction lowers the space temperature and resets the state point on a psychrometric chart to a new position to the left, leaving initial dew point temperature unchanged and resulting in higher humidity. In figure 5.6, after implementation of thermal energy source control, an initial hot and humid air mass at state point A is reset to a final state point A' with the same dew point temperature (26°C) but with a lower temperature (from 35°C to 31°C) and a higher humidity

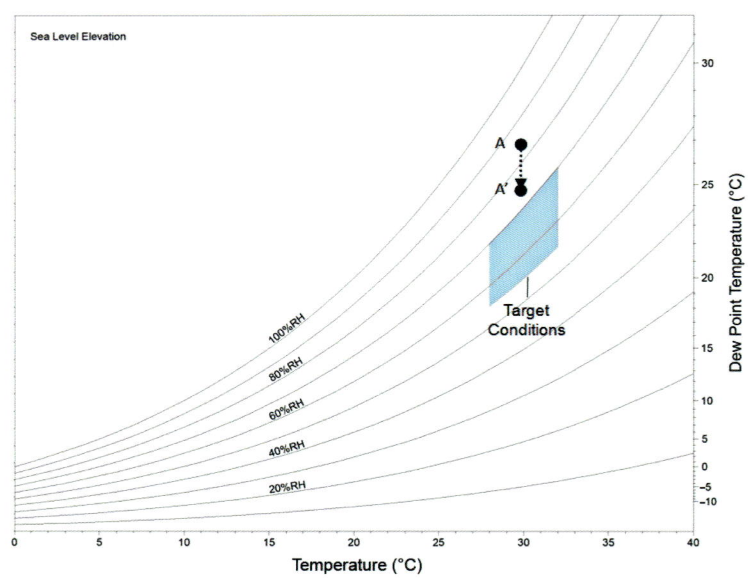

Figure 5.5

Implementation of nonmechanical source moisture control. The dotted line indicates a nonmechanical process.

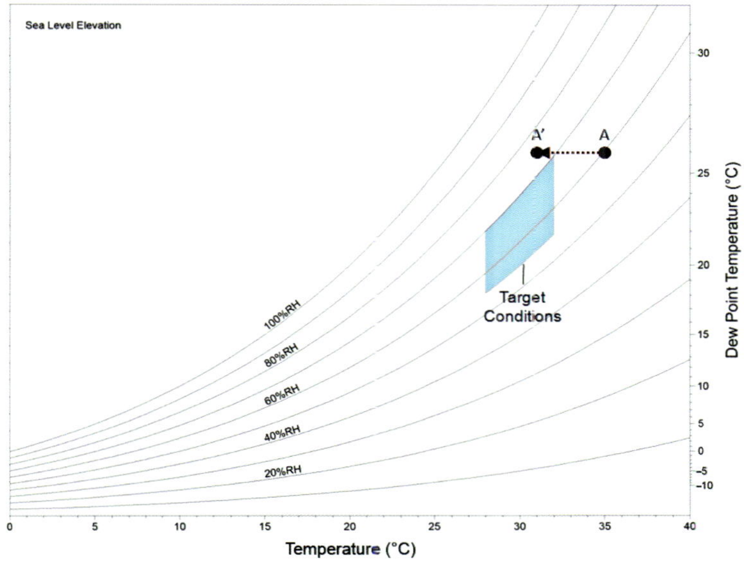

Figure 5.6

Implementation of nonmechanical source thermal energy control. The dotted line indicates a nonmechanical process.

(from 60% to 75% RH). Thus, this nonmechanical strategy resets the initial state point to a position closer to the target range.

Mechanical strategies

Mechanical strategies include the four basic mechanical processes—dehumidification, heating, cooling, and humidification—used in an environmental management system to modify the interior environment. Though not a psychrometric process, the mechanical process of ventilation is also described because of its ability to affect the interior environment. Unlike nonmechanical strategies, a mechanical strategy must consume energy to shift the initial state point closer to, or within, the boundaries of the target environmental conditions.

Dehumidification

Dehumidification removes moisture from the air, thereby reducing its dew point temperature, and represents a key mechanical strategy for environmental management in hot and humid climates. As a consequence of dehumidification, humidity is decreased while temperature remains constant. Dehumidification is represented as a downward process on the psychrometric chart.

In figure 5.7, the A–A'–B process line shows the application of mechanical or desiccant dehumidification (A'–B) following nonmechanical source moisture control (A–A'). In this example, the dehumidification process shifts a hot and humid intermediate state point (A') to a final state point (B) within the environmental target range. While final state point B remains at the same temperature (30°C), dew point temperature is reduced (from 25°C to 22°C), affecting humidity (from 75% to 62% RH). As evidenced by the process from the initial state point C (temperature, 34°C; dew point temperature, 27°C; and humidity, 67% RH) to the final state point D (temperature, 34°C; dew point temperature, 22°C; and

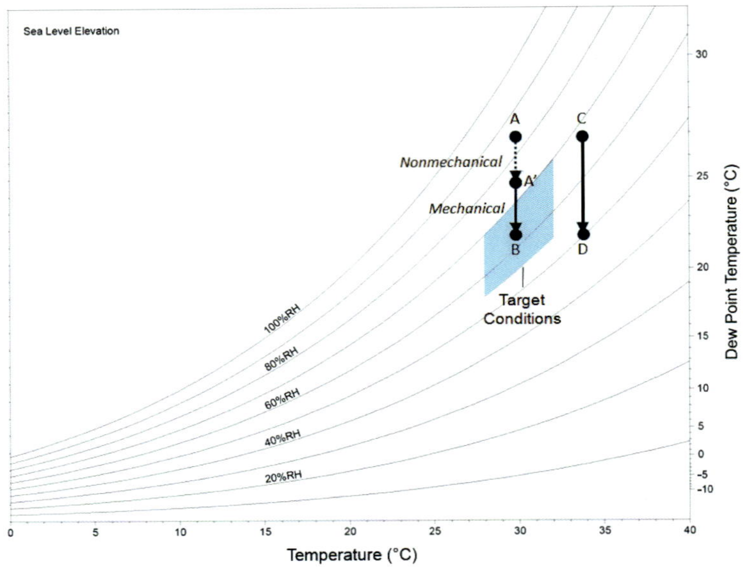

Figure 5.7
Implementation of mechanical dehumidification.

humidity, 50% RH), the sole use of dehumidification for extremely warm air masses may be insufficient to process to a specific environmental target range, necessitating the use of either additional psychrometric strategies or a modified target range of conditions.

It should be noted that hot-humid air requires the removal of a larger amount of moisture than cool-humid air to achieve similar reductions in humidity—this is reflected on the psychrometric chart in the widening of the isohumes (lines of constant humidity) with increased temperature. Although figure 5.7 shows the dehumidification process as a straight-line state change without temperature change, the actual mechanical process path is more complex; it will be further discussed in chapter 7.

Heating

The mechanical process of heating warms the interior air and, as a consequence, reduces humidity. This psychrometric relationship is sometimes used purposely to reduce humidity, a technique termed *conservation heating* or *humidistatic heating*. Since heating only serves to raise temperature, the process to the right, the dew point temperature of the air mass remains unchanged.

In figure 5.8, the A–B process line shows the heating process applied to interior air (A) that is warm and humid. Heating processes the initial state point to the final state point (B) within the environmental target range by increasing temperature (from 26°C to 30°C) and reducing humidity (from 84% to 66% RH) with no change in dew point temperature (23°C). While this psychrometric strategy may be counterintuitive in a hot climate, the use of limited heating can significantly lower humidity and protect against damage from microorganisms. Though heating also results in the shift from the initial state point C (temperature, 28°C; dew point temperature, 27°C; and humidity, 94% RH) to the final state

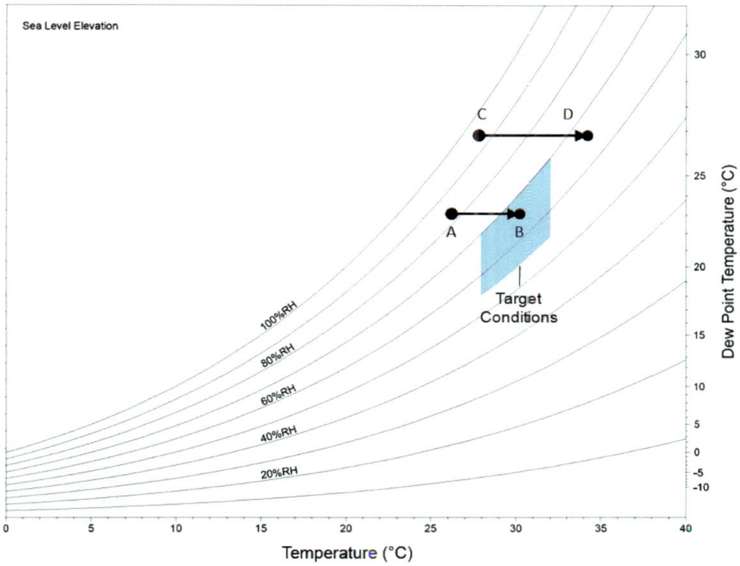

Figure 5.8

Implementation of mechanical heating.

point D (temperature, 34°C; dew point temperature, 27°C; and humidity, 67% RH), the elevated dew point temperature of the air mass renders the heating process insufficient to bring the final state point into the target range of conditions. Thus, the use of additional psychrometric strategies or modified target conditions will be required.

Ventilation—dilution, or conservation heating

Ventilation is a process by which outside air is added to an inside space, displacing a portion of the existing air mass. Though not a psychrometric process, the mixing of interior and exterior air masses at different psychrometric states results in an interior air mass with environmental conditions between those of the initial air masses.

The ventilation process can be further categorized by its effect upon the interior environment. *Dilution ventilation* refers to the addition of drier (lower moisture content) air to a space in order to lower interior moisture content—in this case, the exterior and interior air masses have the same temperature but different dew point temperatures. *Conservation heating ventilation* refers to the addition of warmer air to a cooler space in order to reduce humidity—the two mixing air masses have different temperatures but the same dew point temperature. While dilution ventilation and conservation heating ventilation should be illustrated on the psychrometric chart as vertical and horizontal lines, respectively, these exact circumstances rarely happen in practice, and the process lines will slope depending on differences in the secondary variable of each.

Ventilation is often achieved by mechanical means, such as fans, but it can also be accomplished nonmechanically by using natural ventilation, through either cross ventilation or the stack effect. With either mechanical or nonmechanical strategies, the opportunity for dilution ventilation to lower the moisture content of interior air occurs when the exterior air has a lower dew point temperature than the interior air. The

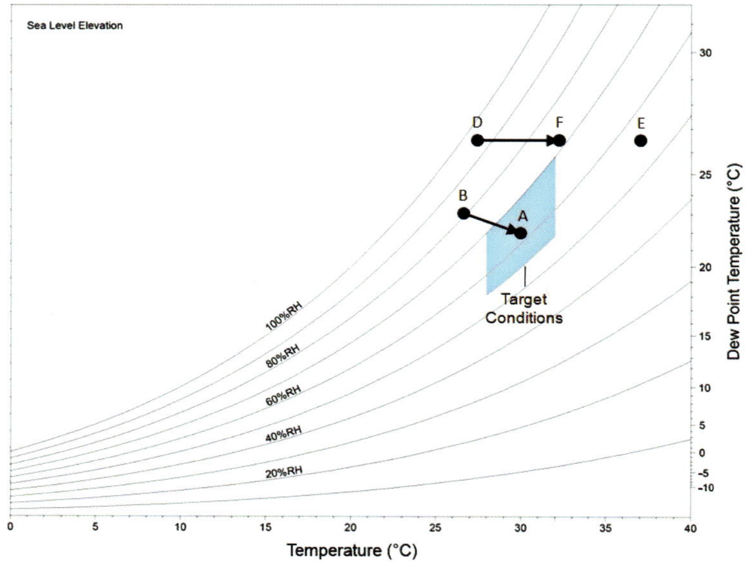

Figure 5.9

Implementation of ventilation. State points A (final) and E represent outside conditions, while initial state points B and D represent initial interior conditions. Final state point F indicates the interior condition after 1 air change from state point E.

opportunity for conservation heating ventilation often occurs during the afternoon on clear warm days.

Figure 5.9 illustrates the effects of ventilation. Using ventilation, an average environmental condition between the original indoor condition at initial state point B (temperature, 26.5°C; dew point temperature, 23°C; humidity, 81% RH) and the outdoor condition at initial state point A (temperature, 30°C; dew point temperature, 22°C; humidity, 62% RH) is achieved following 1 air change. With the further addition of exterior air into the interior space, the indoor condition will asymptotically approach the outdoor condition, matching the exterior environment after 6 to 8 air changes. Rates of natural ventilation in a well-ventilated room can range from 10 to 20 air changes per hour (ACH), while rates of mechanical ventilation range from 4 to 12 ACH, depending on design requirements.

Alternatively, the D–F process line in figure 5.9 indicates the potential limitations of ventilation as an environmental management strategy. In this example of conservation heating by ventilation, equal mixing (or 1 air change) of initially warm and humid interior air (initial state point D: temperature, 27°C; dew point temperature, 26.5°C; humidity, 95% RH) with warmer and less-humid exterior air (initial state point E: temperature, 37°C; dew point temperature, 26.5°C; humidity, 55% RH) results in a final state point (point F: temperature, 32°C; dew point temperature, 26.5°C; humidity, 72% RH) that remains outside the environmental target range because of the high dew point temperature or moisture content of the two air masses. Continued air changes would eventually shift the interior environment at state point F to state point E, reducing humidity at the expense of high temperature.

The limited use of conservation heating ventilation can result in a final state point that will require less subsequent mechanical dehumidification and cooling to achieve the target range of conditions than would be required without the use of ventilation. The effectiveness of ventilation is dependent on the seasonal or diurnal availability of exterior air with conditions appropriate for processing toward the target range of conditions.

Cooling

Mechanical cooling processes lower the temperature and raise humidity. Cooling plays an important role in improving occupant thermal comfort, particularly in hot and humid climates. However, in the absence of concurrent dehumidification, overcooling of an interior environment must be avoided, as this can raise humidity above target levels.

Cooling will shift an initial state point to the left on the psychrometric chart, maintaining dew point temperature until saturation (100% RH) is reached. At saturation, continued cooling of an air mass moves the state point to the left along the saturation line, lowering the dew point temperature of the air mass (process of dehumidification by cooling) until the desired temperature is reached.

In figure 5.10, the A–A'–B process line shows the reset of initial interior conditions by nonmechanical source control of thermal energy (A–A') followed by the application of mechanical cooling (A'–B), which brings the final state point within the environmental target range. In this example, mechanical cooling reduces the temperature from 33°C to 30.5°C, increases humidity from 56% to 64% RH, and maintains dew point temperature at 23°C.

As discussed previously for heating, the sole use of a cooling strategy for air masses with high dew point temperature may be able to meet the target temperature range but unable to arrive within the target humidity range in hot and humid climates. In figure 5.10, the application of cooling processes the initial state point C (temperature, 37°C; dew point temperature, 27°C; humidity, 57% RH) to an intermediate state point D (temperature, 31°C; dew point temperature, 27°C; humidity, 79% RH). Further cooling of the intermediate state point D to the final

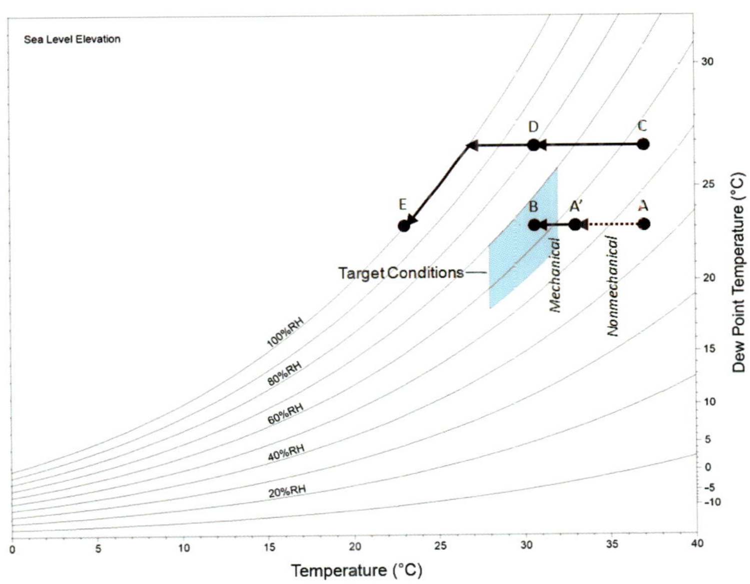

Figure 5.10

Implementation of mechanical cooling.

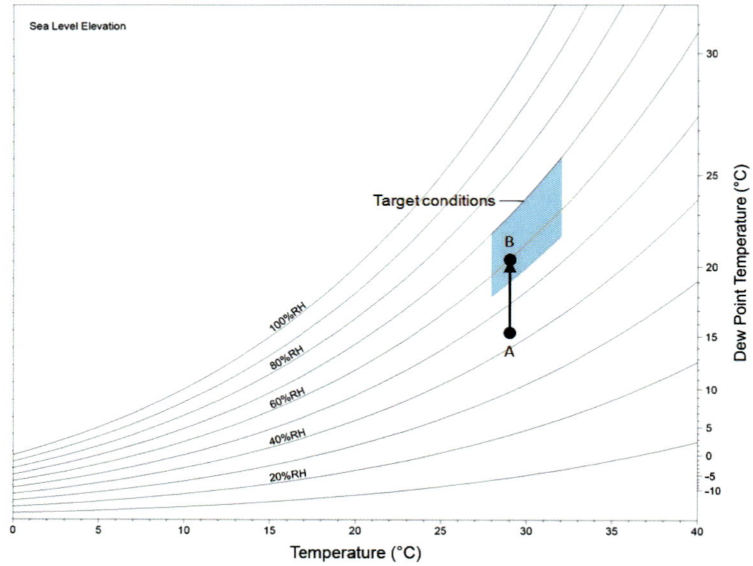

Figure 5.11
Implementation of mechanical humidification.

state point E (temperature, 23°C; dew point temperature, 23°C; humidity, 100% RH) will result in dehumidification as moisture condenses from the air mass along the saturation curve, maintaining the air mass at 100% RH.

Humidification

Though not generally required in hot and humid climates, the mechanical process of humidification adds moisture to the air, increasing its humidity and dew point temperature while maintaining the same temperature. Humidification can be of particular importance during drier winter months in mixed climates, when heating can result in low humidity conditions that may cause embrittlement or dimensional change in objects or result in occupant discomfort. On the psychrometric chart, this process results in the upward process from an initial state point.

In figure 5.11, humidification processes the initial warm air mass (initial state point A) to the final state point B within the environmental target range that is elevated in both humidity (from 43% to 60% RH) and dew point temperature (from 15°C to 20.5°C) while maintaining the same temperature (29°C). In reality, the process of humidification will also affect the temperature of the air mass—evaporative humidification systems will slightly depress temperature, while steam humidification will slightly elevate temperature.

Summary

This chapter showed how a psychrometric chart is used to depict a given environment as well as target ranges of environmental conditions for collections conservation and thermal comfort. This chapter also examined basic psychrometric processes—heating, cooling, dehumidification, and humidification—used to alter the interior environment of buildings housing collections, and it reviewed nonmechanical and mechanical environmental management strategies. Through the graphic presentation of these processes and strategies on the psychrometric chart, one can visualize how the strategies and their resulting change in environmental conditions relate to a nominal set of target ranges of conditions. Additionally, viewing these strategies on the psychrometric chart allows one to grasp the interdependence of the various environmental parameters and to distinguish which parameters must be managed to achieve the target ranges.

We have thus set the stage for the discussion of specific nonmechanical strategies (chap. 6) and mechanical strategies (chap. 7) that can be employed to manage interior environments in hot and humid climates.

Reference

ANSI [American National Standards Institute] and ASHRAE
2013 *Standard 55-2013—Thermal Environmental Conditions for Human Occupancy*. Atlanta: ASHRAE.

Chapter 6

Nonmechanical Strategies for Environmental Management

Chapter 5 described a number of psychrometric strategies for managing the thermal energy and moisture loads, or gains, that are imposed on buildings in hot and humid climates. This chapter discusses nonmechanical strategies for implementing these psychrometric strategies, while chapter 7 presents mechanical strategies.

This chapter is directed to architects, engineers, and conservators. It provides architects and engineers with an overview on how the site and building envelope can be used to moderate thermal and moisture loads on a building, reducing the need for, or size of, mechanical systems. Architects and conservators will learn how the proper selection and assignment of building spaces for collections storage and exhibition can reduce the influence of climatic factors on collections. This chapter closes with a review of the importance of maintenance, operation, and documentation of nonmechanical strategies.

In this chapter, we cover the following topics on nonmechanical strategies for interior environmental management in hot and humid climates:

- site influences on the interior building environment
- building envelope and hygrothermal performance
- influences of the spatial organization of a building
- source moisture control strategies
- thermal energy management strategies
- convection strategies
- control strategies for particulates and insects
- space use strategies
- layered enclosure strategies—secondary envelopes, micro-environments, and collections housing
- occupant load strategies
- maintenance and sustained implementation of nonmechanical strategies

Overview

Nonmechanical strategies can be combined with mechanical strategies for environmental management, effectively reducing energy consumption while providing an acceptable collections conservation environment that is thermally comfortable. For example, a thermally massive building envelope can slow the rate of change in the interior environment if operation of the mechanical system is interrupted by an electrical power outage. A thermally massive envelope may also slow the effects of extreme low or high exterior temperatures on the interior environment. If nonmechanical strategies are used to reduce the total cooling, dehumidification, or heating loads on a space as well as their dynamic range, the capacity of the mechanical system equipment can be reduced; as a result of the smaller loads and load range, the system equipment can operate closer to its peak efficiency point for a greater percentage of the time.

Nonmechanical strategies can improve the stability of conservation environments and occupant thermal comfort and can potentially significantly reduce system installation and operating costs; therefore, nonmechanical strategies should be considered the first step in managing the interior environment. The nonmechanical approach is particularly advantageous in historic buildings that predate the widespread use of mechanical systems for thermal comfort. The envelope construction and spatial organization of these buildings may already incorporate features to mediate the influence of the exterior climate on interior conditions. For new buildings, nonmechanical strategies should be integrated into the building design.

Site Influences on the Interior Building Environment

Chapter 1 described the moisture, thermal energy, and convective characteristics of hot and humid climates. These characteristics are statistically representative of each climate zone but will vary within each zone because of local influences and site-specific factors. Local or situational influences may moderate or attenuate the typical climatic factors related to collections risks or the comfort of building occupants. The site and contextual influence of thermal and moisture loads on a building are summarized in figure 6.1 and discussed below.

Moisture

Locally, the available moisture in liquid or vapor states can be influenced by:

- topography, water permeability of site soils, surface and subsurface hydrology, and roof drainage runoff from impervious pavements;
- vegetative cover and moisture uptake, and the tree canopy and its effects on soil drying by evaporation;
- irrigation systems, fountains, and overspray on building surfaces;
- pools, ponds, and adjacent bodies of water such as rivers, lakes, and oceans;
- leaks from underground piping.

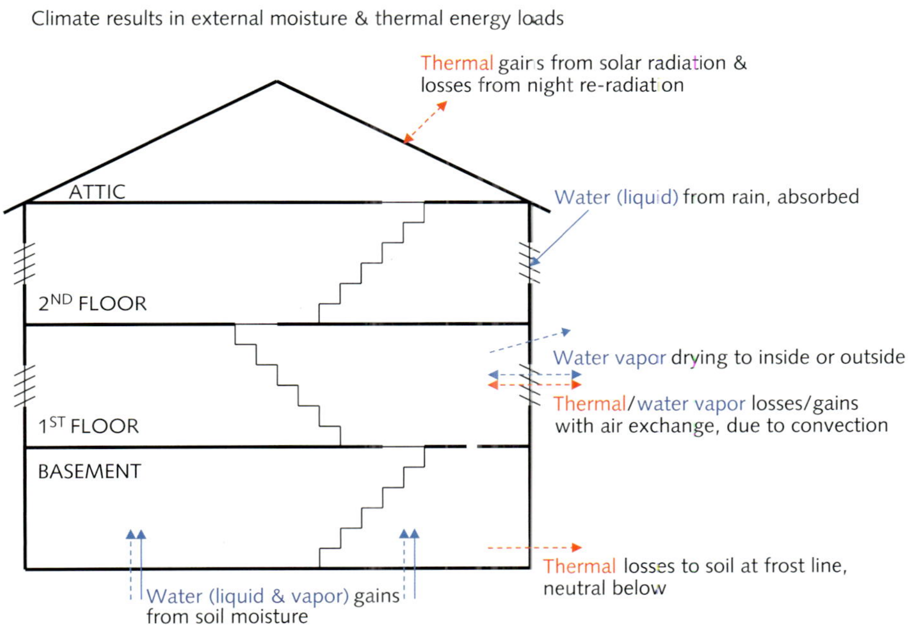

Figure 6.1
Thermal and moisture influences on a typical building. Illustration: Michael C. Henry.

Sunlight

Direct and reflected sunlight is the dominant thermal energy source, and the effects may be locally influenced by:

- building orientation, surface area, and geometry;
- shading from topography, vegetation, or adjacent buildings (fig. 6.2);
- reflected sunlight from adjacent buildings and pavement surfaces;
- thermal energy absorbed by surroundings and reradiated at night;
- temperature of soil in contact with the portions of a building below ground.

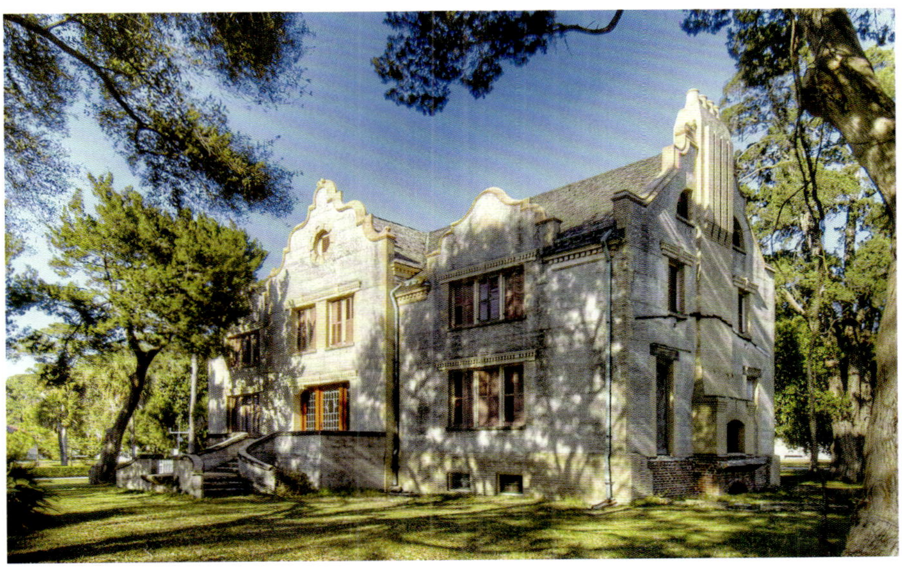

Figure 6.2
Shade from adjacent trees at Hollybourne Cottage, Jekyll Island Historic District, Georgia, United States, reduces thermal gain and drying of the porous concrete (case study, chaps. 9 and 10). Photo: Harlan Hambright, courtesy of Jekyll Island Authority.

Air movement

Air movement transports atmospheric moisture and thermal energy around and through the building. Wind-driven rain will increase moisture uptake by porous building surfaces and at breaks in the continuity of the envelope.

Locally, air movement around the building is influenced by topography, vegetation, and adjacent buildings, as well as the orientation, surface area, and geometry of the building. Air movement through the building is influenced by windows, doors, and openings through the envelope, as well as by the organization of interior spaces.

Ground contact

The influences of the surroundings on a building include the portion of the building that is set in the ground. The influence of ground moisture, salts, and geothermal effects can occur at a basement, at a shallow crawl space, or at the floor if it is set directly on ground. In Warm-Humid (3A) to Very Hot-Humid (1A) climate zones, traditional buildings have been typically elevated above the ground on piers or supports for protection from flooding and from animals or to improve natural ventilation. With elevated buildings, the piers or supports limit ground contact, creating a microclimate in the space between the building and the ground (fig. 6.3). In special circumstances, some buildings, such as vaults, may be located partially or entirely in the ground (fig. 6.4).

Building Envelope and Hygrothermal Performance

The building envelope is multifunctional and provides shelter, physical security, aesthetics, and information about the use and occupancy of the building. The building envelope mediates and/or buffers the exchange of water vapor, liquid water, thermal energy, and air between the exterior environment and the building interior.

The building envelope consists of the surfaces and materials of the building that separate the interior spaces from the surrounding environment, above and below the ground:

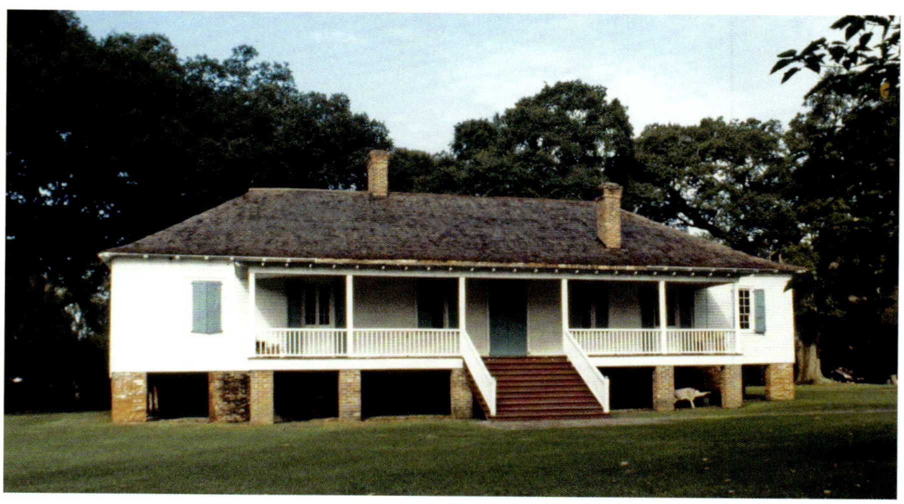

Figure 6.3
Traditional raised Creole cottage in Louisiana. Photo: Michael C. Henry.

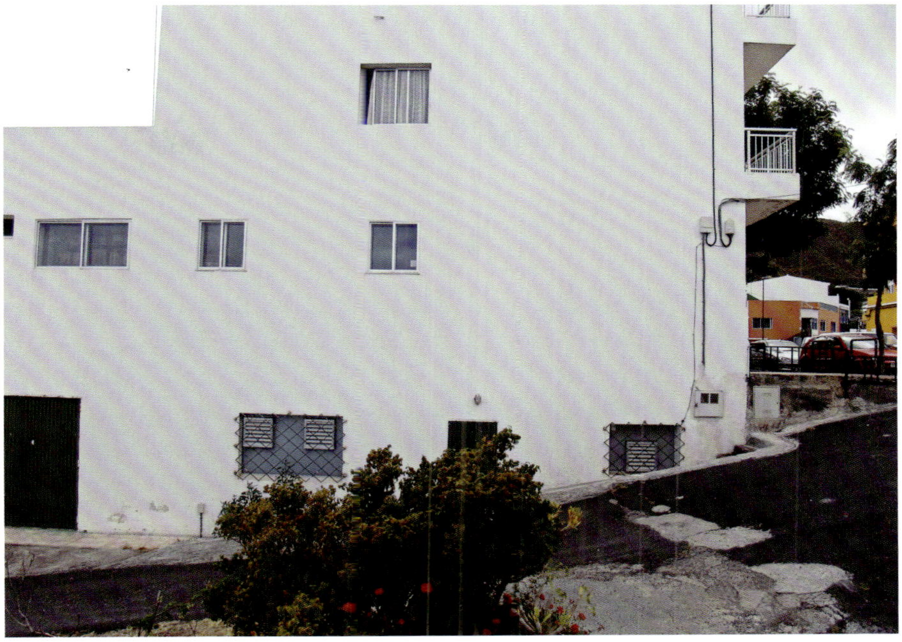

Figure 6.4
Building set into the ground at Valle Guerra museum storage, Tenerife, Spain (case study, chap. 12). Photo: Shin Maekawa. © J. Paul Getty Trust.

- roof planes and roof appurtenances and projections such as eaves, soffits, chimneys, and parapets;
- floors situated below the ground, placed directly on the ground, or elevated above the ground on supports;
- walls above and below the ground;
- openings through the above surfaces and operable closures such as window sashes, shutters, doors, hatches, and flue/ventilator dampers.

Figure 6.5 illustrates the above elements.

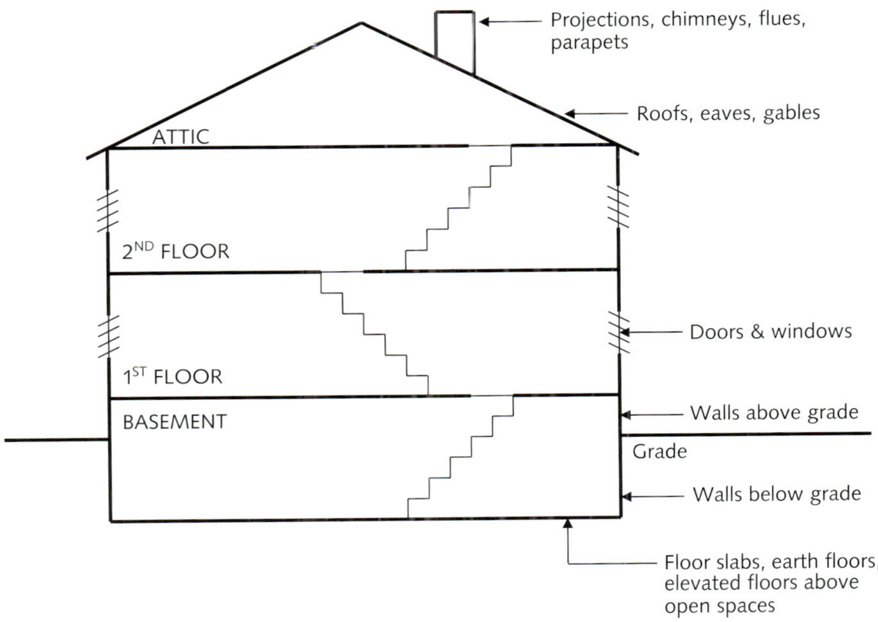

Figure 6.5
The building envelope and typical components. Illustration: Michael C. Henry.

Moisture sources and transport

Moisture transport into or through the building envelope requires a moisture source, a driving force, and material properties or geometries that are favorable to one or more of the transport mechanisms. In the real world, these requirements are often easily satisfied. Water, in its liquid or vapor states, can move by one or more processes—the common processes being gravity flow, capillary flow, diffusion, and convection. Gravity flow is the movement of liquid due to differences in elevation. Capillary flow is the movement of liquid in a porous solid due to adhesion between the liquid and the solid; it is influenced by surface tension of the liquid, density of the liquid, and the size and interconnectedness of the pores in the solid. Diffusion is the movement of moisture vapor due to differences in concentration, and it may occur in air as well as in porous materials. Convection is the movement of liquid or vapor due to pressure differences, as with forced convection from a fan or pump, or due to density differences caused by temperature, as with natural convection.

The interactions among moisture sources, the building and its materials, and the environment are illustrated in figure 6.6. The diagram indicates: moisture sources in the liquid and vapor states, the processes by which moisture is transported from the sources to the building materials (referred to as the "moisture storage"), and the wetting processes by which moisture is taken up by the building materials. The diagram then shows how stored moisture is released into the environment

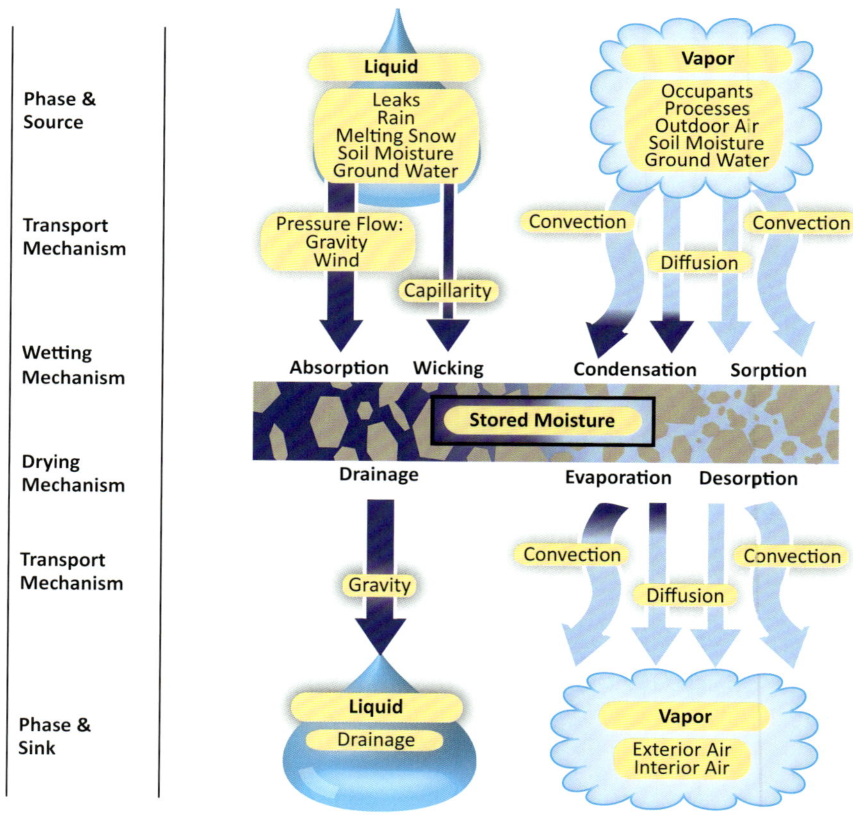

Figure 6.6

Interactions among moisture sources, the building and its materials, and the environment. Illustration: Christine Beckman. Adapted from Straube 2002.

("moisture sinks") from the building materials and indicates the associated drying and transport processes. It is important to note that although moisture can enter building materials by several transport processes, moisture leaves the materials by fewer processes—the dominant process being convection of desorbed or evaporated water vapor. In hot and humid climates, the rates of desorption and evaporation are limited by the high humidity.

Thermal energy transfer

Thermal energy is transferred into and through the envelope by conduction, convection, and radiation when there are temperature differences among the exterior, the envelope, and the interior. Conduction heat transfer occurs between adjacent solid molecules in contact within a nonporous solid material or at contact surfaces between two solid materials, such as the soil-wall interface. Convection heat transfer occurs between a solid and a moving liquid or gas; movement is caused by density differences (natural convection) or pressure differences created by a pump or fan (forced convection). Radiation heat transfer occurs between two bodies without an intermediary medium or direct contact; it is dependent on temperature difference and emissivity and reflectance of the material surface.

Effectiveness of the building envelope

With older buildings, the rate at which moisture and thermal energy are transported through the building envelope will change with time because of deterioration of the building materials and envelope components, particularly if maintenance is lacking. Moisture and thermal energy transport may also be affected by functionally incompatible repair or replacement materials; the removal of original operable envelope features, such as shutters or awnings; or the addition of new materials such as insulation, vapor retarders, or low-permeability coatings.

Ultimately, the effectiveness of the building envelope in mediating the exterior climate and the interior environment limits the interior conditions that can be realistically and economically achieved.

Influences of the Spatial Organization of a Building

The spatial organization of a building—the arrangement of interior spaces, their proximity to the exterior envelope, and the horizontal and vertical connections between interior spaces—will affect the transport of thermal energy and water vapor within the building, as well as the entry of pollutants and particulates with the infiltration of outside air, as illustrated in figures 6.7 and 6.8.

For example:

- Spaces bounded by the building envelope are strongly influenced by exterior conditions; windows will allow solar gain and infiltration of outside air.

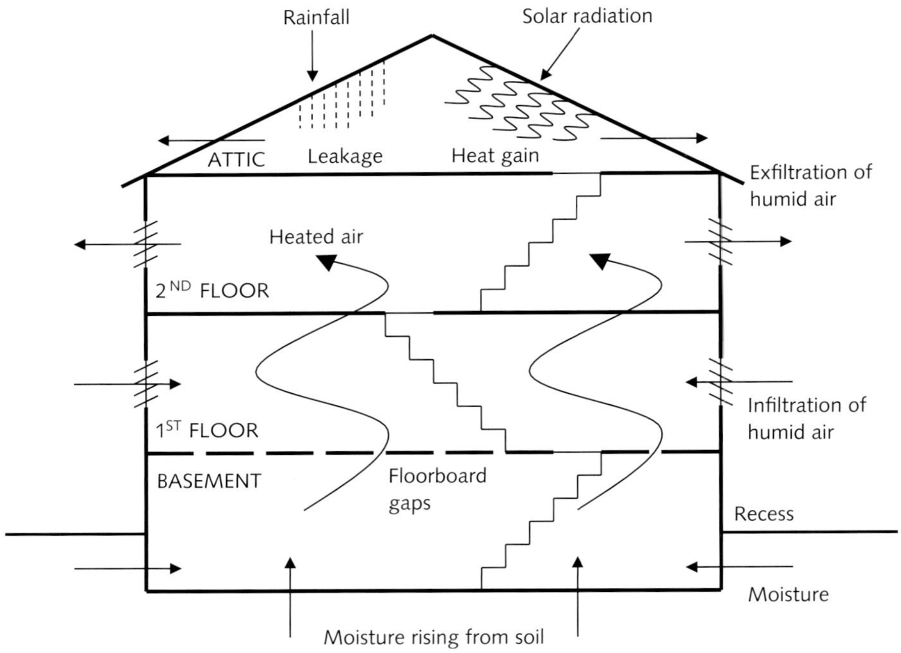

Figure 6.7
Schematic showing sources of moisture and thermal energy affecting Hollybourne Cottage.

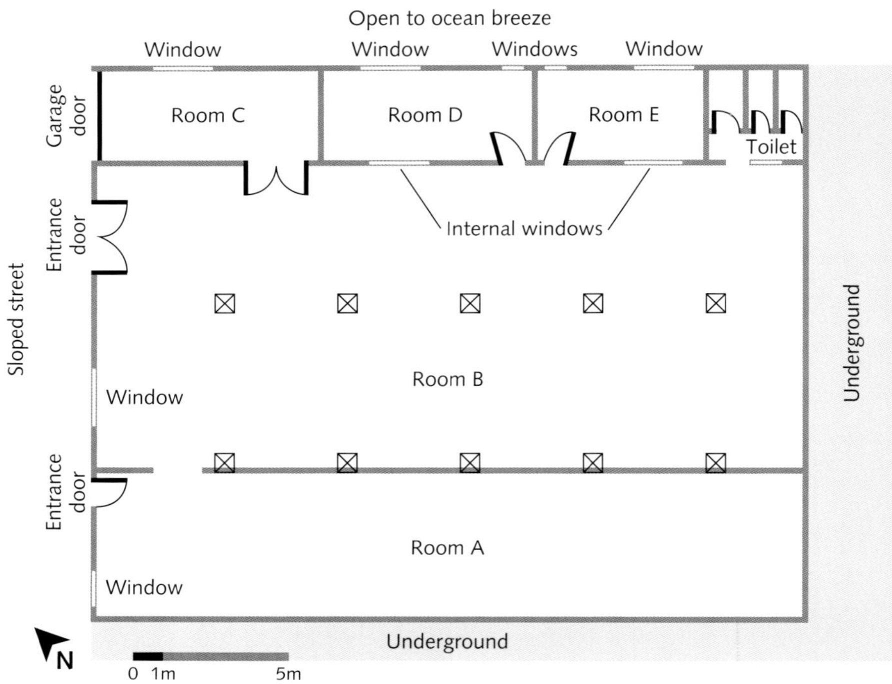

Figure 6.8
Floor plan showing window locations and walls that are in contact with ground soil at the Valle Guerra museum storage, Tenerife, Spain.

- Spaces located directly below roofs or unconditioned attics will be influenced by radiant thermal gains.
- Spaces on opposite sides of the building, and connected horizontally or vertically by interior spaces, will be influenced by pressure differences from wind.
- Vertically connected spaces will be influenced by the stack effect, the vertical convection of air and vapor because

of density differences between floor levels with different temperatures; where the building envelope is perforated by chimneys, windows, and doors, the stack effect can lead to significant convective air exchange with the exterior, even in the absence of wind.
- Underground spaces will be strongly influenced by the temperature of the surrounding soil mass. Soil temperature is typically lower than air temperature in warm to very hot climates; as a consequence, interior humidity in underground spaces will be elevated compared with the warmer spaces above the ground level.

Source Moisture Control Strategies

Hot and humid climates provide ample water vapor and liquid moisture for uptake by building materials and interior environments. In these climates, the sources of exterior moisture include high humidity, rainfall and resultant surface water runoff, and elevated soil moisture.

As shown in figure 6.6, removal of absorbed moisture from building materials is dominated by evaporation and convective drying—processes that are slow in humid climates. Therefore, limiting the uptake of moisture by building materials through the implementation of source moisture control is an essential nonmechanical strategy for managing interior humidity.

Source control strategies for gravity transport of exterior water include interception and diversion of:

- rainwater with gutters and downspouts and with wall projections and drip edges;
- surface water with slopes and berms;
- groundwater with perimeter foundation drains;
- water entering through penetrations, perforations, joints, and cracks in the envelope with the application of joint sealants and weather stripping.

Convective transport and infiltration of water vapor into the building can be reduced by using air barriers in walls and by sealing penetrations of the envelope by such means as weather-stripping windows and doors. Capillary transport of water through building materials can be reduced with either physical or chemical capillary barriers. However, these barriers may elevate moisture content in the material on the source side of the barrier and should be implemented only after careful study.

In any climate, interior sources of moisture include: water from plumbing, piping, and equipment; condensate from dehumidification equipment; and water vapor from occupants and occupant activities such as cooking, showering, and drying laundry. Source moisture strategies for interior moisture should include early detection and repair of leaks in plumbing systems, use of local exhaust fans to remove water vapor from appliances and occupant activities in kitchens and bathrooms, and installation of drip pans and drains under equipment containing water.

Source moisture control should also take into account moisture sources from construction or restoration activities. Building materials such as masonry, plaster, concrete, and paint can initially contain significant amounts of water, which is released during the drying or curing process. Strategies for management of construction moisture should identify moisture in building materials and schedule completion of construction with sufficient time for drying before occupancy or installation of collections. Permanent mechanical systems for environmental management should not be used for construction dehumidification; such use can introduce construction contaminants and particulates into the equipment and ductwork.

After source moisture control measures have been implemented, the removal or drying of remaining water in building materials can be improved by increasing ventilation when atmospheric moisture levels are low.

Thermal Energy Management Strategies

In many hot and seasonally hot climates, thermal energy management strategies are used to lower interior temperature. However, in humid climates, lowering temperature without dehumidification will result in unacceptably high humidity. Therefore, in hot and humid climates, the objectives for humidity and thermal management for collections conservation may negatively impact human comfort.

The primary sources of thermal energy in buildings include:

- solar radiation absorbed by the building envelope and immediate surroundings;
- solar radiation penetrating openings in the building envelope at windows;
- thermal conduction through the building envelope;
- infiltration of hot outside air;
- internal loads such as waste heat from lights, appliances, computers, and equipment and dissipated heat from occupants and their activities.

Strategies for managing thermal energy gains through the envelope include the use of lighter exterior colors or the use of high-emissivity coatings for nighttime reradiation of daytime thermal gains. Solar radiation gains through windows and wall openings can be controlled by shading openings with exterior awnings or shutters or by interior shutters, roller shades, blinds, or window treatments. Solar radiation gains through roofs and walls may be reduced by the application of reflective radiant barriers. The infrared component of natural light passing through the window glazing can be reduced by light-filtering glazing or by application of light-filtering film to existing plain glass.

Thermal conduction through the envelope may be reduced through the strategic application of insulation to exterior walls or roofs; however, the benefit of thermal insulation in high-thermal-mass buildings in hot climates without mechanical cooling may be marginal at best. Infiltration of hot exterior air can be reduced by improving the fit of windows and doors in their frames. With any heritage building, the potential negative impact of envelope improvements, such as appearance change or reduced durability, must be assessed and evaluated.

Internal thermal gains from lighting and other equipment can be reduced by using lower-wattage incandescent bulbs or by refitting light fixtures with higher-efficiency lamps, such as light-emitting diode (LED) and compact fluorescent lamps. Internal thermal gains related to building occupancy can be reduced by managing the occupant census so that peak occupancy does not coincide with peak external thermal loads.

Convection Strategies

The convection or movement of air within a building may benefit the conservation and occupant-comfort environments and can result from:

- cross ventilation and/or vertical ventilation due to exterior wind channeled through interconnected spaces by open windows and doors,
- gravity or stack ventilation due to temperature gradients within a space or a series of connected spaces,
- fans or similar devices for moving air.

When used as a strategy for suppressing microbial activity, air movement must occur near the surface of the material at risk, but it must not disturb the object or abrade the material surface with particulates. Therefore, convective strategies should provide low-velocity, low-turbulence airflow that is evenly distributed throughout the room or space, including the walls, floor, and ceiling.

Control Strategies for Particulates and Insects

Particulates

In any climate, particulates can damage collections through abrasion; in hot and humid climates, reactive particulates can lead to accelerated chemical deterioration of surfaces. Particulates may also lead to deposition of nutrients necessary for microbiological growth. Unless particulates can be removed by mechanical ventilation and filtration, nonmechanical particulate control strategies are necessary. Examples of particulate control strategies include:

- walk-off mats, shoe coverings (booties), or carpets along visitor routes to remove particulates on shoes (fig. 6.9);

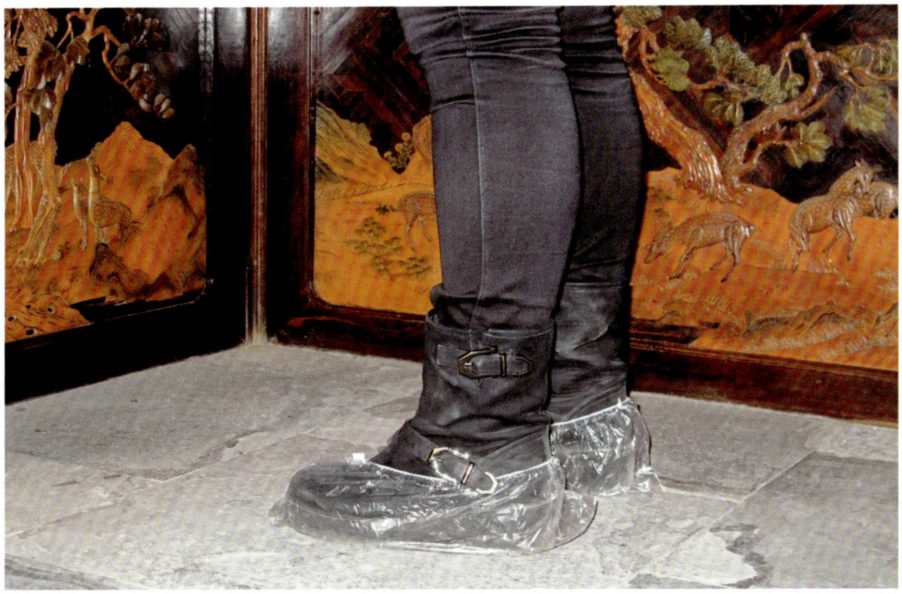

Figure 6.9
Disposable shoe covers used to control dust at Juanqinzhai, Beijing (case study, chap. 15). Photo: Shin Maekawa. © J. Paul Getty Trust.

- water-soluble tackifiers (chemical compounds used in formulating adhesives to increase the tack, or stickiness, of a surface) sprayed on window screens to improve particulate capture;
- calcium chloride dust-control sprays on nearby stone- or earth-surfaced pedestrian walkways and vehicular driveways, and reduced vehicular speeds or restricted vehicular access on unpaved roads near the building;
- cessation of ventilation when airborne pollen counts are high;
- management of landscape maintenance activities to reduce dispersal of particulates from soil and small engine exhaust (fig. 6.10);
- rigorous housekeeping procedures to remove accumulated dust from interiors;
- use of entry vestibules.

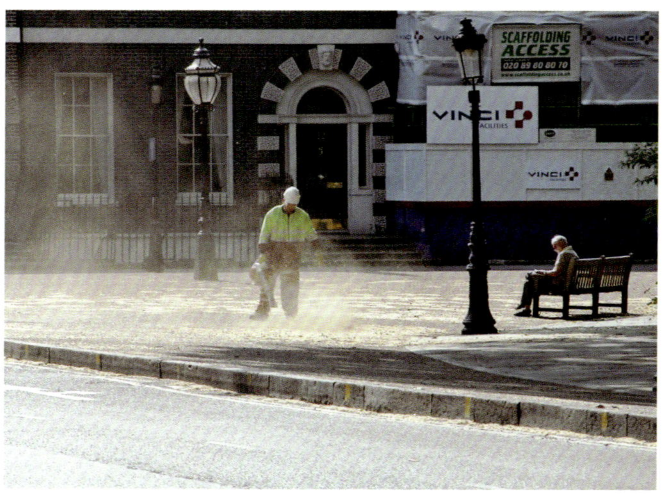

Figure 6.10
Leaf blower being used for street cleaning—and the resultant airborne dust. Photo by permission of Jim Murray, The Bloomsbury Associates.

Insects

When open windows and doors are used to ventilate buildings in hot and humid climates, insect entry must be controlled with tight-fitting screens. Integrated Pest Management, discussed in a sidebar in chapter 3, must also be used throughout the building.

Space Use Strategies

The envelope and spatial configuration of a building incorporate features and systems for occupant comfort and the environmental needs of the contents, as well as for aesthetics and functional considerations, such as the architectural program, security, and cost.

The conversion of an existing, and possibly historic, building into a museum containing collections is often a significant change from the original intentions for the building (fig. 6.11). In order to successfully adapt an existing or historic building for museum use, one must understand the original design intent and construction of the building. It is also important to understand the environmental comportment, or behavior, of the building and its spaces.

When determining which spaces are best suited for collections, it is useful to assess the building, its spaces, and their interior environmental behavior, and to consider:

- suitability of the unmodified interior environmental conditions of the space for collections;
- adequacy of the space, its configuration, and location to accommodate the collections without congestion and handling risks;
- location of the space relative to exterior walls, where thermal and moisture loads are high;

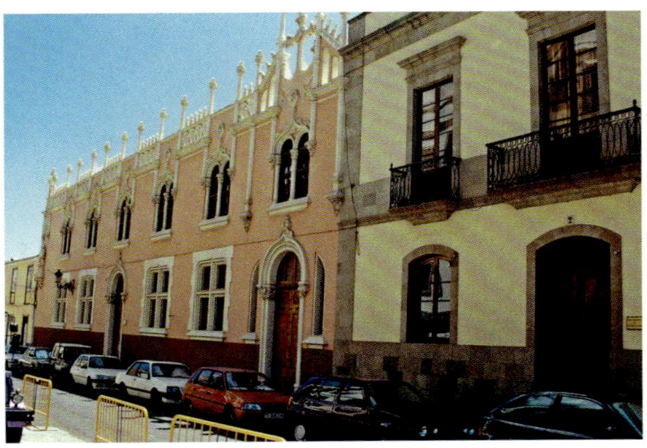

(a)

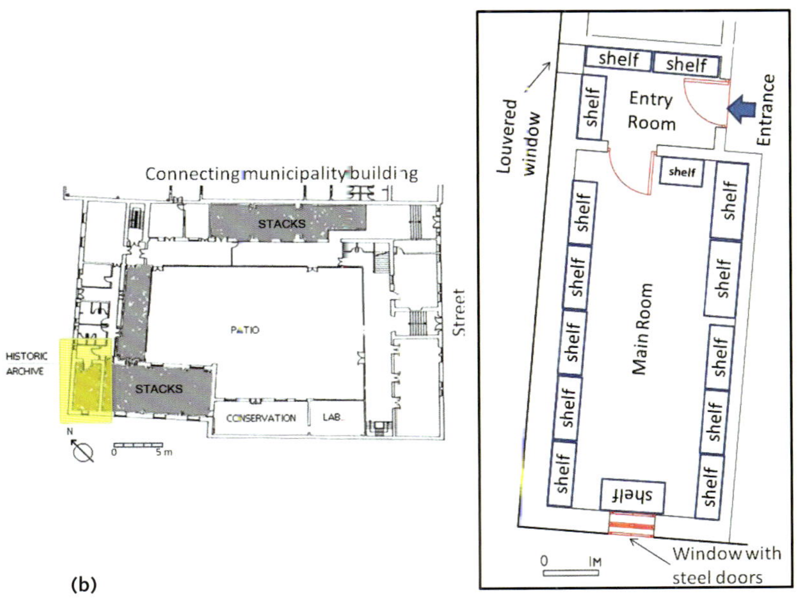

(b)

Figure 6.11

Building facade (a) and floor plan (b) of the Municipal Archive in San Cristóbal de La Laguna, Tenerife, Spain. The building was originally a convent (case study, chap. 11). Photo: Shin Maekawa. © J. Paul Getty Trust.

- adjacency of the collections space to potential hazards, such as kitchens, restrooms, and mechanical rooms;
- proximity of the collections spaces to unsupervised or controlled building entrances and exits;
- separation of staff/private spaces from public spaces, including circulation control;
- proximity to building emergency exits for evacuating or salvaging the collections in the event of a disaster.

The spaces in a building can be grouped into zones that are defined by the level of security, interior environmental conditions, and proximity to utilities and services. This approach simplifies the task of separating collections from obvious risks and can lead to identification and placement of collections in zones that present the lowest aggregate risk to the collections. Separating spaces containing collections from non-collection spaces such as offices will simplify mechanical environmental management strategies by eliminating the competing and conflicting objectives for occupant comfort and collections conservation. With heritage buildings, evaluation of the possible allocation of spaces and uses must also include the potential impact on aesthetics and historic building fabric that may result from changes in use.

Layered Enclosure Strategies—Secondary Envelopes, Microenvironments, and Collections Housing

With some collections, it is necessary to reduce the range and frequency of temperature and humidity variations. This conservation objective can be effectively accomplished with nonmechanical strategies using layered enclosures, ranging from room-sized secondary envelopes to object-sized microenvironments.

The layered enclosure strategy creates layers of resistance to the transport of thermal energy and moisture between the environment immediately surrounding the collections and the larger environment of the surrounding room or the exterior. Layered enclosures divide the total difference between the object and nearby conditions into a set of small thermal, pressure, and moisture gradients. The division of a single steep gradient into multiple gradients across several layers effectively reduces the influence of fluctuations in exterior temperature and humidity on the environment immediately surrounding the collection.

Layered enclosures also reduce the volume of the environment surrounding the collections; the smaller volume can then be more easily managed with either passive control materials, such as silica gel, or smaller mechanical systems.

Examples of layered enclosure strategies include:

- Creating perimeter corridors or circulation spaces between exterior walls and interior collections spaces. This is particularly effective in heritage buildings, where it is important to retain the fenestration and material fabric of the exterior walls but where the thermal and moisture performance of the exterior walls cannot maintain the desired conservation environment. This also constitutes effective protection for the historic building fabric where the desired interior conditions might result in the deterioration of the exterior wall if there were no buffer space. A new interior partition enclosing the collections zone and erected inside of the exterior wall can be constructed to provide the necessary thermal and moisture performance for the collections zone. The perimeter space may be occupied, such as a corridor, or it may merely be a narrow space wide enough for housekeeping and maintenance of the exterior wall.

- Exhibiting collections in closed cases or gasketed cabinets (fig. 6.12).
- Storing collections in boxes (fig. 6.13).

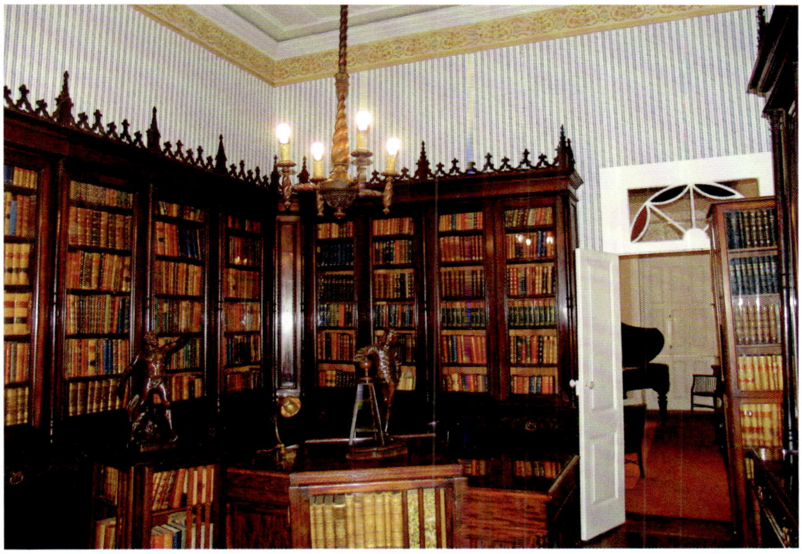

Figure 6.12
Book collections kept in custom-built book cabinets with covers at the Museu Casa de Rui Barbosa, Rio de Janeiro, Brazil (case study, chap. 14). Photo: Shin Maekawa. © J. Paul Getty Trust.

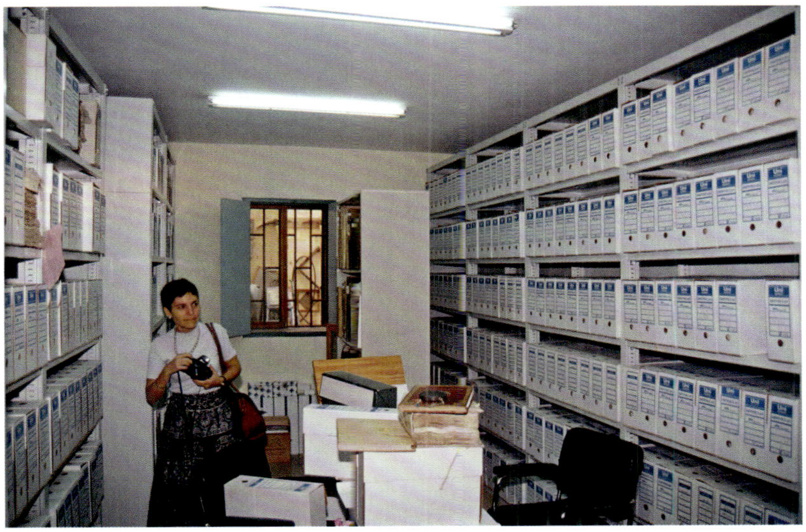

Figure 6.13
Archival documents kept in acid-free archival boxes in the Municipal Archive of San Cristóbal de La Laguna, Tenerife, Spain. Photo: Shin Maekawa. © J. Paul Getty Trust.

Occupant Load Strategies

Occupants can be a significant source of heat and moisture in a building; these gains are proportional to the number of occupants and the duration of their occupancy. The amount of outside air that must be conditioned and introduced into occupied spaces to maintain an acceptable indoor air quality increases with the number of building occupants. For most buildings containing exhibited collections, the occupant census varies throughout an individual day and over the course of a year.

In order to reduce the need for mechanical cooling for thermal comfort, the occupant census can be managed to shift peak occupant thermal gains to times of the day when thermal loads are lower, such as morning hours. Managing occupant census and visitor routes in collections areas provides opportunities to reduce visitor congestion and the associated physical risks to collections, as well as to improve visitor comfort and the viewing experience.

Strategies for managing occupant census include:

- timed ticketing or a visitor reservation system to limit the number of people during the periods of day when thermal gains are greatest,
- discounted entry prices to encourage visitation during periods when thermal gains are low,
- scheduling group tours during periods when thermal gains are low,
- scheduling opening hours to correspond with times when thermal gains are low.

Maintenance and Sustained Implementation of Nonmechanical Strategies

In order to realize the full benefits of nonmechanical strategies, their assumed effectiveness and performance must be clearly documented and integrated into the design of the mechanical strategies. Like mechanical systems, nonmechanical strategies require systematic preventive maintenance to maintain performance. Furthermore, some nonmechanical strategies, such as operating window shades or solar control devices, may require active adjustment by occupants. The benefits of some nonmechanical strategies, such as use of spaces or occupant load management, can be lost when decisions to reallocate use of space to increase visitation or occupancy are made without considering the ramifications on interior environmental management.

In the long term, nonmechanical strategies are well served by an operating and maintenance (O&M) manual. The O&M manual should include the design intent, the importance of the strategy to environmental management, and the routine measures required to maintain performance at an acceptable level. In addition to periodic inspection, examples of activities that might be required to maintain the performance of nonmechanical strategies are:

- clearing of all drainage systems and removal of debris and litter that slow flow or reduce capacity;
- replenishment of the soil around buildings to maintain a positive grade to direct rainwater away from building foundations;
- adjustment of windows and doors for ease of operation and airtight closure;
- adjustment of operable features for solar gain control, such as shades and shutters, especially those mounted on the building exterior;
- spot repair or touchup of breaches in protective coatings;
- trimming and pruning of vegetation near the building and removal of vegetative litter that may serve as food/nutrients for pests or microorganisms;
- verification of the effectiveness of the nonmechanical strategy through monitoring and evaluation.

Maintenance of nonmechanical strategies should be systematically documented in the same manner as is mechanical systems maintenance. See chapter 8 for additional discussion of maintenance.

Summary

Nonmechanical strategies reduce thermal and moisture loads on the interior environment of a building by using building or site features, whereas mechanical strategies use mechanical processes to modify the interior environment by removing heat and moisture gains. This chapter described a range of nonmechanical strategies that may potentially reduce the energy required to condition the interior environment of a building by mechanical means.

The next chapter will address mechanical strategies, with the underlying assumption that nonmechanical strategies have been implemented first.

Reference

Straube, J. F.
2002 Moisture in buildings. *ASHRAE Journal* 44 (1): 15–19.

Chapter 7

Mechanical Strategies for Environmental Management

The nonmechanical strategies for environmental management described in chapter 6, such as source moisture control and envelope improvements, are highly effective at reducing the amount of cooling and dehumidification needed in hot and humid climates. Mechanical strategies, however, are generally needed to bring interior conditions within the desired ranges of humidity and temperature for collections conservation and thermal comfort.

This chapter is directed at readers, including architects and museum staff, who may need a general review of the various types of mechanical equipment and devices that can be used to implement the psychrometric strategies of dehumidification, heating, cooling, and humidification identified in chapter 5 for collections environments. It emphasizes simple equipment and devices that can be utilized in hot and humid climates, as well as in marine climates with cool-humid winters, where availability, reliability, maintainability, and low energy consumption are major operational objectives.

This chapter describes:

- mechanical equipment for psychrometric and related processes of dehumidification, ventilation, heating, cooling, and humidification
- nonpsychrometric mechanical equipment necessary for the implementation of mechanical strategies, including control systems and filtration

This chapter also provides an overview of the issues of configuration and risks related to:

- centralized and noncentralized mechanical systems in historic buildings
- mechanical strategies in general

Overview of Air-Conditioning

The term *air-conditioning* was first used in 1906 by Stuart Cramer to describe his integrated, automatically controlled approach to cleaning, humidifying, and distributing air within large textile mills and factories

(Cooper 1998). Since then, the engineering professional's use of the term has expanded to include the mechanical processes and control of dehumidification, ventilation, filtration, cooling, heating, and humidification.

The conventional application of air-conditioning in hot and humid climates relies on mechanical cooling to depress the air temperature (sensible cooling) for human comfort; the cooled air is supplied to the conditioned space; as a result, the space temperature is lowered, and thermal comfort is improved. If the air temperature is not reduced below the dew point temperature during the cooling process, the resultant humidity actually increases, because the moisture content (indicated by the dew point temperature) of the air has not been reduced by mechanical cooling. If, as part of the cooling process, the air temperature is lowered below the dew point temperature, dehumidification (latent cooling) will occur as moisture is removed from the air by condensation. In this case, the cooled air must be heated (in a process called *reheat*) before it is supplied to the conditioned space to prevent overcooling and to depress the humidity; if reheat is not used, the air in the conditioned space must be sufficiently warm that, when mixed with the cooled air, the desired space temperature and humidity of the cooled air will be achieved (fig. 7.1).

The process of cooling air below its dew point temperature followed by reheating for dehumidification and temperature control is utilized by many museums and cultural institutions for collections conservation and occupant comfort. However, cooling and then reheating the air is an energy-intensive process. This approach can be problematic when dehumidification loads are much greater than cooling loads, as in hot and humid climates, particularly at night or during rainstorms. The same problem occurs in marine climates during the cool-humid winter. The cooling-then-reheat approach is also problematic when humidity fluctuations require tighter control than temperature fluctuations, as set out in chapter 3, where conservation environments in hot and humid climates

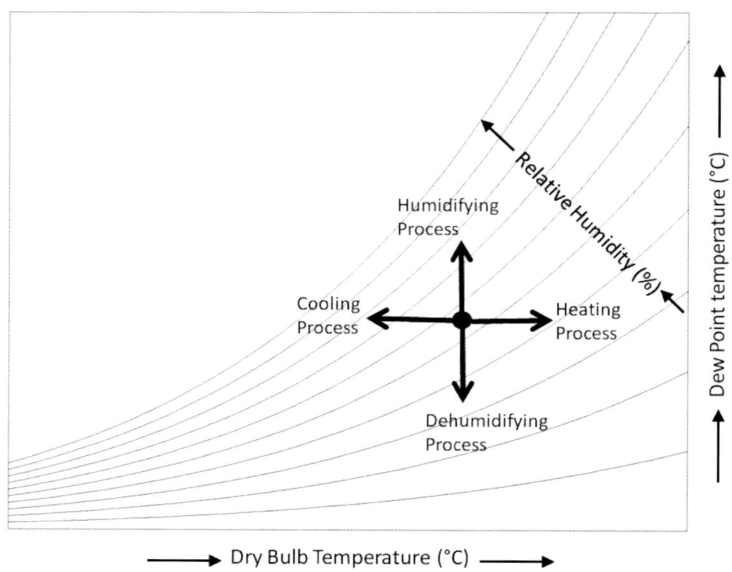

Figure 7.1
The cooling, dehumidification, and reheating processes demonstrated on a psychrometric chart.

and marine climates prioritize humidity control over temperature control. In these instances, alternative mechanical approaches for environmental management are needed.

Alternative strategies for environmental management employ the same basic psychrometric processes as conventional strategies based on air-conditioning. However, the primary objective is to economically produce and maintain a conservation environment for both collections and historic interiors that is focused on humidity control using the simplest and most reliable combination of mechanical equipment. Occupant comfort can also be addressed with alternative strategies, but the conservation environment for the collections remains the first priority.

Mechanical Equipment for Psychrometric Processes

The following introduction to mechanical equipment is in the order of importance of the basic psychrometric processes for alternative environmental management strategies in hot and humid climates: dehumidification, heating, cooling, and humidification. It concludes with a review of nonpsychrometric processes such as ventilation.

Dehumidification

The process of dehumidification removes water vapor from air, lowering the dew point temperature of the air. Dehumidification can be performed mechanically by condensing water vapor from the humid air or chemically by using desiccants to adsorb water vapor. Dehumidification cannot be achieved solely by heating air, such as conservation heating, which depresses the humidity by raising the air temperature but does not change the moisture content of the air. Conservation heating is discussed in the section on heating.

Portable mechanical dehumidifiers

Portable mechanical dehumidifiers, such as the one shown in figure 7.2, utilize the refrigeration cycle (see sidebar "Refrigeration Cycle") to maintain the temperatures of the evaporator (cold coil) and the condenser (hot coil). Humid room air is taken through a dust filter to the evaporator where the air is cooled below its dew point temperature. As a result, water vapor from the air condenses on the evaporator and drains to the water receptacle. The dehumidifier then reheats the cooled and dehumidified air by passing it over the condenser before returning it to the room. Note that this is a departure from the dehumidification achieved by a refrigerant-based mechanical cooling process, in which the condenser rejects the absorbed heat to the outside (see "Cooling," p. 122). The temperature of the dehumidified air is increased by waste heat from the compressor and the fan motor. The net result is that the returned dry air is warmer than the entering moist air. For the same amount of thermal energy, the temperature rise in dry air will be greater than the temperature rise in moist air (fig. 7.3).

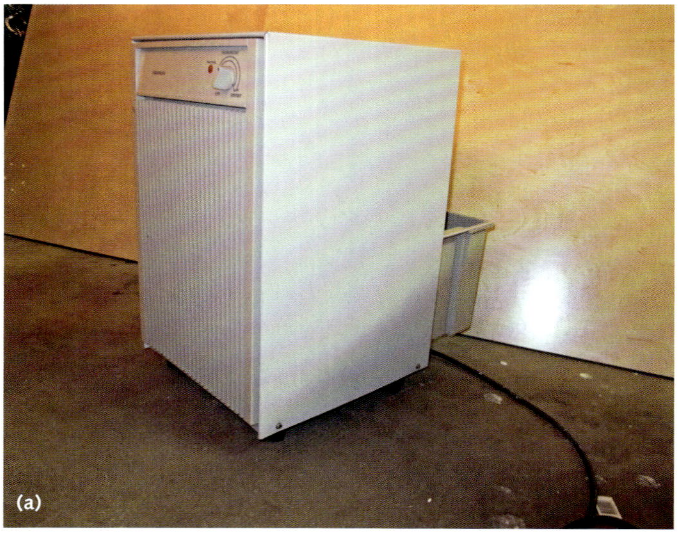

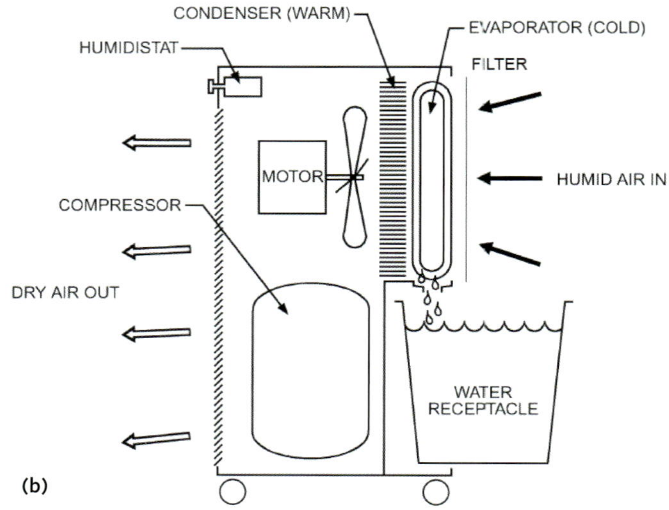

Figure 7.2

Image (a) and schematic (b) of a typical portable mechanical dehumidifier. Modified from fig. 24.2, chap. 7.39, ASHRAE 2008. Photo: Shin Maekawa. © J. Paul Getty Trust.

Refrigeration Cycle

The refrigeration cycle is also referred to as the vapor compression cycle. Natural (thermodynamic) law states that heat flows from warm space to cool space, but not vice versa. The refrigeration cycle makes use of this principle. The figure below describes the mechanical processes of the refrigerant in the cycle. Gaseous refrigerant is compressed mechanically, a process that raises the gas temperature. The compressed refrigerant gas is cooled as it passes through a condenser, where the heat of compression is dissipated to the outside air or to a liquid cooling medium. As a result, the refrigerant gas condenses to the liquid phase. The pressurized liquid refrigerant then passes through an expansion valve, expanding back into the gas phase, which lowers the refrigerant temperature to the coldest state in the cycle. The cold misty refrigerant gas next flows through an evaporator coil, where it is heated by room air that is forced over the coil by a fan; as the refrigerant is heated, the room airstream is cooled. The absorption of heat converts the refrigerant completely into the gas phase, and it returns to the compressor, completing the cycle.

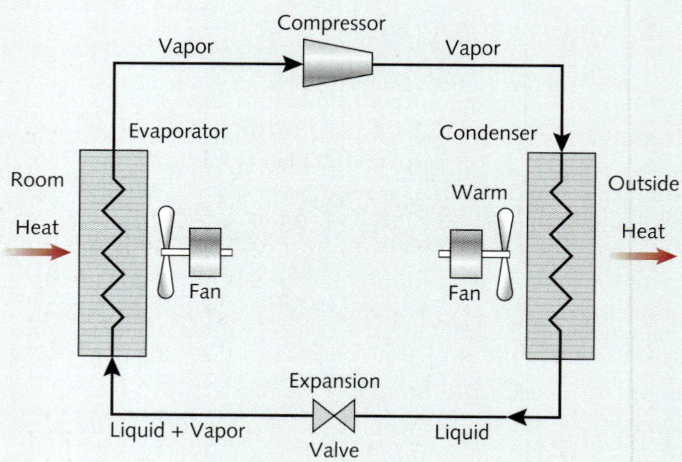

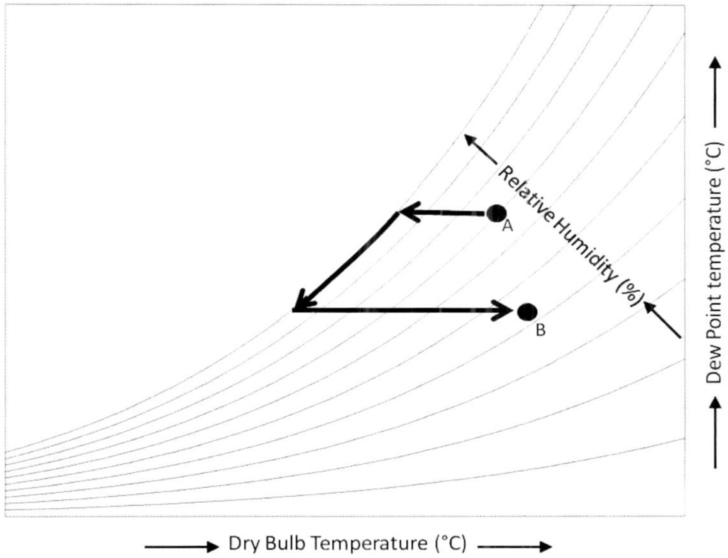

Figure 7.3

Psychrometric process of mechanical dehumidification. State point A represents the room air entering the dehumidifier. The air is cooled and reaches its dew point temperature. The air is further cooled to below its dew point temperature, and moisture is removed. The cooled and dehumidified air is heated to state point B, where the dehumidified and reheated air exits the dehumidifier.

The effectiveness and durability of mechanical dehumidifiers is well proven, particularly in hot and humid climates. However, portable dehumidifiers are not without potential problems:

- The efficiency of mechanical dehumidifiers decreases in cool or dry (low dew point temperature) environments due to ice formation on the evaporator coil.
- If the filter is clogged or if the fan motor fails and reduces airflow, ice may form on the evaporator coil and reduce or fully block airflow, resulting in low efficiency or shutdown.
- The condensate receptacle must be emptied regularly unless the dehumidifier is fitted with a condensate pump or gravity drain for continuous removal of the condensate. Most dehumidifiers are equipped with a sensor that shuts down the unit when the condensate receptacle is missing or full; however, this may result in loss of dehumidification if the manually drained bucket is not attended to—for example, over a weekend or holiday. Failure of the sensor can result in an overflowing condensate receptacle and resultant damage to flooring and adjacent furniture. If the dehumidifier shuts down with a receptacle full of water, room humidity will rise as the water evaporates.
- Noise from the compressor, the fan motor, the fan, or the vibrating housing may be intrusive, exceeding 60–70 dB. Excessive noise in interpreted spaces can diminish the quality of the visitor experience, although acoustically insulated enclosures can provide limited noise reduction.
- The low-cost humidistats supplied with portable dehumidifiers are not sufficiently accurate to control collections environments, although a technician can resolve this issue by bypassing the standard humidistat and replacing it with

a separate (more accurate) aftermarket RH sensor and control device.
- Mechanical dehumidifiers (both portable and duct-mounted units) require diligent filter maintenance to prevent the coil from clogging.

The performance capacity of small portable dehumidifiers is specified by the airflow rate of the fan and the amount of water removed per day (or per hour); the rate of water vapor removal may be expressed in terms of either weight or volume. The temperature and humidity of the air in the conditioned space will affect the water removal rate of the unit, so stated removal rates should be listed alongside the environmental conditions under which the test was conducted. Reputable manufacturers will indicate the water removal rate for several temperature and humidity combinations, often as a graph. Condensate capacity (for receptacles) or removal flow rate and head pressure (for condensate pumps) are also important parameters. Most portable mechanical dehumidifiers can operate with power consumption on the order of 1.75 kW/hr and can be powered from standard domestic electrical receptacles and circuits. Portable dehumidifiers were utililized for the case studies at Hollybourne Cottage, part B, and the Emílio Goeldi Museum (chaps. 10 and 13).

Duct-mounted mechanical dehumidifiers

A duct-mounted mechanical dehumidifier (fig. 7.4) uses the same mechanical process as does a portable unit, but the duct-mounted types typically have larger dehumidification capacities and receive and discharge air through ductwork, allowing them to be located nearby, rather than within, the room they serve. The controls for duct-mounted dehumidifiers may be of higher quality than those on portable units. A humidity sensor is typically mounted in the space served or in the return duct, rather than on the unit. Duct-mounted units normally use gravity or a small condensate pump to discharge condensate to a nearby drain, allowing for continuous unattended operation.

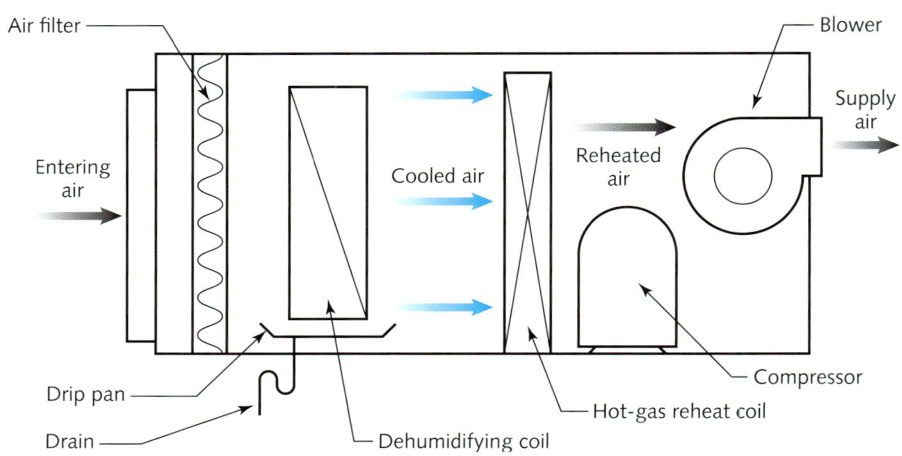

Figure 7.4

Schematic of a typical packaged in-duct dehumidifier. Modified from fig. 24.2, chap. 7.39, ASHRAE 2008.

Issues with duct-mounted mechanical dehumidifiers include:

- The internal condensate collection systems can fail or overflow if the unit is not installed plumb and level.
- Drip pans need to be routinely checked for debris and occasional small vermin that can block drain lines.
- A spill protection pan must be fitted below the air-handler unit to protect the building and collections against accidental overflow in the event of condensate tray leakage, pump failure, or gravity drain blockage. The spill protection pan should be fitted with a separate oversize drain and water sensors, the latter connected to locally audible alarms or to an alarm indicator in the environmental management control system. Water in the spill protection pan should also trigger automatic shutdown of the dehumidifier.
- Noise and vibration of in-duct dehumidifier units can be controlled through good duct design, installation of in-duct noise suppression devices, and vibration isolation mounts.
- Mechanical dehumidifiers (both portable and duct-mounted units) require diligent filter maintenance to prevent the coil from clogging.

In recent years, duct-mounted dehumidifiers with hot-gas reheat and bypass (fig. 7.5) have entered the market and are capable of providing cooling for limited thermal comfort. This type of dehumidifier is

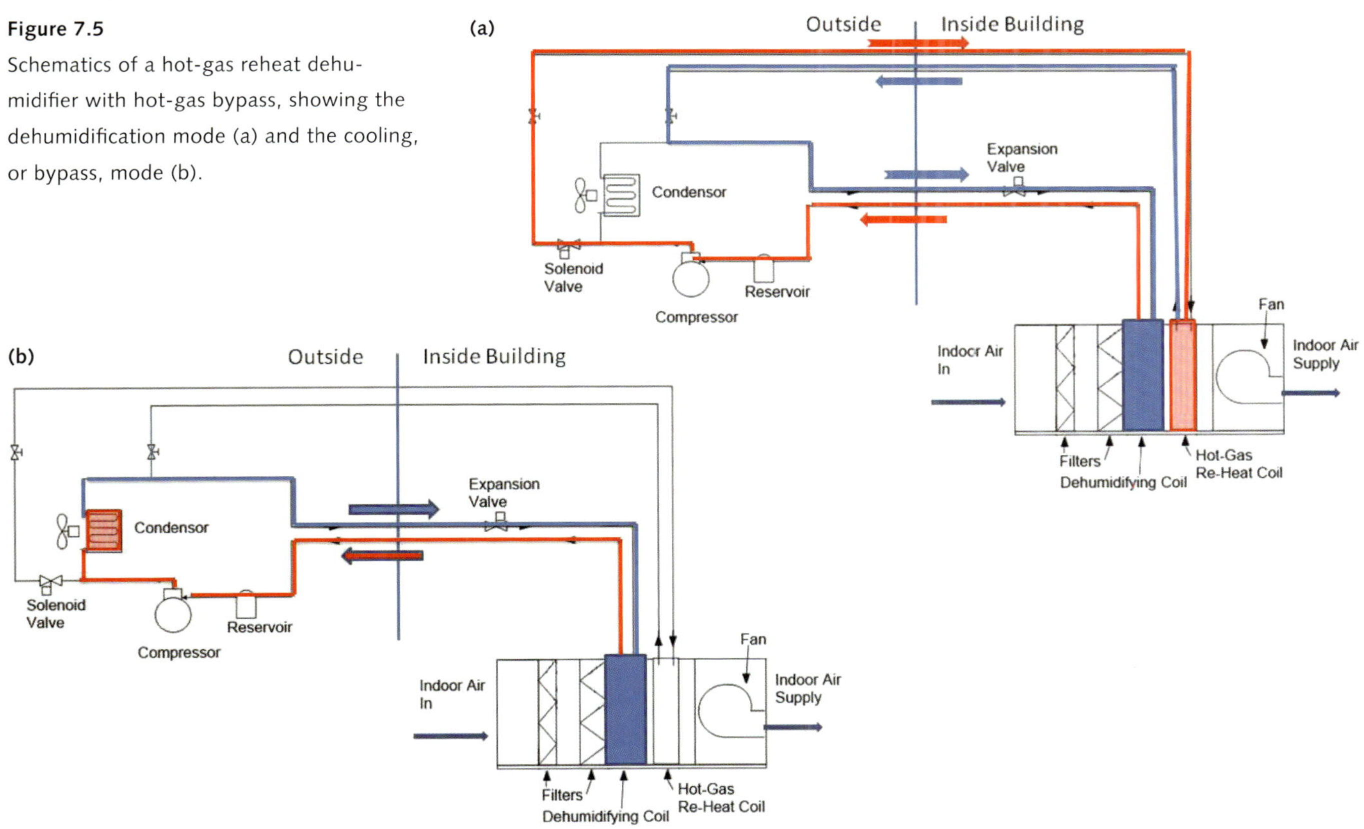

Figure 7.5
Schematics of a hot-gas reheat dehumidifier with hot-gas bypass, showing the dehumidification mode (a) and the cooling, or bypass, mode (b).

equipped with two condenser coils, one downstream of the dehumidifying coil (evaporator) and the other remotely mounted on the exterior, similar to a conventional direct expansion (DX) air conditioner (see the discussion of cooling below). When the interior temperature is below the set point (normally 25°C–28°C), the unit operates in conventional dehumidification mode, and the refrigerant is cycled between the dehumidifying coil and the hot-gas reheat coil, both located in the conditioned airstream. In this mode the unit supplies dehumidified air, near to or slightly above the room temperature, to the conditioned space. When the temperature of the return/room air exceeds the set point temperature, the unit switches from dehumidification-reheat mode to cooling mode, and hot refrigerant is directed to the exterior condenser coil to dissipate the heat, bypassing the reheat coil in the duct or air handler. In cooling mode, cool air is returned to the conditioned space with only incidental dehumidification in order to reduce the space temperature. During the cooling mode, the system does not control humidity, and it cannot be operated on the humidity priority. This system has been used in Juanqinzhai, China (case study, chap. 15).

Desiccant dehumidifiers

Desiccant (chemical) dehumidifiers can also be used to control the moisture content of air. In desiccant-based dehumidifiers, moist room air is passed over a desiccant bed that chemically adsorbs water vapor from the moist airstream, drying the air. In closed environments, such as tight display cases or vitrines, simple desiccants or humidity buffers can be either regenerative (reusable) or nonregenerative (nonreusable). The desiccant media can include:

- silica or alumina gel (regenerative),
- activated alumina (regenerative),
- molecular sieves (regenerative),
- lithium chloride salt or salt solution (regenerative),
- glycol solution (regenerative),
- hygroscopic salts such as calcium chloride, urea, and sodium chloride (nonregenerative).

Desiccant dehumidification can either work passively or with continuous regeneration. In passive dehumidification, the regenerative media must be periodically removed and regenerated by heating in a separate oven to remove the adsorbed moisture; the nonregenerative media must be removed, disposed of, and replaced. Passive dehumidification is not effective where large rates of air change or moisture generation are required, because in such instances the desiccant media will rapidly become loaded with moisture.

Where the air change rate is high and for HVAC applications, continuous regeneration (reactivation) of the desiccant media is required. A typical continuous regeneration dehumidifier (fig. 7.6) uses a rotating wheel (cassette) loaded with desiccant media. The moisture-laden space air passes through the dry section of the desiccant wheel (normally about 75% of the desiccant surface area), where the water vapor is adsorbed by

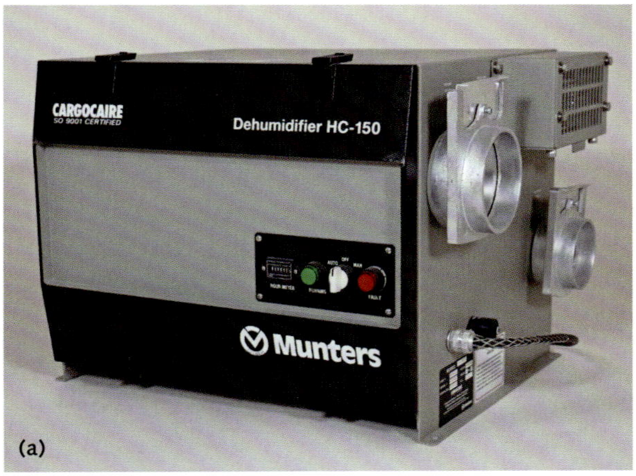

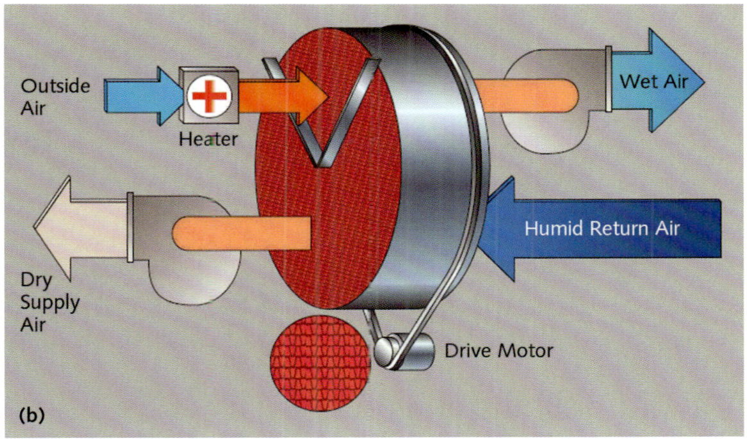

Figure 7.6
A typical desiccant wheel dehumidifier (a) and its operational schematic (b). Photo: Vincent L. Beltran. © J. Paul Getty Trust.

the media; simultaneously, the portion of the wheel containing moisture-loaded desiccant media is regenerated (reactivated) by hot air from a gas burner or electrical resistance heater, which removes the moisture from the media and discharges the hot humid air to the exterior. Slow rotation of the wheel continuously cycles the desiccant between the moisture adsorption and regeneration sections of the dehumidifier, the sections being separated by airtight seals (fig. 7.7).

In continuous regeneration dehumidifiers, the desiccant media will retain some heat from the reactivation process as it returns to the adsorption section. As a result, the processed air returns to the space at a slightly higher temperature.

Requirements for continuous regeneration dehumidifiers include:

- If mounted indoors, these desiccant dehumidifiers require three or four separate air ducts, one of which must be open to the exterior for the exhaust of humid regeneration air.

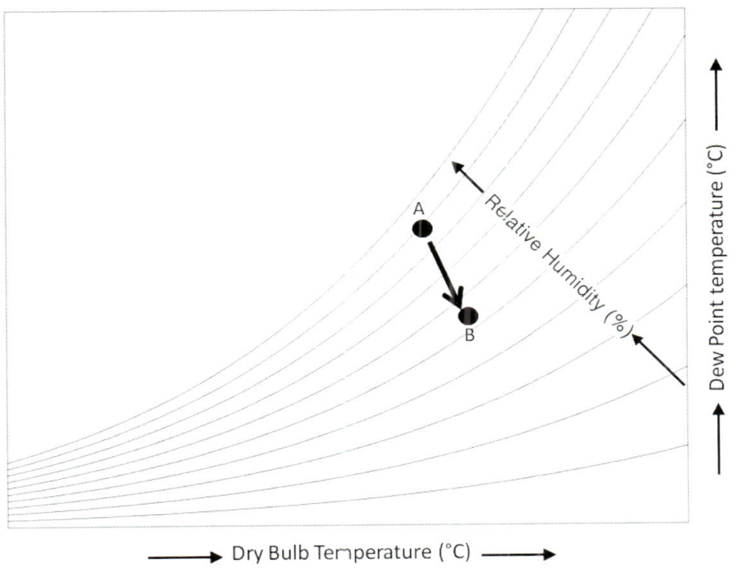

Figure 7.7
The psychrometric processes in desiccant dehumidifiers. State point A is the air entering the dehumidifier; state point B is the dehumidified air exiting the dehumidifier.

These ducts, particularly the regenerative air ducts, must be carefully designed and routed to avoid condensation and space depressurization problems.
- The conditioned space air and the supply air for regeneration must be filtered to prevent the contamination of the desiccant media with particulates. In addition to a rigorous schedule for filter replacement, these desiccant dehumidifiers should be checked routinely for the integrity of the radial and circumferential seals of the desiccant cassette. The desiccant cassette wheels on some dehumidifiers are driven by belts rather than gears; it is important to replace the belts according to the manufacturer's schedule.
- The regenerative heaters in these dehumidifiers require electrical power or natural gas. Small-capacity desiccant dehumidifiers have been available since the 1990s. Small units may operate at domestic grade single-phase power, but electric regenerative heaters and fans in larger units will require two- or three-phase power.

Although the use of continuous regeneration desiccant dehumidifiers is increasingly common, many service and facilities personnel are not familiar with them. They provide more efficient dehumidification than the incidental dehumidification achieved with refrigerant-based or chilled-water mechanical cooling systems (discussed under cooling below), especially when low dew point temperatures (<5°C) are required, but they have higher initial and operating costs than mechanical dehumidifiers.

Ventilation

Ventilation is the process of replacing air in a space to improve indoor air quality; this may involve the addition or removal of heat and water vapor or the dilution of indoor air contaminants (fig. 7.8). The air move-

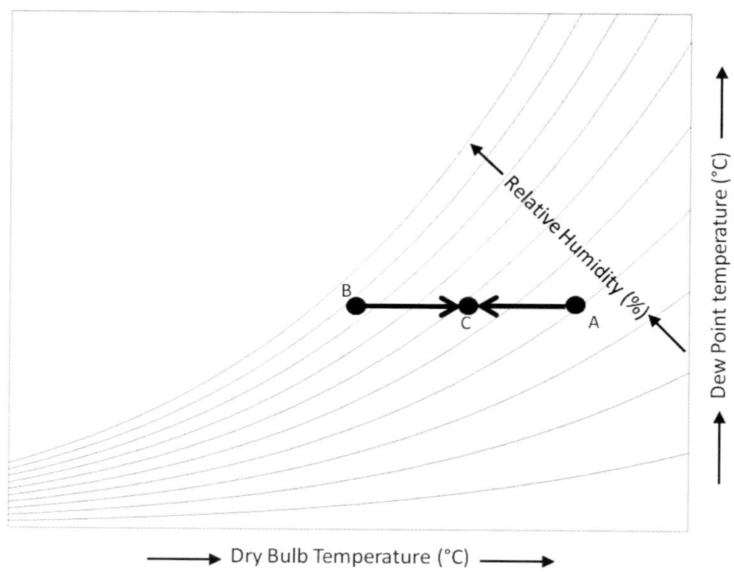

Figure 7.8

Psychrometric process of dilution ventilation. State point A is the room air that requires ventilation; state point B is the outside air that is taken into the room, and state point C is the resulting room air following 1 air change.

ment in the space resulting from ventilation can improve thermal comfort in hot or humid conditions by enhancing the exchange of moisture and heat between skin and space air. It can also inhibit fungal growth by eliminating localized cool or highly humid areas by mixing the air and equilibrating collections temperature and humidity with the environment (see chap. 3). While this chapter focuses on mechanical ventilation, ventilation can also be accomplished nonmechanically, as discussed in chapter 6.

Whole-house or whole-room ventilation

Mechanical ventilation involves a system for supplying and/or exhausting air from a building. A supply ventilation system usually includes an air intake, filters to remove particulates, a fan, ducts to distribute the air to various spaces, and registers to regulate the volume of air supplied to each space. The location of an intake supply ventilation system is very important and should be arranged away from sources of contaminants or pollutants, dust, or landscape debris. The supply air may be conditioned by heating, cooling, humidification, or dehumidification; the equipment for these processes is described in related sections of this chapter.

A system that supplies air only, primarily used for ventilation and offering minimal conditioning, may rely on displacement of existing space air by exfiltration (air leakage due to pressurization generated by the supply fan) through the building envelope, including open windows and doors. If substantial conditioning of the air is needed, the space air should be recirculated to the fan and conditioning equipment via return air ducts; the return air is mixed with a desired amount of outside air to supply ventilation or to offset losses from exfiltration, then mixed air is filtered and conditioned and returned to the space via the supply air system.

Exhaust ventilation systems provide air removal on a building-wide or local basis. The exhausted air must be replaced by outside air; the replacement outside air may be delivered as part of the supply air system (as described above) or by outside air that infiltrates the building envelope, either intentionally or unintentionally. In the absence of a system of conditioned supply air, operation of an exhaust air system will result in the space air temperature and moisture conditions approaching those of the exterior air.

Local exhaust ventilation

Local exhaust ventilation systems remove concentrated heat and contaminants near their source; removing undesirable materials where they are concentrated is more efficient than removing diluted materials as part of whole-building exhaust. Local exhaust ventilation systems are commonly used in laboratories, restrooms, and kitchens. In laboratories, local exhaust ventilation normally consists of fan suction that will draw a contaminant stream into a collection hood through a duct system and air cleaning device, exhausting the air directly to the outside. Local exhaust ventilation systems require high air velocities in the hood and duct to prevent diffusion of the contaminants into the space as well as to prevent the deposition of contaminants in the duct system.

Fans

Fans are the main components that move air in a ventilation system; they produce pressure differences that cause the air movement through ductwork that, along with associated control devices, directs and distributes the moving air to the desired delivery points. There are three types of fans: axial, centrifugal, and mixed type. Axial fans move the air along the duct axis and are designed to produce higher-volume flows, although they do not produce much pressure across the devices. Centrifugal fans move the air radially and work more efficiently to produce lower-volume flows, while they produce higher pressure than axial fans. Mixed-type fans produce fan performance between those of the axial and centrifugal fans. The sizing and selection of fans depends on system flow rate and system pressure, which are determined by the size, routing, and flow resistance of ducts, as well as the flow resistance of any filtration or conditioning equipment.

Small fans can be operated with domestic grade single-phase power. However, larger units require higher voltages and two- or three-phase power. Energy costs for operating fans are relatively small in comparison to those of cooling and heating devices, provided that the fans do not have to overcome high pressures in ductwork or large pressure differentials across filters or desiccant media. However, as heating, cooling, and dehumidification equipment become more efficient, the relative cost of fan operation becomes more important.

Fans in simple supply or exhaust systems operate at constant speed. Air distribution in constant volume systems is controlled by resistance devices such as louvers or dampers in the supply air ducts. Recently, variable frequency drive motors have become available, providing the opportunity to adjust airflow rates to meet ventilation or conditioning requirements and reduce the energy consumption of fans—but at increased capital cost and increased complexity of controls. The use of supply and exhaust ventilation systems is described in the case studies for Hollybourne Cottage, the Historic Archive of San Cristóbal de La Laguna, the Valle Guerra storage, the Goeldi Museum, and Casa de Rui Barbosa (chaps. 9–14).

Although not used for ventilation, fans that simply circulate room air can be beneficial in hot and humid climates and cold-humid marine climates for suppressing germination of fungal spores, mixing stratified room air, and eliminating microenvironments. Circulation fans can also provide improved thermal comfort in hot and humid climates. Their use is discussed in the case studies for the Historic Archive, Valle Guerra, and the Goeldi Museum (chaps. 11–13).

Heating

The elevation of air temperature through the addition of heat is a fundamental process in mechanical systems in climate zones where heating is necessary for thermal comfort. A wide range of equipment for production and transfer of thermal energy to conditioned spaces has been developed to meet this need in various climates and buildings.

Conservation heating

Conservation heating depresses humidity by raising the air temperature but does not change the dew point temperature, which indicates moisture content of the air. In the conservation heating strategy, air temperature is typically raised by an electrical resistance element or by the combustion of fuel, such as natural gas or fuel oil. Radiant and conduction heaters, which deliver heat directly to objects or humans, are not effective in raising air temperature. Wall- or floor-mounted convection heaters without fans will heat the surrounding air, but temperature distribution in the space will likely be uneven, since natural convection must provide the air movement; in these cases, low-speed portable or ceiling-mounted fans should be used to improve air mixing, mitigating uneven temperatures and the stratification of air. Forced air heaters and furnaces with integral electrical fans can provide even distribution of the heated air, as long as the supply air and return air ducts are configured to effectively mix the room air.

Museums and cultural institutions have used conservation heating to depress humidity in cool-humid and cold-humid climate zones, such as the United Kingdom, eastern Canada, and portions or seasons of mixed-humid and mixed-marine climate zones, such as the northeastern United States. As shown in table 7.1, in a cold-humid climate at 5°C, 80% RH, humidity can be reduced to 61% RH by increasing the temperature by 4°C to 9°C, without changing the dew point temperature.

In hot and humid climates, heating loads for thermal comfort should be small or nonexistent, but heating the air in a collections space can be part of a humidity management strategy. In hot and humid climates, the use of conservation heating has been limited owing to concern over raising space temperatures beyond the limits considered safe for collections or comfortable for occupants. As shown in table 7.1, in a 28°C, 80% RH environment, humidity can be reduced to 60% by increasing the temperature by 5°C to 33°C—again, without changing the dew point temperature.

Conservation heating can also be achieved without heating equipment, through the ventilation of warm and dry outside air into the building using supply and exhaust ventilation. This ventilation and heating strategy raises the temperature of the interior environment, collections, and interior building surfaces. In hot and humid climates, warm and dry outside conditions will typically occur during the midday period of specific seasons. All case studies in part 2, except that of Juanqinzhai (China), utilize conservation heating by ventilation.

Table 7.1

Humidity reductions by conservation heating in cold-humid and hot-humid climates.

	Cold and Humid Climates						Hot and Humid Climates					
Temperature (°C)	5	6	7	8	9	10	28	29	30	31	32	33
Humidity (% RH)	80	75	70	65	61	57	80	75	71	67	64	60
Dew point temperature (°C)	1.8						24.2					

Note: Values are estimated at sea level.

Heaters

Wherever possible, high-temperature electric heating elements, combustible fuels, furnaces, and boilers should be isolated from the collections space by fire-resistant construction. Combustion heaters must have a source of outside air so that combustion and exhaust gases must be properly vented to the exterior. With a high-temperature heat source located outside the conditioned space, a fluid—such as oil, glycol, water, or steam—is used to transfer the heat from the heating device into the space; the fluid heats the air in the space through a radiator or heats the heating coil in an air handler as part of a supply air system.

Low-temperature heat sources can be located directly in the conditioned space since the lower operating temperature is likely to present a lower risk of fire. Such devices include self-contained, oil-filled heaters of the electric resistance type that heat a fluid in a closed-loop system, transferring heat to the space air by natural or forced convection.

Electric heaters are mechanically simple, small, quiet, and rugged and can be operated with minimum maintenance, especially where heating loads are small and heating is only used for humidity management. Although the initial cost of a typical heater is low in comparison to dehumidifiers, the operating energy cost can be significantly higher. A high-voltage or two- or three-phase power line should be installed for safely operating higher-capacity (>1.5 kW) heaters. Forced air electrical heaters were used in the Hollybourne Cottage, Historic Archive, and Valle Guerra Storage case studies (see chaps. 9–12).

Cooling

Cooling lowers the air temperature by removal of heat. If the air temperature is lowered below its dew point, dehumidification will occur. In dry climates, where atmospheric moisture is low, cooling can be accomplished by the addition of moisture to the dry air. Evaporative coolers (so-called swamp coolers) pass hot, dry air through a water mist or through a water-saturated medium; the water evaporates to the dry air, absorbing heat from the air and lowering the air temperature, resulting in cooler air with a higher dew point temperature. However, in hot and humid climates where atmospheric moisture is high and the reduction of humidity is important, evaporative cooling is not effective, and other approaches are needed.

Cooling with incidental dehumidification

Conventional air-conditioning systems cool the air by passing it through a cooling coil (evaporator), which is maintained cold (4°C–13°C) by a constant flow of a cold liquid. The removal of heat lowers the air temperature and, if the temperature of the cooling medium and the coils is below the dew point temperature of the incoming air, dehumidification also results as moisture condenses on the cooling coil. Upon leaving the cooling coil, the temperature of the air remains near the dew point temperature, resulting in high humidity. In order to lower the humidity, a heating (reheat) coil downstream of the cooling coil raises the air temperature; the reheat can be achieved by an electric resistance heater or by a coil supplied with hot water or steam. Cooling with reheat is effective in controlling both temperature and humidity in hot and humid climates, but it can be an energy-intensive process. However, reheating may not be necessary in hot climates, where the heat gain of buildings and rooms is large. In this case, reheat can be achieved by mixing the conditioned cold and humid air from the cooling coil with existing room air; room temperature is thereby decreased, and humidity is reduced.

Cooling systems that utilize the refrigeration cycle are referred to as direct expansion (DX) systems because a refrigerant (vapor expansion/compression) is used as the cooling medium in the coil. DX systems are well suited for smaller applications and, when sized correctly, are capable of dehumidification because they operate at low coil temperatures without risk of freezing. DX cooling systems pose a low risk to collections and buildings because refrigerant escaping from leaks or breaks in the piping system quickly evaporates at room temperature. Where large-capacity cooling is required, chilled water may be the preferred cooling medium; chilled water systems (described below) operate at more constant fluid temperature than DX systems, but they require piping, valves, and pumps for distribution of the chilled water. It can cause major damage to collections and building interiors when overhead chilled water pipes break or leak.

Packaged terminal air-conditioning

Packaged terminal air-conditioning (PTAC) systems are typically used where there is little or no space available for air distribution ductwork. PTAC systems (fig. 7.9) incorporate DX cooling for energy efficiency. PTACs include a wide variety of equipment configurations:

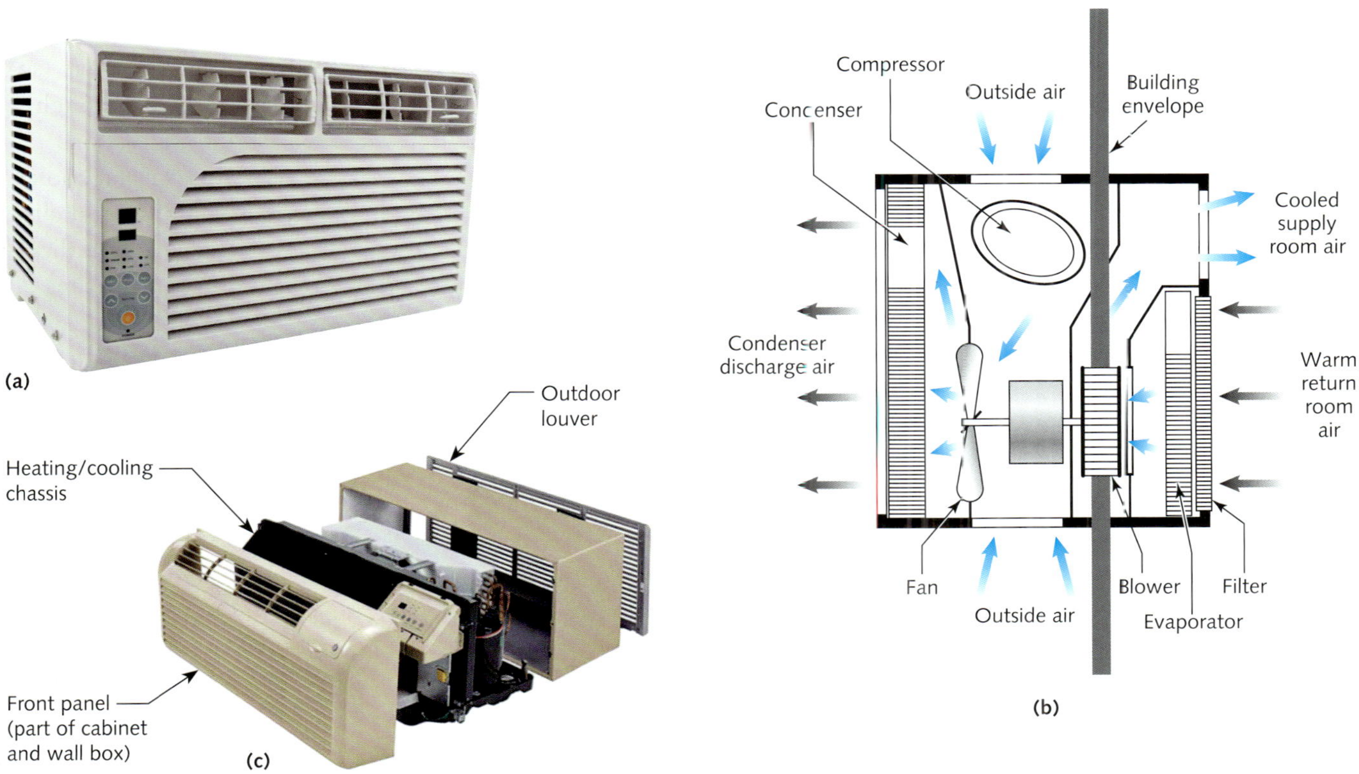

Figure 7.9
Images and schematic of packaged terminal air-conditioning (PTAC) units: a typical window-mounted unit (a and b) and a wall-mounted unit (c). Photo (a): © iStock.com/George Peters.

- window- and wall-mounted units, the most common packaged cooling device and prevalent in all climate regions (see fig. 7.9)
- split units and multisplit units (fig. 7.10)
- freestanding floor units
- cassette units
- ducted units
- variable refrigerant flow (VRF or VRV) units

For window-mounted or wall-mounted PTAC units, the equipment housing is divided into two compartments. The room-side compartment contains a fan that draws room air through a grille, passing the recirculated room air through a filter and an evaporator coil and discharging the cooled air through another grille into the room. A shallow pan under the evaporator collects the condensate and drains it to the outside drain. The exterior compartment is open to the ambient environment and shields the refrigerant compressor, condenser, and a second fan motor from rain and objects. This second fan in the exterior compartment draws outside air over the compressor and the condenser coil to remove the heat of compression as well as remove the heat absorbed by refrigerant at the evaporator coil.

Window- and wall-mounted units must penetrate the building envelope and may allow excessive air and water vapor infiltration between the equipment and the opening in the window or wall; therefore, they are mainly used for cooling and are not necessarily effective for dehumidification, as the infiltration of humid outside air can be faster than the dehumidification provided by the unit. Noise is a significant drawback of window and through-wall units because of poor acoustic isolation of the compressor from the interior, as well as the transmission of mechanical vibrations from the unit to the surrounding structure.

Split-system cooling units

Split-system cooling units physically separate the interior and exterior equipment functions but retain the installation convenience of a compact interior wall- or ceiling-mounted unit for the evaporator and fan assembly. This configuration resolves the problems inherent in a PTAC of significant envelope penetration—the through-wall connection is much narrower—as well as noise.

In the split-system arrangement (fig. 7.10), the compressor, condenser, and a fan and its motor are contained in a remotely located separate exterior unit, and the evaporator and a fan are located in a separate interior unit; refrigerant piping and wiring connect the two units. This scheme allows flexibility in locating both the interior and the exterior units, and the wall penetration only needs to be large enough for refrigerant lines, electrical cables, and condensate drains, which are easier to seal than the penetration for a wall- or window-mounted air conditioner. The separation of the two parts reduces noise intrusion from the compressor and condenser fan, even with larger-capacity units. Although wall-mounting of the indoor evaporator unit is most common, variations are avail-

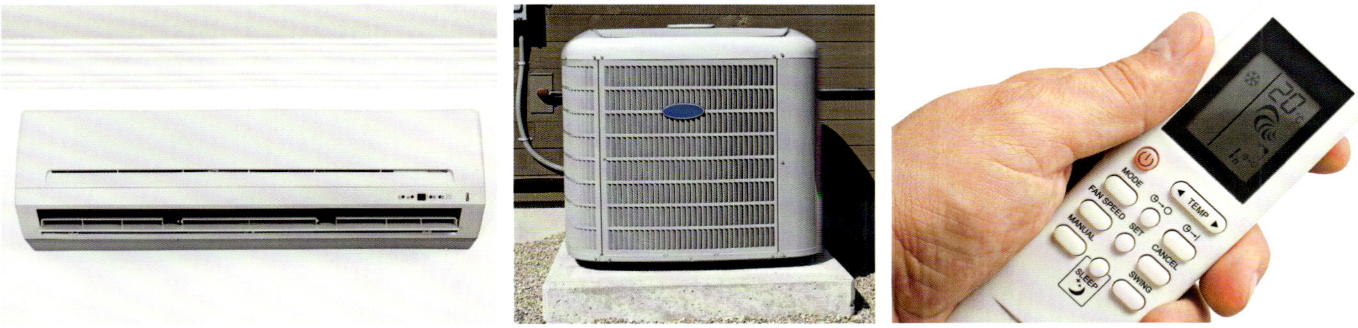

(a)

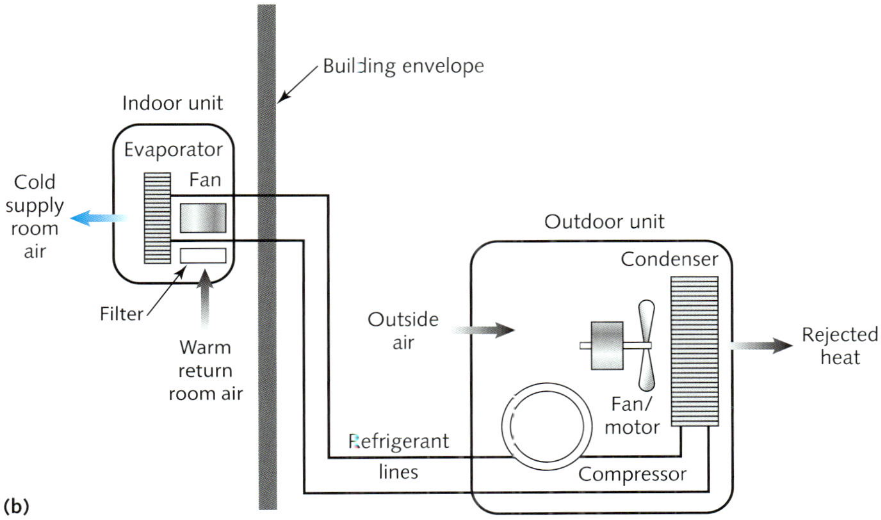

(b)

Figure 7.10
Images (a) and schematic (b) of split air conditioning. The photos (a) show the interior unit (left), the exterior unit (middle), and the wireless control (right). Photos (a): left, © iStock.com/Fuatkose; middle, © iStock.com/gmnicholas; right, © iStock.com/amete.

able, including freestanding floor units and ceiling-mounted units, also known as cassettes. Some manufacturers produce models in which one exterior compressor and condenser unit can serve several interior evaporators, providing the opportunity to have different temperature zones in a building without multiple exterior units. Typically split-system units provide incidental dehumidification but are not equipped with reheat for controlling the humidity of the supply air to the room. As a result, split systems may produce a large humidity variation in the indoor space. Placement of the interior evaporator unit must take into account the need for condensate drainage and drain line routing, especially if gravity drainage is used.

Central air-conditioning systems

Central systems for ventilation and air-conditioning generally incorporate cooling. Typically these systems consist of a centrally located air handler containing a fan for filtration, ventilation, and recirculation of air within a given building area; ducts distribute supply air and return air to and from the conditioned spaces (fig. 7.11). Cooling and dehumidification are accomplished by a cooling coil, and heating is performed by a heating coil (if positioned downstream of the cooling coil, the heating coil can perform the reheat function). The cooling medium may be a refrigerant (direct expansion) or chilled water, and the heating coil may use hot water, steam, or electricity.

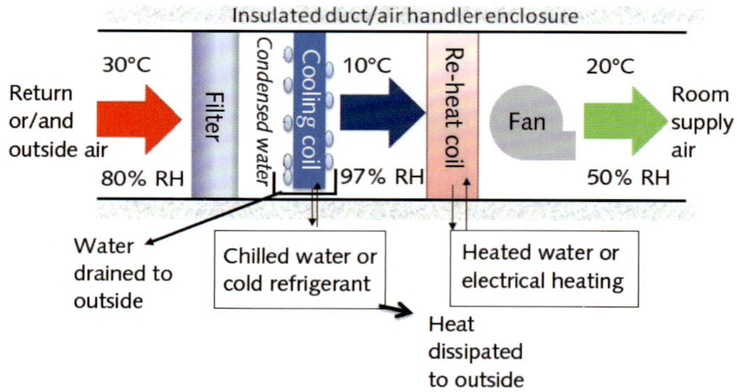

Figure 7.11

Schematic of a typical central air-conditioning system.

The refrigerant-based system employs an evaporator coil for cooling and a remotely mounted refrigerant compressor and condenser coil. This type of system was installed at the Museu Casa de Rui Barbosa (see case study, chap. 14). In smaller-capacity systems, the condenser coil is air cooled; it is water cooled for larger-capacity systems. The water-cooled units must employ some method of cooling the condenser water, using ambient air with wet or dry cooling towers or using the earth through groundwater cooling loops or wells. Advanced cooling systems modulate or stage the capacity of the compressor to match the variations in cooling load. Other advanced units come with variable refrigerant flow (VRF or VRV) technology, in which the flow rate of refrigerant is varied to match the cooling load. The variable flow system allows for improved control of the evaporator coil temperature while providing very high efficiencies.

Chilled-water cooling

In chilled-water cooling, cold water is provided by either a closed- or open-loop water supply system. The open-loop water supply can be either well water or another cold-water reservoir; however, in hot and humid climates, the supply temperature of these water sources will be too high for effective dehumidification. Closed-loop systems utilize various types of packaged water chillers to provide a chilled water supply and to remove heat from the return water. Water chillers usually utilize a water-based or refrigerant-based system for chilling the water supply, but they require air-cooled condensers or a water-based cooling tower to expel the absorbed heat into the atmosphere. Some chilled-water supply systems expel absorbed heat into the ground through a system of piping and/or wells. Chilled-water systems efficiently produce a chilled-water supply in large volume for large cooling loads, and they are normally associated with centralized HVAC systems with multiple zones. However, their cost and complexity are difficult to justify for smaller applications. As with any water-based system, there is a risk of leaks and consequent damage to the building and collections.

Humidification

As described in chapter 1, some hot and humid climates, especially in Hot-Humid (2A), Warm-Humid (3A), and Mixed-Humid (4A) climate zones, have three to six months of a relatively cool and dry, or cold and

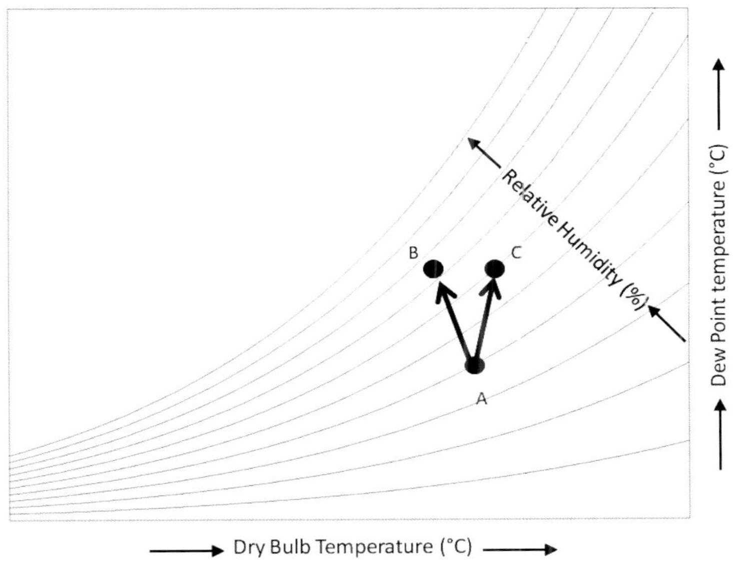

Figure 7.12
Psychrometric process of humidification. State point A is the room air to be humidified; state point B is the cooled and humidified air from an ultrasonic humidifier; state point C is the heated and humidified air from a steam humidifier.

dry, season. Humidification may be needed during these periods for conservation environments. There are three methods for humidifying the air:

- evaporative humidification using a wetted element or wick
- ultrasonic mist generation
- steam generating humidification (fig. 7.12)

In evaporative humidifiers, dry air flows through a porous medium (wick) that is constantly wetted by water and humidifies through the evaporation of the water. The wetting element can be configured as a single element, rows of elements, or a rotating disk, depending on the design.

Ultrasonic humidifiers produce mists by vibrating metal diaphragms in water tanks/baths at ultrasonic frequencies. The mist is mixed with the airstream, generated by a small internal fan, passing through the unit and into the conditioned space.

In both evaporative humidifiers and ultrasonic mist devices, impurities—such as minerals, salts, and bacteria in the water supply—can become concentrated in the reservoirs or wetted media, both of which are at room temperature. Therefore, a purified water source must be used to avoid mineral deposits and the prospect of biological growth in the humidifier—and subsequent contamination of the conditioned airstream and ducts.

Steam-generating humidifiers heat water in small tanks to produce hot pressurized steam, which is introduced through nozzles into the airstream to be conditioned. Although steam humidifiers do not allow impurities in water to become airborne, salts and minerals in the water supply become concentrated in the steam humidifier. If the hot steam is not thoroughly mixed with the supply airstream, the steam can condense on cool ducts, on supply air registers, or on cold walls near supply air diffusers, leading to corrosion, water damage, or in some instances biological growth. To avoid this, the ducts downstream of the steam humidifier must be carefully designed to maximize mixing and to collect and drain any steam condensate. Water quality is important with steam-generating humidifiers; these must be cleaned regularly to control mineral and salt deposits. Poorly maintained steam humidifiers can leak or fail because of corrosion.

Nonpsychrometric Mechanical Equipment

Mechanical systems must also execute nonpsychrometric processes in order to operate; these functions include filtration and control.

Filtration

HVAC filters remove particulate or gaseous chemical contaminants from the recirculated airstream or from the outside air used for ventilation. Particulates include dusts, viruses, tobacco smoke, pollens, fungal spores, bacteria, and viruses. Chemical filters are used to remove noxious or toxic gaseous pollutants such as nitrogen oxides and ozone from car and truck exhaust.

Particulate filtration

Particulate filters are made of woven or nonwoven fibers and are configured into flat panels, pleated sheets, or bags (fig. 7.13). Some filters are reusable/washable, but the majority of filters and filter media are disposable.

The minimum efficiency reporting value (MERV) is used to identify and rank filtering capacities of HVAC air filters according to a testing procedure in ANSI/ASHRAE Standard 52.2 (ANSI and ASHRAE 2007) (see sidebar). The MERV rating indicates a filter's ability to capture and hold particulates of specific size ranges (table 7.2). A larger MERV number means better filtration and removal of a wider range of particulate sizes. However, better filtration results in a higher pressure drop across the filter media, requiring higher system pressure to maintain airflow. The MERV rating is now an international industry standard administered by the ASHRAE Standard 52 Committee. Air filters must be

ANSI/ASHRAE Standard 52.2-2007

ANSI/ASHRAE Standard 52.2-2007, *Method of Testing General Ventilation Air-Cleaning Devices for Removal Efficiency by Particle Size* (ANSI and ASHRAE 2007), establishes a test procedure for evaluating the performance of air cleaning devices as a function of particle size.

Standard 52.2 was developed so that both users and suppliers can compare products, predict a uniform level of verified performance under operating conditions, and determine appropriate air cleaner efficiencies for specific situations.

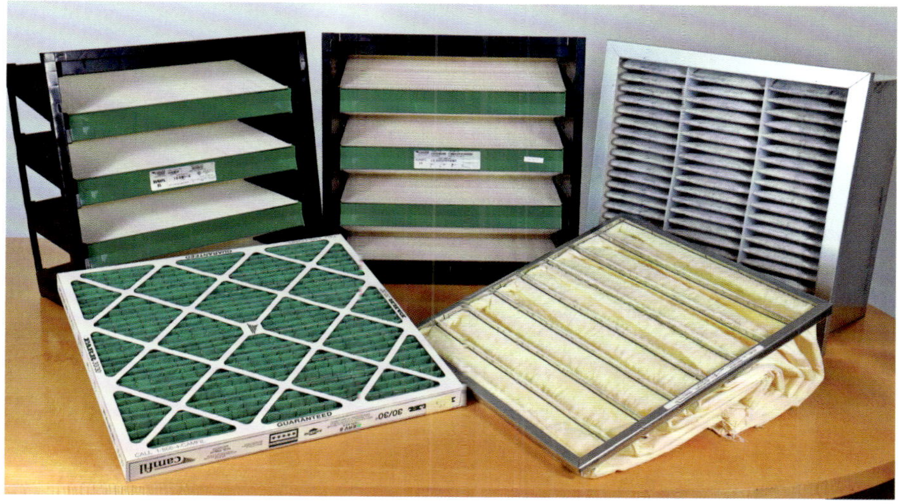

Figure 7.13

Images of various types of HVAC particulate and chemical (upper right) filters. Photo: Vincent L. Beltran. © J. Paul Getty Trust.

Table 7.2

MERV ratings, average particle sizes, typical contaminants, efficiency, and typical filter configurations. Modified from table E-1, ANSI/ASHRAE Standard 52.2-2007.

MERV Rating	European Efficiency Class	Average Particle Size (μm)[1]	Typical Contaminant	Efficiency[2]	Typical Air Filter Configuration
1–4	G1, G2	>10	Pollen, dust mites, sanding dust, spray paint dust, textile fibers, carpet fibers	<20%	Electrostatic Washable aluminum mesh Throwaway filters
5–8	G3, G4	3.0–10	Mold spores, cement dust	20%–60%	Pleated filters Cartridge filters Throwaway filters
9–12	F5, F6	1.0–3.0	Legionella, humidifier dust, lead dust, coal dust, auto emissions, welding fumes	40%–85%	Box filters Bag filters
13–16	F7, F8, H11	0.3–1.0	All bacteria, most smoke, cooking oil, copier toner, paint pigments	70%–98%	Box filters Bag filters
17–20 H10–H14	U13, U14, U15	≦0.3	Viruses, carbon dust, sea salt, combustion smoke	99.97% 99.99% 99.999% 99.9995%	HEPA/ULPA filters

[1] Note that a human hair is 20–40 μm in diameter.
[2] Dust spot method.

changed on a regular schedule to prevent clogging and reduction of system airflow. In extreme cases, clogged filters may fail structurally, allowing unfiltered air to bypass them completely.

HVAC filters in typical residential applications have MERV ratings of 1 to 4. These disposable panel-type filters are used primarily for preventing buildup of debris on mechanical devices and coils rather than for indoor air quality. They are inadequate for museums and other cultural institutions because they do not capture particles smaller than 10 μm. Filters with MERV ratings of 5 to 8 are commonly found in commercial applications and will collect particles as small as 3 μm. Filters with MERV ratings of 9 to 12 are used in commercial and industrial applications and will capture particles in the 1–3 μm range. Filters with MERV ratings of 13 to 16 eliminate bacteria, smoke, and copier pigments (0.3–1 μm).

High efficiency particulate air (HEPA) and ultra-low penetration air (ULPA) filters use dense filtration materials and remove more than 99.7% of airborne particles that are 0.3 μm or larger, which is significantly more effective than a pleated paper or fiberglass filter. Depending on efficiency, HEPA and ULPA filters are classified from H10 through H14. These filters are extremely efficient, but a significant air pressure is needed to maintain airflow through the filter. A typical HVAC system will not be designed to provide enough air pressure to use these filters.

An electrostatic precipitator (ESP), or electrostatic air cleaner, is another particulate filtration device. ESP cleaners impart an electrostatic charge across the airstream, precipitating fine dust particles onto collection surfaces that are periodically cleaned. ESP cleaners are highly efficient in removing particles with low airflow restriction, and they have minimal affect on fan size. However, the discharge of high-voltage electricity ionizes the air, creating ozone—a highly reactive gas known to oxidize pigments and deteriorate synthetic materials. Therefore, the ESP air cleaners are not recommended for cultural institutions.

Gaseous filtration

Chemical filters use absorbent media such as activated charcoal to capture gaseous pollutants (fig. 7.13). Some media use reactive chemical compounds to absorb and remove target gases from the airstream. Chemical filters are particularly useful for trapping low- and moderate-concentration gaseous air pollutants, such as sulfur dioxide, nitrogen oxides, and ozone. Activated charcoal filters are competent at trapping carbon (organic chemicals) and chlorine-based pollutants but do not absorb many other chemicals, such as sodium and nitrates. Potassium permanganate (Purafil) can be used for trapping other gas contaminants. Chemical filters must be replaced or regenerated when all of the bonding sites are filled by trapped pollutants. Chemical filters require high system pressure, resulting in larger fans and greater energy consumption. Because of operating costs and replacement filter costs, chemical filtration is justified only when the monitoring of gaseous pollutants indicates a clear risk to collections.

Activated charcoal filters and potassium permanganate filters are generally found in large urban institutions with sufficient resources to install and properly service chemical filters. In many instances, the need for chemical filtration can be decreased by placement of outside air intakes to avoid gaseous pollutant sources such as automobile and truck engine exhausts.

Control and control equipment

The separate pieces of mechanical equipment described above require controls in order to operate as a system. The control system is an integrated set of devices that communicate to enable, disable, and modulate the equipment performing the psychrometic processes, resulting in the desired conditions for collections conservation or thermal comfort. Some packaged mechanical equipment incorporates sensors and control devices; these are sold and installed as a complete turnkey system. However, systems that incorporate specialized equipment from several vendors will require custom-designed control systems.

Open- and closed-loop control schemes

Environmental management systems can be operated by a closed-loop or open-loop control scheme. Most environmental management systems operate with a form of closed-loop control. In the closed-loop (feedback) control scheme, a sensor in the controlled space measures the interior condition and sends a signal to a controller, which in turn modulates, enables, or disables devices (such as fans, valves for cold and hot water, and air dampers) to adjust the system in response to a deviation of the interior conditions from a desired set point.

In a closed-loop system, control may be executed through simple two-state actions (high/low or on/off), through modulating actions (0%–100% capacity), or combinations of the two types of actions. The simplest and least expensive is the two-state action for which devices can be set up to simply turn high and low, or on and off through transducers or relays. The deadband, or differential, between the two settings defines the on/off cycle of the equipment (fig. 7.14). For dehumidification,

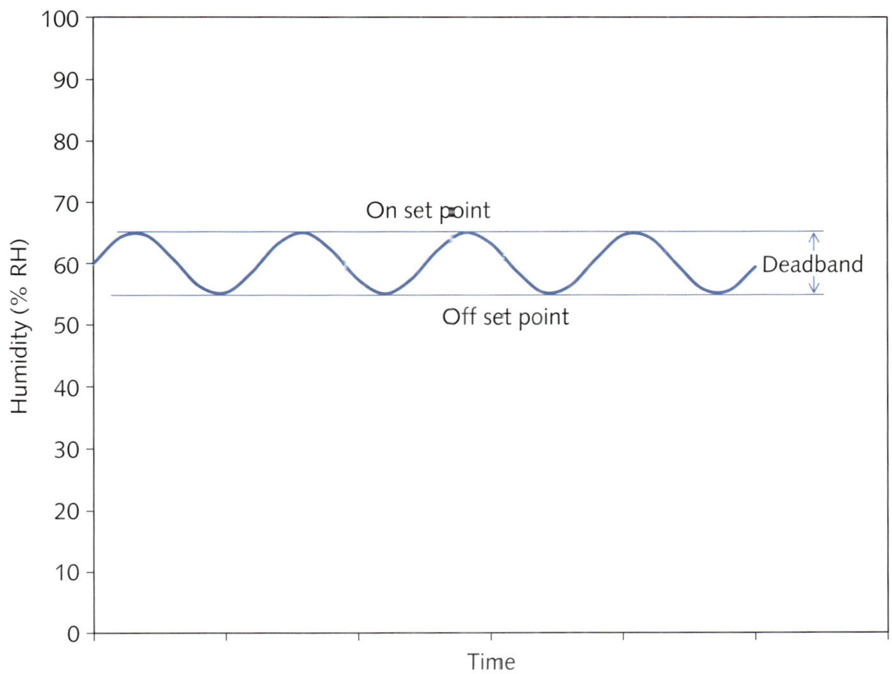

Figure 7.14
Schematic of set points and deadband.

the dehumidifier will be enabled at the set point humidity (e.g., 65% RH) and disabled at the set point RH minus its deadband (i.e., 55% RH, if the deadband is set to 10% RH). For modulating control, both the controller and the activation device must have capabilities to modulate signals and function, respectively, from 0% to 100% capacity. For example, if the room temperature is slightly higher than the set point temperature, a modulating control system would open the chilled water valve a small amount to increase supply to the cooling coil and progressively decrease the valve opening as the temperature difference decreased; for a larger difference in temperature, the initial response in opening the valve would be greater.

With open-loop control, devices are activated or deactivated by a human operator based on an external condition or timer, such as an exterior temperature or the time of day. There is no feedback loop in which the interior condition is measured, and a controller adjusts the device operation. The designer of an open-loop system must anticipate a fixed relationship between the interior condition and the exterior condition or the time of day. Open-loop systems include ventilators and air conditioners that are continuously operated or are enabled during particular hours (e.g., operational or occupied hours). The equipment operation is not modulated, and capacity is based on a fixed operational scenario, such as typical exterior conditions and/or occupant load. Open-loop systems provide a simple and inexpensive way to operate equipment, but their use will result in wide variations of interior environmental conditions as exterior and occupant cooling or dehumidification loads vary.

Control devices for simple closed-loop control systems. Simple closed-loop control systems (see sidebar "Control Devices") can be analog or digital and generally consist of devices like thermostats or humidistats. Analog thermostats were commonly used in the past and remain a viable option, especially where the technical capacity of service contractors is limited. Simple digital controllers, such as programmable thermostats, offer features that allow automatic adjustment of settings based on time, day of week, or other predictable factors.

Pneumatic Control Devices

In the past, the majority of controllers and control devices were pneumatic devices, which work with compressed air. The devices were introduced early in the twentieth century and widely applied in industrial environments. They are reliable and inexpensive and are therefore still in use in various systems, even with the current predominance of electronic controls. Pneumatic controllers and pneumatically driven control actuators and devices can generate and transmit very high forces to modulate large devices through minimal and inexpensive tubing between the controller and the devices. Since there is no appreciable airflow between the sensor transmitter and the controller at a steady state condition, the system is very accurate. Sensors are normally mechanical devices, and they convert the mechanical forces to pressure changes by the use of diaphragms.

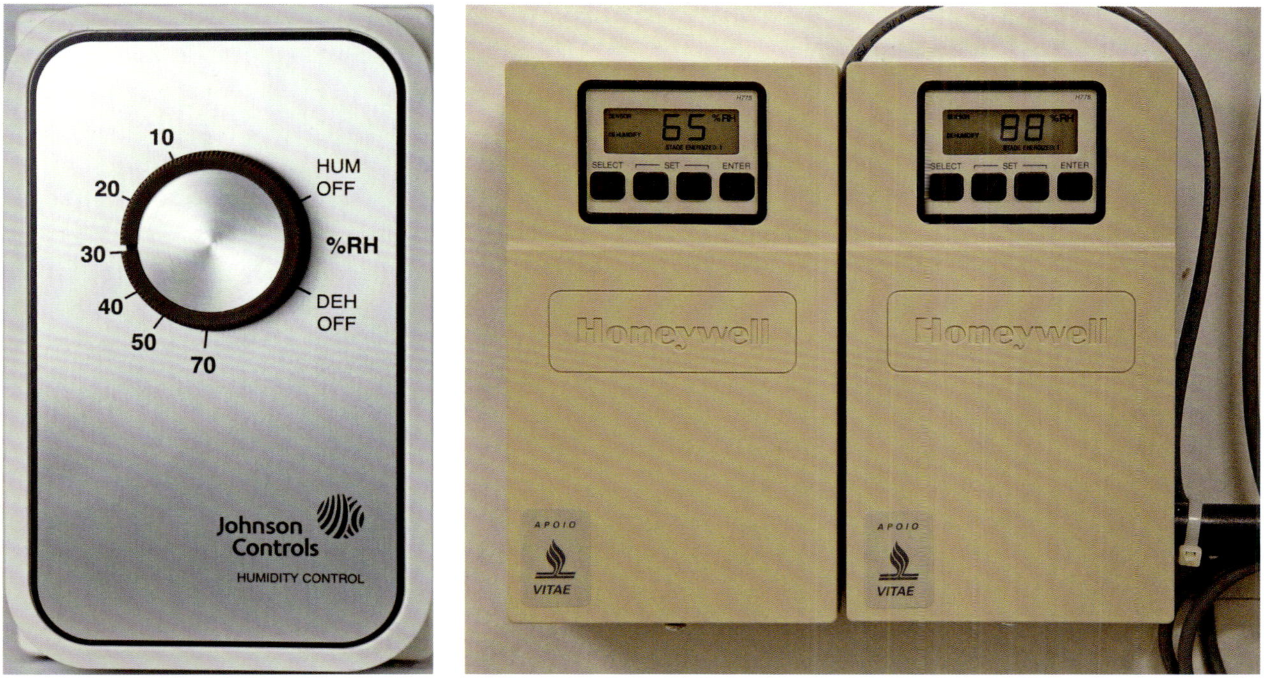

(a) (b)

Figure 7.15
Analog (a) and digital (b) humidistats. Photo (b): Shin Maekawa. © J. Paul Getty Trust.

A humidistat consists of a humidity-sensing element or remote humidity sensor, a logic circuit for processing the sensor signal and adjusting set point, and relay switches for controlling power to another device or a piece of equipment. Humidistats control humidification and/or dehumidification equipment by on/off regulation of equipment such as humidifiers, dehumidifiers, ventilators, or air conditioners (fig. 7.15). The user manually selects a humidity set point at which the appropriate equipment is enabled by turning "on" the electrical power; the equipment is disabled when a satisfactory condition is achieved and the humidistat turns "off" the electrical power. Some humidistats allow the selection of the "on" set point, with the "off" value being the result of a preset or adjustable differential (deadband). Analog devices operate in a similar fashion but incorporate mechanical linkages instead of electronic logic.

A simple closed-loop control system might utilize two humidistats: a room humidistat that controls the operation of equipment depending on the room humidity, and an outside humidistat that controls the selection of equipment for dehumidification, ventilators or heaters, depending on the outside humidity (fig. 7.16). Table 7.3 lists the operational control conditions for such a system.

A thermostat (fig. 7.17) controls temperature by enabling/disabling power for heaters, air conditioners, and ventilators. Similar to a humidistat, an electronic thermostat includes a temperature sensor, an electronic logic circuit, and switching relays. The thermostat operates with set points and deadbands in the same way as a humidistat.

Control system devices for more complex closed-loop systems. Humidistats and thermostats have simple on/off control logic and usually incorporate a humidity sensor, logic circuitry, and switching relays in a single

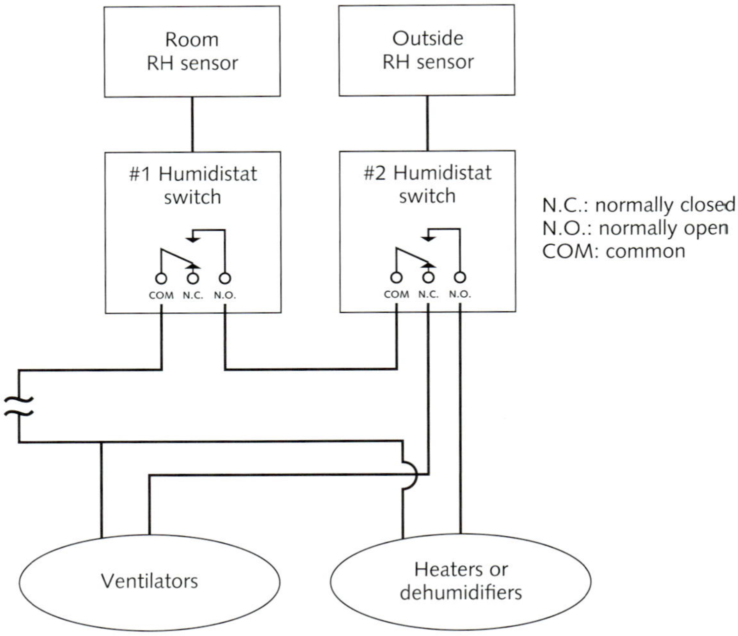

Figure 7.16

Wiring diagram of the two-humidistat system that operates ventilators, heaters, or dehumidifiers (listed in table 7.3) based on control conditions.

Table 7.3

Operational control conditions for a simple two-humidistat closed-loop control system.

Psychrometric strategy	Enable	Disable
Ventilation	$H_{interior} > 75\%$ RH and $H_{exterior} \leqq 70\%$ RH	$H_{interior} \leqq 70\%$ RH
Conservation heating	$H_{interior} > 75\%$ RH and $H_{exterior} > 75\%$ RH	$H_{interior} \leqq 70\%$ RH
Idle	$H_{interior} \leqq 70\%$ RH	$H_{interior} > 70\%$ RH

small enclosure. Complex closed-loop control systems incorporate multiple sensors, complex control logic that can be reprogrammed, and multiple devices and equipment that must be controlled by on/off switching or modulation; for complex closed-loop control systems, the sensors,

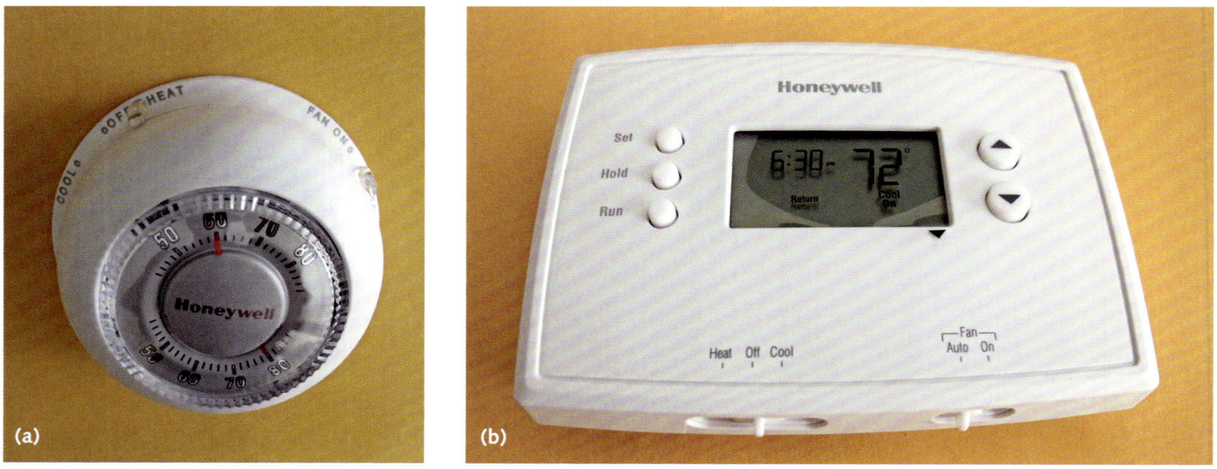

Figure 7.17

Analog (a) and digital (b) thermostats. The analog unit's interior mechanisms include a bimetal coil (mechanical) temperature sensor and a mercury switch (red arrow). The digital unit has an electronic sensor and relay switch and is programmable for both the temperature range and the control schedule (internal clock). Photos: Shin Maekawa. © J. Paul Getty Trust.

control logic, and switching/modulating activation functions are performed by separate devices.

Typical sensors

Sensors produce or modify an electrical signal that varies in proportion to the measurement of a particular environmental parameter; the signal is sent to a particular controller or control system. Sensor performance specifications include power requirements, operating range, output signals (analog or digital), accuracy and repeatability, response time, drift over time, calibration frequency, and the ambient environmental conditions within which the sensor can operate.

Typical sensors used in environmental management systems measure temperature, humidity, pressure, flow rate, indoor air quality, occupancy, or visible light levels. Indoor air quality sensors, such as CO_2 sensors, are used as inputs for control of ventilation in order to provide sufficient fresh air for occupants. Occupancy sensors may rely on thermal sensing to detect the presence of people in a space, or they may employ motion sensing to detect occupant movement. These sensors are commonly used in energy conservation applications, such as the control of artificial lighting.

Sensors are not limited to the conditioned space. Environmental management systems utilize a wide array of sensors to control the various processes, components, and equipment that constitute an HVAC system in order to maintain the desired conditions in the space. Overall, system operation is determined by the temperature and humidity conditions of a particular space, so that the sensors measuring temperature and humidity (and from which dew point temperature is typically calculated) are most commonly encountered by the building owner or occupant. Humidity and temperature sensors are discussed in detail below.

Space sensors, such as humidity and temperature sensors, for controlling the environmental management system, are remotely positioned either on interior wall surfaces in the middle of the conditioned space or in the main return duct—locations where they can sense the average conditions of the space. Other considerations for sensor placement are the security of the sensors and their connections to the system equipment, as well as ease of maintenance and calibration.

Humidity sensors. Relative humidity sensors measure humidity in % RH, and the signal is typically based on the variation of electrical capacitance or impedance of a hygroscopic polymer in response to changing humidity in the air surrounding the sensor. Impedance sensors are less accurate and respond more slowly than capacitive types. The main disadvantages of both sensors include sensitivity to chemical pollutants, salts, or sustained or repeated condensing conditions; these sensors may have reduced accuracy over wide temperature ranges, or they may drift with time, necessitating routine calibration. Sensor performance degradation is typically related to the contamination of the sensor material.

There are many low-cost RH sensors available on the market, but environmental management systems for collections conservation require RH sensors with high accuracy, reliability, and stability. Based on the authors' experience, the sensor measurement range should be 0% to 100% RH with an accuracy equal to or better than ±3% RH between 10% and 90% RH when measured at 20°C. The repeatability (or precision) of measurement should be equal to or less than ±1% RH. Response time should be less than 60 seconds to register 90% of the humidity change at 20°C or with a fixed air speed. The long-term drift should be less than 1% RH/yr (see sidebars "CMOS Sensors" and "Chilled-Mirror Hygrometers").

RH sensors should be protected from direct sunlight, physical damage, and rain but must be directly exposed to the air because the sensor material reacts with the air by adsorbing/desorbing small quantities of water vapor. However, this exposure subjects the sensor to potential contamination from airborne particulates or gaseous pollutants. Such contamination can affect the long-term stability and sensitivity of RH sensors,

CMOS Sensors

Recently, digital RH and temperature sensors that are based on complementary metal-oxide-semiconductor (CMOS) technology were introduced. The digital sensors, calibrated during the manufacturing process, cannot be recalibrated. They need to be replaced annually, so it is more economical to use the CMOS sensors if many RH sensors are installed.

Chilled-Mirror Hygrometers

Chilled-mirror, or condensation, hygrometers are the most precise instruments for humidity measurement. All national standard institutions throughout the world use them as the primary national standard. In these devices, the dew or frost point temperature of the air is determined by optical sensing of the presence of a dew or ice layer on the surface of a temperature-controlled mirror. In combination with precise temperature measurements, very accurate humidity measurement can be achieved. However, these hygrometers have two significant disadvantages: (1) each measurement system is many times the cost of a high-quality RH sensor, and (2) any contamination of the mirror surface will cause measurement errors.

Relative Humidity Sensor Calibration

RH sensors can be calibrated with certain saturated salt solutions. Lithium chloride produces 11.5% RH, magnesium chloride produces 33.3% RH, and sodium chloride produces 75.5% RH. The sensor humidity values can be adjusted while the devices are exposed to the humidity environments produced by these salt solutions.

and therefore all RH sensors should be calibrated at least once a year (see sidebar "Relative Humidity Sensor Calibration").

Temperature sensors. Electrical temperature sensors detect a change in an electrical property in response to a change in temperature. Thermistors, resistance temperature devices, and thermocouples are the most common electrical temperature sensors.

Thermistors are semiconductors that exhibit large changes in electrical resistance with small changes in temperature. However, the resistance change of a thermistor is not proportional to the temperature change; therefore the sensor output must be electronically processed and converted to a linear and proportional indication of temperature. Thermistors are widely used because of their low cost and high resolution.

Resistance temperature devices change resistance in proportion to temperature change. These sensors are durable, accurate, precise, stable, and interchangeable, and they are frequently used in laboratory and industrial processes. They are more expensive than other temperature sensors.

Thermocouples are also used in industries and laboratories because of their low cost and their accuracy and durability. These sensors consist of two dissimilar metals joined together at one end, called the junction. When the junction of the two metals is heated or cooled with respect to the other ends, a voltage is produced. Thermocouples are available in different combinations of metals, with the four most common ones known as J, K, T, and E types. Each thermocouple type has a different temperature range. Disadvantages of thermocouples are their low-voltage output signal and the fact that the temperature of the other end or junction has to be known for reference.

Accurate and reliable temperature measurement is essential for the operation of an environmental management system. Based on the authors' experience, the accuracy of a temperature sensor, normally measured at 20°C, should be equal to or better than ±0.3°C within the range of expected temperatures. The repeatability should be less than 0.1°C. The temperature dependence of repeated measurements should be less than ±0.02°C per degree Celsius. Long-term drift should be equal to or less than 0.1°C over a year, and response time should be less than 15 seconds at 20°C with a static condition or with a fixed air speed.

Controllers

Controllers receive and process one or more sensor signals using input/output conditioners. Logic circuits or decision programs within the controller respond to the sensor signals by modulating equipment or devices or by switching equipment or devices on or off. The control logic may be programmed into the controller, or it may be adjustable by use of software and a personal computer.

Microprocessor-based controllers perform multiple and complex (both Boolean and arithmetic) processes on sensor input and output operational signals (fig. 7.18). Input and output (I/O) modules receive signals from sensors and transmit them to the controller for processing. The controller transmits output processing signals to actuating devices, such as switches, fans, compressors, vanes, and valves, through the

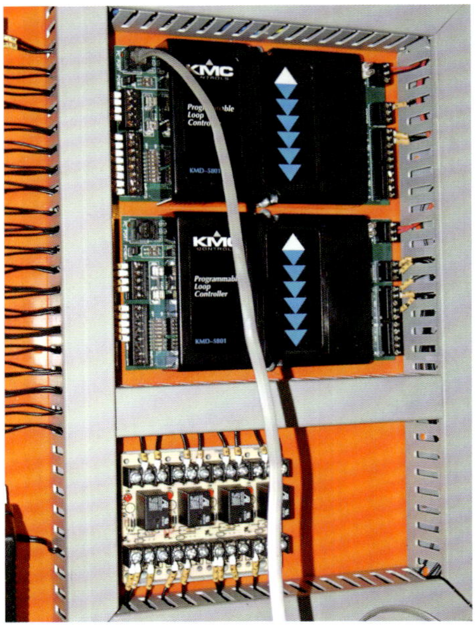

(a)

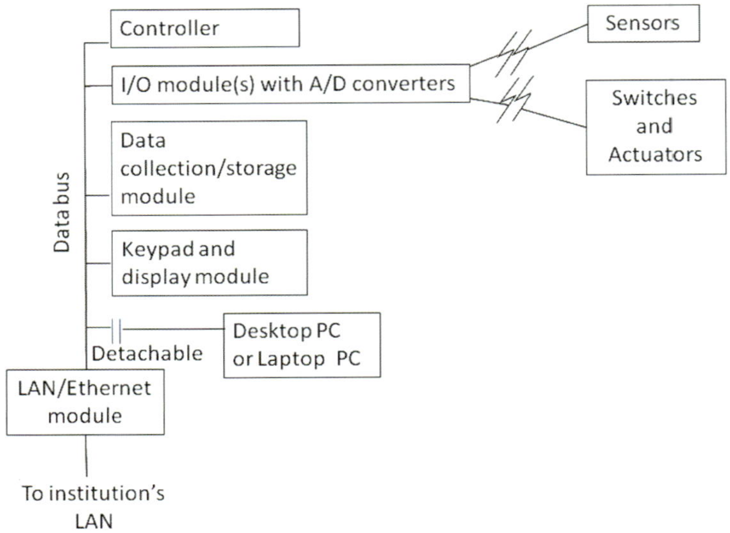

(b)

Figure 7.18
Programmable logic controller control panel (a) of a centralized climate control system, and schematic of a modular-based control system (b). Photo: Shin Maekawa. © J. Paul Getty Trust.

I/O modules. Data storage modules record environmental data and operational data that can be downloaded for analysis and diagnostics of system performance. A keypad and display unit or a computer is normally installed with the system for inputting or modifying control values in the program, as well as for displaying operating conditions of the system for diagnostics.

Control programs are typically prepared, compiled, and downloaded to the controllers with a personal computer. Among the numerous advantages of digital controls are their flexibility, interchangeability, low cost, and ability to easily integrate into a larger existing building automation system (BAS) or building management system. Modern controllers, such as programmable logic controllers (PLCs), can be equipped with a LAN/ethernet communication module for connecting to a local area network and/or a telephone modem for remote access and alarm notifications via email, phone call, SMS/text message, and/or audible alarm.

Activation devices

Valves, dampers, pump, compressors, and fans are typical activation devices in the control of environmental management systems. Valves and dampers are controlled through actuators that receive electrical or pneumatic signals from controllers via a variety of transducers and relays. Valves control the flow of liquids such as refrigerants as well as cold and hot water, while dampers and fans control airflow. Pumps and compressors move fluids and gases through the system.

Monitoring

Although the sensors in control systems can also be used for monitoring the collections environment, it is preferable to separate the monitoring function from the control function so that curatorial and conservation staff can readily access information on environmental conditions and independently confirm that the desired environmental conditions are being maintained (fig. 7.19). Appendix 1 discusses the analysis of data from monitoring.

Although using separate sensors to monitor and control systems introduces the possibility of differing readings, coordination of the two efforts can minimize these differences through sensor selection and sensor placement. Sensor placements will typically differ for monitoring and control functions, since it is often important to monitor conditions nearest a specific collections object, and such placement is impractical for conventional wall-mounted or duct-mounted HVAC systems devices used for control.

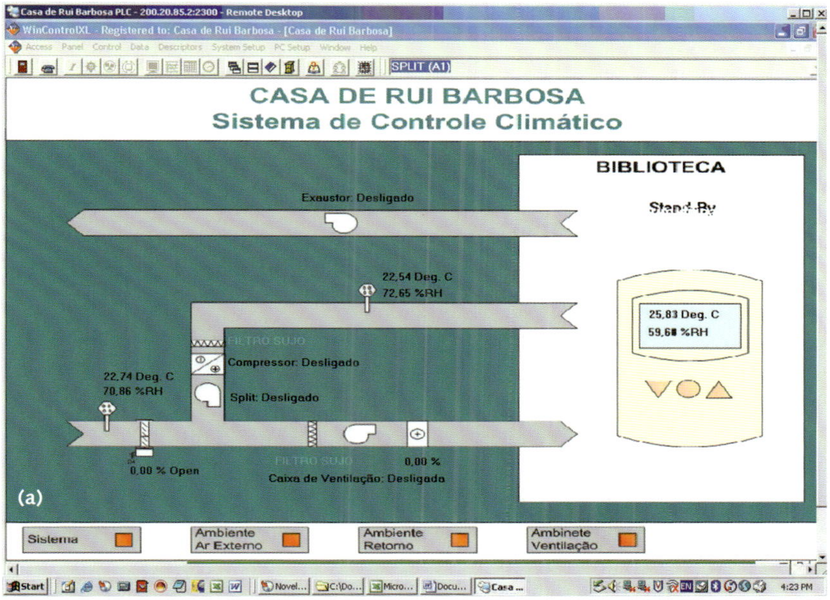

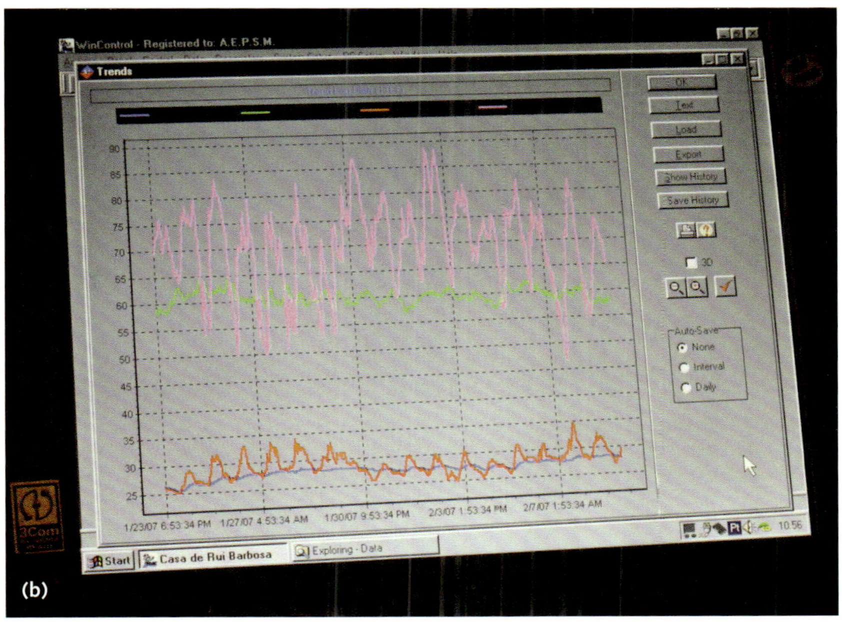

Figure 7.19

Online real-time monitoring screen (a) and a trend plot for performance review (b) displayed on a PC connected to the climate control system at the Museu Casa de Rui Barbosa (case study, chap 14). Sensors for controlling both the environmental management system and monitoring are displayed on the monitoring screen. Photo (b): Shin Maekawa. © J. Paul Getty Trust.

Additional Topics Related to Mechanical Strategies

The implementation of mechanical strategies for conservation environments for collections must address three additional considerations:

- finding and allocating space to accommodate the mechanical equipment and related distribution piping and ductwork
- determining system distribution—centralized and noncentralized systems
- identifying and reducing the risks of unintended consequences that can result from operation or failure of the mechanical systems

Spatial requirements for mechanical systems

The introduction of a mechanical system for environmental management will require space for equipment and distribution piping and ducts; an existing building may have no provisions for HVAC equipment and other requirements, such as:

- evaporators, condensers, compressors, air filters, fans, and supply and return ducts;
- insulated piping for hot and chilled water and insulated tubing for refrigerants and condensate piping (much of this piping must be installed at a proper slope to promote gravitational drainage);
- conduits and wiring for electrical power and control;
- associated electrical devices;
- clearance space for service access and major maintenance.

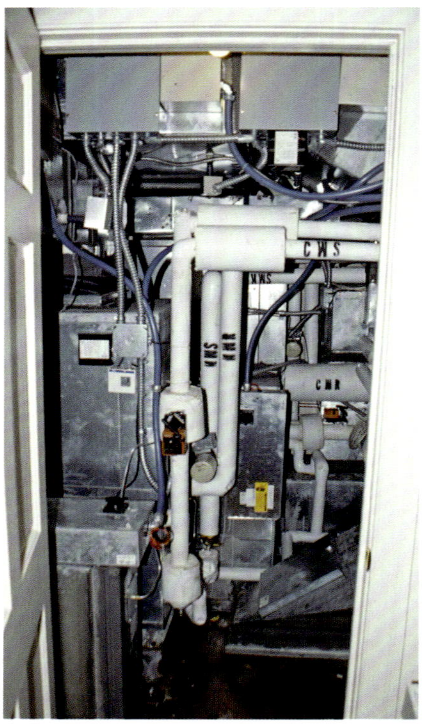

Figure 7.20
Mechanical components crowded into a small closet, an example of inadequate space. Photo: Michael C. Henry.

This mechanical equipment must be moved into and through the building for installation, as well as for later replacement at the end of its functional service life. For maximum service life and best performance, the equipment and devices must be easily accessed for maintenance, repairs, and periodic adjustments. Ducts, pipes, and cables need to penetrate walls, floors, and ceilings, and some will have to be insulated or protected in conduits. Many electrical components have clearance requirements for the safety of operation as well as for access by servicing personnel. The need for sufficient space around the equipment for maintenance cannot be overemphasized; if adequate space is not provided, maintenance will be difficult to perform, and the equipment will eventually fail to operate properly, often prematurely (fig. 7.20).

The spatial requirements for mechanical systems for environmental management are often underestimated or inadequately anticipated during the early stages of design; interpretive spaces, visitors' amenities, and staff offices have strong and vocal constituencies compared to equipment and facilities operation. As a result, architects and engineers are forced to resort to a variety of equipment placement strategies, including:

- locating mechanical equipment outside the building at ground, on the building roof, on exterior walls, in a nearby building, in a new vault or room below ground, or on balconies (fig. 7.21);

- creating new interior cavities to accommodate mechanical systems by erecting false ceilings, false floors, or new walls (fig. 7.22);
- installing the mechanical equipment within the room being served, usually in the form of freestanding or self-contained units placed near windows (fig. 7.23);
- repurposing existing spaces/rooms for mechanical systems (fig. 7.24).

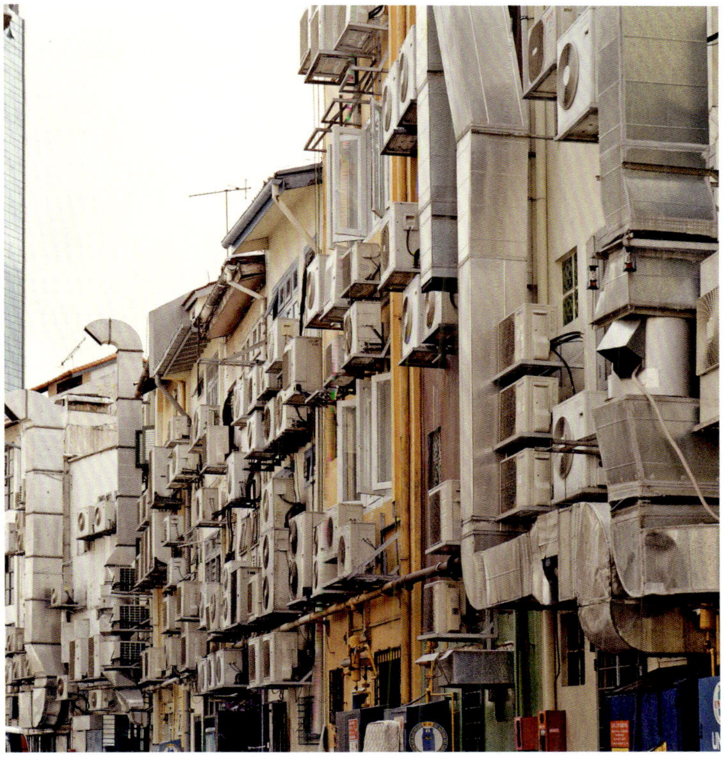

Figure 7.21
Air conditioners are mounted on the facade of a building in Singapore. Photo: Shaun Higson colour/Alamy.

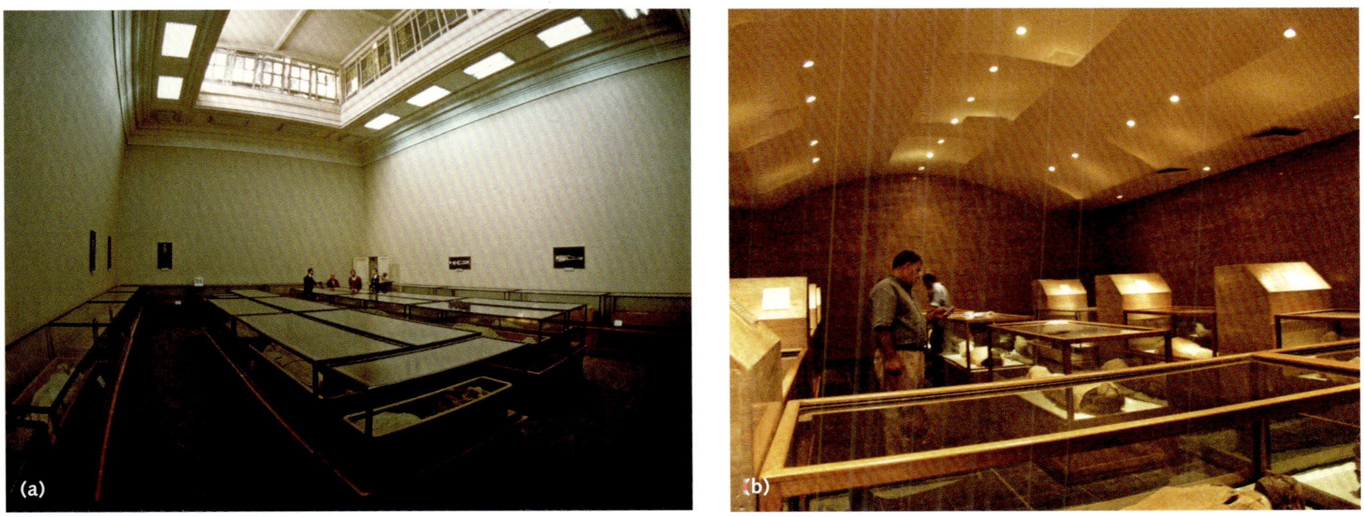

Figure 7.22
The Royal Mummy Room in the Egyptian Museum, Cairo, before (a) and after (b) installation of an HVAC system. Note the reduced height of the new ceiling in the postinstallation view. Photos: Shin Maekawa. © J. Paul Getty Trust.

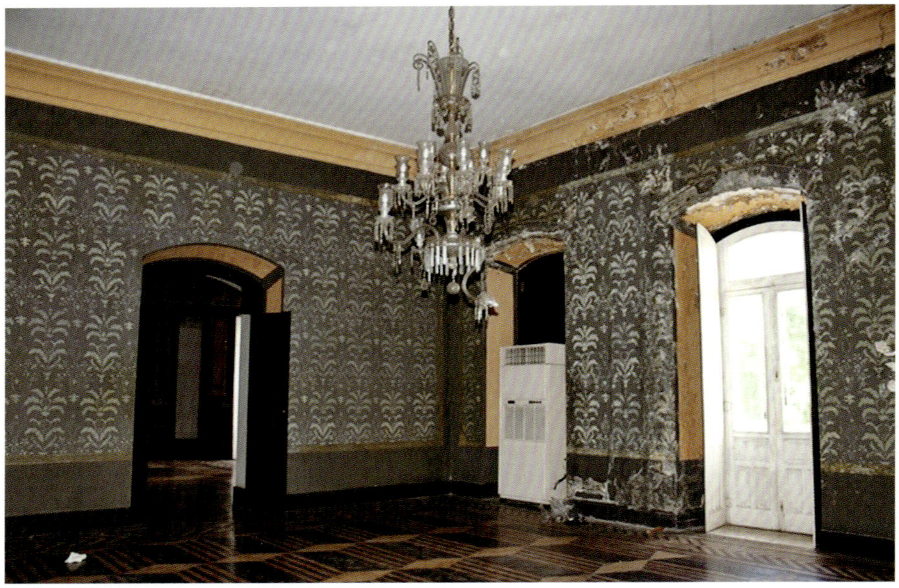

Figure 7.23
Self-contained air-conditioning unit located within the walled-off window space of a nineteenth-century palace building that now houses the Museu do Estado do Pará, Brazil. Photo: Franciza Toledo.

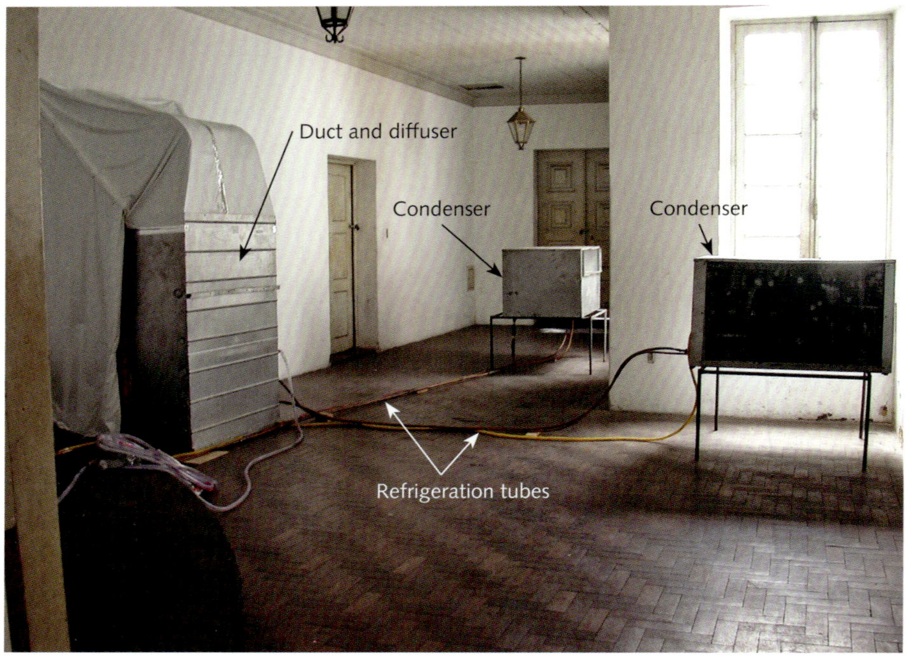

Figure 7.24
A split air-conditioning unit installed in a side room connected to the main chapel of the Madre de Deus church in Recife, Brazil. Photo: Franciza Toledo.

Noncentralized and centralized mechanical systems

Mechanical equipment may be configured in either noncentralized or centralized systems, depending on the size and complexity of the building to be conditioned. For noncentralized systems, individual environmental management equipment, such as heaters, fans, dehumidifiers, and packaged air-conditioning units, can be placed in various locations within a building to condition the local environment. A noncentralized system typically features small-capacity equipment with low initial cost that is simple to install. Space requirements can be minimized, easing installation of equipment in existing or historic buildings. Noncentralized systems, with small-capacity equipment located throughout a building, can easily address a variety of interior environmental objectives on a room-by-room basis.

Centralized mechanical systems use fewer pieces of equipment but require more extensive networks of piping and ducts. The major mechanical equipment and components are centralized in one location, either inside or outside the building, and the conditioned air is distributed throughout the building through ductwork. Most recently constructed commercial and residential buildings utilize centralized systems because the space required for centralized equipment and for the distribution of conditioned air can be designed into the building.

Advantages and disadvantages of noncentralized and centralized environmental management approaches are summarized in table 7.4.

Table 7.4

Comparison between centralized and noncentralized environmental management systems.

Quality	Centralized System	Noncentralized System
Initial capital cost	High	Low
Maintenance costs	Lower due to fewer pieces of equipment with longer service life and easier service access	Higher due to multiple stand-alone systems, often packaged as units; tends to have shorter service life and difficult-to-access components
Equipment availability	Project-specific combinations of equipment, sometimes from several different manufacturers	Off-the-shelf package units typically available from a single manufacturer
Equipment delivery time	Longer delivery time due to design, assembly and transport of system equipment	Short delivery times for standardized off-the-shelf units
Equipment location	One central location of major equipment	Multiple locations in/near rooms to be conditioned
Service access	Accessed in central location	Equipment is often located in scattered, unrelated and often small spaces
Space needs	Requires suitable dedicated space for equipment plus a network of connected spaces for distribution ductwork and piping	Equipment usually positioned in/near rooms to be conditioned, leading to lots of visual clutter
Air distribution	Requires network of distribution ductwork and piping between equipment and spaces served	Equipment typically close to room served, minimizing ductwork and piping
Cooling/dehumidification capacity	High	Low
Conditioning flexibility	Typically conditions air to lowest necessary temperature and humidity, then reheats and/or humidifies air as needed to maintain different space conditions	Conditions air to meet specific needs of the space served, such as higher or lower temperature/humidity than the surrounding spaces
Environmental stability and space fluctuations	Large systems can provide more consistent supply air conditions, minimizing temperature/humidity fluctuations in the space	Cycling of small local equipment units results in more frequent variations in supply air conditions
Malfunction result	Will affect all spaces served by centralized system	Limited to spaces served by failed equipment
Unoccupied spaces	Modular heating and cooling equipment is needed to accommodate wide working range of loads due unoccupied spaces	Systems serving unoccupied spaces may be individually turned off without affecting other units/spaces
Control logic	May be complex	Simple
Filtration	Numerous options for particulate and gaseous filtration. Fan sized to match filtration pressure drop	Equipment is normally designed with low-capacity fans that do not allow higher capacity filters; equipment configuration and air inlets may not conform to standard filter sizes
Noise and vibration	Equipment spaces are usually separate from occupied spaces so structural and duct transmission of vibration noise is low	Equipment noise is in close proximity to visitor/work space; may be difficult to isolate vibrations or absorb unwanted noise

Risks associated with the operation of mechanical strategies

The operation of mechanical strategies in previously unconditioned buildings in hot-humid climates and cold-humid marine climates may introduce new risks and unintended consequences, typically resulting from problems with humidity control, including:

- high infiltration of humid exterior air,
- intermittent system use,
- insufficient dehumidification capacity.

High infiltration of humid exterior air

An example of high infiltration of humid exterior air was seen by one of the authors in the Document Archive of the Institute of Brazilian Studies at the University of São Paulo, Brazil. The air-conditioning system was thermostatically operated to maintain the storage temperature at 19°C–20°C, resulting in daily fluctuations in humidity in the archive ranging from a minimum near 55% RH to a maximum near 80% RH (fig. 7.25). From the trend plot, one can clearly see that the fluctuations in humidity were driven largely by fluctuations in dew point temperature (10°C–17°C) rather than by the small variation in air temperature. Cooling (DX compressor) typically operated for 18–20 minutes, followed by 18–20 minutes of an off-cycle period; the cycling on and off of cooling is evident in the fluctuation in dew point temperature as the evaporator coil alternately cools and warms the air. Although the exterior dew point temperature was not measured at the site, it was estimated to be between 15°C and 18°C owing to its proximity to a major river and heavy vegetation. This dew point temperature is significantly higher than the typical value for São Paulo in August, 13°C (www.weatherbase.com). Small fluctuations of temperature in the room indicated a comparatively small heat gain and an overcapacity of the installed cooling unit for the space.

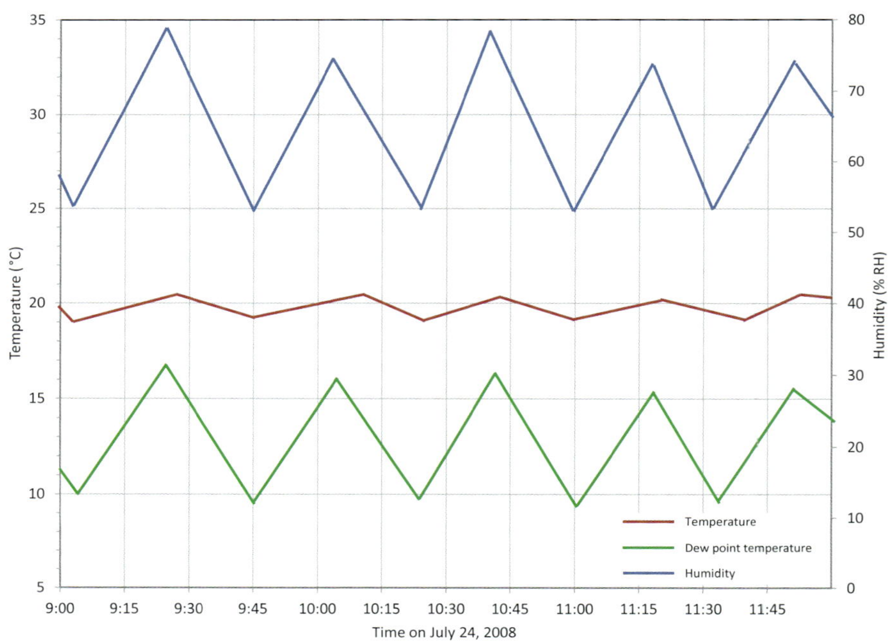

Figure 7.25

Temperature and humidity in the Document Archive of the Institute of Brazilian Studies of the Federal University of São Paulo on July 24, 2008, during the low-rainfall season. Fluctuations of 55% to 75% RH are seen at a 30–40 minute cycle as the temperature was thermostatically maintained at 19°C–20°C. Data courtesy of Franciza Toledo.

However, the rapid rebound of room humidity as the unit cycled off indicated a large dehumidification load. Since interior moisture sources—such as wet surfaces, occupants, ventilation, domestic loads, and vapor permeation—were not found in the room, it was concluded that the cause was the infiltration of a large volume of the humid exterior air; this finding was supported by the discovery of significant air leaks around windows and doors.

Intermittent system use

The introduction of air-conditioning is driven by the need to provide a comfortable environment, primarily with respect to temperature, during periods of occupancy. This objective is distinct from the goals for the conservation of collections or building fabric, which prioritizes humidity management. As a result, many institutions mistakenly operate air-conditioning systems only during visiting or working hours when people are present, and turn the systems off during nights and days when the museum and offices are closed. The intermittent use of air-conditioning may also be driven by the need to reduce energy consumption and operational costs. This operational pattern, however, can lead to the "rebound," or sharp increase in humidity, when the air-conditioning system is turned off. If the air-conditioning equipment is the sole means of dehumidification, high humidity may result from high moisture contents of collections and interior building fabric and/or the infiltration of nighttime humid air when cooling loads are low and there is no demand for air-conditioning or the system is off. Though short in duration, these periods of spiking humidity may cause mechanical stresses in objects or lead to condensation on cool surfaces, such as glass and metals.

An example of the risks associated with intermittent use of an air-conditioning system was seen by one of the authors at the Eva Klabin Foundation, a house museum dating from the 1950s, in Rio de Janeiro, Brazil. The museum operated a centralized air-conditioning unit in the house only when visitors were present. In figure 7.26, visiting periods are

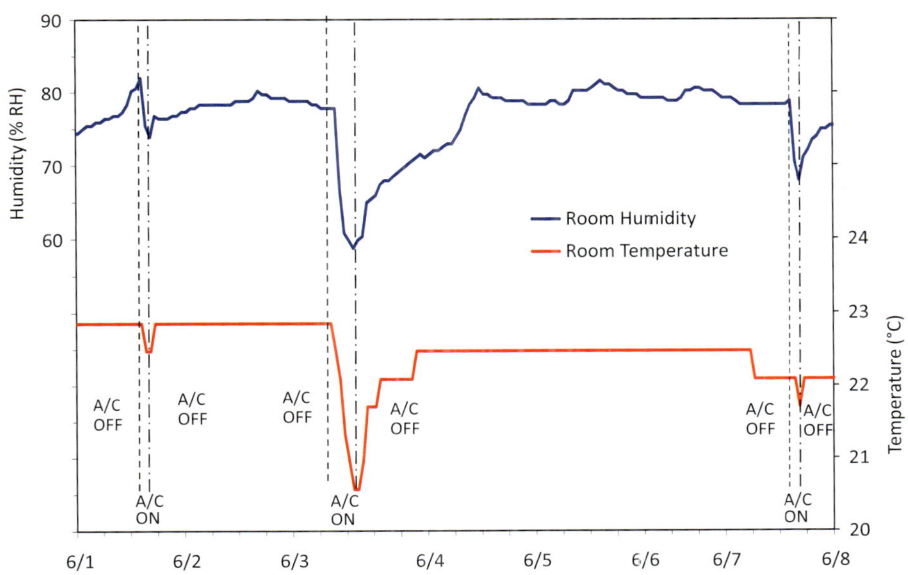

Figure 7.26

Temperature and humidity values in the main exhibition room of the Eva Klabin Foundation, Rio de Janeiro, Brazil, during the rainy season (June 2005). Data courtesy of Franciza Toledo.

indicated by sharp drops of humidity, corresponding to operation of the cooling system. During visiting hours, the air-conditioning system cooled and dehumidified the air, and humidity and temperature were reduced to 50%–60% RH and 24°C –25°C. After visiting hours, the system was turned off, and interior humidity levels rebounded to and remained at 70%–80% RH depending on outside dew point temperature. The museum suffered from fungal odors and fungal activity in its textile and paper collections, and fungi were suspected in air-conditioning ducts.

Since the early 2000s, the US conservation community has been examining the use of off-hours "idling" of environmental management systems as a means to save energy (Lev-Alexander 2011). For buildings with large thermal mass and low infiltration rates, such as national archive buildings and underground storages, this practice is effective for energy savings and safe for collections conservation. However, the situation is different in hot and humid climates, where humidity approaches saturation (100% RH) during the evening and early morning hours. To operate the system effectively in off-hours idling mode in these climates, the infiltration (air change) rate of the building envelope must be extremely low and well controlled (<0.3 air changes/hr), and the building must be comparatively impermeable to moisture vapor entry.

Insufficient dehumidification capacity

An example of insufficient dehumidification capacity is seen in the exhibition room at Casa das 11 Janelas Museum, in Belém, Brazil, which was subjected to daily activation and deactivation of the environmental management system (fig. 7.27). Located in a very hot-humid climate zone near the Guamá River, the museum is housed in a restored eighteenth-century governor's palace, a massive stone building with numerous windows on

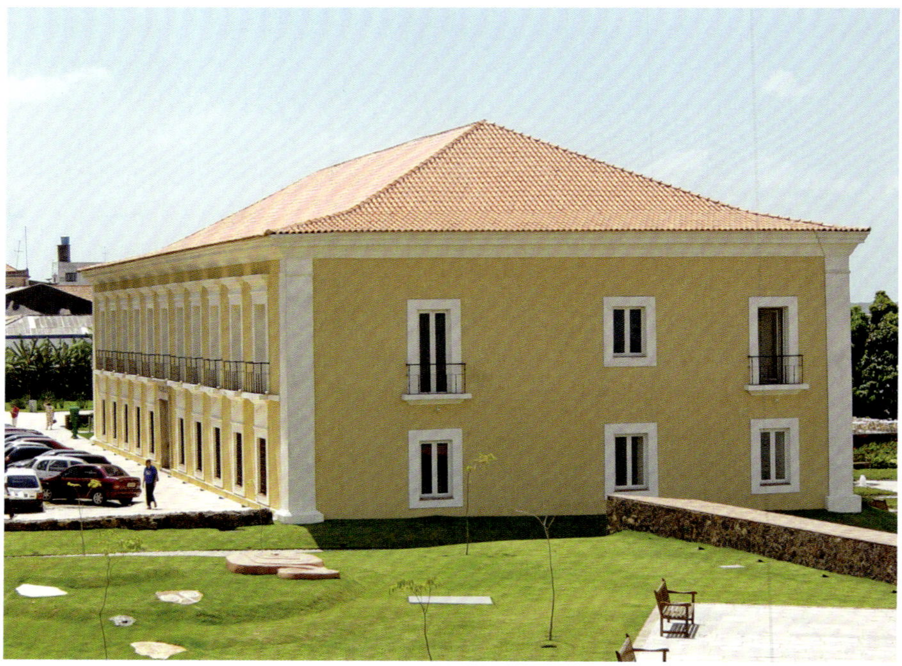

Figure 7.27
Building of Casa das 11 Janelas Museum, formerly a governor's palace in Belém, Brazil. Photo: Franciza Toledo.

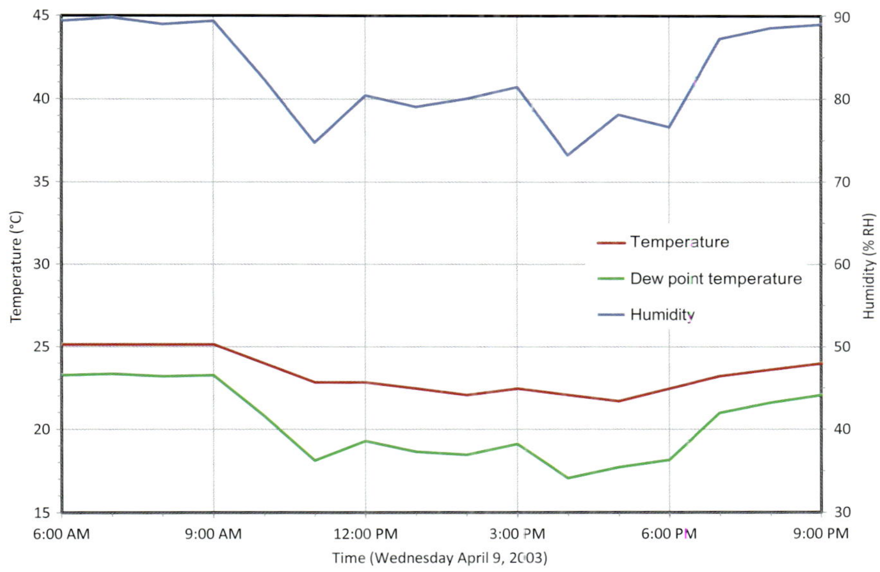

Figure 7.28
Temperature and humidity recorded in Casa das 11 Janelas Museum, Belém, Brazil, during April (the rainy season). Data courtesy of Franciza Toledo.

its front and back facades. Figure 7.28 shows temperature, humidity, and calculated dew point temperature of the environment in the building during the rainy season in April. The air-conditioning system was operated only Monday through Friday during working hours (9:00 a.m.–5:00 p.m.) and was turned off each weekday evening and for the entire weekend. Humidity remained at 80% RH or less only during the operation of air-conditioning, and it immediately rebounded to about 90% RH one or two hours after the air conditioner was turned off.

The air-conditioning system cooled the environment to 21°C–23°C from about 25°C and dehumidified to 75%–80% RH from 90% RH by reducing the dew point temperature from 23°C to 17°C–19°C. Although the outside condition was not recorded, temperature typically ranges from 23°C to 30°C daily, and dew point temperature remains stable at about 23°C during April, during the dominant rainy season in Belém.

The data in figure 7.28 suggest the following possibilities:

1. Impaired performance of the mechanical system,which may have been affected by (a) the temperature of the cooling coil exceeding the dew point temperature needed for sufficient incidental dehumidification, or (b) inadequate airflow volume due to undersized ducts, fouled coils, low fan speed, clogged air filters, or blocked supply/return ductwork or registers.
2. Inadequate source moisture control due to (a) large moisture source(s) present inside or outside the building, or (b) excessive infiltration of humid outside air.

During an on-site investigation, the temperature of the cooling coil was found to be higher than 15°C during operation of the air-conditioning system, limiting its capacity for dehumidification.

Summary

This chapter outlined the types of equipment for dehumidification, ventilation, heating, cooling with incidental dehumidification, and humidification that are suitable for mechanical strategies for alternative environmental management for collections and heritage buildings in hot and humid climates and in cold-humid marine climates. Other mechanical equipment necessary for completing management systems, such as control devices and sensors, were described. The chapter also presented the advantages and disadvantages of centralized and noncentralized systems, especially in heritage buildings, as well as the risks associated with the operation of air-conditioning systems in hot and humid climates. For maximum effect, the mechanical strategies discussed here are intended to be applied in combination with the nonmechanical strategies set out in chapter 6.

Chapter 8 will present a process and methodology for the commissioning, design, and implementation of environmental management strategies, and it will discuss various issues commonly encountered when applying the process. It will also address the evaluation and selection of mechanical and nonmechanical strategies.

References

ANSI [American National Standards Institute] and ASHRAE
2007 *Standard 52.2-2007—Method of Testing General Ventilation Air-Cleaning Devices for Removal Efficiency by Particle Size.* Atlanta: ASHRAE.

ASHRAE
2008 *ASHRAE Handbook—HVAC Systems and Equipment* (SI edition). Chapters 2, 3, 20–24, 27–29, 44, and 49. Atlanta: ASHRAE.

Cooper, G.
1998 *Air-Conditioning America: Engineers and the Controlled Environment, 1900–1960.* Baltimore: Johns Hopkins University Press, 19.

Lev-Alexander, N.
2011 Energy savings trials in the Library of Congress Adams Building. Paper presented at the 25th Annual National Archives Preservation Conference, Hyattsville, MD, March 17, 2011.

Chapter 8

Design and Implementation of Environmental Management Strategies

This chapter, which is directed at museum staff, conservators, and architectural and engineering consultants, describes a process and methodology for selecting, designing, implementing, and commissioning an integrated nonmechanical and mechanical environmental management strategy that is specific to a collection, building, and site. This process and methodology also call upon the information provided in the preceding chapters, particularly during steps concerning assessment and strategy development. While generally applicable to environmental management projects in all climate types, for the purposes of this book, this chapter illustrates the issues commonly encountered in alternative environmental management projects in hot and humid climates.

Overview

Environmental management strategies in archives and museums must address numerous considerations including: collections vulnerabilities, conservation objectives, staff and visitor comfort, the budget available for equipment installation and operation, the performance and spatial limitations of the building, and, in many instances, the conservation of heritage building fabric. The development of a strategy that is specific to the collections, building, and site is best accomplished through the engagement and active collaboration of all project stakeholders—financial and administrative personnel, as well as technical professionals from several disciplines and specialties.

Five Steps to a Successful Environmental Management Strategy

A five-step process is recommended for implementation of a successful strategy to manage conservation and thermal comfort environments in museums and archives. These steps, illustrated in figure 8.1, are:

- team building
- assessments
- strategy development
- design and implementation
- operations, maintenance, and continued evaluation and improvement

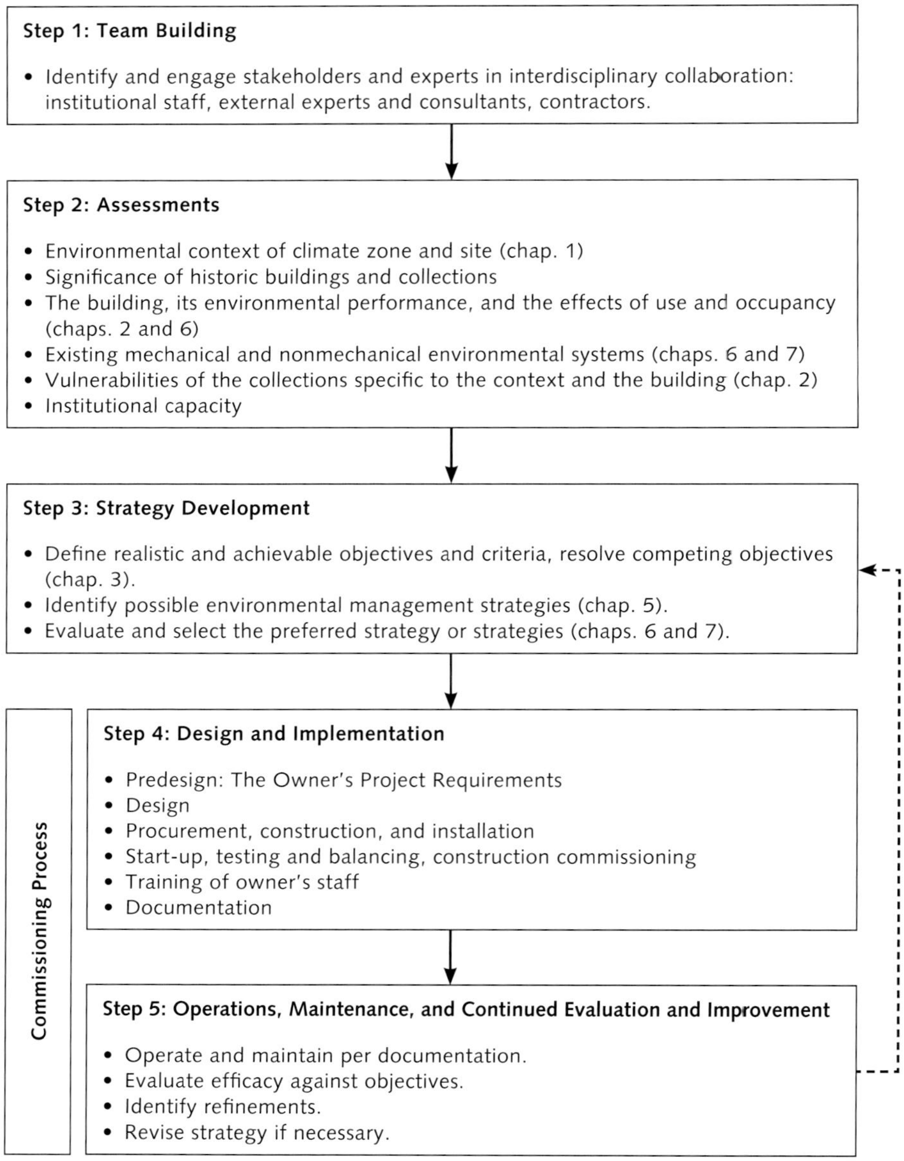

Figure 8.1

Process for designing and implementing an environmental management strategy.

Step 1: Team Building

An effective environmental management strategy must balance the multiple needs, objectives, and capacities of an institution and the collections for which the institution is responsible. Finding this balance requires a problem solving approach that crosses organizational boundaries and technical disciplines. It requires collaboration and team building. Over the life of the project, the team will involve all institutional departments—facilities/estates, conservation, curatorial, education, interpretation, and finance—as well as design professionals, specialists, equipment vendors, and construction contractors; the project will span planning, design, procurement, construction, and postoccupancy phases.

Typically, project participants will include: a collections conservator; the mechanical or environmental systems engineer; the architect; museum curators for collections, interpretation, and education; the commissioning specialist or agent; museum facilities and maintenance staff and/or service contractor; fire and security consultants or staff; the museum director and chief financial officer; and key board members. Ordinarily the core members of the team are the architect, conservator, engineer, facilities manager, and commissioning agent; the architect is the overall project manager and, in the case of heritage buildings, is likely to be a specialist in such buildings.

An example of team building is seen in the case study for Museu Casa de Rui Barbosa (MCRB) in Rio de Janeiro, Brazil. The team included:

- MCRB staff members: deputy director (representing the director), the staff architect/project manager, curator, furniture conservator, and book conservator;
- representative of an outside funding organization for the project;
- consultants for environmental management and for mechanical engineering;
- contractors for installation and for commissioning.

Step 2: Assessments

A number of parameters will affect what improvements are possible for collections environments at an institution. These parameters include: the climate; the site context; the architectural significance of the building; the environmental performance of the building and its envelope and systems; uses of the building; vulnerabilities of the collections to environmental factors; and the institutional capacities with respect to finances, technology, and human resources. These parameters must be assessed for each project in order to identify objectives, opportunities, and the potential limitations of possible future environmental improvements.

Assessments of the building, collections, and existing environmental management should be performed by an architect/conservation architect, collections conservator or manager, and an engineer.

Assessment of the environmental context of climate and site

Chapter 1 described various differences between climate zones and a method of determining climate zone from weather data. These climatic risks can be attenuated by the site context and microclimate at the building.

General climatic risks result from thermal energy and available moisture present in a climate zone. Site-specific climate risks can be identified through assessment of at least one year of on-site exterior weather data; however, more than one year of data collection is needed to get a representative dataset. Multiyear datasets of up to thirty years or more are useful but should be carefully examined for indications of recent trends that may result from global climate change. The exterior environmental variables of greatest interest typically include air temperature, dew point temperature, and humidity; data on local rainfall and wind direction and speed may also be useful.

In the United States, historical weather data can be obtained through the National Climatic Data Center (NCDC); data for other countries may be available through national organizations of meteorological study. Bear in mind that climate data are sensitive to location and may be strongly influenced by local geographical features such as topography or vegetation. If local factors are influential, it may be necessary to establish a weather station at or near the project site (fig. 8.2) and institute an environmental monitoring program (see appendix 1, "Environmental Monitoring for Alternative Climate Management"). If establishing a local weather station is not possible, then it is important to understand the contexts of the remote data collection point and the building location so that site-specific anomalies can be identified.

Regardless of the monitoring location and duration, weather data must be supplemented by direct observation of the site and the building at different times of day, during weather events such as intense rain-

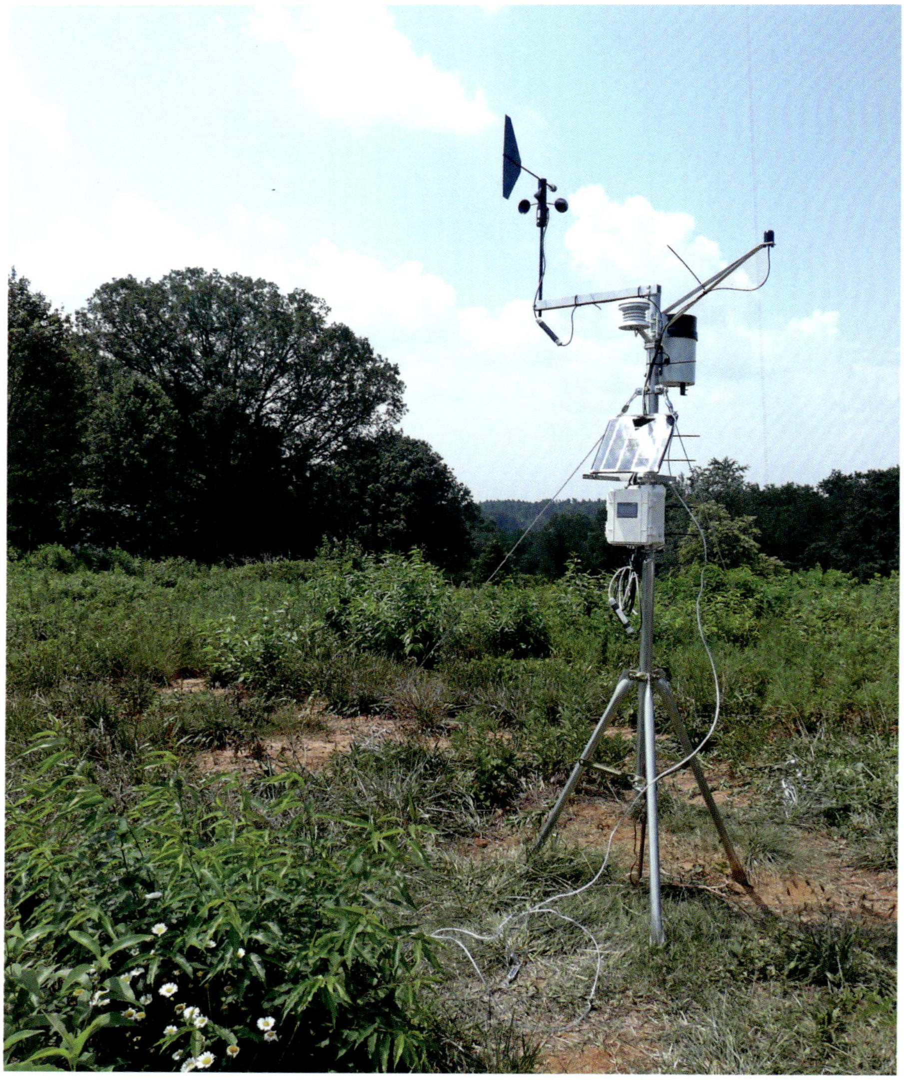

Figure 8.2

A site-specific weather station can help identify how site conditions differ from data collected at regional weather stations. This solar-powered weather station provides real-time access to exterior conditions via a mobile phone uplink to the Internet. Photo: Michael C. Henry.

fall, and throughout the seasons, so that time-dependent local factors may be fully understood.

Assessment and understanding of historically significant buildings

Heritage and architecturally important buildings pose special challenges for both conventional and alternative approaches to environmental management. Chapter 2 described typical risks to heritage buildings and architectural fabric that can result from interior environmental management, particularly for conservation environments. In order to avoid damaging heritage and architecturally important buildings, the first actions are to assess and identify:

- The significance of the building, usually based on its architectural, historical, or cultural importance.
- The character-defining physical features that give the building its aesthetic identity or are linked to its historical or cultural significance. For example, case study Hollybourne Cottage (see fig. 6.2; see also chaps. 9 and 10) is architecturally significant for the early use of exposed, unfinished concrete for a residential structure. This significance is amplified by the designer's intent to mimic vernacular "tabby" concrete with the inclusion of oyster shell fragments and the frank expression of the board formwork in the exterior surface. The concrete is inherently porous, but its intrinsic material significance prevents the application of surface treatments that would reduce water absorption and moisture vapor adsorption. Character-defining features of the building include the concrete, the terracotta accents, the wood doors and fenestration, and the stylistic expression of the gable ends and parapets.
- The state of integrity or authenticity of the building, generally including materials, components, assemblies, and finishes that constitute the original fabric of the building and relate to its significance. An example is the paper mural that covers the walls and ceiling of the theater area in Juanqinzhai (fig. 8.3), an eighteenth-century building located in the Forbidden City in Beijing (see case study, chap. 15). Because the mural is aesthetically and culturally important, it limited the options for placing an opening for delivery of conditioned air.

Most of the above information may be available in the institution's master plan, in reports on the building such as a heritage structure report or conservation plan, or in the documentation submitted as part of the nomination of the building for listing as a historic site. If the information is not available or compiled, it must be developed.

The next action in assessing the historical significance of buildings is to identify the risks to the building and its architectural fabric that might result from possible changes in the interior environment or

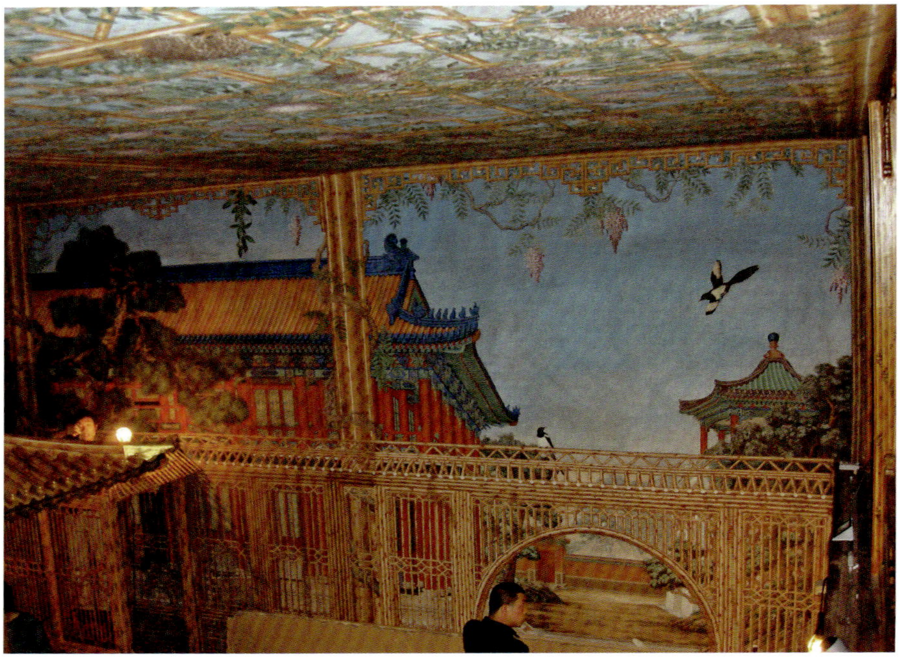

Figure 8.3
An aesthetically and culturally important paper mural covers the walls and ceiling of the theater area in Juanqinzhai, in the Forbidden City, Beijing (case study, chap. 15). Photo: Shin Maekawa. © J. Paul Getty Trust.

from installation and operation of environmental management systems in the building. The potential for these building-specific risks must be identified and assessed in each project.

Assessment of the building, its environmental performance, and effects of use and occupancy

The building envelope moderates the effects of exterior climate on the interior environment and the collections. The envelope may accomplish this with a fixed element, such as a roof assembly, or with an active, non-mechanical element, such as an operable window and shutters. The building envelope influences the range of possible interior conditions that are realistically achievable. For example, vernacular buildings in hot and humid climates were often constructed with large window and door openings to enhance natural ventilation; as a result, these buildings are likely to have excessive infiltration and exfiltration, which will pose a challenge for managing interior humidity.

Assessment of the building and its environmental management performance should include:

- determination of the effectiveness of the building envelope as a moderator of the exterior climate, including its overall condition and the condition of operable features, such as windows and shutters;
- assessment of the condition of the building as an indicator of past conservation efforts or problems and as an indicator of institutional capacity for conservation;
- assessment of the configuration and interconnection of spaces within the building and their effect on natural (convective) ventilation either horizontally or vertically;

- identification of spaces that may be inherently suitable for conservation environments;
- understanding of how the building is used and how the use influences the interior environment owing to moisture-generating activities, such as cooking, bathing, or housekeeping, and interior heat sources such as office equipment and lighting;
- determination of how occupant use and census affect the interior environment of the building, including the variability of occupant census and the possible coincidence of peak occupant census with maximum exterior thermal or moisture loads;
- identification of how building use and occupant census will influence occupant expectations for thermal comfort (this topic is described in chap. 4);
- determination of whether it is possible to engage building occupants in the operation of building features such as windows and shutters or awnings;
- if mechanical strategies are likely, identification of space that will be available for environmental management equipment and systems, with the clearances and access necessary for maintenance and repair.

The present performance of the building (and systems, if functioning) can be established through a program of environmental monitoring of the interior spaces and the exterior conditions (see appendix 1). Typically, such a program would use temperature and humidity dataloggers (fig. 8.4). The resultant data can be analyzed to determine the building's effectiveness in moderating the thermal and moisture loads from the exterior climate. In the absence of mechanical environmental

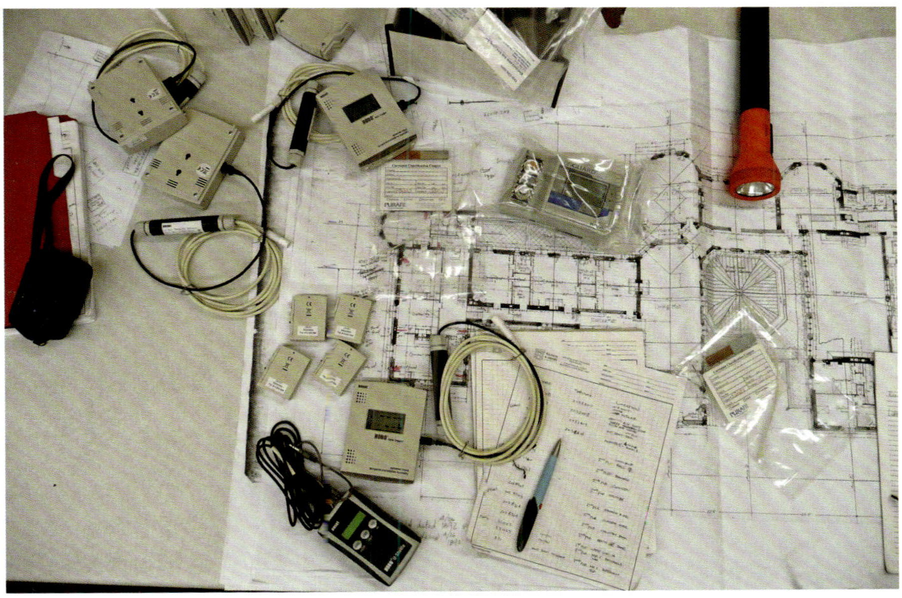

Figure 8.4

An environmental monitoring program that uses dataloggers, such as those shown here—along with thoughtful deployment and diligent attention to data downloads and data file management—can provide credible data for analysis. Photo: Michael C. Henry.

management systems, the monitoring data can indicate the overall comportment, or response, of the interior spaces to exterior conditions. Analysis of the interior data for temperature, dew point temperature, and humidity can indicate the extent that seasonal extremes or fluctuations pose risks to collections, which will inform development of seasonally specific strategies for reducing these risks. The analysis should include classification of the existing environmental performance of the building based on the Conservation Environment Classification–HH set out in chapter 3.

The influence of visitors on the interior environment can be understood by monitoring one or more factors, such as the frequency and duration of the opening of entry and exit doors, which are often significant factors in infiltration of outside air into small buildings. Visitor census may be correlated with interior carbon dioxide concentration unless ventilation or infiltration of outside air is significant. For example, at Our Lord in the Attic museum in Amsterdam, a converted seventeenth-century Dutch merchant residence, increases in carbon dioxide concentration were coincident with visitor entry counts, especially during winter months. The museum uses a centralized radiator-based heating system throughout the building in winter, and the supply of fresh outside air is limited to the uncontrolled infiltration of outside air through the opening and closing of entrance and exit doors as well as through small gaps and cracks in the building envelope (Maekawa 2007).

Dosimeters, such as Blue Wool Standards for light damage and datalogger-based dosimeters for gaseous pollutants, can assess other environmental risks to collections by indicating changes in a sacrificial or proxy material that has been exposed to the collections environment.

Monitoring data is essential in the prioritization and development of environmental management strategies. Monitoring data on existing conditions prior to the implementation of environmental management strategies provides a benchmark for assessing the effectiveness of strategies after implementation.

Assessment of existing nonmechanical and mechanical environmental management systems

Existing nonmechanical strategies (chap. 6) and mechanical environmental management systems (chap. 7) used for managing the interior environment of the building should be assessed. This assessment includes the identification of building features and mechanical equipment, design intents, and operating conditions. Once the nonmechanical strategies are identified by visual inspection and/or documentation, it is essential to understand the principles of their operation. As building use evolves over time, these nonmechanical features may be modified, blocked, or disabled due to various reasons, such as security requirements, fire protection/regulations, assignment of the space to other use, or a mechanical system installation. Thus, the assessment of existing nonmechanical systems should document their current working conditions and the operability of their features.

Similarly, a building's mechanical environmental management systems can deteriorate, undergo modifications, or be disabled over the course of their service life. While a mechanical system is typically accompanied by a service log, a visual inspection of the existing mechanical equipment is required in the event that no service log is available or changes to the system were not recorded. The operational principles of the mechanical system should be identified, and its mechanical components—including working condition, capacity, and layout—should be documented. Spaces allocated for existing mechanical equipment, as well as resulting modifications to the building interior and envelope, can be reused for a future system. The operational sequence of the system should also be identified and the integrated sensors of the mechanical system checked for accuracy and precision. In addition to analysis of available building environmental performance data, the current performance of the existing mechanical environmental management system can be evaluated by the use of strategically placed temperature/humidity dataloggers and a handheld sensor for spot measurements.

Assessment and identification of collections vulnerabilities specific to context and building

Accurate assessment and evaluation of the collections vulnerabilities to environmental risks is essential to developing strategies for alternative environmental management (this topic is addressed in chap. 2). Collections will differ from one institution to another in many respects, including: type, material, value, size, condition, past history of environmental exposure, and possibly past intervention and repair. Within an institution's collection, significance may vary, and not all objects will warrant equal efforts for conservation.

In theory, the number of environmental risks to a collection may be very large, but in practice, the number of probable risks may be smaller for a particular object or collection. Identification of the collection-specific probable risks is informed by a variety of factors, including the analysis of environmental context and building performance, as described above.

The specific vulnerabilities of collections to these risks must also be taken into account, since the existence of probable risks does not necessarily result in deterioration of or damage to collections. Understanding the present condition of the collections is important, since the objects are "information rich"; as the physical manifestations of the effects of past uses and prior environments, they contain the history of what has gone before. Combining data on the former environments to which collections have been exposed with the current physical state of the collections enables understanding of their vulnerabilities to current and future environmental conditions.

Analysis of the collections' vulnerabilities should include:

- identification of the current conditions and the recent rates of change in condition, which can indicate active deterioration mechanisms;

- determination of whether the observed conditions correlate with recent risks identified in the environmental analysis;
- identification of past environmental exposure and risks to the collections and determination of whether present conditions correlate to exposure to these past risks;
- assessment as to whether the collections may have been "proofed" by exposure to past environmental extremes;
- identification of past conservation treatments or interventions, including resupport, and determination of whether these actions may have compromised the object's ability to respond to variations in environmental conditions;
- assessment of the significance of the collections and determination as to whether all objects warrant the same level of environmental management.

A broad assessment of all potential collections vulnerabilities and risks should also be done. This assessment will include:

- determination of the proposed service life of the environmental management systems and building improvements, which will define the time frame for the remaining analyses;
- estimation of how the collections might change in material and size;
- assessment of whether recent building performance can be maintained;
- understanding of how local or regional climate are expected to change, recognizing that short-term exterior monitoring data will not indicate the rate and direction of future climate change;
- assessment of the capacity, reliability, and resiliency of the external infrastructure, including local water, sewer, electrical power, and information systems necessary to operate systems and controls. Protective and environmental systems for cultural heritage typically depend on the availability of some level of built and social infrastructures. Power and fuel must be delivered, staff must be nearby and available to service and maintain systems, and there must be firefighters and police to respond to new emergencies.

Assessment of institutional capacity

The success of an environmental management strategy will depend on the institution's capacity to implement and maintain the strategy.

Mechanical and nonmechanical environmental management strategies will require maintenance and periodic adjustment in order to perform within the specified outcomes. As the equipment and components age, the frequency of repairs and maintenance will increase, and as the end of the expected service life is approached, replacement will be needed, requiring a new input of capital funds.

The premature failure of mechanical environmental systems is often traceable to institutional inability to maintain them. Examples include:

- insufficient dehumidification and low airflow caused when cooling coils become clogged by debris and particulates because of inadequate filter replacement;
- unbalanced distribution of conditioned air caused by changes made to supply air register settings by building occupants or caused by the blockage of return air grilles by furnishings or exhibit panels;
- incorrect space conditions caused by sensors or controls that have not been periodically calibrated;
- "mystery" systems that no one knows how to operate—situations that occur when sensors and controls are replaced or modified without coordination or integration with the original system design; they can also result from lack of documentation or when the single person with a working knowledge of the operation leaves the institution without a trained replacement (fig. 8.5).

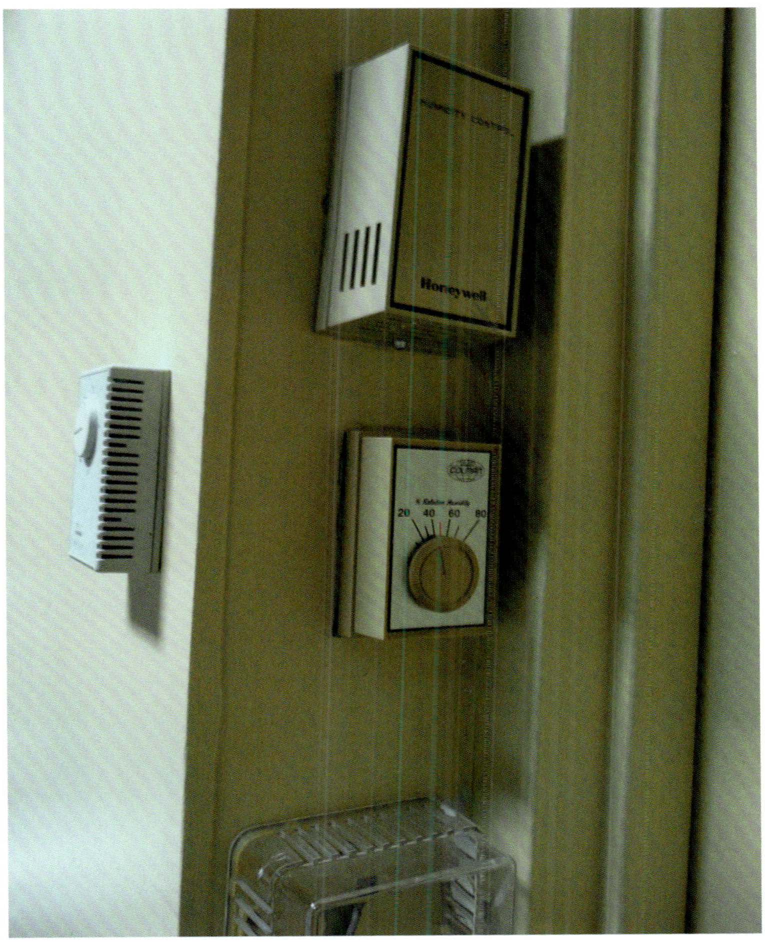

Figure 8.5
"Mystery" controls resulting from undocumented replacements and interventions. Occupants were unaware which of these controls actually control the environmental management system of the space. Photo: Michael C. Henry.

To avoid such problems, a nonmechanical or mechanical environmental management strategy should be matched to the institution's long-term capacity to maintain and operate the systems designed to execute the strategy. To this end, an institution's financial, human resource, technological, and organizational capacities should be assessed, including:

- access to funds for capital investment for building and systems improvements;
- authority to make decisions with clear accountability for actions taken or not taken;
- capacity to generate annual contributions to a building reinvestment fund for systems replacement at the end of their service life;
- reliability of annual income for recurring expenses for software and hardware upgrades, fuels, power, and preventive maintenance contracts for mechanical and nonmechanical systems;
- availability of reserve income for unplanned increases in energy costs or for increases in maintenance costs as the envelope or systems age;
- number of in-house or contracted personnel with the technical knowledge and skills needed to operate and maintain the building, equipment, and controls.

Step 3: Strategy Development

The assessments phase in the five-step process for a successful environmental strategy frames the constraints and opportunities of the environmental management problem, the solution to which may lie in a number of possible strategies. The next step is development of the possible environmental management strategies, which leads to the evaluation and selection of a strategy that is the best fit for the specific climate, building, collections, and institution. The approach is described below.

Define realistic and achievable objectives and criteria and resolve competing objectives

Objectives, and the criteria for their fulfillment, form the basis for evaluating potential environmental management strategies and ultimately become the basis for design and evaluation of postimplementation performance.

Based on the assessment results (see "Step 2: Assessments" in fig. 8.1)—climate, historic significance, the building's environmental performance, occupancy, existing nonmechanical and mechanical strategies, vulnerabilities of the collections, and institutional capacity—opportunities and objectives for improving the interior environment should be developed, consistent with vulnerabilities of the collection, building performance, and thermal comfort. However, successful environmental management for collections conservation is defined by more than a set of allowable temperature and humidity fluctuations and their extremes; issues such as reliability, resilience, ease of maintenance, cost, emissions, energy consumption, fire risk, and expected service life should be addressed. For heritage buildings, disturbance of historic architectural fabric will be a significant issue, as will be the service life expected before replacement and future disturbance is required.

The objectives for reliability and resilience of an environmental management system must be considered. The *reliability* of a system is an indication of the rate of equipment failure over a period of time, while *resilience* is an indication of a system's ability to recover and function following an incident such as a hurricane or an earthquake. A high degree of reliability for an environmental management system may be achieved with design and construction quality and routine maintenance and testing; high reliability does not necessarily require redundancy. Nonmechanical strategies for building envelopes and collections housing can delay or buffer the effects of short-term environmental excursions that may result from interruption of systems operations. These nonmechanical strategies may be targeted for materials or material assemblies that are vulnerable to short-term environmental excursions.

Therefore, it is important to consider which environmental conditions resulting from a major incident pose the greatest risks to a collection. Critical features of the environmental management system can therefore be identified and designed to provide system resilience. Similarly, contingency plans must be developed for severe storms or seismic activity, so that essential conservation conditions for the collection can be maintained during periods when external utilities might not be available.

The objectives for environmental management are multifaceted, so it is appropriate to involve all of the stakeholders in defining and prioritizing the project objectives. A participatory workshop is an effective method for engaging the stakeholders; the workshop process is both informational and collaborative, and it should lead to a consensus regarding the realistic and achievable project objectives.

Neutral facilitation is important to workshop success, and it should be provided by a professional experienced in environmental management for collections conservation. Since each stakeholder represents a specific point of view, some stakeholder objectives may be competing or conflicting and possibly even mutually exclusive—for example, environmental conditions for thermal comfort versus collections conservation. A facilitator can lead the stakeholders in resolving competing or conflicting objectives, and resolution should occur early in the project.

An example of competing or conflicting objectives is seen in the case study at Museu Casa de Rui Barbosa in Rio de Janeiro (see chap. 14). In that case, the goal of minimizing environmental risk to the collection and the goal of visitor comfort in the interpreted historic spaces are mutually exclusive, and the conflict could only be solved by the provision of separate environments for each (fig. 8.6).

Once the objectives for environmental management are resolved and prioritized, specific criteria must be developed against which the effectiveness of possible strategies is measured. Clarity and simplicity

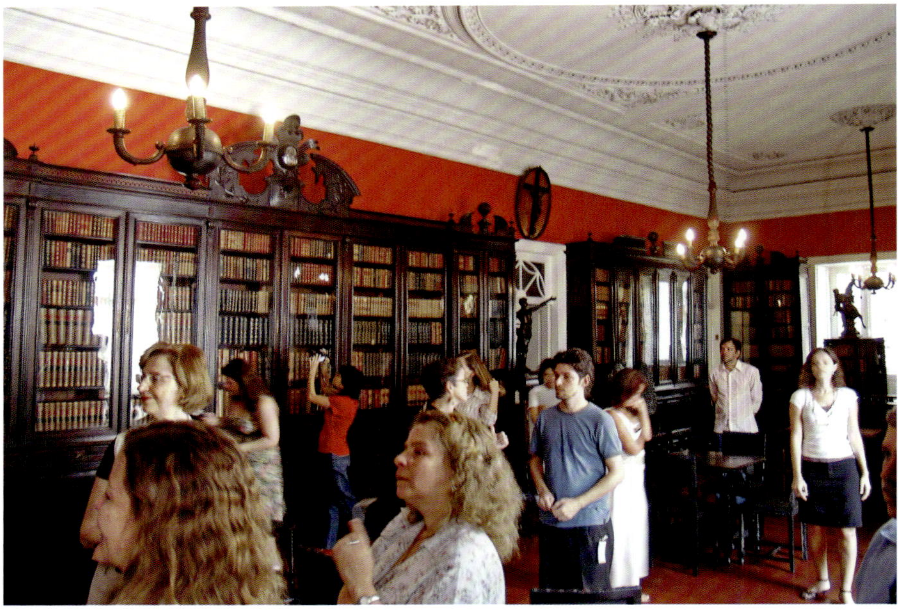

Figure 8.6
The conservation objectives for the book collections and the thermal comfort objectives for visitors to the Library of Museu Casa de Rui Barbosa were resolved by providing microenvironments for the bookcases and by delivering conditioned fresh air along the tour route. Photo: Shin Maekawa. © J. Paul Getty Trust.

of the resultant objectives and criteria are essential for the design and successful implementation of environmental management strategies.

Examples of objectives and possible criteria, with references to the case studies, include:

- Require a certain level of environmental management that is based on the unique vulnerabilities of a particular object or class of objects in the collections (Juanqinzhai, Beijing; chap. 15).
- Require system survivability long enough to maintain interior conditions during an incident, such as loss of electrical power (Museu Paraense Emilío Goeldi, Belém, Brazil; chap. 13).
- Conform to specific local codes or regulations, such as building, electrical, mechanical, and fire and life safety codes (all case studies).
- Reduce energy consumption using current energy consumption as a baseline, or achieve energy efficiencies required by codes or regulations. The environmental sustainability of a collections management strategy is best based on three factors: energy consumption relative to conventional systems, efficacy of mitigation of environmental risk factors to collections, and use of materials/equipment that are nontoxic and maximally recyclable (Museu Paraense Emilío Goeldi, Belém, Brazil; chap. 13).[1]
- Provide for occupant comfort in specific areas of the building for certain large-occupancy events (Museu Casa de Rui Barbosa, Rio de Janeiro; chap. 14).
- Limit the placement of systems equipment in certain rooms or spaces based on an aesthetic, architectural, or historical

ranking of the significance of the spaces within the building (Museu Casa de Rui Barbosa and Juanqinzhai; chaps. 14 and 15).

The objectives and criteria for an environmental management strategy should be included in the "Owner's Project Requirements" document (also known as the project's design brief), which is described in the "The Commissioning Process" section below.

Identify possible environmental management strategies

Nonmechanical measures

Identification of possible strategies for interior environment management should be based on clear objectives and criteria for environmental management, after which possible psychrometric strategies, discussed in chapter 5, can be identified. The selection of strategies begins with nonmechanical measures to mitigate thermal and moisture gains, as discussed in chapter 6. Some of the nonmechanical measures used in the case studies are described below:

- Source control of bulk moisture, including roof and site drainage measures, as well as building envelope repairs such as for roof, doors, and windows: Hollybourne Cottage, Jekyll Island, Georgia, United States (chaps. 9 and 10).
- Reestablished functionality of original nonmechanical architectural features, such as window shutters for control of light and solar gain, or operable skylights for gravity ventilation: Museu Casa de Rui Barbosa, Rio de Janeiro (chap. 14).
- Envelope improvements, such as repairing the roof and reestablishing attic space; filling gaps and unintended openings on the building envelope due to wood shrinkages and warping; secondary glazing for windows; and window shades: Juanqinzhai, Beijing (chap. 15).
- Collections segregation by level of significance and by environmental vulnerability: Valle Guerra museum storage, Tenerife, Spain (chap. 12).
- Locating collections in spaces away from exterior building surfaces or segregating exterior thermal and moisture conditions by creating interior enclosures with archival boxes—the "box within a box" approach: Historic Archive of San Cristóbal de La Laguna, Tenerife, Spain (chap. 11).

Mechanical strategies

After nonmechanical strategies for reducing thermal and moisture loads have been identified, mechanical strategies, such as those described in chapter 7, can be developed. Mechanical strategies may range from incremental improvements through the use of simple devices to the introduction of new, energy-efficient alternative mechanical systems. Design considerations for mechanical strategies include:

- calculation of maximum dehumidification and/or cooling loads,
- selection of a centralized or noncentralized system to implement the strategies,
- identification of proposed placement for mechanical equipment and components.

Examples of mechanical systems used in the case studies in part 2 include:

- nonducted small-scale conservation heating and mechanical ventilation (Historic Archive of San Cristóbal de La Laguna; chap. 11);
- portable mechanical dehumidification and ducted ventilation (Museu Paraense Emilío Goeldi; chap. 13);
- ducted mechanical dehumidification with heat dissipation for the maximum temperature limit, and HEPA filter for particulate removal (Juanqinzhai; chap. 15).

Evaluate and select the preferred strategy or strategies

The best combination of nonmechanical (chap. 6) and mechanical strategies (chap. 7) for environmental management for a particular building and collections is one that satisfies the project objectives.

The decision-making process by which the "best" combination of strategies is evaluated and selected may be simple, or it may be formal and complex. The objectives and their criteria may be qualitative and quantitative, including cost considerations. The process of evaluation should be objective, transparent, and well documented. Individual or professional biases or other influences should be avoided in the evaluation process.

In practice, formal methods of evaluation may not account for the qualitative aspects of cultural heritage care, and formal structured evaluation methods may be too cumbersome for implementation on small-scale projects.

For small projects, a direct, expedient, and reasonably accurate approach, suitable for the purposes of most environmental management decision making, might be to evaluate the proposed strategies against the objectives and criteria using one of three grades (*meets*, *does not meet*, or *neutral*). Equivalent to a "straw poll," this method provides a rapid estimation of whether or not one strategy is better suited than the others. If the results are inconclusive, a simplified variant of one of the formal evaluation methods might be warranted, such as the structured decision-making methodologies described in appendix 4.

Ultimately, the evaluation method should result in consensus-based decisions by the stakeholders, since stakeholder commitment and engagement in the selected strategy are essential to long-term support and success.

Regardless of the decision-making method used, it is important to document the objectives, criteria, candidate strategies, and the

> **Unique Approach at Hollybourne Cottage**
>
> The case study conducted at Hollybourne Cottage, reported in chapters 9 and 10, was unique with respect to its experimental nature. Unlike most projects that implement a chosen environmental strategy following analysis of environmental data, the work undertaken at this historic house tested various sets of environmental strategies over a multiyear period, then made an environmental and economic comparison of the different sets. This approach was possible because the structure had limited occupancy and because there was no collection stored therein—the building's historic interior was the primary conservation focus.

evaluation method for selection of the preferred strategy; this information will be useful in the future when the success of the implemented strategy is assessed. See the sidebar "Unique Approach at Hollybourne Cottage" for an unusual decision-making and evaluation tactic.

The Commissioning Process

The commissioning process comprises steps 4 and 5 of the five-step process for implementation of a successful strategy to manage conservation and thermal comfort environments in museums and archives. Step 4—design and implementation—and step 5—operations, maintenance, and continued evaluation and improvement—are discussed below.

Commissioning is an integrated process that addresses systems and the building envelope and spans the life of the project, from predesign through occupancy and operations, including "continuous commissioning" during operation. ASHRAE Guideline 0-2005, *The Commissioning Process* (ASHRAE 2013) notes that:

> Due to the integration and interdependency of facility systems, a performance deficiency in one system can result in less than optimal performance by other systems. Implementing the Commissioning Process is intended to reduce the project capital cost through the first year of operation and also reduce the lifecycle cost of the facility. Using this integrated process results in a fully functional, fine-tuned facility, with complete documentation of its systems and assemblies and trained operating.

ASHRAE Guideline 0-2013 (ASHRAE 2013) defines the commissioning process as a "quality-oriented process for verifying and documenting that the performance of facilities, systems, and assemblies meets defined objectives and criteria." The fundamental objectives of the commissioning process are to:

- clearly document the Owner's Project Requirements (OPR) in the predesign phase;

- provide documentation and tools to improve the quality of deliverables;
- verify and document that systems and assemblies perform according to the OPR;
- verify that adequate and accurate system and assembly documentation is provided to the owner;
- verify that operation and maintenance personnel and occupants are properly trained;
- provide a uniform and effective process for delivery of construction projects;
- deliver buildings and construction projects that meet the owner's needs, at the time of completion;
- utilize high-quality sampling techniques to detect systemic problems, as such sampling provides high value, efficient verification, accurate results, and reduced project costs;
- verify proper coordination among systems and assemblies and among all contractors, subcontractors, vendors, and manufacturers of furnished equipment and assemblies.

Thus, the commissioning process is more than the "start-up" of the system. The extent and level of detail of the commissioning process will be determined by the complexity of the project.

Step 4: Design and Implementation

Predesign—the Owner's Project Requirements

The selection of the preferred strategy for environmental management should result in a statement called the "Owner's Project Requirements," as described in ASHRAE Guideline 0. The OPR would include:

- project schedule and budget;
- project documentation requirements for submittals, training materials, reports, and the systems manual;
- applicable building codes or, alternatively, minimum standards for materials and construction;
- energy-efficiency, environmental, and sustainability goals;
- building occupant load and schedules, and requirements for human comfort, health, safety, and barrier-free access;
- noise and vibration limits, as well as seismic performance and fire-risk requirements;
- security and data/telecommunications compatibility;
- operational and maintainability requirements, and minimum warranty periods.

In addition to the above, the OPR must identify the following considerations unique to cultural heritage buildings and collections:

- criteria for acceptable interior environmental performance consistent with the class of control specifications based on the climate; for hot and humid climates, the conservation objectives and environmental criteria classifications discussed in chapter 3 should be applied, and for temperate or heating-dominated climates, the classification criteria for conservation environments as documented in ASHRAE's "Museums, Galleries, Archives and Libraries" (2011) are applicable;

- performance metrics to be used to ascertain compliance with the OPR;
- conservation requirements for the heritage building fabric with respect to potential damage from installation and/or operation of the proposed improvements;
- criteria for resilience of the building and systems in maintaining an interior environment in the event of equipment failure or loss of supporting infrastructure.

Design

Selection of the mechanical and nonmechanical environmental management strategies and completion of the OPR will enable the design team of architects and/or engineers to design the system and building improvements.

The design procedure is well established for larger systems. Heating and cooling loads for a particular building must be calculated, based on information on building envelope construction and details, climatic data on thermal loads and moisture loads, interior thermal and moisture loads due to use and occupancy, and the interior conditions of temperature and humidity to be maintained. Equipment is sized and selected, based on the heating/cooling load calculations. A whole-building energy simulation model, such as EnergyPlus,[2] can be used to model heating, cooling, lighting, ventilation, other energy flows, and water consumption and to fine-tune design decisions, including equipment sizing, as well as to identify potential energy efficiency improvements.

In hot and humid climates it is important to distinguish between thermal loads and moisture loads, especially if dehumidification is accomplished independent of cooling. As discussed in chapter 6, in hot and humid climates, the largest dehumidification load comes from outside humid air entering the building.

Also part of the design step is the selection of sensors and instrumentation for monitoring the performance of the environmental management system and development of the sequence of operations for mechanical and nonmechanical strategies.

Procurement, construction, and installation

The system design is physically realized through the procurement, construction, and installation of equipment. Conformance with the design intent and the OPR is confirmed through a variety of methods, including: review and acceptance of contractor's technical submittals for materials and equipment to be supplied; in-factory acceptance of complex equipment before shipment; qualification and certification of tradespersons for certain critical installation activities such as welding; and field observation and inspection of equipment and systems during installation.

Construction quality, including cleanliness of systems during construction, is critical; all incomplete piping and ductwork should be kept closed or sealed during installation to prevent the introduction of dust and debris.

Start-up, testing and balancing, and construction commissioning

Start-up, testing, and balancing are performed once an environmental management system is installed. In preparation for start-up, preliminary testing and cleaning is conducted as follows:

- *Design conformance:* The installed system must be checked for conformance with the design intent and the OPR. Specifically, the leak-tightness of air and hydraulic systems must be verified by pressure testing; pump and fans checked for proper rotation; valves and control devices checked for correct actuation/response; and the electrical continuity and proper polarity of electrical wiring and connections must be confirmed. Sensors, instrumentation, and control devices must be checked for correct calibration and signal/response.
- *Cleaning:* Hydraulic systems must be flushed-cleaned using start-up strainers and air systems operated with construction filters; both operations must continue until cleanliness requirements are met, as indicated by the amount of construction-related debris and particulates captured in the strainers and filters.
- *Start-up:* Operation of each piece of equipment must be initiated in accordance with a start-up procedure provided by the equipment manufacturer. It is essential to adhere to the manufacturer's start-up procedures, since the manufacturer's warranty period begins with the initial power-up of equipment; failure to follow start-up instructions can void the warranty.

Upon completion of preliminary testing, cleaning, and start-up of individual components, equipment, and assemblies, the start-up sequence and shut-down sequences for the entire system must be verified before the system can be operated. Upon successful system start-up and shut-down, the system can be balanced for operation, and construction commissioning can take place.

- *Balancing:* The balancing phase consists of measuring system airflow and fluid flow, making adjustments, and balancing the flows to match design flow rates. This may require an adjustment of the airflow using dampers in ducts and/or pulleys or belts at the fans. Similarly, water flow rates in hot or chilled water systems must be measured, adjusted, and balanced. Both cooling and reheating capacities may also need to be adjusted by refrigerant compression or by regulation of cooling or heating fluid flow rates, to produce designed heating, cooling, or dehumidification. Electrical loads from equipment must also be verified.

- *Construction commissioning:* After testing and balancing, construction commissioning takes place, during which phase the performance of the system is checked against the OPR and verified. In climates with wide ranges of thermal and moisture loads over four seasons, performance verification may take up to twelve months; in climates with more consistent thermal and moisture loads throughout the year, performance verification might be accomplished in six months.

Training of owner's staff

The facilities staff at the building where the environmental management system is installed should be familiar with the OPR and knowledgeable about design intent, operation, maintenance, and basic troubleshooting of the system. In-house knowledge of the OPR will help protect the collections, minimize unnecessary service calls, maintain system operation within performance specifications, and help prevent premature failures. After system start-up, testing, and full commissioning, the facilities staff and building maintenance personnel must be trained in the operation and maintenance of the environmental management system. The training should start with the fundamentals of the design intent for the system, followed by explanation of the operations and maintenance manual prepared for the system. Training should also include:

- hands-on practice by facilities staff with each of the necessary service/maintenance operations and control systems adjustments,
- basic troubleshooting procedures for the system,
- operation of the monitoring features of the environmental management system and early identification of performance issues.

Documentation

Complete documentation of the environmental management system from inception through start-up is a product of the commissioning process and is provided to the facility owner. The system documentation should include the OPR, design intent of the environmental control system, design documents, performance specifications, technical submittals, inspection reports, calibration records, testing and balancing reports, a detailed sequence of operation, the start-up and shut-down procedures, a maintenance schedule for each of the system's components, and simple diagnostic and/or troubleshooting procedures.

Step 5: Operations, Maintenance, and Continued Evaluation and Improvement

The operation of any environmental management system requires a program of preventive maintenance, ongoing performance monitoring, and periodic performance assessment and evaluation for overall effectiveness in collections conservation and in energy use.

If changes in operational parameters, collections conditions, or energy efficiency are noted, the current system's performance should be revisited as per the steps described previously in the "Strategy Development" section:

- define realistic and achievable objectives and criteria, resolve competing objectives;
- identify possible environmental management strategies;
- evaluate and select the preferred strategy or strategies.

Once a new strategy development phase is completed, the process continues through (a) the design and implementation phase and (b) the operations, maintenance, and continued evaluation and improvement phase. This repeat cycle of actions is shown in figure 8.1. Repeating this process makes it possible for the system to operate at optimal performance and at the lowest energy consumption.

Summary

Drawing on the information given in preceding chapters, this chapter provided a methodology and framework for identifying, selecting, designing, and implementing a strategy for alternative environmental management that is specific to a collection, building, and site. The methodology utilizes assessments of the collections, the building and its interior environment, and the institutional capability to implement and maintain an environmental management system as the basis for setting realistic and achievable environmental management objectives, while integrating both nonmechanical and mechanical approaches into the environmental management strategy. Although the technical considerations are specific to collections conservation in hot and humid climates, the general principles of the methodology in this chapter are also applicable to systems for conservation environments in all climates.

Part 2 of this volume presents seven case studies of alternative environmental management systems for collections conservation in hot and humid and marine climates that were referenced throughout part 1. Although these case studies evolved as research projects rather than as commercial installations, they illustrate the application of the methodology elaborated in this book for arriving at a suitable strategy for alternative environmental management that is specific to a collection, building, and site.

References

ASHRAE
2011 Museums, galleries, archives and libraries. Chapter 23 in *ASHRAE Handbook—HVAC Applications*. Atlanta: ASHRAE.

———.
2013 *Guideline 0-2005—The Commissioning Process*. Atlanta: ASHRAE.

Maekawa, S.
2007 Investigation into impacts of large numbers of visitors on the collection environment at Our Lord in the Attic. In *Museum Microclimates: The Copenhagen Conference, November 19–23, 2007*, ed. Tim Padfield and Karen Borchersen, 99–106. Copenhagen: National Museum of Denmark.

Notes

1 Two major programs are available to evaluate the environmental sustainability of a building, and thus improve its energy efficiency: (1) LEED, in the United States (http://www.usgbc.org/leed), and (2) BREEAM, in the United Kingdom (http://www.breeam.org). Although neither program is suitable for evaluating the performance and energy consumption of collections management strategies, they can inform choices with respect to environmentally friendly building materials and systems.

2 EnergyPlus is a building energy simulation program developed by the US Department of Energy for use by architects and engineers in designing energy-efficient buildings. The EnergyPlus software is free and can be downloaded at http://apps1.eere.energy.gov/buildings/energyplus/. However, the program needs third-party interface software if the building design data are to be input directly from proprietary design software.

PART 2

CASE STUDIES

Introduction to Part 2

Case Studies of Alternative Approaches to Environmental Management in Hot and Humid Climates

The seven case studies presented here provide examples of collections environment management projects that illustrate application of the strategies and methods set out in part 1 of this book. These case studies provide insights into the successes and potential issues of alternative environmental management strategies in hot and humid climates as well as in some warm-humid and marine climates with similar conservation issues. Readers are reminded that each case study is the result of a unique set of factors and constraints; therefore, the studies should be used as guidance —not recipes—for their projects.

The projects, ordered chronologically, are:

- Hollybourne Cottage, a vacant historic house museum, originally the late nineteenth-century winter resort residence of a wealthy family, in Jekyll Island Historic District, Georgia, United States (two separate case studies are reported on this project);
- The Historic Archives section of a municipal archive, set in the rehabilitated ground-floor level of a former nineteenth-century convent, in the Municipality of San Cristóbal de La Laguna, Tenerife, Spain;
- Museum storage in a contemporary building of the Organismo Autónomo de Museos y Centros of Tenerife in Valle Guerra, Tenerife, Spain;
- Ethnographic storage in a contemporary office building of the Museu Paraense Emílio Goeldi, in the equatorial city of Belém, Brazil;
- Library of Casa de Rui Barbosa, a furnished and interpreted historic house museum in Rio de Janeiro, Brazil;
- Juanqinzhai, a restored eighteenth-century Chinese emperor's retirement studio in the Forbidden City, Beijing, China.

The projects at Hollybourne Cottage in the state of Georgia (United States) and at the Historic Archive and the Valle Guerra museum storage on Tenerife Island (Spain) were initiated in 1999 as field trials of the GCI Collections in Hot and Humid Environments research project to test the viability of the various environmental management approaches discussed in part 1. The remaining projects were initiated and funded by partner organizations and utilized local engineers and architects; for these the GCI provided conceptual design and technical support. The final project, at

Juanqinzhai in Beijing, concluded in approximately 2010. GCI researchers Shin Maekawa, Franciza Toledo, and Vincent Beltran were involved in the various case studies. Project participants along with their contributions are listed at the end of each case study.

For ease of reference, the case studies are summarized in two tables—table 1 for project background information and table 2 for results. To avoid repetition, the "Case Study Terminology" sidebar defines key terms used in the case studies.

Case Study Format Overview

For ease of comparison, the case studies are presented in a standardized format. Additional discussion can be found in journals and conference proceedings; relevant publications are listed at the end of each case study.

Table 1

Project background information for case studies.

Case Study	ASHRAE Climate Zone	Building Characteristics & ASHRAE Building Class	Collection	Occupancy
Hollybourne Cottage, Historic House, Jekyll Island, GA (United States)	Hot-Humid (2A)	Cast-in-place concrete Class IV Two-story historic Basement, 1st & 2nd floors, and attic: 279 m^2 each/attic: 214 m^2	Historic interior only; no collection	Limited infrequent tours
Historic Archive, San Cristóbal de La Laguna, Tenerife (Spain)	Mixed-Marine (4C)	Heavy masonry Class IV Two-story historic Whole building: 825 m^2 (footprint) Historic Archive: 25.4 m^2	Bound and unbound archival documents made of papers and parchment	Regular access by archivists
Museum Storage, Valle Guerra, Tenerife (Spain)	Hot-Humid (2A)	Light masonry Class III Four-story modern 260 m^2 (one floor)	Mixed collection of folk artifacts such as agricultural/small industry objects, textiles, pottery, basketry, and furniture—materials include organics (e.g., sapwood, paper, leather, hide, sticks, leaves) and metals	Periodic light work by conservators, later partly occupied by registrars
Goeldi Museum Ethnographic Storage, Belém (Brazil)	Very Hot-Humid (1A)	Hollow brick Class III Single-story modern Whole building: 920 m^2 Portion of storage: 270 m^2	Mixed ethnographic collection consisting of objects used in agriculture, fishing/hunting, food processing, ceremonies, and celebrations—materials include plant fibers, wood, seeds, and feathers	Limited access by conservators and scholars
Library at Casa de Rui Barbosa, Rio de Janeiro (Brazil)	Very Hot-Humid (1A)	Heavy masonry Class IV Partially two-story historic Whole building: 880 m^2 Library portion: 165 m^2	Historic interior, book collection, artwork, and furniture	Regular supervised tours
Juanqinzhai, Heritage House Museum, Beijing (China)	Mixed-Humid (4A)	Timber post-and-beam assembly and heavy masonry (hybrid) Class IV/III Partially two levels with unfinished attic Historic 224 m^2 (footprint)	Interior decorative finishes (polished ironwood framing, silk embroidery, bamboo marquetry, wood relief, paintings and calligraphy mounted on wallpaper, mural paintings on silk, painted wood surfaces), conserved and restored interior elements	Limited infrequent tours

Table 2

Case study results: environmental management strategies, achieved risk classes, and energy metrics.

Case Study	Nonmechanical Strategies	Occupant Load Strategy	Mechanical Strategy	Conservation Environment Classification (Postinstallation)	Energy Intensity	Energy Use Intensity
Historic house, Jekyll Island, GA (United States)	Repaired building envelope/ sealed gaps in roof/ improved drainage of area surrounding building	Infrequent guided visits	Ventilation and heating (part A)	Basement Microbial risk: C Mechanical risk: b Chemical risk: + Attic Microbial risk: F Mechanical risk: a Chemical risk: 0	Basement: 83 W/m² Attic: 3.5 W/m²	28.3 kWh/m²/yr
			Ventilation and dehumidification (part B)	Basement Microbial risk: E Mechanical risk: b Chemical risk: + Attic Microbial risk: E Mechanical risk: b Chemical risk: +	Basement: 11 W/m² 1st and 2nd floors: 1.8 W/m² Attic: 38 W/m²	26.6 kWh/m²/yr
Archive, Tenerife (Spain)	Installed vinyl curtain at entrance/ removed internal door/filled window and door gaps	Limited access by archivists	Ventilation and heating	Microbial risk: C Mechanical risk: b Chemical risk: +	77 W/m²	10.8 kWh/m²/yr
Museum Storage, Tenerife (Spain)	Sealed external windows or converted for ventilator use/sealed gaps in doors/installed airtight door in restroom/removed all internal doors and windows	Limited access	Ventilation and heating	Room D Microbial risk: E Mechanical risk: a Chemical risk: +	45 kW/m²	53–83 kWh/m²/yr
Museum Storage, Belém (Brazil)	Vestibule area/ ceiling insulation/tightening of envelope/paved area surrounding/ perforated compact shelving	Limited access	Ventilation and dehumidification	Microbial risk: C Mechanical risk: b Chemical risk: –	277 W/m²	24 kWh/m²/yr
Library of historic house, Rio de Janeiro (Brazil)	Repaired building envelope/closed windows and doors/improved microenvironment (repaired bookcase doors)	Guided visits and modified visitor pathway	Ventilation, dehumidification, and limited cooling	Microbial risk: B Mechanical risk: b Chemical risk: +	148 W/m²	524 kWh/m²/yr
Historic house, Beijing (China)	Thermal insulation/ shaded double-glazed windows and UV filters into original window grilles/ sealed gaps in doors and attic space	Guided tours	Seasonal dehumidification and limited cooling	Return air Microbial risk: A Mechanical risk: b Chemical risk: 0	47 W/m²	139 kWh/m²/yr

Case Study Terminology

To avoid repetition, we provide here a review of terms common to all case studies.

HDD and CDD: Heating and Cooling Days

The demand for energy needed to heat or cool a building, as defined in the ANSI/ASHRAE/IES Standard 90.1-2013 climate classification system (see chap. 1), is proportional to the value of heating degree days (HDD) or cooling degree days (CDD) at the location. HDD and CDD are defined relative to Standard 90.1-2013 base temperatures—the outside temperature below which a building needs heating (18°C) or above which a building needs cooling (10°C). In the case studies, these base temperatures are included when writing out the two terms, e.g., CDD10°C and HDD18°C. A daily CDD10°C value is determined by the difference between the daily average temperature and the CDD base temperature, with a zero value given if the base temperature exceeds the daily average. Similarly, a daily HDD18°C value is the difference between the HDD base temperature and the daily average temperature, with a zero value given if the daily average exceeds the base temperature. Annual CDD10°C and HDD18°C values are calculated by summing daily temperature differences between the daily average and the base temperature over a year.

CDD and HDD values are calculated from actual temperature data. Websites, such as http://www.degreedays.net/#, are available to assist energy professionals to evaluate CDD and HDD for specific base values at specific geographical locations.

Damage Risks to Cultural Heritage

The Conservation Environment Classification–Hot and Humid (HH) protocol (see chap. 3) defines a building interior's environmental risk for microbial, mechanical, and chemical damage to cultural heritage objects in hot and humid climates. This protocol is displayed in a table format (see table 3.3) showing the three risk categories, the specific risk in each category, the humidity and temperature criteria, and the resulting risk classes, which are identified by letters or characters as shown in the figure.

PMV and PPD: Occupant Thermal Comfort

In the case studies, human thermal comfort in a building is based on the heat balance model (see chap. 4). This model arrives at the average comfort vote of occupants based on the predicted mean vote (PMV), which is a seven-point index of thermal comfort, and the predicted percentage dissatisfied (PPD), indicating the percentage of a population that would consider the environment unacceptable. PMV and PPD values were evaluated using the air speed model in the ASHRAE Thermal Comfort Calculator.

Dry-Humid Index

The case studies refer to the Dry-Humid Index, which is a comparison formula, based on a location's annual mean precipitation and annual mean temperature, to determine if the location's moisture climate zone should be characterized as Dry (B) or Humid (A)—this assumes exclusion of a marine classification when the moisture criteria are applied (see chap. 1).

This Dry-Humid Index is arrived at with the following formula:

$$\text{Dry-Humid Index} = 20 \times (\text{annual mean temp. in } ^\circ\text{C} + 7)$$

Boxplots

In addition to line plots and psychrometric charts, the case studies use boxplots (also called box-and-whisker plots) to show the monthly distribution of humidity, temperature, and dew point temperature data. Note that for datasets extending over a calendar year, data for multiple months (e.g., data from January 2004 and January 2005) are combined when determining monthly statistics for the plots. The figure below is an example of a boxplot.

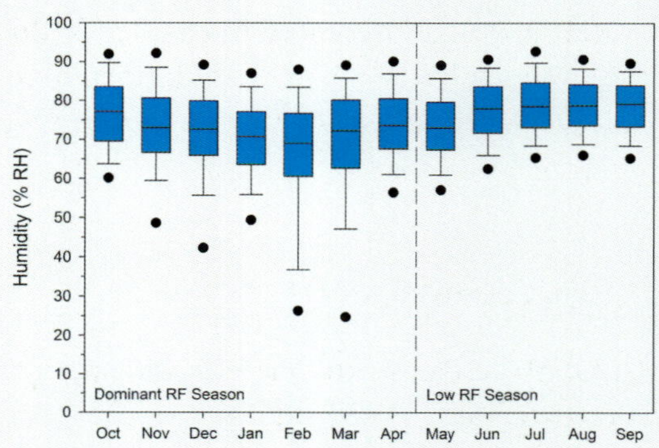

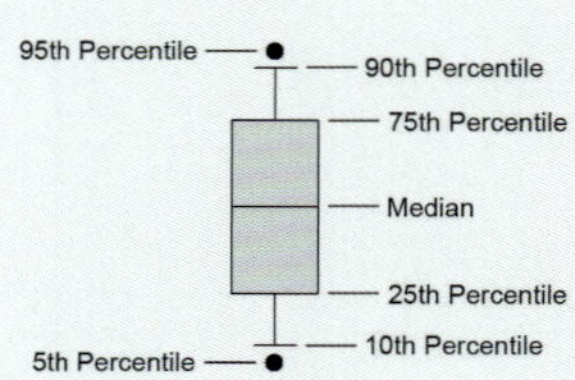

Each case study begins with a short project summary and introduction providing contextual information, such as an overview of the project or its background. This is followed by a discussion of:

- *Climate*, including geographical information, historical climate data from a nearby reference site, the area's Köppen-Geiger classification, and the ANSI/ASHRAE/IES climate zone designation (chap. 1).
- *Building characteristics*, including historical use, construction, architectural and envelope description, nonmechanical environmental management features, and the ASHRAE building classification (chap. 3).
- *Collections information*, including collection types, materials, and volume of collections displayed or stored.

Preinstallation Condition Assessments

A series of assessments characterized each project before the alternative environmental management strategies were applied:

- *Building envelope assessment.* Typically performed by a staff architect of the cultural institution or by a contracted preservation architect, this assessment addressed site influences on the interior building environment, the envelope performance for infiltration and source moisture control, and the influences of the building's spatial organization (chap. 6).
- *Exterior and interior environmental assessments.* These assessments describe the site climate—humidity, temperature, and dew point temperature—and the structure's preinstallation interior environment. Both pre- and post-installation environmental data are depicted in boxplots, which describe the distribution of the data.
- *Collections condition assessment.* Performed by the collections manager, this assessment describes the types of collections at the site, their materials, and indications of past environmental vulnerabilities of the collection (chap. 2).
- *Preinstallation risk class assessment.* Based on the Conservation Environment Classification–Hot and Humid (HH) protocol presented in chapter 3, this assessment classifies the collection's preinstallation environment (humidity and temperature) with respect to microbial, mechanical, and chemical risk.

Environmental Management System

The environmental objectives of the project, the approach taken, and details of the alternative environmental management strategies applied at the site are broken down into the following subjects.

Psychrometric strategies

This section gives an overview of the combination of nonmechanical and mechanical strategies that will be implemented to address the desired outcomes.

Nonmechanical strategies

These strategies cover a wide range of actions to protect collections—for example, source moisture control, particulate and externally generated pollutant control, collections housing, and occupant load management (chap. 6).

Mechanical strategies

This topic covers the following:

- *Environmental management strategy*, describing the target environmental conditions sought through installation of mechanical systems for conservation heating using ventilation, conservation heating using heaters, dehumidification, and/or cooling.
- *Mechanical system description*, describing the major installed equipment and its specifications, along with its locations and arrangement (chap. 7).
- *Operational control sequence*, describing the control sequence for enabling and disabling specific equipment for the various operational modes required to achieve the psychrometric goals.

Postinstallation System Performance

This section compares the interior environmental conditions—humidity, temperature, and dew point temperature—between the preinstallation and postinstallation periods via boxplots and psychrometric charts (chap. 5). In the section on postinstallation risk class assessment, the Conservation Environment Classification–HH protocol for hot and humid climates (chap. 3) is again used to describe the microbial, mechanical, and chemical risks of the postinstallation collections environment based on the system's humidity and temperature performance; pre- and postinstallation risk levels are compared.

Other postinstallation data might include:

- results of microbe or insect population monitoring, if conducted (see sidebar "Monitoring of Microbes, Insects, Particulates, and Gaseous Pollutants");
- measures of particulate deposition and gaseous pollutants, if conducted (see sidebar "Monitoring of Microbes, Insects, Particulates, and Gaseous Pollutants");
- thermal comfort values for building occupants (chap. 4);
- noise levels at various locations or distances from mechanical components (chap. 4).

Monitoring of Microbes, Insects, Particulates, and Gaseous Pollutants

Microbes

Microbial activity in both the environment and on documents was monitored monthly before and after installation of the environmental management system. To test environmental samples for bacteria and fungi contamination, 300 L of room air was directed onto the surface of several agar plates, which were then placed in a laboratory incubation chamber for 14 days. Counting the number of bacterial and fungal colonies on the agar plates indicated the level of microbe activity. To test surface samples for microbes, surfaces of various books and documents were wiped with cotton swabs moistened with 0.1% sodium chloride solution. The swabs were transported to a laboratory in sealed containers and rinsed with the same sodium chloride solution, which was then transferred to nutrient media and incubated for 14 days. Again, colonies were counted to determine microbial activity.

Insects

Insect activity was monitored using small sticky traps distributed throughout a space. Following a period of deployment, the traps were recovered, and insect species were identified and counted for each trap.

Particulates

Dust deposition rates were measured using the quartz crystal microbalance technique. In this process, a preweighed quartz plate is exposed to the environment for one to two months and then weighed, with the pre- and post-exposure weight difference indicating particulate deposition over the time period.

Airborne particulates were measured using an optical airborne particle counter. This unit assessed particle counts per cubic foot of air for two size fractions: 0.3–1 μm and 1–5 μm in diameter. The instrument manufacturer's proprietary algorithm allowed for conversion to micrograms per cubic meter of air for each size fraction. Sampling error is estimated at ±10% of calibration aerosol.

Gases

Gaseous pollution was measured using passive samplers. These units assessed time-weighted average (approximately 100 hr exposure time) pollutant concentrations of ozone, sulfur dioxide, nitrogen dioxide, and nitric oxide. The passive samplers utilize precoated pads specific to each pollutant. Postexposure analysis was conducted at the manufacturer of the samplers using flow injection analysis. Sampling and analytical measurement error was estimated to be a combined ±10% of the pollutant concentration (personal communication, Donald Schaeffer, Ogawa & Co.).

Maintenance

This section discusses the installed system's maintenance, as well as any issues that developed during the project.

Energy Metrics

Because costs associated with individual projects are subject to numerous factors unrelated to the mechanical system itself—such as local labor costs and taxes—project costs cannot be reliably compared among the case studies. As an alternative, the following comparable energy metrics for the installed equipment are provided: the *energy intensity* (installed total capacity per unit floor area in watts per square meter) and *energy use intensity* (annual energy use of the environmental management system in kilowatt-hours per square meter per year).

Closing Discussions

This section focuses on various topics, depending on the particulars of the individual case study, among them:

- *conclusions*, a general summary of the project's successes and challenges with respect to its conservation goals,
- *lessons learned*, describing the expected and unexpected findings from the project,
- *postscript*, reporting any developments following the project's completion.

Chapter 9

Hollybourne Cottage, Historic House on Jekyll Island, Georgia, United States: Part A—Basement Ventilation (Conservation Heating) and Heating, and Attic Ventilation (Dilution)

Climate zone 2A (Hot-Humid)

Objectives

- Reduce moisture and heat source in the cottage.
- Maintain less than 75% RH in the cottage to prevent microbial deterioration during the hot-humid season.
- Reduce heat accumulation in upper-floor spaces during the hot-humid season.
- Avoid placement of mechanical equipment in the first- and second-floor (interpreted) spaces.
- Install low-cost, robust, and low-energy system.

Summary

This case study considers the application of a conservation-focused environmental management system in the historic Hollybourne Cottage, located on Jekyll Island, Georgia, United States, to address high-humidity events in the basement and high temperatures in the attic during the hot-humid season in a Hot-Humid (2A) climate zone.

Following implementation of nonmechanical strategies, such as source moisture control and the reduction of envelope infiltration, mechanical strategies of ventilation and heating were applied in the basement to maintain humidity below 70% RH, while mechanical ventilation alone was used in the attic to lower the maximum daily temperature.

During operation of the environmental management system, monthly mean humidity in the basement was maintained between 60% RH and 70% RH, and maximum monthly mean attic temperatures were lowered by Δ3°C. Humidity conditions on the structure's first and second floors were also reduced, though these intermediate spaces were not actively managed by the mechanical environmental management system. Installation and operational costs of the environmental management system were significantly less than costs associated with conventional air-conditioning systems.

Background

Hollybourne Cottage, a two-and-one-half-story building located on the Atlantic coast of the southeast United States in a Hot-Humid (2A) climate zone, had suffered extensive biological damage to its interior architectural fabric as a result of moisture. The project objective at Hollybourne Cottage was to slow the decay of the remaining interior architectural woodwork and finishes by improving the interior environment through the implementation of nonmechanical and mechanical alternative environmental management strategies.

Hollybourne Cottage represented a unique experimental opportunity to evaluate the efficacy of multiple system configurations over an extended time. The first of six experimental systems tested is described in this chapter, designated part A; the following case study, part B (chap. 10), employed the fifth experimental strategy. The two systems reported in this book are:

- basement ventilation/heating and attic ventilation (part A),
- whole-house ventilation, basement dehumidification, and attic heating (part B).

Full descriptions of the overall project and of all the experimental systems can be found in Maekawa and Toledo 2002 and Maekawa, Beltran, and Toledo 2007.

Climate

Roughly equidistant between Savannah, Georgia, to the north (130 km) and Jacksonville, Florida, to the south (110 km), Jekyll Island (31°3'40" N, 81°25'19" W) is a barrier island situated along the Atlantic coast of the southeastern United States (fig. 9.1). The nearest mainland city is

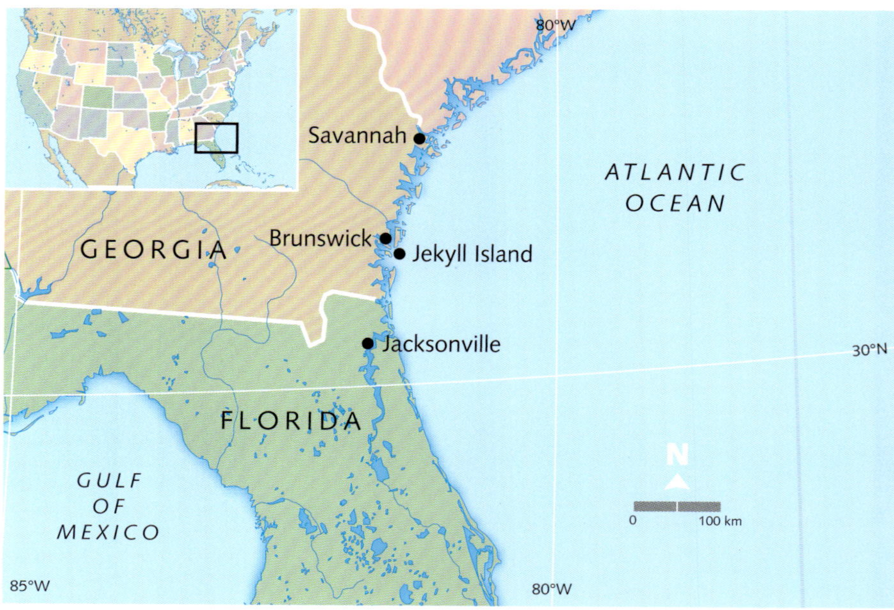

Figure 9.1
Map of the southeast Atlantic coast of the United States showing the location of Jekyll Island.

Brunswick, 12 km to the northwest. The island is 11 km in length (north–south direction) and has a width of 2.4 km.

Historical climate data collected from 1973 to 1996 at the Malcolm McKinnon Airport (31°9'0" N, 81°23'21" W) in Brunswick, Georgia, shows wide seasonal shifts over the year (AFCCC and NCDC 1999). The summer months are the warmest (maximum monthly mean of 28°C in July), while the cooler winter months (minimum monthly mean of 11°C in January) are relatively mild owing to the island's proximity to the Atlantic Ocean (fig. 9.2). Monthly average dew point temperatures range from 6°C in January to 23°C in July. Annual means of temperature, dew point temperature, and relative humidity (calculated from temperature and dew point temperature) were 19.9°C, 15.2°C, and 74% RH, respectively. While rainfall is recorded throughout the year, average monthly rainfall from June through September ranges between approximately 125 mm and 175 mm and constitutes a substantial percentage of the total annual precipitation (1247 mm); these high-rainfall months also have frequent summer thunderstorms as well as the highest average monthly temperatures.

Based on the Brunswick/Malcolm NCDC historical dataset, two seasons were identified for the locality:

1. The hot-humid season, occurring from June through September, was defined by average monthly rainfall exceeding 100 mm, and monthly average temperatures and dew point temperatures above 25°C and 20°C, respectively.
2. The cool-humid season, occurring from October through May, had relatively lower average monthly rainfall (50–100 mm), and monthly average temperature and dew point temperature ranged from between 10°C and 20°C and 5°C and 18°C, respectively.

During the first three years of this study (1999–2001), conditions were warmer and drier than the NCDC historical averages. In contrast, the climate during this study's final four years (2002–5) was generally cooler and wetter than NCDC historical values.

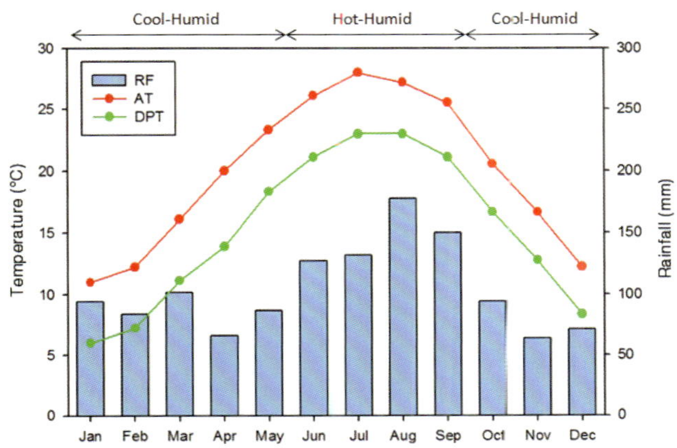

Figure 9.2

Mean monthly historical (1973–96) data for temperature (AT), dew point temperature (DPT), and rainfall (RF) recorded at the Brunswick/Malcolm station (WMO no. 722137; data from *Engineering Weather Data* [AFCCC and NCDC 1999]).

Five years (2006–10) of climate data from the weather station at the Brunswick Airport (Weather Channel 2014) produced a CDD10°C of 3954 cooling degree days. This thermal index places Jekyll Island in the Hot (2) thermal zone as defined in table 1.1 (see chap. 1). Excluded from a Marine moisture designation because its warmest month (28°C) exceeds 22°C, the local climate was classified as a Humid (A) type owing to its annual precipitation (1247 mm) being greater than the Dry-Humid index of 538. Because of this combination of thermal and moisture criteria, the climate of Brunswick and the nearby Jekyll Island is classified as Hot-Humid (2A). In the United States, climate zones have already been identified by location, and the above calculation confirms the climate classification found in ANSI/ASHRAE/IES Standard 90.1-2013 (ANSI, ASHRAE, and IES 2013).

Building

Constructed in 1890 for engineer and bridge builder Charles Stewart Maurice (1840–1924) and sold to the state of Georgia in 1947, Hollybourne Cottage is part of the Jekyll Island National Historic Landmark District (fig. 9.3). Jekyll Island came to prominence in 1886 when it became the winter retreat for America's financially elite families, including those of J. P. Morgan, Joseph Pulitzer, William Rockefeller, and William Vanderbilt. Located on the west side of Jekyll Island and facing the back bay and mainland, Hollybourne Cottage sits on sandy soil at approximately 5 m above sea level and 120 m from the shoreline of the back bay, with adjacent buildings roughly 100 m to the south (Crane Cottage) and north (Villa Ospo).

Hollybourne Cottage was constructed for seasonal use as a winter residence; it was not occupied in summer. The house was heated by ten fireplaces when necessary in winter, and it could be cooled by cross ventilation if needed. A central coal-fired furnace in the basement and a network of heating ducts were added at the turn of the twentieth century, but there were no functional mechanical systems at the start of the case study project in 1999. The building has remained vacant since 1947 and does not currently contain a collection, though it is open to visitors on a limited basis.

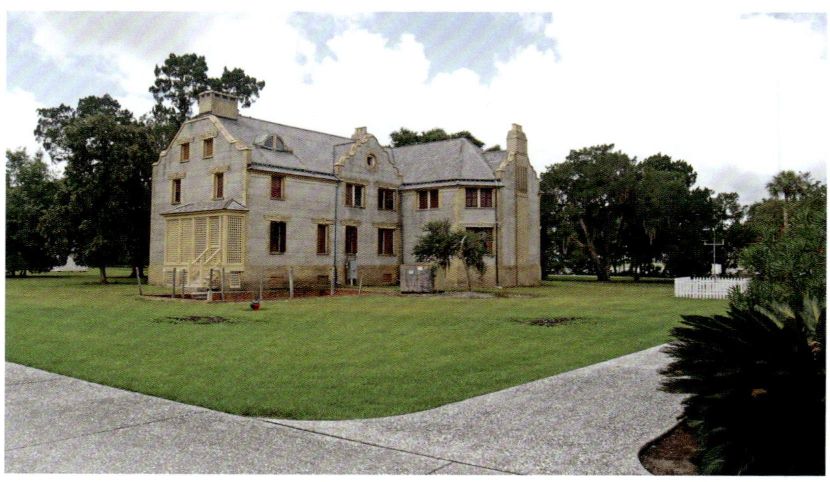

Figure 9.3
Hollybourne Cottage in Jekyll Island Historic District, Jekyll Island, Georgia, United States. Note the weather station to the right of the building. Photo: Shin Maekawa. © J. Paul Getty Trust.

Hollybourne Cottage is a concrete and masonry structure with a wood roof and floor framing. Its load-bearing exterior walls are cast-in-place concrete made of cement, sand, and seashell fragments to resemble traditional "tabby." The exterior surface is weathered concrete, retaining the pattern of wood formwork; on the interior face, the formwork was left in place and used as sheathing/furring for wood and plaster interior finishes. Reflecting the engineering influence of its owner, the loads of the second floor and attic are distributed via two simple trusses to the exterior concrete walls to allow for large open spaces on the ground floor of the north–south wing. Basement walls are composed of two parallel brick walls with a 13 cm space acting as a capillary break, and the basement floor is a concrete slab on soil. Hollybourne Cottage can be classified as an ASHRAE Class IV (heavy masonry) building (ASHRAE 2011).

T-shaped in plan (fig. 9.4), Hollybourne Cottage has an approximate total floor area of 1050 m^2 and a volume of 2550 m^3. The basement, first-, and second-story floor areas are approximately 280 m^2

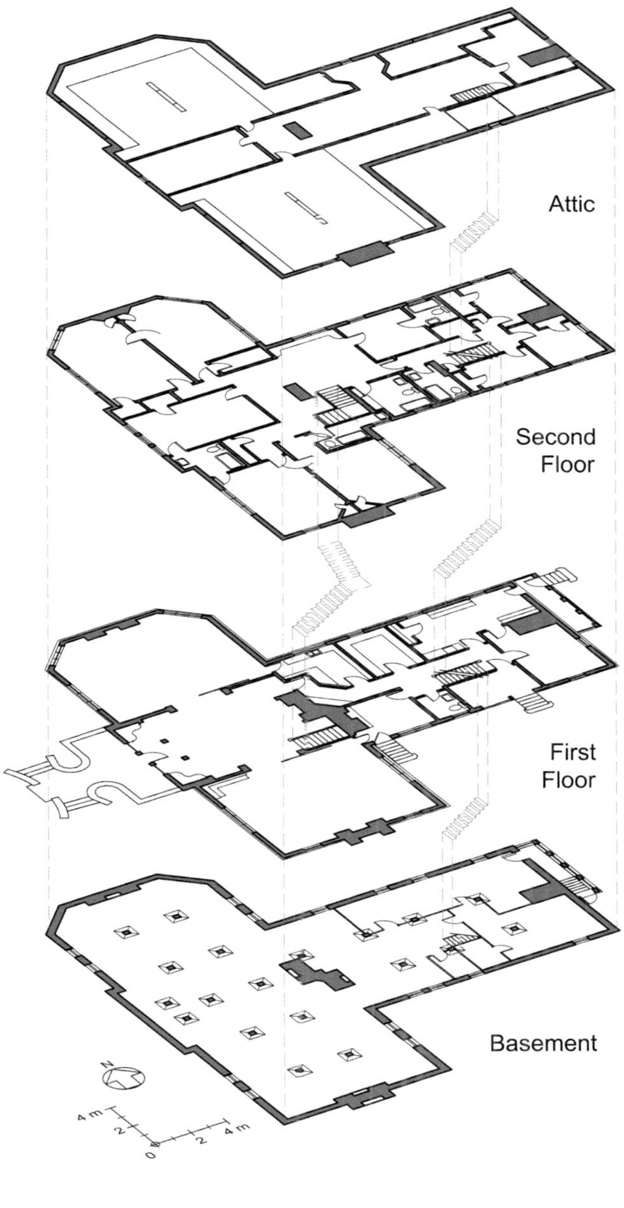

Figure 9.4

Axiometric view of Hollybourne Cottage, from the basement (bottom) to the attic (top), including connecting stairways.

each, enclosing approximately 650 m^3 each. The attic has approximately 215 m^2 of floor space with 400 m^3 of volume. An open main staircase (1.22 m wide) connects the first and second floors at the center of the house, while an enclosed secondary staircase (0.89 m wide) on the southeast side provides a vertical passage from the basement to the attic. A door is located at the base of the secondary stairway, at the entrance to the basement.

The design and architectural details of Hollybourne Cottage indicate an awareness of the local climate. Large window openings on the four facades feature exterior shutters with adjustable wood louvers for cross ventilation while they protect against rainfall and excessive solar heat gain. A covered porch at the end of its south wing, no longer extant, expanded the living space of the parlor and provided shade and fresh air. The high water table and porous soil of the location were addressed by the cavity construction of the foundation walls and by an underground foundation drainage system that discharged to the back bay. Building vacancy and lack of maintenance severely degraded performance of these nonmechanical environmental management features and resulted in extensive deterioration of interior and exterior wood elements.

Preinstallation Condition Assessments

In preparation for the development of the environmental management strategies at Hollybourne Cottage, the building condition and the environment were assessed. The building assessment, which included the site context surrounding the building, the building envelope, and the building interior, was performed by a consulting conservation architectural/engineering firm (Watson & Henry Associates 1998). The assessment of the interior and exterior environments was performed by the Getty Conservation Institute (GCI).

Building envelope assessment

The building assessment highlighted extensive biodeterioration of wood structural members and interior architectural woodwork as the major conservation issue at Hollybourne Cottage. Damage from microorganisms and termites was largely localized to the structural assembly—beams, joists, and the subfloor of the first floor, which are exposed to the basement environment; the interior architectural woodwork, plaster lath, and furring strips on the floors, walls, and ceiling of the first floor—with the worst damage located in the north end of the northwest wing (fig. 9.5). Wood framing members in the attic and second floor did not display obvious cracking or decay, though summer temperatures were elevated, particularly in the attic space.

A potential mechanism for wall damage may have been moisture absorption by the porous exterior concrete wall, with moisture migration into the left-in-place interior wood formwork from the initial construction. Damage may have been further exacerbated by numerous

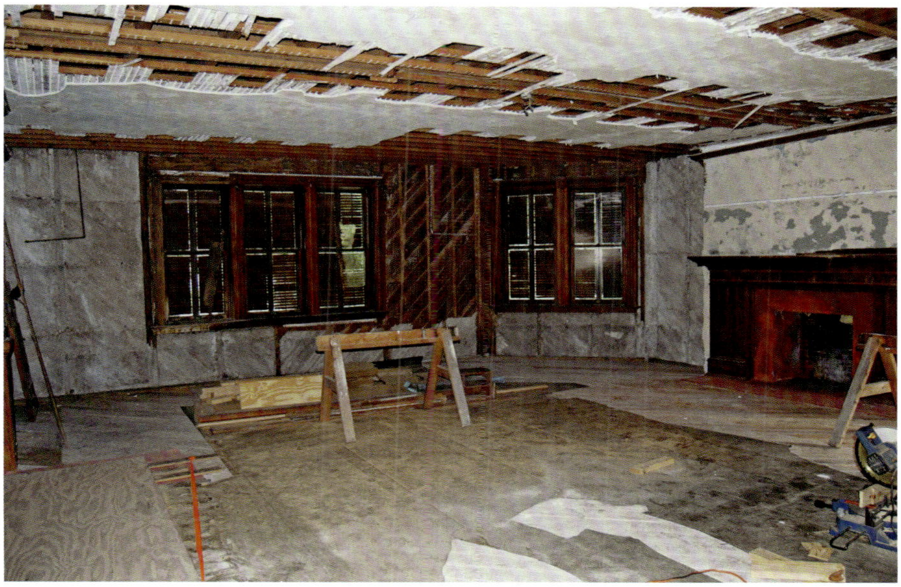

Figure 9.5

First-floor living room of Hollybourne Cottage, with rotted floor boards and lath-and-plaster walls. Photo: Shin Maekawa. © J. Paul Getty Trust.

factors, including: overflowing gutters, lack of air circulation, shading by adjacent overgrown trees, a high water table in conjunction with a partially subterranean basement, poor surface drainage away from the structure, and frequent and heavy seasonal rains. Vacancy of the building, starting in 1947, and the cessation of winter heating required for occupancy would have exacerbated moisture issues and reduced opportunities for the building materials to dry.

Exterior environmental assessment

The plots in figure 9.6 show 24 hour trends of average exterior temperature, dew point temperature, and humidity during the months of January (cool-humid season) and August (hot-humid season). While the daily variation (difference between maximum and minimum) of each variable is similar for both months, the range of each variable was shifted to higher

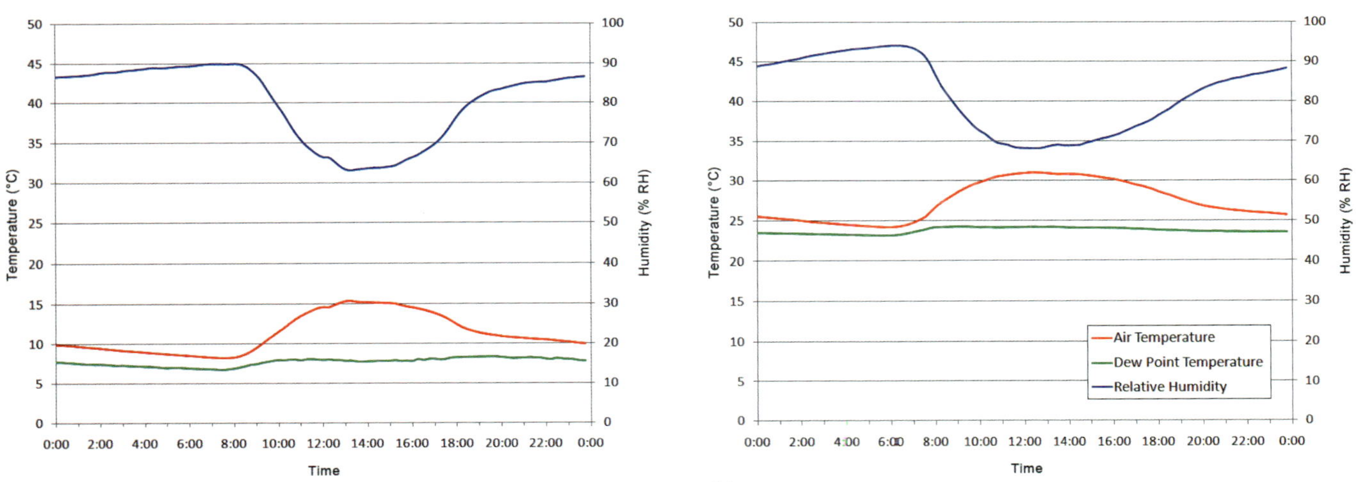

Figure 9.6

Mean daily trends of exterior temperature, dew point temperature, and humidity during January (a) and August (b) for the study period of January 1999 to October 2005.

values in August (hot-humid season) than in January (cool-humid season), particularly for temperature and dew point temperature. In January, the humidity of the outside air was below 70% RH (as low as 62% RH) for roughly 6 hours (11:00–17:30) on an average day, making ventilation a feasible environmental management strategy. In contrast, the minimum average daily humidity in August was only slightly below 70% RH, limiting the use of ventilation to reduce interior humidity.

Interior environmental assessment

Nonmechanical strategies for source moisture control in the building and the site context and for reduction of uncontrolled infiltration of moist air through the building envelope were implemented by January 1999 (see discussion under "Environmental Management System" below). The interior temperature and humidity of Hollybourne Cottage were then monitored from January 1999 through June 2000, prior to installation of mechanical equipment. This monitoring provided a preinstallation baseline against which to assess the effectiveness of the mechanical strategies. Dew point temperature was calculated from paired measurements of temperature and humidity.

In the following sections, humidity, temperature, and dew point temperature data are presented as monthly boxplots to highlight seasonal shifts, with multiyear datasets combining data for the same month.

Humidity

Humid conditions predominated in the lower floor levels of Hollybourne Cottage during the preinstallation period. Humidity levels in the basement and in first-floor spaces exceeded the 75% RH threshold for microbial activities during 78% and 44% of the preinstallation period, respectively (fig. 9.7) Monthly mean basement humidity ranged from 72% RH in January to 96% RH in July, while that of the first floor ranged from 68% RH in December to 81% RH in July.

As expected, the high humidity in the basement and on the first floor of Hollybourne Cottage coincided with areas of major wood decay and damage identified by the building assessment. Ceiling collapse in the north wing of the first floor and corrosion of the small fasteners for the wood lath may be attributed to the effects of high humidity on the lath-and-plaster ceiling. Lower-humidity environments in the second floor and attic coincided with minimal interior damage. Cracked and/or split wood elements, typical of damage due to low moisture content or to a wide range of moisture content, were not observed in the attic. This finding may be attributed to the moderate humidity fluctuations observed in the attic, even during summer, when the attic temperature was very high.

A short-term experiment was conducted on the floor of the basement to examine moisture infiltration. A portion of the concrete basement floor slab, approximately 1.0 × 1.2 m, was covered with vapor barrier film and thermal insulation, and the humidity in the intermediate space was monitored for six months. The humidity in the space between the floor and the covering was approximately 100%, indicating net vapor

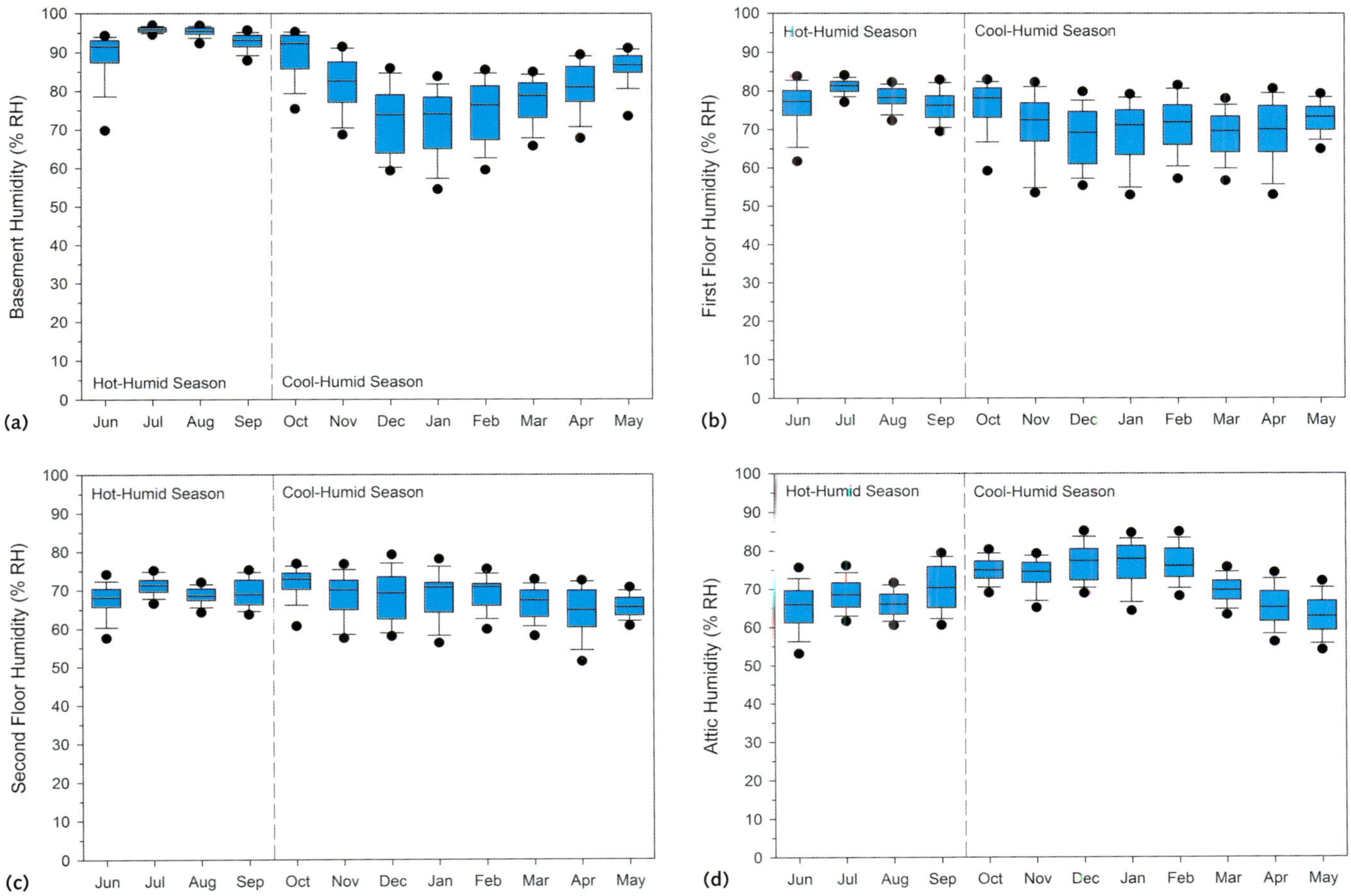

Figure 9.7

Preinstallation humidity conditions from January 1999 to June 2000.

moisture migration from the slab floor into the basement space. However, indications of condensation were not observed in the basement.

Temperature

High temperature in the attic space, particularly during the summer months, was identified as an environmental management and building conservation concern. Preinstallation attic temperatures exceeding 35°C were recorded from May to September, and two months (July and August) exhibited maximum values above 40°C (fig. 9.8). These high temperatures were attributed to heat transfer through the roof and the accumulation of unvented buoyant hot air from lower floors. While the minimum monthly mean temperatures were similar for all floors (~14°C in January), the maximum monthly mean differed considerably between the basement (26.3°C in August) and the attic (33.0°C in August).

The second-floor and attic environments only exceeded the 75% RH threshold during 6% and 22% of the preinstallation period, respectively. The period in which the threshold was exceeded largely coincided with the cool-humid season (see fig. 9.7). Monthly mean humidity for the second floor ranged from 64% RH in May to 72% RH in October, while levels in the attic ranged from 63% RH in May to 77% RH in December.

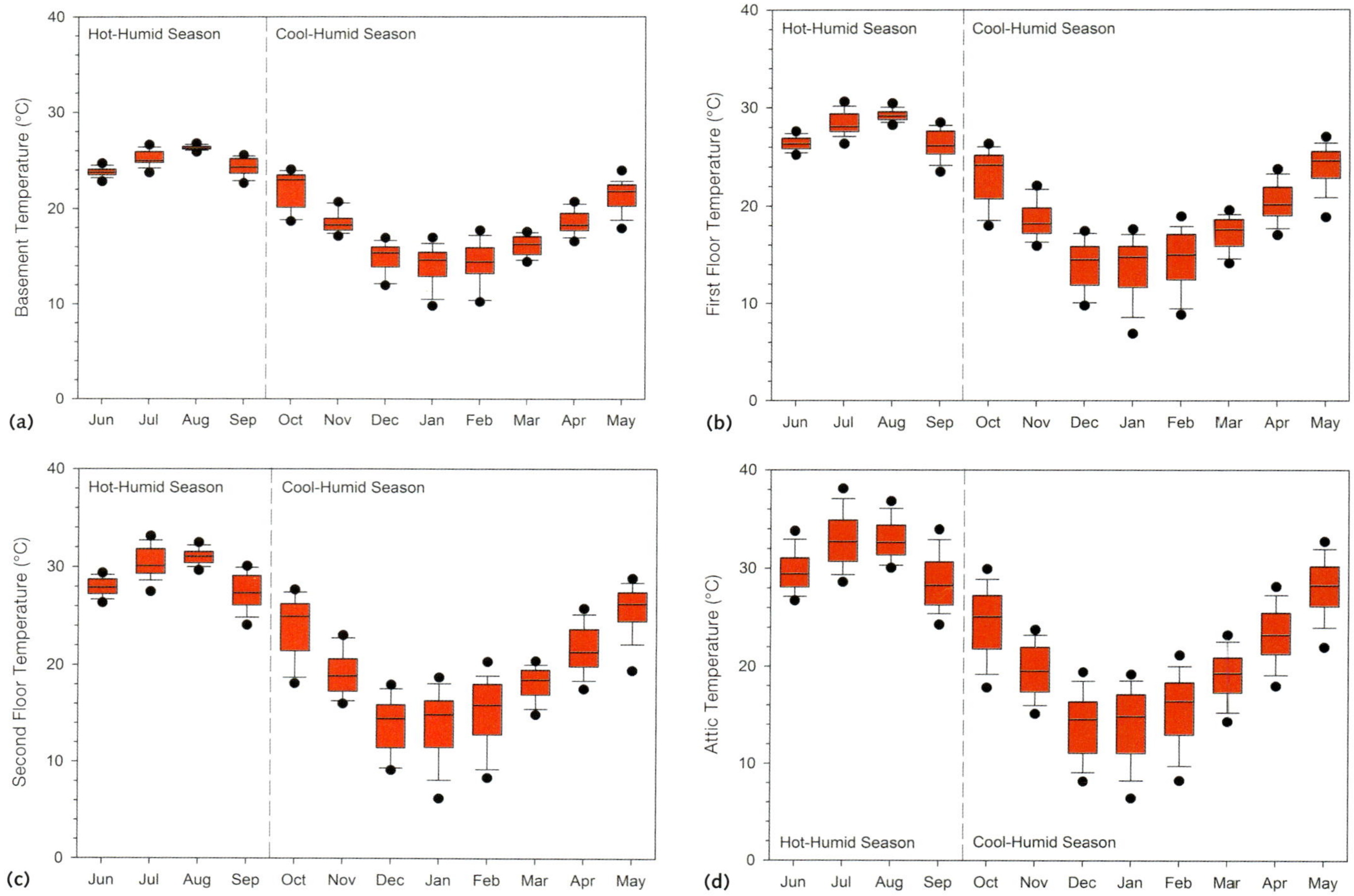

Figure 9.8

Preinstallation temperatures from January 1999 to June 2000.

Figures 9.9a and b

Preinstallation dew point temperatures from January 1999 to June 2000.

Dew point temperature

Dew point temperatures were similar on all floors of Hollybourne Cottage, and interior monthly mean dew point temperature was typically at its lowest (~10°C) during the cool-humid season and peaked (~25°C) during the hot-humid season (fig. 9.9). In addition, the highest monthly variability in dew point temperature was observed during the cool-humid season, with a much narrower range of values seen during the wet-humid season.

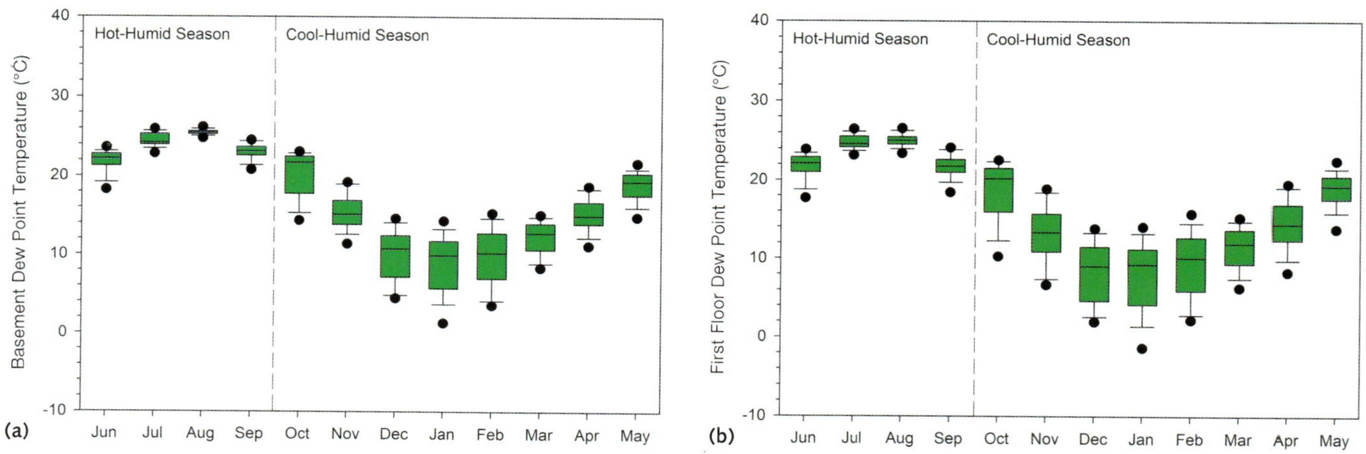

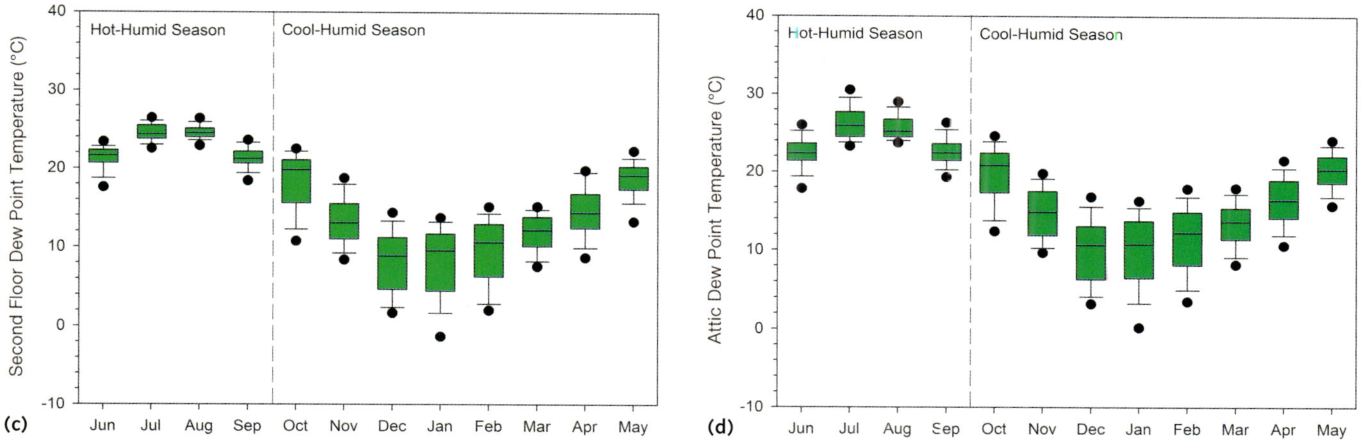

Figures 9.9c and d

Preinstallation dew point temperatures from January 1999 to June 2000.

Preinstallation risk class assessment

The Conservation Environment Classification–HH protocol for the preinstallation interior environment of Hollybourne Cottage is summarized in table 9.1.

Table 9.1

Conservation Environment Classification–HH protocol for the preinstallation environment of Hollybourne Cottage.

Risk			Humidity		Temperature	
Category	Specific Risk (statistic)	Floor	Value (% RH)	Class	Value (°C)	Class
Microbial	Germination threshold (P97.5)	Attic	84	F	—	—
		2nd floor	77	F	—	—
		1st floor	83	F	—	—
		Basement	97	F	—	—
Mechanical	Short-term variation (P95 of rolling 24 hour variation)	Attic	Δ13	b	Δ6.6	b
		2nd floor	Δ13	b	Δ2.6	a
		1st floor	Δ17	b	Δ3.5	a
		Basement	Δ13	b	Δ1.2	a
	Seasonal variation (absolute difference of seasonal means, RH means ≦ 70%)	Attic	Δ4	a	Δ9.8	a
		2nd floor	Δ2	a	Δ10.1	b
		1st floor	Δ7	a	Δ10.6	b
		Basement	Δ14	b	Δ9.9	a
	Deviation from historical mean (absolute difference)	Attic	Δ4	a	Δ3.6	a
		2nd floor	Δ7	a	Δ2 4	a
		1st floor	Δ2	a	Δ1.4	a
		Basement	Δ9	a	Δ0.1	a
Chemical	Deviation from historical mean (difference)	Attic	—	—	Δ3.6	+
		2nd floor	—	—	Δ2.4	0
		1st floor	—	—	Δ1.4	0
		Basement	—	—	Δ0.1	0

Overall mechanical risk for each floor:[1]
Attic: b
2nd floor: b
1st floor: b
Basement: b

[1] Overall mechanical risk for each floor is based on the highest risk classification for each floor in any of the specific humidity- and temperature-based mechanical risks (short-term variation, seasonal variation, and deviation from historical mean).

Environmental Management System

Objectives

The following are the project objectives based on the assessments summarized above:

- Reduce moisture and heat source in the cottage.
- Maintain less than 75% RH throughout the cottage to prevent microbial deterioration during the hot-humid season.
- Reduce heat accumulation in upper floor spaces during the hot-humid season.
- Avoid the placement of mechanical equipment in the first- and second-floor interpreted spaces.
- Install low-cost, robust, and low-energy system.

Psychrometric strategies

Since the site climate of Hollybourne Cottage is typically humid, it was essential to reinstate nonmechanical source moisture control elements to the building envelope to reduce the infiltration of humid outside air. Following implementation of nonmechanical psychrometric strategies, the mechanical psychrometric strategies to be utilized at Hollybourne Cottage were determined. These included heating/ventilation in the basement (use of conservation heating or the transfer of warm-dry outside air into the space) and ventilation in the attic (dilution ventilation) (fig. 9.10). These mechanical strategies focused on environmental management in areas with extreme interior conditions (high humidity in the basement and peak summer temperature in the attic). Because the basement space was isolated from the upper floors by the installation of a door at the base

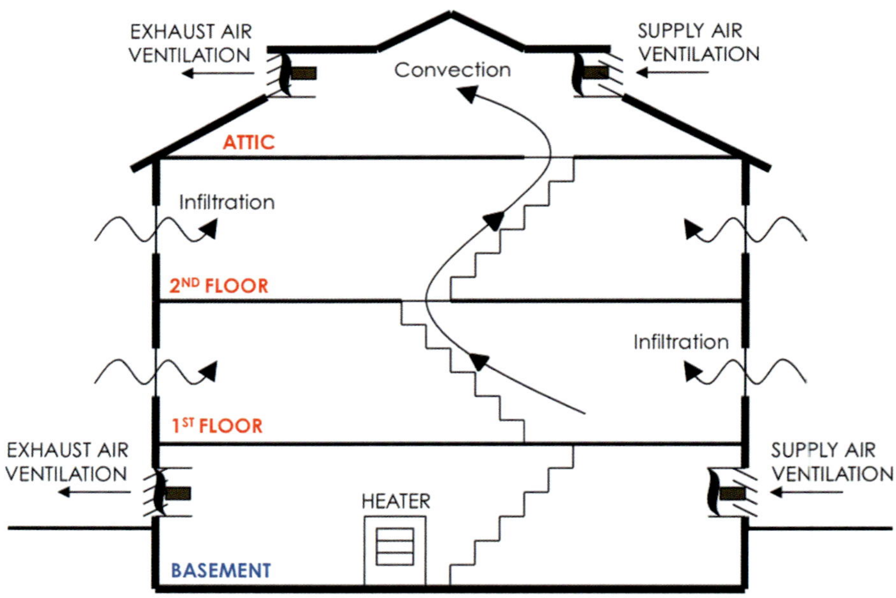

Figure 9.10

Schematic showing basement ventilation and heating, and attic ventilation.

of the secondary stairway, the first- and second-floor environments were not directly affected by the implementation of mechanical psychrometric strategies.

Nonmechanical strategies

The following nonmechanical strategies were implemented at the site and to the building envelope of Hollybourne Cottage:

- reestablishment of the original soil contours immediately adjacent to the building foundation to naturally drain surface water away from the building;
- clearing of debris from underground drain pipes;
- trimming of overgrown trees to allow more air movement around the building;
- removing tree litter from the roof;
- cleaning of roof gutters and downspouts;
- capping of chimneys to reduce uncontrolled infiltration and exfiltration due to the stack effect, and to prevent animal entry and habitation;
- adjusting/repairing windows and doors to close properly;
- temporary closure of missing or broken windows with thick polyethylene sheets;
- temporary closure of holes on the first-floor floor/basement ceiling with plywood;
- temporary replacement of the original entry door with a plywood door with penetrations for power and control cables from the mechanical systems.

Because of the historic significance of the building and the high water absorption and moisture vapor adsorption of the concrete (which contribute to interior humidity), moisture control could not be addressed by applying a sealer to the exterior surface.

Mechanical strategies

Mechanical system description

The basement ventilation and heating system consisted of two sets of propeller-type supply and exhaust fans and three electric resistance convection heaters (fig. 9.11). The two supply fans were placed in existing window openings on the north and south foundation walls of the east–west wing, and the two exhaust fans were placed in the window openings of the west foundation walls of the north–south wing. This placement produces a sweeping airflow at approximately 5 air changes per hour (ACH) from the east wing to the west wing basement. The flow of filtered supply air exceeded the exhaust flow by 0.2 to 0.3 m^3/s to offset infiltration of unfiltered air. Three convective heaters were hung from floor joists of the first floor at the three ends (north, south, and east wings) of the space to uniformly distribute heated air in the basement (fig. 9.12).

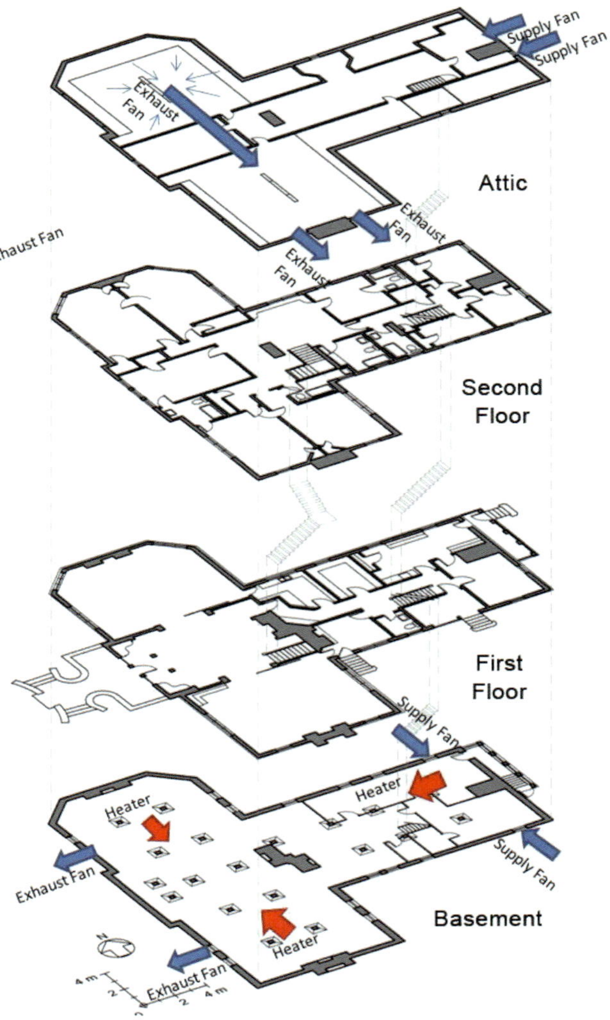

Figure 9.11
Locations of mechanical components in the basement and attic.

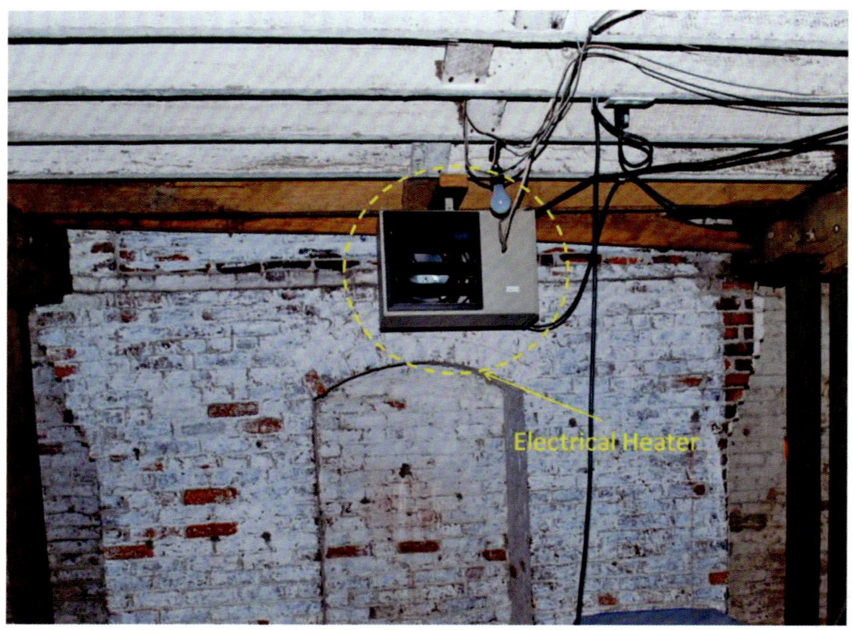

Figure 9.12
Electrical heater installed in the basement. Photo: Shin Maekawa. © J. Paul Getty Trust.

Table 9.2

Performance specifications for mechanical components of Hollybourne Cottage environmental management system (NA = not available).

Unit	Power (kW)	Flow Rate (m^3/s)	Static Pressure (kPa)
Supply fan	0.25	1.130	0.062
Exhaust fan	0.12	0.847	0.062
Extraction fan	0.02	0.052	NA
Heater	7.5	0.271	NA

The attic ventilation system consisted of similar paired sets of supply and exhaust fans sized to provide approximately 10 ACH (see fig. 9.11). Supply fans were placed in window openings in the east gable wall, and exhaust fans were placed in window openings in the south gable wall. Since the north wing of the attic did not have an outside opening, a ducted extraction fan was placed in the middle of this room approximately 2 m above the floor to move air toward the exhaust fans positioned in the south wing. The attic supply flow exceeded exhaust flow by 0.5 to 0.6 m^3/s to eliminate the possibility of uncontrolled infiltration.

Motorized shutters and antimicrobial MERV 7 filters were used in conjunction with supply fans, and exhaust fans were equipped with gravity-operated shutters. Heaters had the capacity to increase interior temperatures by approximately Δ7°C (subject to a programmable maximum temperature threshold), potentially reducing humidity by 30% to 35% RH. Table 9.2 lists specifications for each type of equipment.

Steps were taken to make the installation of mechanical equipment at Hollybourne Cottage simple and unobtrusive, with special attention given to minimizing the visual impact of exterior installations of ventilators and louvers. After a window opening was selected, the window sash was removed and an oversize plywood panel with an opening for the fan was mounted over the window, leaving the original windowsills, frames, and moldings in place. The fan was mounted with screws to the plywood panel and supported by a weight-bearing stand (fig. 9.13). Wood louvered shutters and wooden window frames with insect screens were mounted on the window openings to camouflage the fan from the outside,

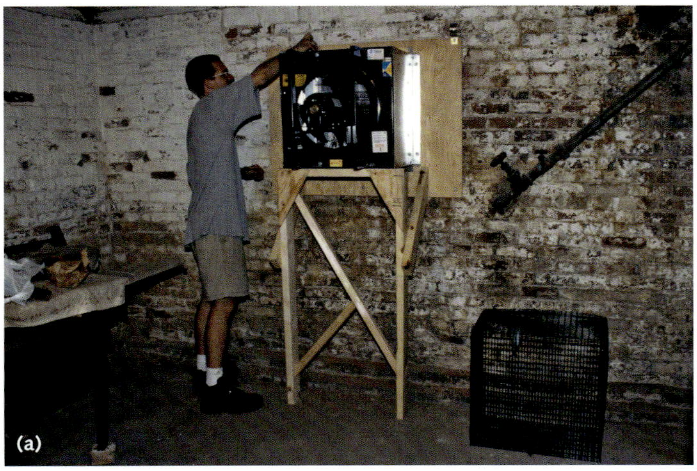

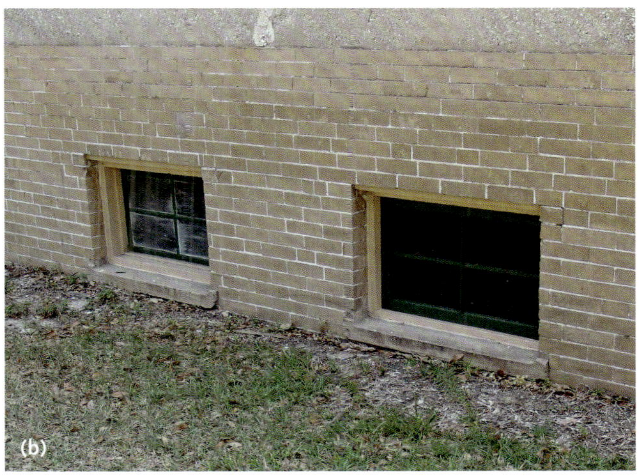

Figure 9.13

Installation of a fan in a basement window opening (a), and an exterior view of basement windows, with an original window on the left and a custom-made wood louver and window frame on the right (b), concealing the fan installation. Photos: Shin Maekawa. © J. Paul Getty Trust.

although these shutters were not original to the specific window openings. The new louvered shutters were fabricated to reproduce the appearance of the original shutters on the first- and second-floor windows.

Operational control sequence

Automated control of the attic and basement mechanical systems was executed by a programmable controller configured for independent operating control schemes. The control schemes were based on comparisons of the interior and exterior temperature and humidity conditions.

The basement controller had three modes of operation: ventilation, heating, and idling. The ventilation mode was enabled when interior humidity was greater than or equal to 65% RH and exterior humidity was less than 65% RH (table 9.3). Ventilation was disabled when either interior humidity was less than 60% RH or exterior humidity exceeded 65%. Basement heating was enabled when both interior humidity and exterior humidity were greater than or equal to 70% RH and ceased when one of the following three conditions occurred: interior humidity dropped below 65% RH, exterior humidity was less than 70% RH, or basement temperature was greater than or equal to 30°C. The temperature limit of 30°C was based on the 99.6 percentile of the recorded temperature for Brunswick, Georgia (Harriman, Brundrett, and Kittler 2001). In general, humidity of 65% RH can be achieved if the building interior is maintained at approximately 15°C or higher during the cool-humid season, and 32°C or higher during the hot-humid season. It should be noted that, during this period, the operational bands for ventilation (65%–60% RH) and heating (70%–65% RH) did not overlap. The idling mode was enabled when the ventilation and heating modes were both disabled.

The attic environmental management system utilized only the dilution ventilation mode. Attic ventilation was enabled when the roof (ceiling) temperature exceeded outside temperature and attic temperature was greater than or equal to 30°C (see table 9.3). Ventilation was disabled when the attic temperature dropped below 30°C.

Table 9.3

Operational control conditions for the environmental management system.

	Psychrometric Strategy	Enable	Disable
Attic	Ventilation	$T_{interior} \geqq 30°C$ and $T_{Roof} > T_{outside}$	$T_{interior} < 30°C$
	Idle	$T_{interior} < 30°C$	$T_{interior} \geqq 30°C$ and $T_{Roof} > T_{outside}$
Basement	Ventilation	$H_{interior} \geqq 65\%$ RH and $H_{exterior} < 65\%$ RH	$H_{interior} < 60\%$ RH or $H_{exterior} > 65\%$ RH
	Heating	$H_{interior} \geqq 70\%$ RH and $H_{exterior} \geqq 70\%$ RH	$H_{interior} < 65\%$ RH or $H_{exterior} < 70\%$ RH or $T_{interior} \geqq 30°C$
	Idle	$H_{interior} < 60\%$ RH or $T_{interior} \geqq 30°C$	$H_{interior} \geqq 70\%$ RH

Postinstallation System Performance

The performance of the environmental management system was assessed by evaluating interior environmental data for system operation from June 2000 to January 2001.

Humidity

Figure 9.14 shows system performance with respect to humidity. Installation of the environmental management system shifted the environment in the basement to drier and warmer conditions throughout the year. With the use of ventilation and heating in the basement, the monthly mean humidity ranged from 59% RH in December to 69% RH in June. The maximum humidity recorded in the basement was 77% RH (June). Occurrences of humidity above 75% RH were reduced to only 0.3% of the postinstallation period (June 2000 through January 2001).

With only dilution ventilation used in the attic, the postinstallation humidity in the attic remained similar to conditions observed during the preinstallation period. The monthly mean humidity ranged from 63% RH (May) to 77% RH (December) during the preinstallation period, and from 66% to 72% RH during the postinstallation period. Similar to preinstallation, the 75% RH threshold was exceeded in the attic during

Figure 9.14

Humidity comparison between preinstallation (January 1999–June 2000) and postinstallation (June 2000–January 2001).

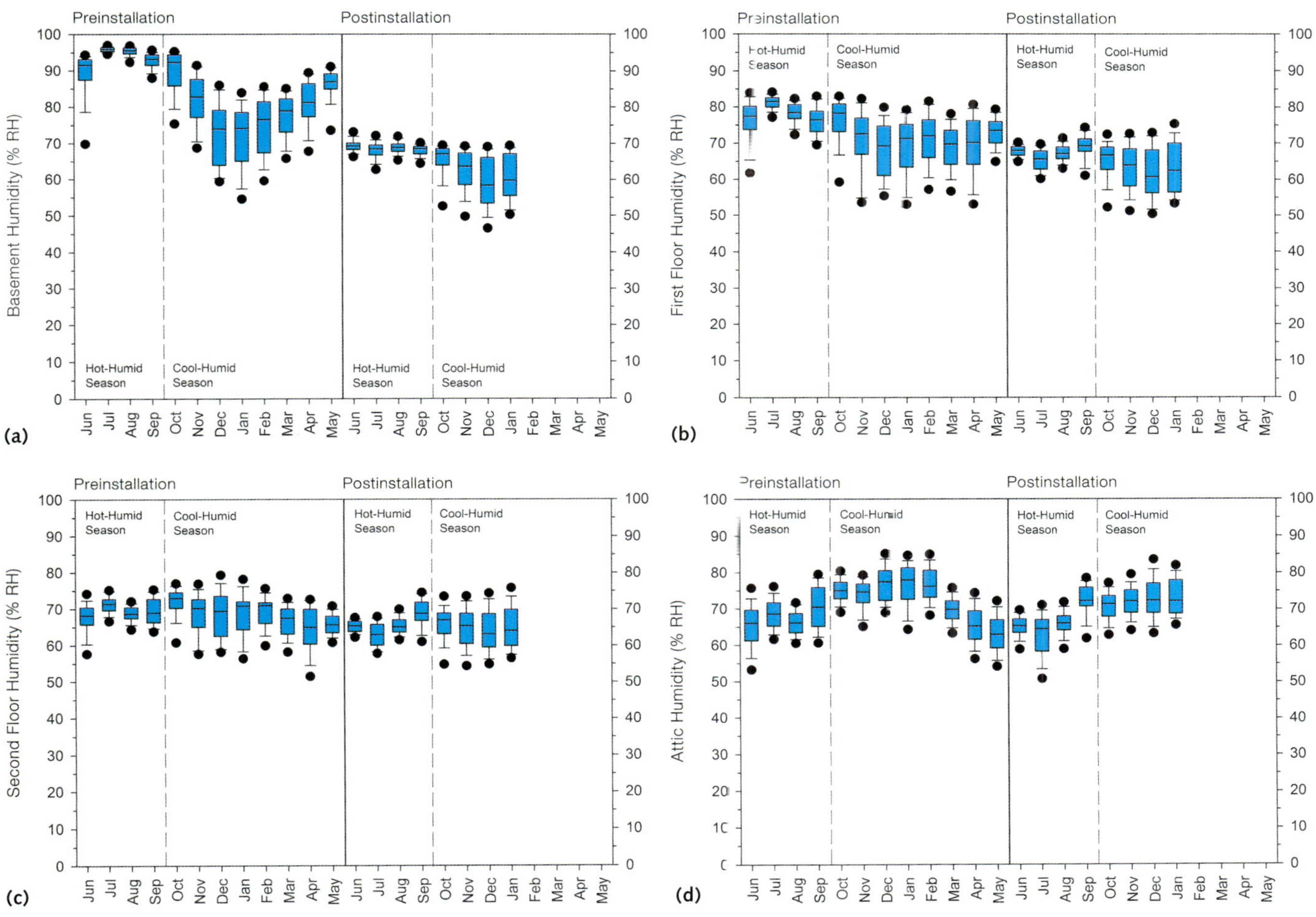

approximately 22% of the postinstallation period, with the majority of high-humidity conditions occurring in evenings of the cool-humid season.

Humidity levels on the first and second floors were both lower than preinstallation values, despite the lack of direct management of the first- and second-floor spaces. Following system installation, monthly mean humidity on the first and second floors ranged from 62% RH (December) to 69% RH (September) and from 62% RH (July) to 69% RH (September), respectively, and the 75% RH threshold was exceeded less than approximately 2% of the period on the first and second floors.

Since the system was actively managing interior humidity during the hot-humid season due to high exterior humidity conditions, narrower humidity ranges were observed for all floors. In contrast, the lack of a low humidity threshold limited system operation during the cool-humid season, resulting in a wider range of humidity conditions for all floors. In general, postinstallation humidity fluctuations were narrower than those observed during the preinstallation period.

Temperature

Figure 9.15 shows system performance with respect to temperature. In the basement, the use of conservation heating and ventilation yielded monthly average temperatures ranging from 11.6°C (January) to 29.5°C (August), with a maximum value of 30.6°C recorded in July.

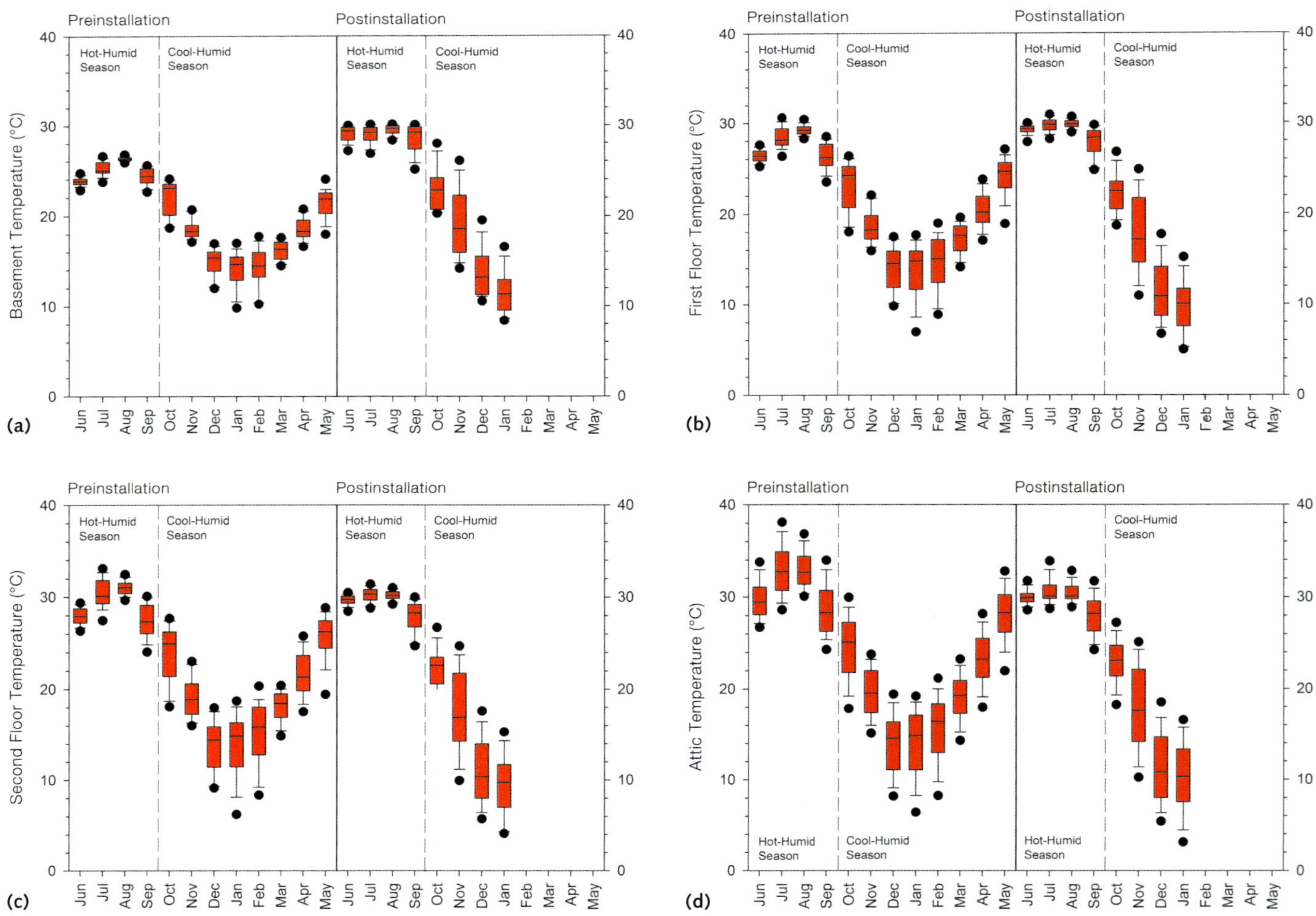

Figure 9.15

Temperature comparison between preinstallation (January 1999–June 2000) and postinstallation (June 2000–January 2001).

In the attic, the ventilation system reduced maximum temperatures compared to preinstallation values. While the maximum preinstallation attic temperature was 37.1°C (July), only one instance above 35°C (August) was recorded following system installation. Monthly mean attic temperature during the postinstallation period ranged from 10.2°C in January to 30.6°C in July. Postinstallation monthly mean temperatures (10°C in January to 30°C in August) in the first and second floors remained similar to preinstallation levels during this period.

Dew point temperature

Figure 9.16 shows system performance with respect to dew point temperature. As was the case during the preinstallation period, dew point temperatures were similar throughout the house during the postinstallation period. Postinstallation monthly mean dew point temperatures ranged from about 4°C in the winter to about 23°C in the summer. Average dew point temperatures did not peak in July and August, as seen in the preinstallation period, remaining relatively flat (22°C–23°C) during the hot-humid season of the postinstallation period. When the system was operating in ventilation mode during the hot-humid season, dew point temperatures in the attic were slightly less than preinstallation conditions.

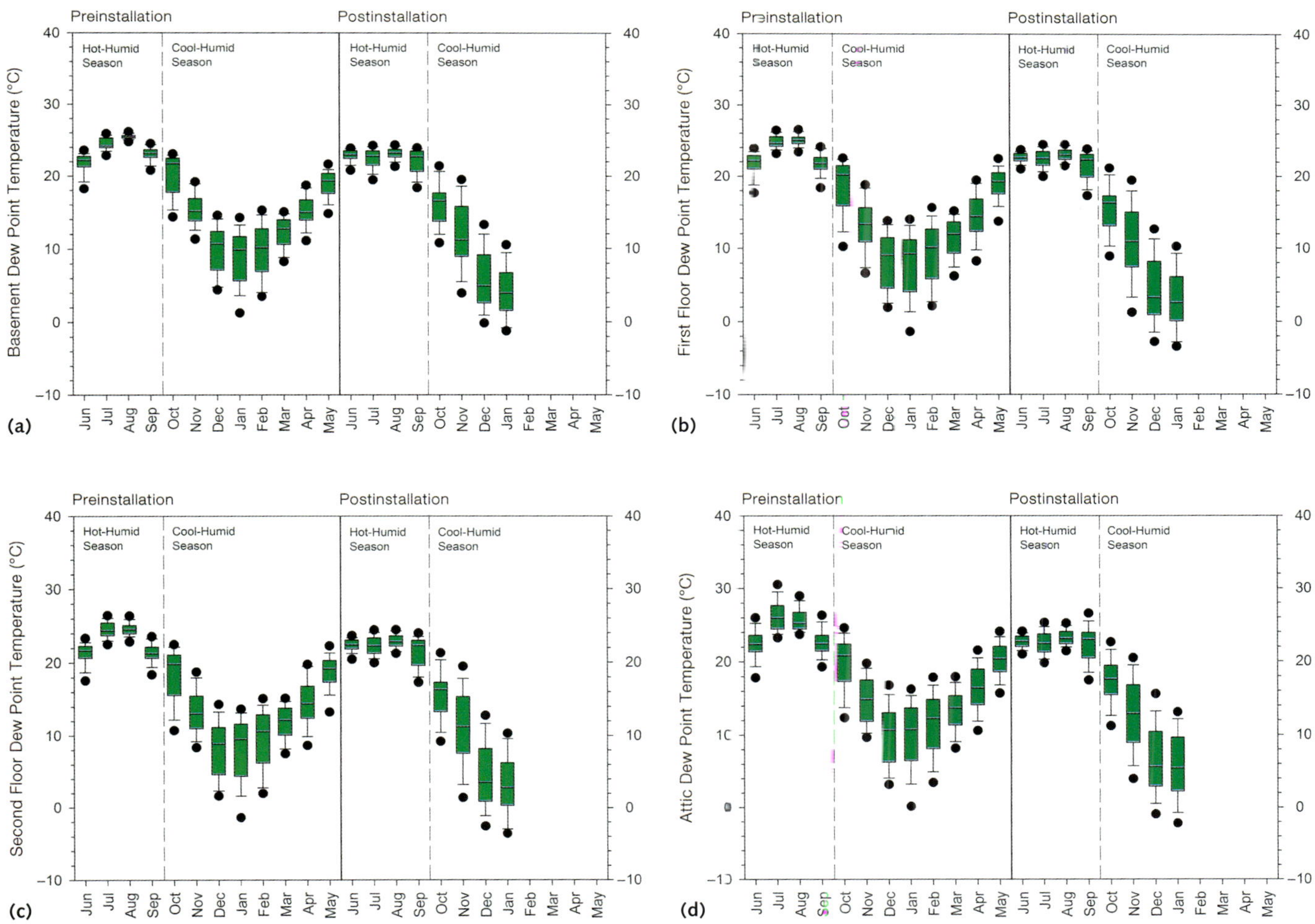

Figure 9.16

Dew point temperature comparison between preinstallation (January 1999–June 2000) and postinstallation (June 2000–January 2001).

Table 9.4

Conservation Environment Classification–HH protocol for the postinstallation environment of Hollybourne Cottage.

Risk			Humidity		Temperature		
Category	**Specific Risk (statistic)**	**Floor**	**Value (% RH)**	**Class**	**Value (°C)**	**Class**	
Microbial	Germination threshold (P97.5)	Attic	81	F	—	—	
		2nd floor	75	C	—	—	
		1st floor	74	C	—	—	
		Basement	72	C	—	—	
Mechanical	Short-term variation (P95 of rolling 24 hour variation)	Attic	Δ18	b	Δ5.5	b	Overall mechanical risk for each floor:[1] Attic: b 2nd floor: b 1st floor: b Basement: b
		2nd floor	Δ12	b	Δ2.6	a	
		1st floor	Δ12	b	Δ3.1	a	
		Basement	Δ11	b	Δ2.8	a	
	Seasonal variation (absolute difference of seasonal means, RH means ≦ 70%)	Attic	Δ5	a	Δ14.2	b	
		2nd floor	Δ1	a	Δ14.2	b	
		1st floor	Δ4	a	Δ13.7	b	
		Basement	Δ7	a	Δ12	b	
	Deviation from historical mean (absolute difference)	Attic	Δ6	a	Δ2.8	a	
		2nd floor	Δ10	a	Δ2.4	a	
		1st floor	Δ10	a	Δ2.4	a	
		Basement	Δ10	a	Δ3.1	a	
Chemical	Deviation from historical mean (difference)	Attic	—	—	Δ2.8	0	
		2nd floor	—	—	Δ2.4	0	
		1st floor	—	—	Δ2.4	0	
		Basement	—	—	Δ3.1	+	

[1] Overall mechanical risk for each floor is based on the highest risk classification for each floor in any of the specific humidity- and temperature-based mechanical risks (short-term variation, seasonal variation, and deviation from historical mean).

Postinstallation risk class assessment

The Conservation Environment Classification–HH protocol for the postinstallation interior environment of Hollybourne Cottage is summarized in table 9.4, and differences from the preinstallation environment classification are discussed in this section. Figure 9.17 also shows on psychrometric charts the preinstallation and postinstallation first-floor datasets, with values pertaining to the conservation environment classification denoted; the boxes indicate the mid-95% of each dataset.

The preinstallation and postinstallation risk classes for Hollybourne Cottage are summarized in table 9.5.

Microbial risk

Compared to the preinstallation Class F (high risk) value for microbial risk for each level, the mechanical strategies implemented by the environmental management system reduced the microbial risk to Class C (moderate risk) for the basement, first floor, and second floor.

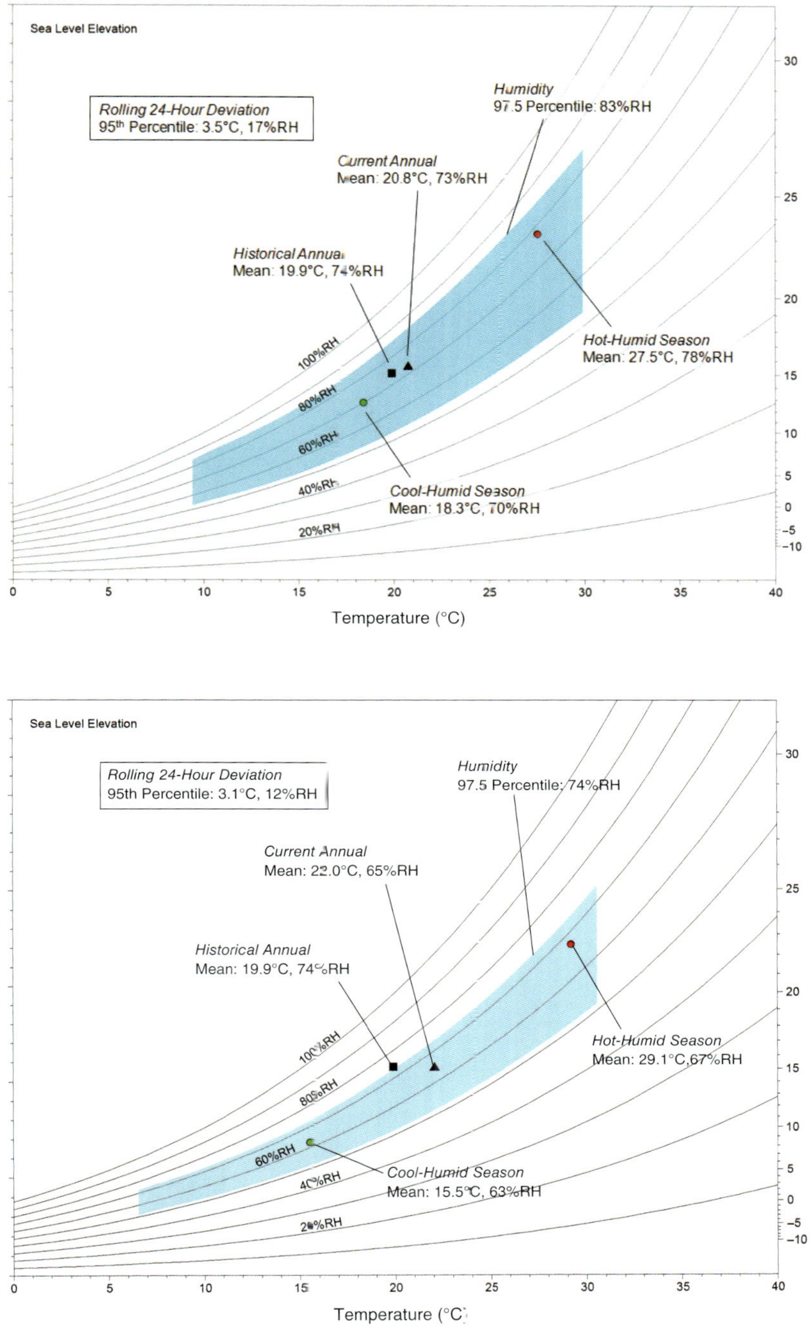

Figure 9.17
Psychrometric charts showing environmental data collected on the first floor of Hollybourne Cottage during preinstallation (a) and postinstallation (b). The boxes are bracketed by the 97.5 and 2.5 percentiles of temperature and humidity.

Mechanical risk

Mechanical risk due to seasonal variations of humidity improved in the basement, from a preinstallation Class b (low risk) to a postinstallation Class a (no risk). Mechanical risk due to seasonal variations of temperature resulted in Class b (moderate risk) on all floors—the preinstallation basement and attic environments were Class a (no risk). However, the *overall* mechanical risk remained Class b for all floors during both pre- and postinstallation.

Table 9.5

Comparison of pre- and postinstallation risk classes at Hollybourne Cottage.

	Microbial Risk		Mechanical Risk														Chemical Risk	
			Humidity						Temperature						Overall[1]			
			Short-term Variation		Seasonal Variation		Deviation from Historical Mean		Short-Term Variation		Seasonal Variation		Deviation from Historical Mean					
Floor	Pre	Post	Pre	Post	Pre	Post	Pre	Post	Pre	Post	Pre	Post	Pre	Post	Pre	Post	Pre	Post
Attic	F	F	b	b	f	f	a	a	b	b	a	b	a	a	f	f	+	0
2nd floor	F	C	b	b	a	a	a	a	a	a	b	b	a	a	b	b	0	0
1st floor	F	C	b	b	f	a	a	a	a	a	b	b	a	a	f	b	0	0
Basement	F	C	b	b	f	a	a	a	a	a	a	b	a	a	f	b	0	+

[1] Overall mechanical risk is based on the highest risk classification in any of the specific humidity- and temperature-based mechanical risks (short-term variation, seasonal variation, and deviation from historical mean).

Chemical risk

For chemical risk, the preinstallation basement environment was Class 0 (same risk as the historic condition), but postinstallation, the environment had deteriorated to Class + (moderately increased risk). This change is the result of increased temperatures during the hot-humid season. Conversely, the chemical risk in the attic was reduced from a preinstallation Class + to a postinstallation Class 0, due to the reduced high temperature conditions during the hot-humid season.

Thermal comfort

Thermal comfort was not a requirement of the environmental management system for Hollybourne Cottage, since the building was not regularly open to visitors. However, the predicted mean vote (PMV) and predicted percentage dissatisfied (PPD) values were evaluated for the mean postinstallation first-floor environment (the area typically accessed when the cottage is open for visitors) during the hot-humid season using the airspeed model in the ASHRAE Thermal Comfort Calculator. Thermal comfort is not a concern during the cool-humid season because transient occupants are likely to be clothed for exterior conditions.

Prior to installation of the environmental management system, the average temperature and humidity on the first floor during the hot-humid season were 27.6°C and 78% RH, respectively. The PMV and PPD values were calculated assuming a mean radiant temperature equal to the room temperature, 27.6°C; an occupant metabolic rate of 1.5 met (combination of standing and walking); a clothing factor of 0.6 clo (typical summer clothing); and a minimum air speed of 0.1 m/s. With these assumptions, the PMV was calculated to be 1.45 (slightly warm), and the PPD was calculated to be 48% (point 1 on fig. 9.18).

After installation of the environmental management system, the average hot-humid season first-floor temperature and humidity shifted to 29.1°C and 67% RH, respectively. Using similar assumptions from the preinstallation analysis, except for an increased air speed of 0.4 m/s due

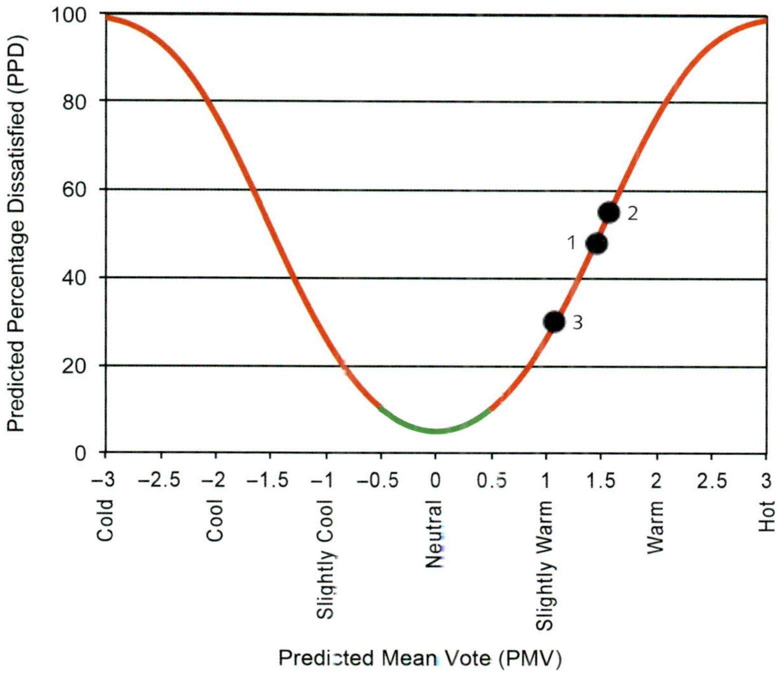

Figure 9.18

Thermal comfort assessment—predicted mean vote (PMV) and predicted percentage dissatisfied (PPD)—of the hot-humid first-floor environment during preinstallation (1), postinstallation with air speed of 0.4 m/s (2), and postinstallation with air speed of 1.2 m/s (3).

to ventilation, the PMV was calculated to be 1.57 (slightly warm), and the PPD was calculated to be 55% (point 2 on fig. 9.18). This decrease in thermal comfort relative to the preinstallation analysis is due to the higher interior temperature. Even if air speed were increased to 1.2 m/s, the upper limit stated by ANSI/ASHRAE Standard 55-2013 (ANSI and ASHRAE 2013), PMV and PPD values would only improve to 1.08 and 30%, respectively (point 3), and thermal comfort would not be achieved.

As stated earlier, Hollybourne Cottage is occasionally the site of special tours lasting approximately thirty minutes. Prior to entering the building, visitors are informed that the interior environment is conditioned for architectural conservation and that it may be too hot or too cold for human comfort. These announcements may psychologically prepare the visitors to anticipate and tolerate environmental conditions outside of the typical thermal comfort zone.

Noise

Noise measurements for the supply fans were 75–90 dBA next to the unit and 66–69 dBA downstream. Exhaust fan noise levels were comparatively lower, at 72 dBA next to the unit and 63 dBA downstream, presumably owing to the lower flow rate.

Noise levels recorded on each floor were influenced by the configuration of the rooms and the building construction materials. The open layout of the basement combined with the hard floor and wall surfaces resulted in noise levels of 58–60 dBA throughout the space. In the attic, the supply and exhaust fans in the similarly open attic space generated noise levels of 65 dBA near the fans and 59 dBA at distance. In contrast, the multiroom configurations of the first and second floors resulted in noise levels of 46 dBA and 48 dBA, respectively—slightly noisier than a typical office background noise of 45 dBA.

Maintenance

A variety of maintenance tasks was necessary to maintain the performance of the environmental management system. These tasks included:

- annual maintenance of nonmechanical passive measures (i.e., clear gutters, good surface water drainage, windows intact, building envelope intact),
- visual inspection of supply fan screens and filters (monthly),
- annual inspection of solenoid switches that trigger operation of fans and heaters,
- check of power cords and insulation for effective fire safety (monthly),
- calibration/functional check of humidity sensor (annually).

Energy Metrics

Energy intensity

The total installed heating capacity (energy intensity) in the basement was 22.5 kW, and the total fan capacity in the basement was 0.74 kW. In the attic, the total fan capacity was 0.76 kW. The total capacity of all mechanical equipment was 24.0 kW, and the total energy intensity of the building was 23 W/m^2.

Energy use intensity

As might be expected, basement heating costs were greater than basement and attic ventilation costs on a per hour basis. Overall, during the postinstallation period, basement heating represented 88% of the operational cost, while the cost of basement and attic ventilation made up only 11% (table 9.6). Based on the total floor area of the building (1050 m^2) and the duration of operation, the energy use intensity for basement ventilation/heating and attic ventilation was 28.3 $kWh/m^2/yr$. It should be noted that these indices reflect operation during less than a full calendar year (June 2000–January 2001); however, operation of this system configuration during the cooler winter and spring months is expected to be minimal.

Table 9.6

Operational use (time operated as percentage of total time for the period from June 2000 to January 2001) of the environmental management system.

		Use of the Monitored Period (%)	Total Energy Used during the Monitored Period (%)
Attic	Ventilation	19	5
Basement	Ventilation	11	6
	Heating	13	88

Conclusions

The dominant factors in preventing microbial growth are limiting prolonged periods of humidity above the 75% RH threshold for germination, and promoting air movement. The basement and first-floor environments of Hollybourne Cottage had high levels of humidity during the preinstallation period, as evidenced by the moisture-related microbial and insect damage to the wood framing and architectural woodwork. Preinstallation data for the attic environment indicated very high temperatures during summer months and seasonally elevated humidity.

Though only the basement and attic environments at Hollybourne Cottage were directly addressed by the nonmechanical strategies and the installation of the basement ventilation/heating and attic ventilation system, the overall environment, including the intermediate floors, benefited.

The major achievement of the environmental management strategy was the significant improvement in the basement and first- and second-floor environments. On those floors, the preinstallation microbial risk, based on the Conservation Environment Classification–HH protocol, improved from a Class F (high risk) to a postinstallation Class C (moderate risk). However, high-humidity events in the attic during the winter kept the microbial risk of the attic environment at Class F.

The environmental management strategy reduced mechanical risk due to seasonal variations by limiting humidity during the hot-humid season, while maintaining the interior humidity near the historical mean exterior humidity. However, the use of conservation heating in the basement resulted in an increased chemical risk due to higher temperatures.

Lessons Learned from Hollybourne Cottage, Part A

- Based on the Conservation Environment Classification–HH protocol, an overall collections risk classification of either Cb0 or Cb+ is achievable in a 2A climate zone with a system of heating and ventilation and humidistatic control.
- Nonmechanical strategies to reduce air infiltration and for source moisture control are essential.
- Active environmental management of basement and attic can indirectly improve conditions in adjacent spaces.
- High-humidity events in the upper floors during cold winter periods may occur unless specifically addressed by the mechanical strategies.
- Ventilation alone cannot maintain an acceptable collections environment.
- Basement humidity can be effectively managed with only heating.

Future Considerations

The basement should continue to be segregated from the upper floors because it is the dominant site for moisture infiltration. In the absence of occupants for visitation or work, the environmental management system for the basement may omit ventilation and rely solely on the heating mode to control humidity. However, without ventilation, the air should be recirculated to further inhibit microbial activity. This reconfiguration of the system would simplify system operation while maintaining a suitable conservation environment for the building fabric.

The attic environment was improved, but the attic remained the highest-risk space for microbial damage in the building. However, no fungal damage has been found in the attic, as the high humidity conditions and coincident low temperatures during the cool-humid season were not sustained for a sufficient duration to allow microbial activity.

The use of attic ventilation effectively reduced maximum and average temperatures during the study period, and this reduction resulted in a reduced risk of chemical damage. Installation of reflective insulation material under the rafters would further limit solar heat gain during the summer and heat loss during the winter. However, the introduction of conventional thermal insulation, such as fiberglass batts on reflective sheets, under the wood roof should be avoided, as it would slow drying and possibly reduce the service life of the wood shingles. Though attic ventilation was primarily directed toward reducing attic temperature, the promotion of air movement throughout the space likely improved occupant comfort during periods of extremely warm attic temperatures.

Thermal comfort was not a requirement for Hollybourne Cottage, since it is an unoccupied building with infrequent tours. Because of elevated temperatures in the building, thermal comfort would not be achieved during the hot-humid season even with an additional air movement of 1.2 m/s. Therefore, adaptive measures are needed for visitors. These may include short tour durations, a briefing prior to the tour on the interior environmental conditions to be expected, and explanations of the environmental management system.

Design efforts for an environmental management system should also address equipment noise, particularly for ventilation. As sound typically registers between 60 and 80 dB at each unit, the visitation experience and work efficiency may be negatively affected near the fans. Though the use of sound-dampening double casing and elbow ducts are not possible because of the lack of ductwork, clever placement of equipment, the positioning of noise reduction forms on wall surfaces, and redesign of the visitor path can mitigate noise perceived by the visitor or worker.

Postscript

Following the successful initial application of ventilation and heating for the environmental management of Hollybourne Cottage described in this chapter—which is part A of the Hollybourne case study—five additional experimental variations or phases of mechanical strategies were subsequently tested in the building.

During the second experimental phase (February–August 2001) of the project, direct management of the first and second floors of Hollybourne Cottage was attempted through the installation of two supply fans positioned in central locations on each floor. Floor openings were also created between the first floor and the basement, allowing for the flow of air between the two spaces. As a result, two environmental control zones were established: (1) the basement and the first floor, and (2) the second floor and the attic. Additional exhaust ventilators were placed in the basement (two) and attic (one) to balance airflow. In the attic, a convective heater was placed in the center of the space to address low temperature–high humidity conditions during the cool-humid season.

The third and fourth experimental phases maintained similar system configurations for the upper floors but deactivated basement ventilation and relied on heating (third phase, August 2001–June 2002) or dehumidification (fourth phase, June 2002–July 2003) as the sole means of managing the basement environment. The basement was segregated from the rest of the structure and represented one environmental control zone. The first floor also remained independent, while the second floor and attic made up the third environmental control zone.

The fifth experimental phase (August 2003–April 2004) is described in "Hollybourne Cottage, Historic House on Jekyll Island, Georgia, United States, Part B: Whole-House Ventilation, Attic Heating, and Basement Dehumidification" (chap. 10). This phase included the reactivation of basement ventilation, the rezoning of the upper floors as one environmental control zone, and the merging of operational bands for each mode.

The sixth and final experimental phase (April 2004–October 2005) of the project maintained the system configuration of the fifth phase but inserted an additional dew point temperature parameter (interior dew point temperature must be greater than or equal to exterior dew point temperature) into the control logic for basement and upper floor ventilation as a means of reducing humidity variations following the ventilation of high dew point exterior air. Results for all experimental phases are summarized in Maekawa, Beltran, and Toledo (2007).

Project Team and Responsibilities

Jekyll Island Museum (Jekyll Island, Georgia, United States)

Warren Murphey (Director)—Local project management

John Hunter (Chief Curator/Assistant Director, operation and museum)

Preservation Department—Envelope improvements, site context improvements, installation of mechanical system, operation, and maintenance

Getty Conservation Institute (Los Angeles, California, United States)

Shin Maekawa (Senior Scientist)—Project management, environmental assessment, system concept, mechanical strategy design, and environmental monitoring design and operation

Franciza Toledo (Assistant Scientist)—Building interior assessment, nonmechanical strategy design, supervision of installation

Michael C. Henry (Principal, Watson & Henry Associates, Bridgeton, New Jersey, United States)—Building envelope condition assessment

Arthur Hasegawa (Huntington Beach, California, United States)—Electrical design

References

AFCCC [Air Force Combat Climatology Center] and NCDC [National Climatic Data Center]
1999 *Engineering Weather Data*. Washington, DC: Air Force Combat Climatology Center. http://purl.access.gpo.gov/GPO/LPS7081.

ANSI [American National Standards Institute] and ASHRAE
2013 *Standard 55-2013—Thermal Environmental Conditions for Human Occupancy*. Atlanta: ASHRAE.

ANSI [American National Standards Institute], ASHRAE, and IES [Illuminating Engineering Society]
2013 *Standard 90.1-2013—Energy Standard for Buildings except Low-Rise Residential Buildings* (SI edition). Atlanta: ASHRAE.

ASHRAE
2011 Museums, galleries, archives and libraries. Chapter 23 in *ASHRAE Handbook—HVAC Applications*. Atlanta: ASHRAE.

Harriman, L., G. Brundrett, and R. Kittler
2001 *Humidity Control Design Guide for Commercial and Institutional Buildings*. Atlanta: ASHRAE, 2001.

Maekawa, S., V. Beltran, and F. Toledo
2007 Alternatives to conventional air-conditioning in coastal Georgia. *APT Bulletin* 38 (2–3): 3–12.

Maekawa, S., and F. Toledo
2002 A climate control system for Hollybourne Cottage, Jekyll Island Historic District, Georgia. In *IAQ 2001—Moisture, Microbes and Health Effects: Indoor Air Quality and Moisture in Buildings, San Francisco, California, November 4–7, 2001*. Atlanta: ASHRAE.

Watson & Henry Associates
1998 Condition assessment survey report for Hollybourne Cottage, Jekyll Island, Georgia. Getty Conservation Institute internal report.

Weather Channel
2014 Weather Underground. http://www.wunderground.com/.

Chapter 10

Hollybourne Cottage, Historic House on Jekyll Island, Georgia, United States: Part B—Whole-House Ventilation, Attic Heating, and Basement Dehumidification

Climate zone 2A (Hot-Humid)

Objectives

- Maintain 70% RH environment throughout the building to protect the historic interior from microbial deterioration.
- Eliminate heat accumulation in the second-floor and attic spaces during the hot-humid season.
- Eliminate greater than 75% RH conditions in the attic during the cool-humid season.
- Provide outside (fresh) air and air circulation to the first- and second-floor spaces for occupant comfort.
- Reduce energy consumption relative to the strategies discussed in part A of the case study of Hollybourne Cottage (chap. 9) using dehumidifiers to replace heaters in the basement.

Summary

Building upon part A of the Hollybourne Cottage case study, part B considers the application of a conservation-focused environmental management strategy in the historic Hollybourne Cottage, located on Jekyll Island, Georgia, United States, to address high-humidity events in the basement (hot-humid season) and attic (cool-humid season) and high temperatures in the attic (hot-humid season) in a Hot-Humid (2A) climate zone.

Part A of the Hollybourne Cottage case study described the application of a mechanical strategy consisting of basement ventilation and heating, and attic ventilation. As detailed in the postscript of part A of this case study, this was the first of six experimental strategies employed at the cottage.

Part B discusses whole-house ventilation, attic heating, and basement dehumidification—the fifth experimental strategy—which reduced humidity levels for all floors relative to preinstallation conditions. However, the introduction of active environmental management to the first and second floors resulted in elevated humidity on these floors compared to part A, even though the part A configuration did not actively manage the first- and second-floor environments. The installation and operational costs of the part B configuration remained significantly less than costs associated with conventional air-conditioning systems.

Overview

Part B of the Hollybourne Cottage case study discusses the installation and operation of whole-house ventilation, basement dehumidification, and attic heating. This system configuration expands upon the lessons learned in part A of the Hollybourne Cottage case study by replacing basement heating with basement dehumidification, adding ventilation to the first and second floors, and adding heating to the attic.

Discussion of Hollybourne Cottage's history, building, climate zone, and preinstallation building and environmental assessment is found in part A of the case study.

Environmental Management System

Objectives

The following objectives were established based on the assessments described in part A of the case study:

- Maintain a 70% RH environment throughout the building to protect the historic interior from microbial deterioration.
- Eliminate heat accumulation in the second-floor and attic spaces during the hot-humid season.
- Eliminate greater than 75% RH conditions in the attic during the cool-humid season.
- Provide outside (fresh) air and air circulation to the first- and second-floor spaces for occupant comfort.
- Reduce energy consumption relative to the strategies discussed in part A of the case study of Hollybourne Cottage (chap. 9) by using dehumidifiers to replace heaters in the basement.

One of the objectives in part A of the Hollybourne Cottage case study was to avoid placement of mechanical equipment in the first- and second-floor spaces. In part B, this objective was relaxed, allowing for placement of systems equipment on those floors.

Psychrometric strategies

As done for part A of this case study, nonmechanical source moisture removal features were reinstated in the building envelope to limit the infiltration of humid exterior air. With respect to the mechanical psychrometric strategies, whole-house ventilation, attic heating, and basement dehumidification were applied to manage the extreme environmental conditions in Hollybourne Cottage (fig. 10.1). In this strategy, conservation

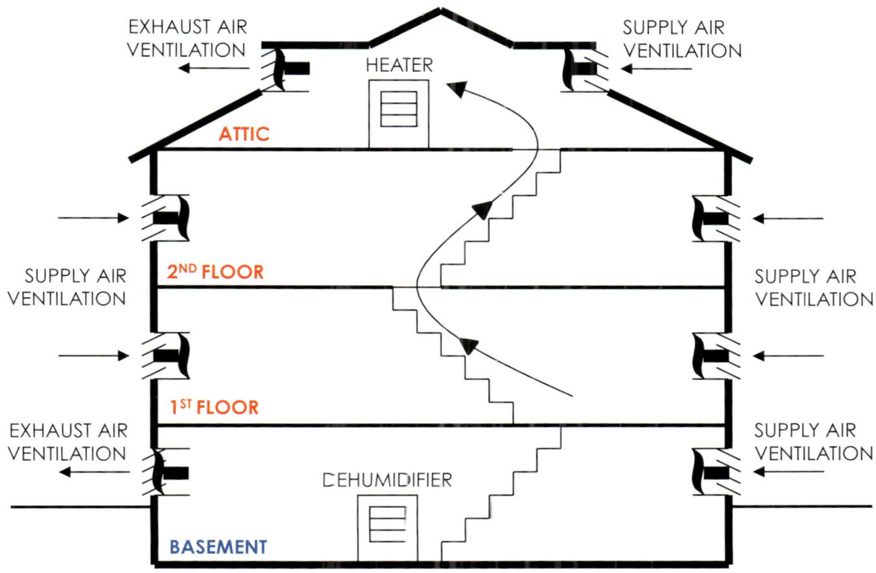

Figure 10.1
Schematic of whole-house ventilation/basement dehumidification/attic heating.

heating by ventilation and dehumidification—replacing heating used in part A—were used to reduce humidity in the basement. In the attic, high temperatures in the hot-humid season were managed by dilution ventilation, and high humidity in the cool-humid season was managed by conservation heating. Conservation heating by ventilation was also introduced into the first and second floors, benefiting the historic interior and providing fresh air for visitor and worker comfort. The basement environment remained separated from the upper-floor environments, effectively dividing the building and the system into two environmental management zones: the basement and the upper floors (first floor, second floor, and attic).

Nonmechanical strategies

Improvements to the building envelope and the surroundings near the building, as described in part A, were maintained throughout part B of the case study. No additional nonmechanical environmental management strategies were applied in part B.

Mechanical strategies

Mechanical system description

The basement ventilation and dehumidification system consisted of propeller-type exhaust and supply fans and mechanical dehumidifiers. Two supply fans were placed in existing window openings on the north and south foundation walls of the east–west wing. Exhaust fans were placed in two window openings of the west foundation wall of the north–south wing and in a window opening in the eastern section of the south foundation wall (fig. 10.2). Ventilation introduced exterior air into the basement near the intersection of the two wings, and air was exhausted at the east and west ends of the basement. The airflow was balanced in the basement. Supply and exhaust fans were sized for approximately 10 air changes per hour (ACH), and they had to fit into existing window and vent openings.

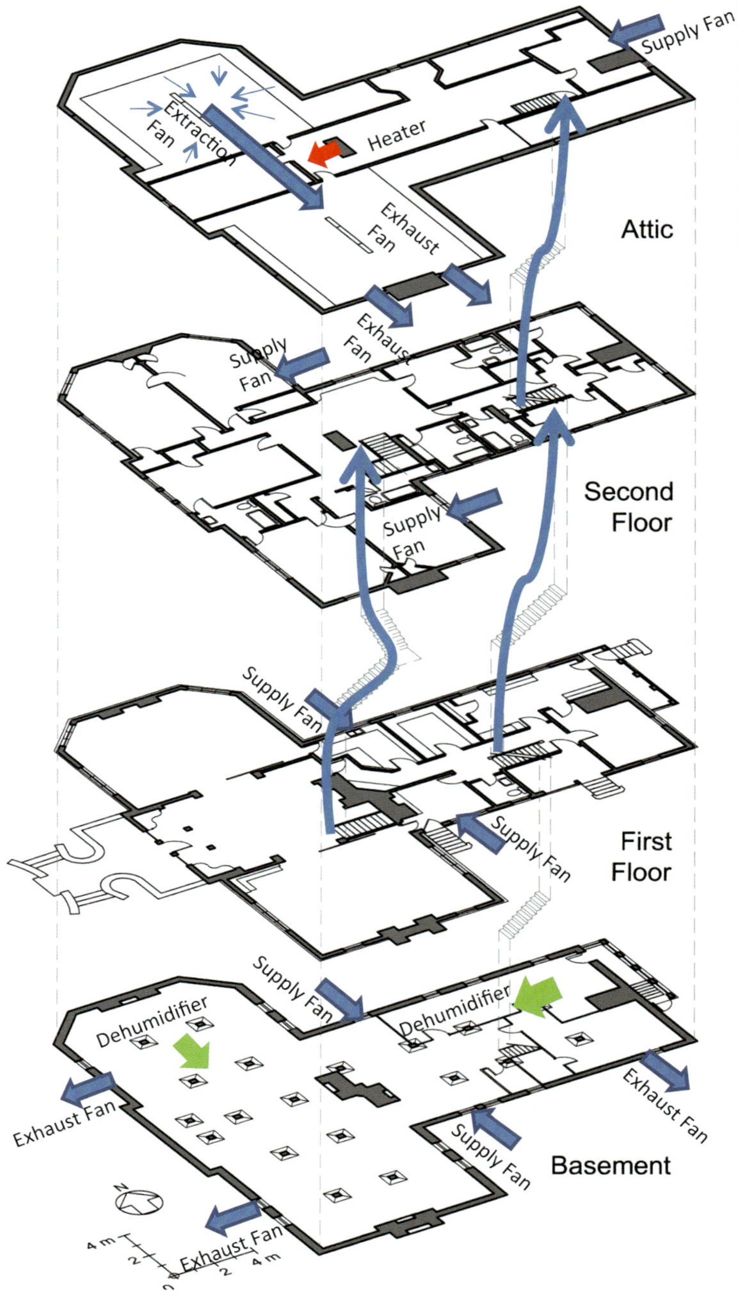

Figure 10.2
Locations of mechanical system components during the case study, part B.

Motorized shutters and antimicrobial MERV 7 filters were fitted to the supply fans, and exhaust fans were equipped with gravity-operated shutters. Two freestanding high-capacity dehumidifiers were installed in the basement, one in the north end of the north–south wing and one in the east–west wing. The dehumidifiers were fitted with condensate pumps that discharged to the exterior. Dehumidifiers in the basement had sufficient capacity to extract 1.5 L/h at 25°C or 1.0 L/h at 28°C.

The ventilation and heating system for the upper floors of Hollybourne Cottage consisted of supply and exhaust fans and a heater. Two supply fans per floor were centrally located on the first and second floors (fig. 10.3). A fifth supply fan was positioned in an east window opening in the attic. Two attic exhaust fans were installed in the south

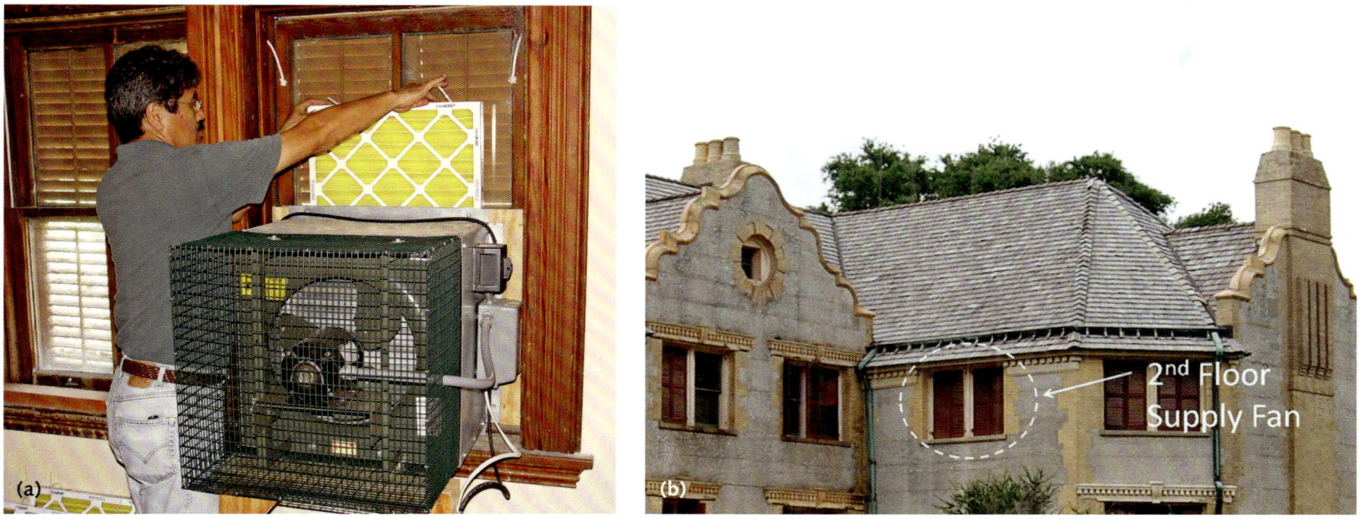

Figure 10.3

Supply fan installed in the second-floor window (a). Supply fans included an antimicrobial filter (yellow pad), an insect screen, and a motorized shutter. Original outside window louvers hide supply fans (b). Photos: Shin Maekawa. © J. Paul Getty Trust.

gable wall window openings, and one exhaust fan was installed in the west wall. A ducted extraction fan was also used in the attic to vent air from an isolated windowless north room. The overall fan configuration theoretically produced an airflow of 3.9 ACH (1.4 m^3/s) from the lower floors to the attic, but in practice, the actual airflow was lower because of exfiltration losses. In the attic, one electric resistance convection heater was centrally placed to reduce high humidity during the cool-humid season. The heater had sufficient capacity to increase interior temperature by approximately Δ7°C, limited by a programmable maximum temperature threshold. Table 10.1 lists specifications for each type of equipment.

Operational control sequence

Automatic control of the mechanical system was achieved with a programmable controller configured for whole-house ventilation, attic heating, and basement dehumidification. The two zones of the environmental management system—(1) basement, and (2) above-ground floors including attic—operated independently; the operational control conditions are summarized in table 10.2 and discussed below.

Above-ground zone: first and second floors and attic. Ventilation for the above-ground zone was enabled by either the attic temperature or the first-floor humidity. When the exterior humidity was below 70% RH, ventilation was enabled when first-floor interior humidity was greater than or equal to 65% RH, or the temperature of the underside of

Table 10.1

Performance specifications for mechanical system components.

Unit	Power (kW)	Flow Rate (m^3/s)	Static Pressure (kPa)
Supply fan	0.25	1.130	0.062
Exhaust fan	0.12	0.847	0.062
Extraction fan	0.02	0.052	
Heater	7.5	0.271	
Dehumidifier[1]	1.1	0.137	

[1] Capacity: 1.5 L/h at 25°C, 1.0 L/h at 28°C.

Table 10.2

Operational control conditions.

Zone	Psychrometric Strategy	Enable	Disable
Above-ground floors (1st floor, 2nd floor, attic)	Conservation heating (attic)	$H_{interior} \geqq 65\%$ RH and $H_{exterior} \geqq 70\%$ RH	$H_{interior} < 60\%$ RH or $H_{exterior} < 70\%$ RH or $T_{roof} \geqq 32°C$
	Ventilation	$H_{interior}$ (1st floor) $\geqq 65\%$ RH and $H_{exterior} < 70\%$ RH, or $T_{Roof} \geqq 32°C$ and $H_{interior} < 70\%$ RH	$H_{interior}$ (1st floor) $< 60\%$ RH and $T_{roof} < 25°C$, or $H_{exterior} \geqq 70\%$ RH
	Idle	$H_{interior} < 60\%$ RH and $T_{Roof} < 25°C$ or $T_{Roof} \geqq 32°C$	$H_{interior} \geqq 65\%$ RH or $T_{Roof} < 32°C$
Basement	Ventilation	$H_{interior} \geqq 65\%$ RH and $H_{exterior} < 70\%$ RH	$H_{interior} < 60\%$ RH or $H_{exterior} \geqq 70\%$ RH or $T_{interior} > 32°C$
	Dehumidification	$H_{interior} \geqq 65\%$ RH and $H_{exterior} \geqq 70\%$ RH	$H_{interior} < 60\%$ RH or $H_{exterior} < 70\%$ RH or $T_{interior} > 32°C$
	Idle	$H_{interior} < 60\%$ RH or $T_{interior} > 32°C$	$H_{interior} \geqq 65\%$ RH and $T_{interior} \leqq 32°C$

the roof was greater than or equal to 32°C. Ventilation was disabled if exterior humidity was above or equal to 70% RH or if first-floor interior humidity was below 60% and the roof (or attic ceiling) temperature was below 25°C.

Conservation heating in the attic was enabled when interior humidity was above or equal to 65% RH and exterior humidity was above or equal to 70% RH, and it was disabled if either of the following conditions were satisfied: interior humidity was below 60% RH, exterior humidity was below 70% RH, or attic temperature was greater than or equal to 32°C. In general, an interior humidity of 65% RH or less can be achieved if the interior of the above-ground zone is maintained at approximately 15°C or higher during the cool-humid season.

The idle mode was enabled when either (1) interior humidity was less than 60% RH and roof (or attic ceiling) temperature was less than 25°C, or (2) roof (or attic ceiling) temperature was greater than or equal to 32°C. The idle mode was disabled when interior humidity was equal to or higher than 65% RH or roof (or attic ceiling) temperature was equal to or less than 32°C.

Basement zone. The basement zone of the environmental management system utilized a ventilation mode and a dehumidification mode. The ventilation mode was enabled when basement interior humidity exceeded or equaled 65% RH and exterior humidity was less than 70% RH. Ventilation was disabled when interior humidity was 60% RH or less, or when exterior humidity was higher than or equal to 70% RH. Basement dehumidification was enabled when interior humidity exceeded or equaled 65% RH and exterior humidity exceeded or equaled 70% RH, and it was disabled when interior humidity was less than 60% RH, or when exterior humidity was less than 70% RH, or when basement temperature exceeded 32°C.

While the control scheme for basement ventilation and heating described in part A of this case study maintained different operational bands (ventilation: 60%–65% RH, heating: 65%–70% RH), the operational bands for the basement ventilation and dehumidification modes overlapped in part B, keeping humidity between 60% and 65% RH in an attempt to firmly maintain the basement humidity at less than 65% RH.

In the control scheme used in part B, the interior temperature threshold for disabling ventilation and dehumidification is 32°C, slightly higher than the 30°C threshold used in part A to disable ventilation and heating. The higher temperature threshold in part B allowed the dehumidifiers to maintain interior humidity between 60% and 65% RH, rather than at the 65%–70% RH range in part A.

The idle mode in the basement was enabled when interior humidity was less than 60% RH—in this case, both the ventilation and dehumidification modes are disabled. The idle mode was disabled when interior humidity was greater than or equal to 65% RH.

Postinstallation System Performance

Whole-house ventilation, attic heating, and basement dehumidification strategies were tested from August 2003 to April 2004, and temperature and humidity were monitored for each floor of Hollybourne Cottage. Dew point temperature was calculated from concurrent temperature and humidity measurements. Since the operational period of this experimental configuration lacked data for a complete hot-humid season, interior environmental data from May to July 2004 was added from the subsequent sixth experimental phase.[1] The added data made it possible to show performance over a full calendar year. Despite a slightly altered control scheme between the fifth and sixth experimental phases, the interior environment remained similar.

Humidity

In the basement zone, the ventilation and dehumidification strategy maintained a less-humid environment compared to preinstallation conditions. Monthly mean basement humidity ranged from 59% RH in December to 66% RH in July (fig. 10.4). Reflecting the overlap of the dehumidification and ventilation operational bands in part B, this range of basement humidity was lower than observed in part A (basement ventilation and heating), which ranged from a monthly mean humidity of 59% RH in

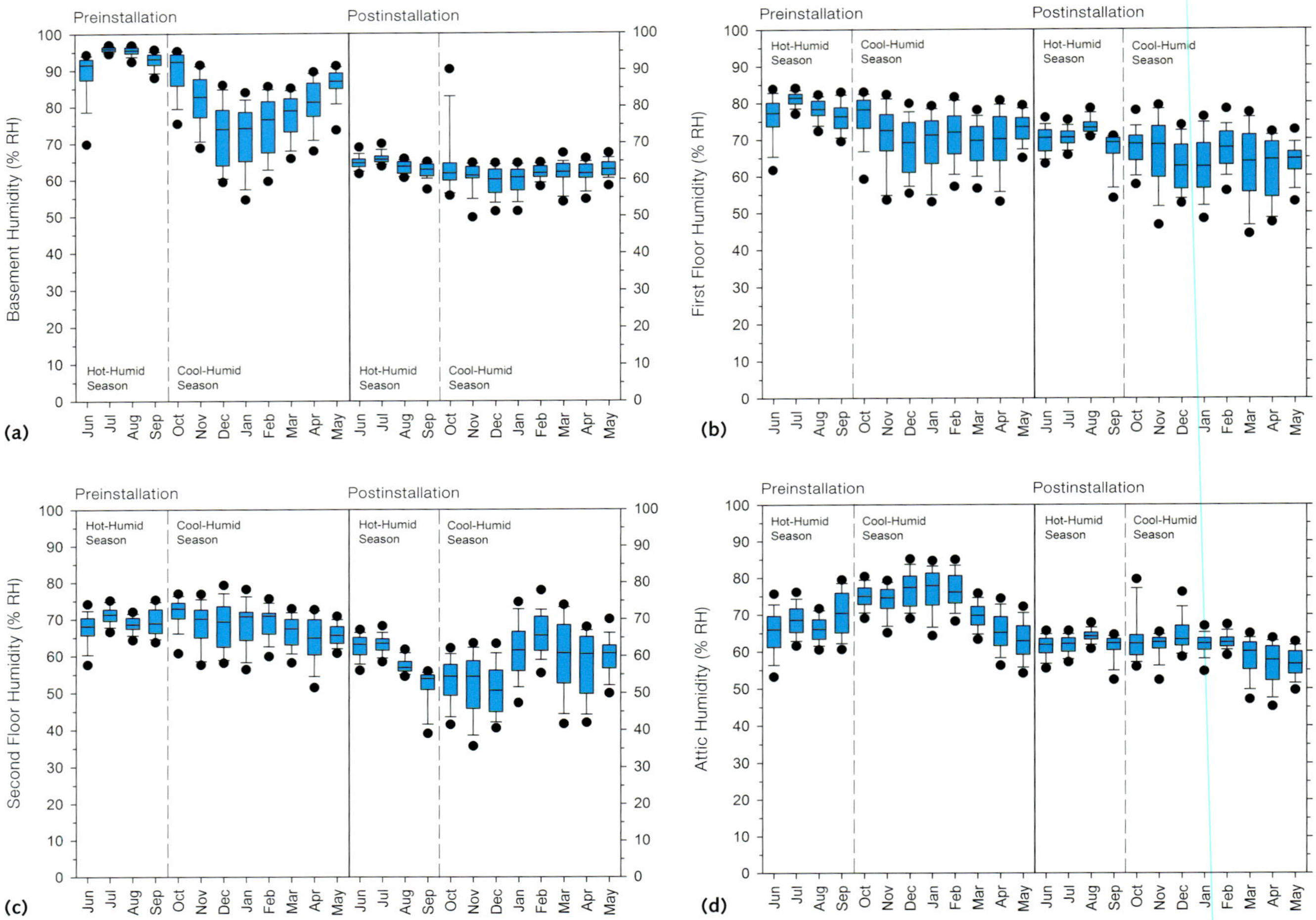

Figure 10.4

Interior humidity comparisons between preinstallation (January 1999–June 2000) and postinstallation (August 2003–July 2004).

December to 69% RH in June. During the operational period of part B, the 75% RH level was exceeded less than 1% of the time, and all high-humidity events coincided with malfunctions in dehumidification or condensate removal.

Compared to preinstallation conditions, the use of supply ventilation in the first and second floors resulted in less-humid conditions. Monthly mean first-floor humidity ranged from 62% RH in April to 74% RH in August, and the 75% RH threshold was exceeded during roughly 8% of that period (fig. 10.4). Monthly average second-floor humidity ranged from 51% RH in December to 66% RH in February, and values above 75% RH occurred during less than 2% of this period. During preinstallation, the first and second floors exceeded 75% RH during 44% and 6% of the period, respectively.

In contrast, compared to the first-floor data of part A (monthly mean ranging from 62% to 69% RH, with 75% threshold exceeded 2% of the period), part B data indicated higher first-floor humidity. The higher levels and wider variability in humidity during the part B operational period may have resulted from intermittent visitation and restoration activities during the hot-humid season—particularly in June, July, and August—that were accompanied by frequent door openings in the first floor and the infiltration of humid air into the interior that did not occur during the part A scheme. In contrast to the first-floor environment, the second-floor environment remained similar during parts A and B.

In the attic, conservation heating lowered interior humidity during the cool-humid season, and the monthly mean humidity ranged from 56% RH in April to 65% RH in December, values that were an improvement over preinstallation monthly means that ranged from 63% RH in May to 77% RH in December. The attic humidity during part B also represented an improvement over part A, whose monthly means ranged from 66% to 72% RH. This improvement was attributable to the elimination of higher than 65% RH conditions in the September–March period by the part B scheme (fig. 10.4). The 75% RH level was exceeded during less than 2% of the period.

Temperature

In the basement zone, dehumidification, which was used frequently during the hot-humid season, increased basement temperature (monthly means ranged from 15.3°C in January and December to 31.4°C in August) compared to preinstallation conditions (14.0°C–26.2°C) (fig. 10.5). The temperature increase was due to rejected heat from dehumidifier operation. Compared to part A, which utilized humidistatically controlled heating, monthly mean basement temperatures during part B were elevated by as

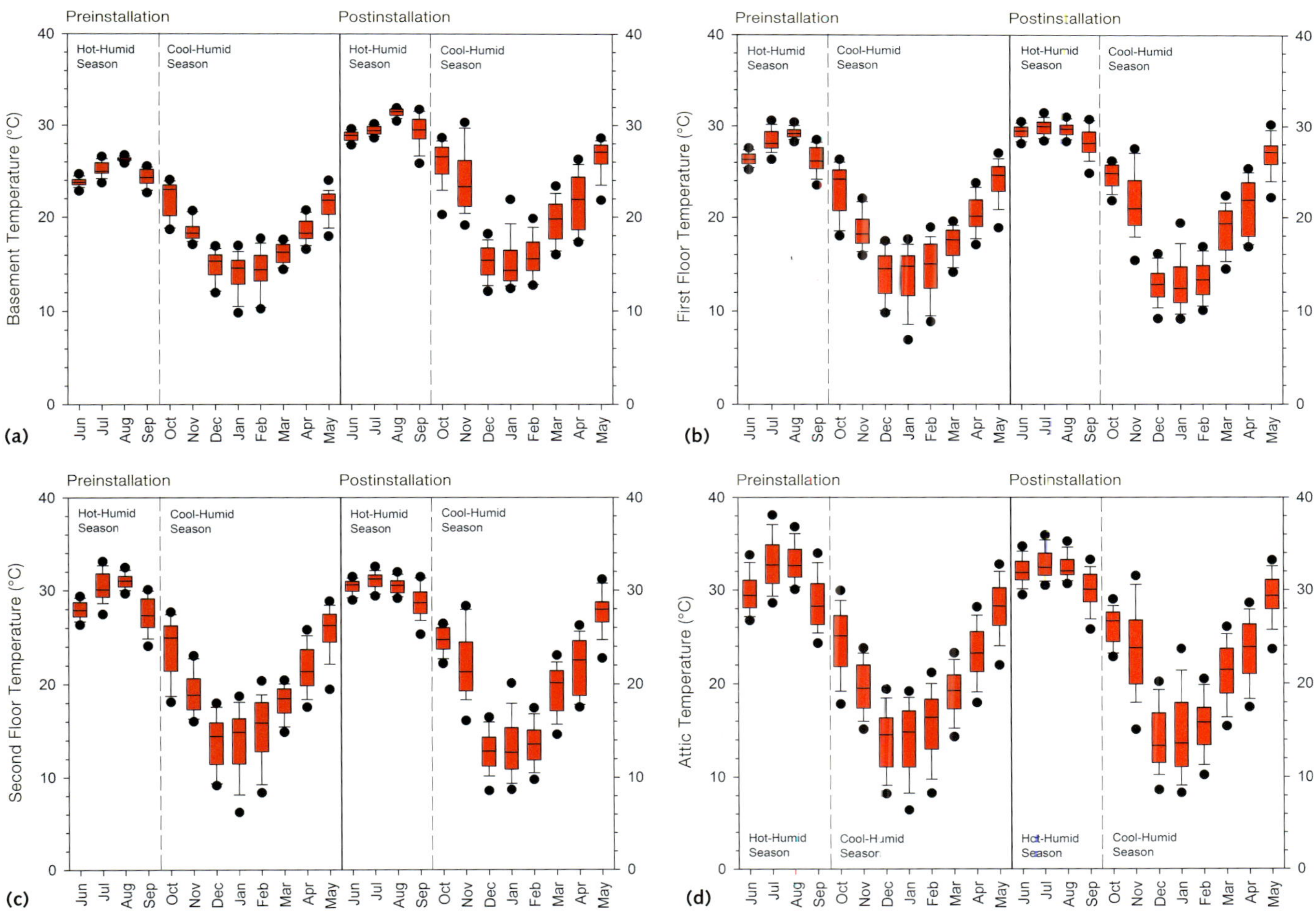

Figure 10.5

Interior temperature comparison between preinstallation (January 1999–June 2000) and postinstallation (August 2003–July 2004).

much as approximately 1.2°C during the hot-humid season and by 2.4°C during the cool-humid season.

In the attic, heating and ventilation resulted in temperature conditions similar to those of preinstallation, but the maximum temperature values were reduced with respect to preinstallation maximums. The monthly mean attic temperature ranged from 14.1°C (December) to 32.8°C (July) (fig. 10.5). Four months exhibited maximum temperature values above 35°C, yet the highest recorded temperature (36.9°C in July) in part B was more than 3°C lower than the preinstallation maximum. Ventilation during warm periods improved thermal comfort for potential short-term occupancy.

Monthly average temperatures were similar for the first and second floors (13°C in December to 31°C in August) and remained consistent with what was observed before system installation (see fig. 10.5).

Dew point temperature

Figure 10.6 shows system performance with respect to dew point temperature. During part B, monthly mean dew point temperature in the basement zone ranged from 7.3°C (January) to 23.5°C (August), which was slightly less than the preinstallation range (8.8°C in January to 25.4°C in

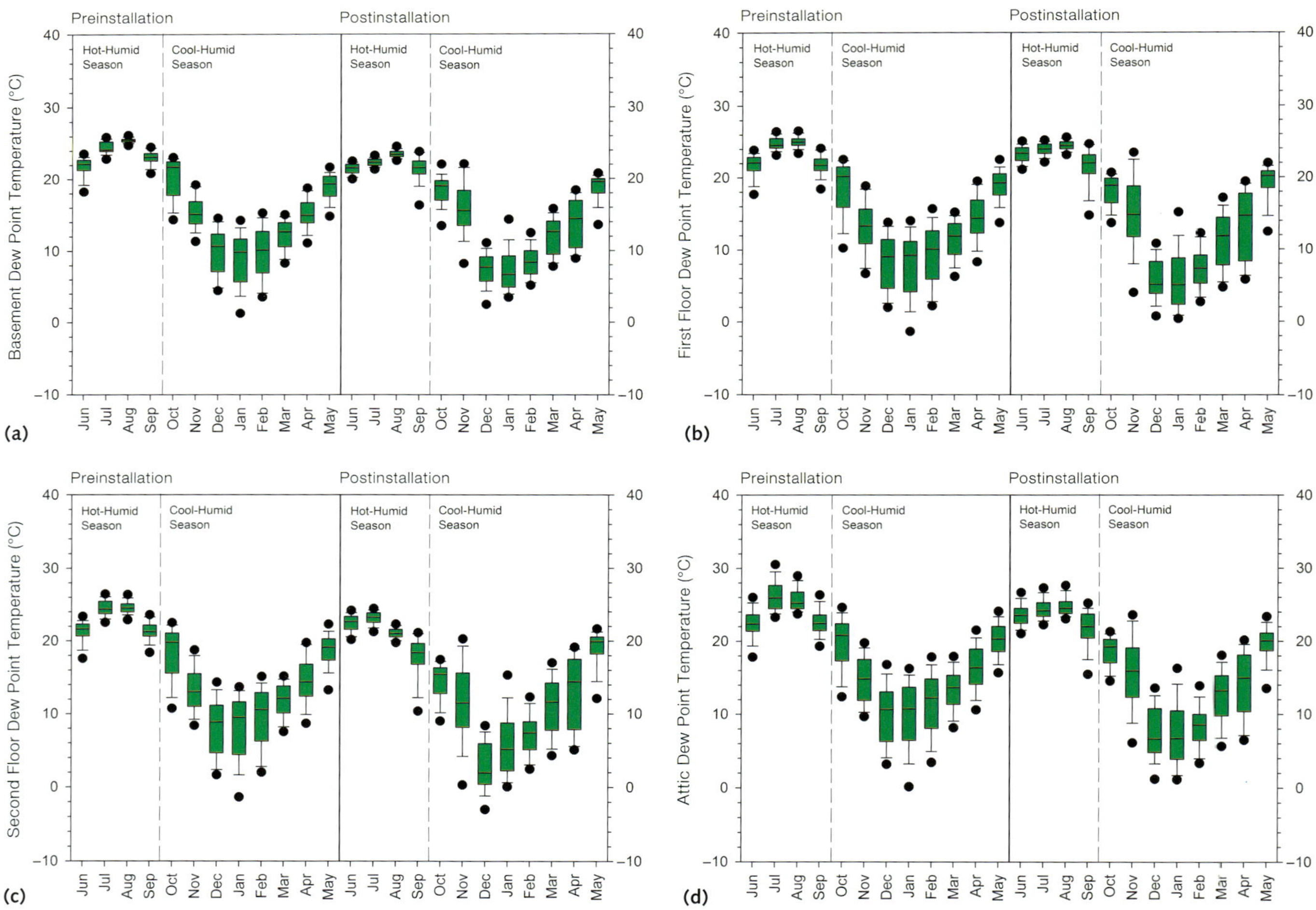

Figure 10.6

Interior dew point temperature comparison between preinstallation (January 1999–June 2000) and postinstallation (August 2003–July 2004).

August). Used heavily during the months of May through August—comprising the end of the cool-humid season and the bulk of the hot-humid season—dehumidifier operation reduced basement dew point temperature by 1°C (August) to 2.4°C (June) relative to exterior values. The monthly mean dew point temperature in the first floor, second floor, and attic ranged from approximately 5°C to 24.5°C, matching the range of exterior dew point temperature.

Postinstallation risk class assessment

Table 10.3 summarizes the postinstallation environment at Hollybourne Cottage, based on the Conservation Environment Classification–HH protocol. Differences from the preinstallation environment classification are discussed below. (Classification of the preinstallation environment is described in part A.) Figure 10.7 also shows the preinstallation and postinstallation first-floor datasets on psychrometric charts, with values relevant to classification denoted; the boxes indicate the mid-95% of each dataset.

Table 10.3

Conservation Environment Classification–HH protocol for the postinstallation environment of Hollybourne Cottage.

Risk			Humidity		Temperature		
Category	Specific Risk (statistic)	Floor	Value (% RH)	Class	Value (°C)	Class	
Microbial	Germination threshold (P97.5)	Attic	70	B	—	—	
		2nd floor	73	C	—	—	
		1st floor	78	F	—	—	
		Basement	69	B	—	—	
Mechanical	Short-term variation (P95 of rolling 24 hour variation)	Attic	Δ15	b	Δ6.6	b	Overall mechanical risk:[1] Attic: b; 1st floor: b; 2nd floor: b; Basement: b
		2nd floor	Δ19	b	Δ3.4	a	
		1st floor	Δ20	b	Δ3.8	a	
		Basement	Δ12	b	Δ3.7	a	
	Seasonal variation (absolute difference of seasonal means, RH means ≦ 70%)	Attic	Δ1	a	Δ10.8	b	
		2nd floor	Δ1	a	Δ10.7	b	
		1st floor	Δ6	a	Δ10.3	b	
		Basement	Δ3	a	Δ9.2	a	
	Deviation from historical mean (absolute difference)	Attic	Δ14	b	Δ4.7	a	
		2nd floor	Δ17	b	Δ3.1	a	
		1st floor	Δ9	a	Δ2.5	a	
		Basement	Δ13	b	Δ3.7	a	
Chemical	Deviation from historical mean (difference)	Attic	—	—	Δ4.7	+	
		2nd floor	—	—	Δ3.1	+	
		1st floor	—	—	Δ2.5	0	
		Basement	—	—	Δ3.7	+	

[1] Overall mechanical risk is based on the highest risk classification in any of the specific humidity- and temperature-based mechanical risks (short-term variation, seasonal variation, and deviation from historical mean).

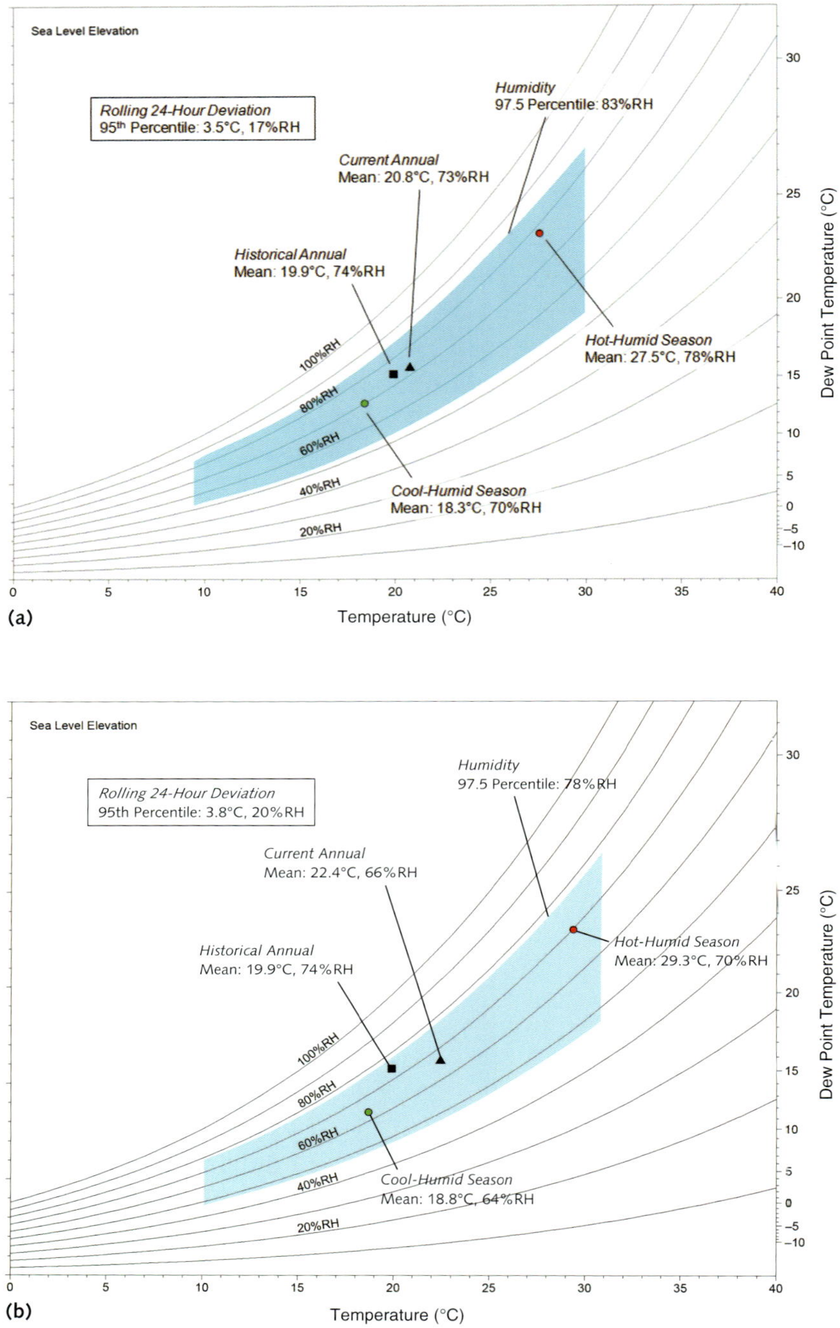

Figure 10.7

Psychrometric charts showing environmental data collected on the first floor of Hollybourne Cottage during preinstallation (a) and postinstallation (b). The boxes are bracketed by the 97.5 and 2.5 percentiles of temperature and humidity.

The pre- and postinstallation (part B) risk classes for Hollybourne Cottage are summarized in table 10.4.

Compared to the preinstallation classification, microbial risk was improved on all floors except the first floor, which remained at Class F (high risk). Mechanical risk due to seasonal humidity variation in the basement, first floor, and attic was decreased from Class f (high risk) during preinstallation to Class a (no risk) during postinstallation. Postinstallation attic temperatures increased mechanical risk owing to seasonal temperature variation from Class a during preinstallation to Class b (moderate risk) during postinstallation. Mechanical risk due to devia-

Table 10.4

Comparison of pre- and postinstallation (part B) risk classes at Hollybourne Cottage.

	Microbial Risk		Mechanical Risk														Chemical Risk	
			Humidity						Temperature						Overall[1]			
			Short-Term Variation		Seasonal Variation		Deviation from Historical Mean		Short-Term Variation		Seasonal Variation		Deviation from Historical Mean					
Floor	Pre	Post	Pre	Post	Pre	Post	Pre	Post	Pre	Post	Pre	Post	Pre	Post	Pre	Post	Pre	Post
Attic	F	B	b	b	f	a	a	b	b	b	a	b	a	a	f	b	+	+
2nd floor	F	C	b	b	a	a	a	b	a	a	b	b	a	a	b	b	0	+
1st floor	F	F	b	b	f	a	a	a	a	a	b	b	a	a	f	b	0	0
Basement	F	B	b	b	f	a	a	b	a	a	a	a	a	a	f	b	0	+

[1] Overall mechanical risk is based on the highest risk classification in any of the specific humidity- and temperature-based mechanical risks (short-term variation, seasonal variation, and deviation from historical mean).

tion from historical mean humidity was increased in the basement, second floor, and attic, going from Class a during preinstallation to Class b during postinstallation. Classification of chemical risk due to deviation from historical mean temperature also indicated increased risk in the postinstallation environments of the basement and second floor, both going from a preinstallation Class 0 (same risk as historic condition) to a postinstallation Class + (moderately increased risk).

Thermal comfort

Assessment of thermal comfort focused on the first floor of Hollybourne Cottage, as this level was the location for visitor tours and restoration activities, both of which took place largely during the hot-humid season. The mean postinstallation temperature and dew point temperature on the first floor were 29.7°C and 22.1°C, respectively. The predicted mean vote (PMV) and the predicted percentage dissatisfied (PPD) were calculated assuming a mean radiant temperature equal to the room temperature, a physical activity of 1.5 met (combination of standing and walking), a clothing factor of 0.6 clo (typical summer indoor clothing), and an air speed of 0.8 m/s. With these assumptions, the PMV was calculated to be 1.38, and the PPD was calculated to be 44% (slightly warm and not in compliance with ANSI/ASHRAE Standard 55-2013 [ANSI and ASHRAE 2013]) (point 2 on fig. 10.8). This was a slight improvement over the thermal comfort assessment using mean hot-humid conditions during preinstallation—27.6°C, 78% RH, and 0.1 m/s (point 1 on fig. 10.8).

Noise

Noise levels recorded in the first-floor Gun and Pantry Rooms, which contained supply fans, were approximately 68 dBA, which may represent an uncomfortably high level. This value dropped to 59 dBA in the hallway between the two rooms, and to between 50 and 54 dBA in the living room and family room. The second floor exhibited values similar to those of the first floor, except for even lower noise levels recorded in the east end of

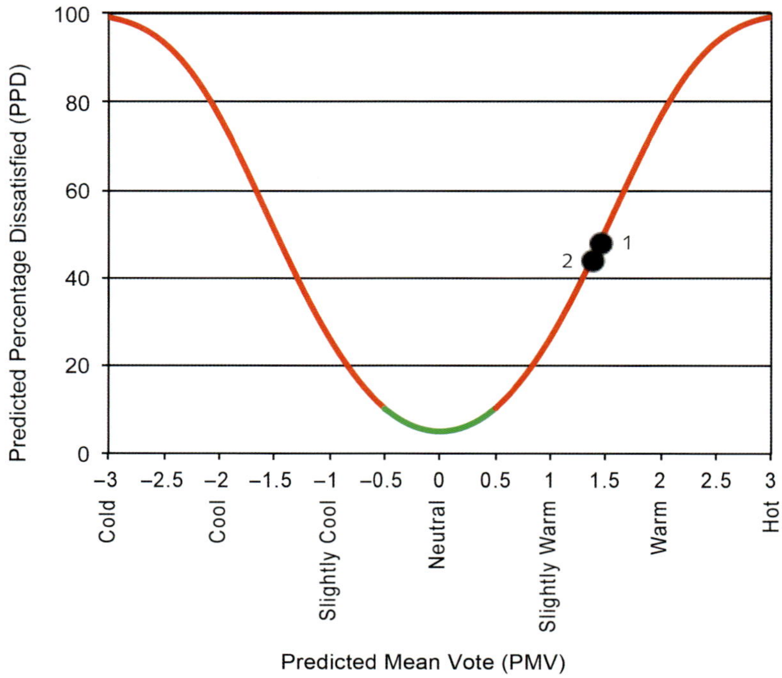

Figure10.8

Assessment of thermal comfort on the first floor of Hollybourne Cottage during (1) preinstallation and (2) postinstallation periods.

the building (45 dBA, which is a typical background office noise) due to greater compartmentalization and more partitions on the second floor.

Design of the part B environmental management system at Hollybourne Cottage focused on fan capacity and appropriate location to create the highest air change efficiency throughout the structure. Due to minimal visitation, design issues such as visitor routing, aesthetics, and noise reduction were not high priorities. However, it is expected that noise levels can be significantly lowered by informed equipment selection and incorporation of noise reduction techniques.

Maintenance

As maintenance for part B was similar to that of part A, please refer to the part A discussion.

Energy Metrics

Energy intensity

In the 280 m^2 basement the total capacity of the two dehumidifiers was 2.2 kW, with an energy intensity of 7.9 W/m^2, which was almost one-tenth that of the three heaters used in the part A experimental configuration (81 W/m^2). The two supply and three exhaust fans in the basement had a total fan capacity of 0.86 kW, and a combined energy intensity of 3.1 W/m^2. The total capacity of the basement system was 3.06 kW with an energy intensity of 10.9 W/m^2. Two supply fans, at 0.25 kW each, were used in the first and second floors, resulting in an energy intensity of 1.8 W/m^2 for each floor. The fans and heater installed in the 215 m^2 attic resulted in a total capacity of 8.0 kW and an energy intensity of 37.3 W/m^2.

Table 10.5

Operational use between August 2003 and April 2004 for whole-house ventilation, basement dehumidification, and attic heating.

Floor		Hours of Use (%)	Energy Use (%)
Attic	Heating	40	63
	Ventilation	3	2
	Idle	57	
2nd Floor	Ventilation	3	1
	Idle	97	
1st Floor	Ventilation	3	1
	Idle	97	

The energy rating of all floors in Hollybourne Cottage was 12.07 kW, and the total energy intensity of the 1055 m^2 building was 11.4 W/m^2 (87.4 m^2/kW), which is half the total energy intensity from part A (23 W/m^2).

Energy use intensity

Analysis of operational use and cost was focused on data collected from August 2003 to April 2004, the period encompassing the operation of the system described in part B. Based on operational use (hours operated as percent of total time), the environmental management system utilized the heating mode over the ventilation mode by a factor of roughly 10 (table 10.5). In the attic, the heating mode was used for 40% of the period, while the ventilation mode was used only 3% of the period. Similarly, the ventilation mode was operated only 3% of the time on both the first and second floors. In the basement, operational use indicated that the dehumidification mode was used during 34% of the period, and ventilation was used during only 3% of part B.

Equivalent energy use of the basement dehumidification and attic heating modes represented 95% of total operational use. Based on the total floor area of the building (1055 m^2) and the operational duration, the energy use intensity for whole-house ventilation, attic heating, and basement dehumidification was 26.6 $kWh/m^2/yr$. It should be noted that this energy use reflects a dataset lacking from May to July, a period that is typically hot-humid. Comparing the energy use of basement dehumidification and basement heating (described in part A) also showed that the energy use of three convective heaters was roughly three times that of two dehumidification units.

Conclusions

In addition to reducing humidity compared to preinstallation conditions, the use of ventilation on the intermediate floors was important from the perspective of visitor comfort, as it promoted air movement and introduced fresh air to the most actively used spaces.

Though the use of dehumidifiers in the basement was initially more expensive and required increased operational time compared to the use of heaters, two dehumidifiers consumed 90% less power than three convective heaters. A calculation of simple payback based on operational costs, in which only heating or dehumidification was used in the basement, indicated that the capital investment in dehumidification would be recouped in approximately one year. The increased maintenance costs of dehumidifiers relative to heaters, however, were not taken into account in the calculation.

While the merging of operational bands of interior humidity for the ventilation and dehumidification modes suggested that further reductions might be possible, lower humidity levels may not be desirable because of the possibility of salt-related damage to the exterior walls of the building and a larger departure of interior humidity from the historic annual mean condition. Described in part A, a covered floor experiment demonstrated that a significant amount of moisture vapor was permeating from the soil through the basement slab floor. However, the surface of the concrete basement floor remained dry throughout system operation. During the study, periodic efflorescence appeared on the basement floor (likely sodium chloride, which deliquesces at 75% RH), indicating that the target humidity level was sufficiently low to allow crystallization of the salt on the slab surface. If the target humidity levels were too low in the basement, subefflorescence could occur within the basement floor slab or the more fragile brick piers, resulting in deterioration. Salt-related damage would be a lower risk for the foundation walls due to the cavity break between the interior and exterior surfaces of the below-grade walls.

Indicating protection against microorganism collection damage, microbial risk Classes B (low risk) or C (moderate risk) were achieved for all floors, except for the first floor, with the installation of whole-house ventilation, attic heating, and basement dehumidification. The first-floor performance may have been compromised during summer by extended periods when windows and doors were open and the first floor was occupied for restoration activities. The part B scheme also provided protection against the mechanical risk from short-term variations, seasonal variations, and deviations from historic mean. While elevated temperatures did not result in higher mechanical risk relative to the preinstallation period, these temperature increases did increase chemical risk for collections, particularly in the basement environment.

Examining the resulting environmental control classes following application of the different strategies described in parts A and B of the case study of Hollybourne Cottage shows that both experimental configurations generally improved the conservation environment in the building compared to the preinstallation environment. In part B, installation of an attic heater eliminated high-humidity conditions during the cool-humid season. However, elevated temperatures in the attic and second floor may have been avoided by the use of dehumidification. Although postinstallation PPD/PMV values did not improve significantly over those of preinstallation, staff members of the Jekyll Island Authority expressed their feeling of improved thermal comfort due primarily to increased interior air velocity. The staff members also noted the elimination of musty mold odors from the first and second floors that were present during preinstallation. Noise levels of part B (whole-house ventilation and the use of dehumidifiers) were higher than in part A throughout the cottage.

Lessons Learned from Hollybourne Cottage, Part B

- Microbial risk Classes B and C are achievable in a Zone 2A climate with a simple system of heating, dehumidification, and ventilation using humidistatic control.
- Replacement of convective unit heaters with dehumidifiers will reduce operational and energy cost.
- High-humidity events in the attic during the cool-humid season can be managed by the placement of heaters.
- Ventilation was used for only 3% of the available hours for maintaining the collection environment.
- Ventilation alone cannot maintain a collection environment in a hot-humid climate.
- Comparing the approach used in parts A and B shows that although ventilation in the first- and second-floor spaces provided the benefits of general humidity reduction, improved indoor air quality, and thermal comfort, ventilation produced higher humidity variations in the space.
- Basement humidity can be effectively managed using only heating or dehumidification.
- Good succession planning for personnel is required to maintain operation.

Future Considerations

Though the environmental conditions of each space were actively addressed during this operational period, the environmental management system configuration for Hollybourne Cottage should focus on the needs of each individual floor. The basement should continue to be segregated from the upper floors, as it is the dominant site for moisture infiltration and remains a cool space during the hot-humid season. The choice between heating and dehumidification will balance the reduced energy costs of mechanical dehumidification versus the lower capital and maintenance costs of heating. In either case, the humidity set point of heating or dehumidification should be maintained at from 60% to 65% RH in order to limit energy use and the possibility of salt-related damage.

If there is no visitation or work activity in the space, basement environmental management can omit ventilation and rely solely on the heating or dehumidification mode, simplifying the system configuration for establishing a conservation environment for the building fabric. Though basement heating or dehumidification was paired with ventilation in parts A and B, other experimental configurations of the environmental management system successfully employed basement heating or dehumidification as its sole psychrometric strategy. Resulting data can be found in "Testing of climate control scheme alternatives to conventional air-conditioning at Hollybourne Cottage, Jekyll Island Historic District, GA" (Maekawa, Beltran, and Toledo 2007).

From a building conservation standpoint, direct mechanical ventilation of the first and second floors of Hollybourne Cottage is of limited value, as these floor spaces were shown to be effectively protected by the controlled environments of the basement and attic in part A of this case study. However, separate mechanical ventilation of these spaces may be necessary for improved occupant thermal comfort and indoor air quality, particularly on the first floor. This can be accomplished by manually activating the ventilation system for comfort and fresh air. While such actions would likely result in higher interior humidity, short-term elevations will not cause significant microbial activity. The use of air recirculation with particulate filters, although it would not introduce fresh air, might be another strategy for addressing occupant comfort.

The present alternative environmental management strategies applied at Hollybourne Cottage yielded a 67% savings in operational costs over those of a conventional HVAC system. Furthermore, the capital cost was approximately 5% of that of a conventional air-conditioning system (estimate based on values presented by Ernie Conrad 2006).

System design efforts should also address noise generated by equipment, particularly for ventilation and dehumidification. As sound typically registers between 60 and 80 dB at each unit, visitation experiences and work efficiency may be negatively impacted by equipment proximity. While double casing and elbow ducts are typically used when ductwork is present, clever placement of equipment and design of the visitor path, as well as the positioning of noise reduction forms on wall surfaces, can mitigate noise in historic structures.

Postscript

At the end of the research project in late 2005 the environmental management system in Hollybourne Cottage was handed over to the Jekyll Island Authority. Prior to the transfer, the control/operational software was modified to include automated phone calls (call-outs) notifying staff members of the Authority's facilities department in the case of abnormal environmental conditions in the building. While the Authority continued to operate the system, maintenance was not performed, as personnel changes followed the hand-over period. The system stopped operating prior to 2010. In August 2012, a collaborative effort between the Jekyll Island Museum and the Getty Conservation Institute restored the environmental management system. Actions included the replacement of fan filters, adjustments of fans and shutters, and repair of dehumidifiers. The environmental management system currently maintains a conservation environment in the building.

Project Team and Responsibilities

Jekyll Island Museum (Jekyll Island Historic District, Georgia, United States)

Warren Murphey (Former Director)—Local project management

John Hunter (Current Director)—Local project management

Preservation Department—Installation of mechanical and electrical systems, and operation/annual and seasonal maintenance

Getty Conservation Institute (Los Angeles, California, United States)

Shin Maekawa (Senior Scientist)—Project management and concept and engineering design for the environmental management strategies and environmental monitoring system

Franciza Toledo (Assistant Scientist)—Building interior condition assessment, architectural design, supervision of installation

Vincent Beltran (Assistant Scientist)—Management of environmental data and programming of control system

References

ANSI [American National Standards Institute] and ASHRAE
2013 *Standard 55-2013—Thermal Environmental Conditions for Human Occupancy*. Atlanta: ASHRAE.

Conrad, E.
2006 Understanding mechanical systems that support preservation environments. Paper presented at the 20th Annual NARA Preservation Conference, "Beyond the Numbers: Specifying and Achieving an Efficient Preservation Environment," Washington, DC, March 16, 2006.

Maekawa, S, V. Beltran, and F. Toledo
2007 Testing of climate control scheme alternatives to conventional air-conditioning at Hollybourne Cottage, Jekyll Island Historic District, GA. Paper presented at the 2006 annual conference of the Association for Preservation Technology International (APT 2006), Atlanta, and accepted for publication in *APT Bulletin*, 2007.

Note

1 The sixth experimental phase, April 2004 to October 2005, maintained the same equipment configuration described here in part B, but it slightly modified the control conditions for ventilation by requiring that the interior dew point temperature be greater than the dew point temperature of the outside air.

Chapter 11

Historic Archive of San Cristóbal de La Laguna, Tenerife, Spain: Ventilation and Conservation Heating

Climate zone 4C (Mixed-Marine)

Objectives

- Eliminate events/conditions above 75% RH.
- Maintain a stable humidity close to the historic average of the site.
- Limit high temperatures to below 30°C and provide outside air to maintain a good indoor air quality.
- Install a low-cost, robust, and low-energy environmental management system.
- Limit placement of mechanical equipment that can obstruct the daily activities of archivists.

Summary

Located in a nineteenth-century masonry building in San Cristóbal de La Laguna, Tenerife, Spain, the Historic Archive of San Cristóbal de La Laguna consists of two connected rooms with a total of 30 m^2 of floor area in the larger Municipal Archive; the historic building contains historically important documents of the Canary Islands. Situated in a Mixed-Marine (4C) climate zone, the archive suffered from a cool and damp environment that resulted in frequent outbreaks of mold on the documents. Portable mechanical dehumidifiers were placed throughout the building to limit high-humidity conditions. However, this environmental management strategy was unsuccessful, as dehumidifiers were operated improperly, particularly over long weekends and holidays, owing to a lack of coordination between the conservation, custodial, and maintenance departments.

The goal of the project was to install and operate a simple, low-cost, and robust environmental management system in the Historic Archive. This system concept would serve as a pilot study and, if successful, would be expanded to the entire Municipal Archive.

Several nonmechanical strategies were implemented at the Historic Archive. Document shelving was reorganized to allow cross ventilation, and a vestibule area was created to control infiltration of humid air from adjacent uncontrolled areas of the Municipal Archive. The mechanical strategy for establishing a conservation environment consisted of humidistat-controlled ventilation and heating, designed to maintain the humidity below 75% RH to prevent microbial damage to the collection. Environmental monitoring following installation of the environmental management system confirmed an improvement in the collections environment.

Background

Located in the historic district of San Cristóbal de La Laguna, the Municipal Archive is situated on an artificially filled native wetland in the highlands of eastern Tenerife Island, Spain. In spite of its proximity to the Tropic of Cancer, the surrounding area is known for a cool and humid climate as a result of its 500 m elevation. Moldy odors, fungal infestation, and foxing of paper are common problems in historic buildings in the region (fig. 11.1). A paper conservator in charge of collections for the larger Municipal Archive attempted to combat these issues through the use of periodic open-window (natural) ventilation and portable mechanical dehumidifiers. Natural ventilation was problematic because of issues of security, insect entry, and dust intrusion. Reliable operation of the portable mechanical dehumidifiers was hindered by temperamental safety switches for condensate reservoirs. These switches would shut off the dehumidifiers when the reservoirs were erroneously positioned or shifted from their positions when the dehumidifiers were accidentally moved or bumped. Even if reservoirs were positioned properly, they quickly filled to capacity, leading to humid conditions in the archives, particularly following weekend periods when dehumidifiers were automatically shut down because of filled reservoirs.

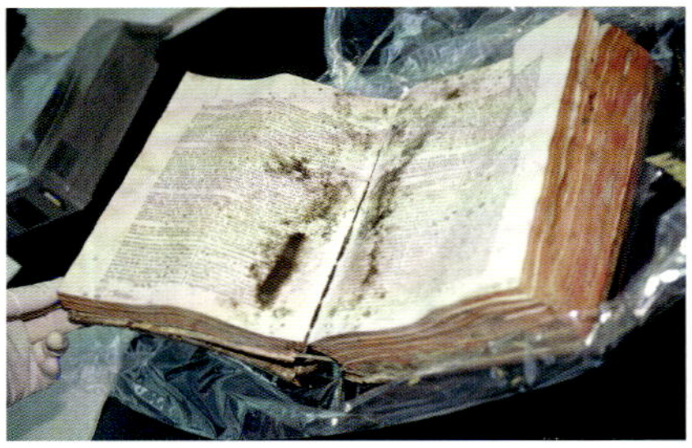

Figure 11.1
Mold-infested book in the collection of the Historic Archive of San Cristóbal de La Laguna. Photo: Shin Maekawa. © J. Paul Getty Trust.

An assessment of the fungal and bacterial infestation in the collection by a staff conservator and a consulting microbiologist confirmed the need for a more reliable means of protecting the collection from microbiological damage. The first proposal for improvements to the collection environment was based on a conventional environmental control system with air-conditioning to control both temperature and humidity for the small but vitally important Historic Archive section of the Municipal Archive. However, consultation with a local HVAC engineer highlighted the high installation and operational costs of the proposed system and the limited space available in the Historic Archive for the necessary HVAC equipment.

Climate

The Historic Archive is housed in a building located in the city and municipality of San Cristóbal de La Laguna (28°48' N, 16°31' W) on the northeast highland of Tenerife Island in the Canary Islands, Spain (fig. 11.2). Designated a World Heritage Site by UNESCO in 1999, this historic city sits at an elevation of 505 m above sea level and is less than 2 km from the Tenerife North Airport (Los Rodeos Airport). The highest elevation on Tenerife Island is the summit of El Teide volcano (3718 m above sea level) in the southwest portion of the island.

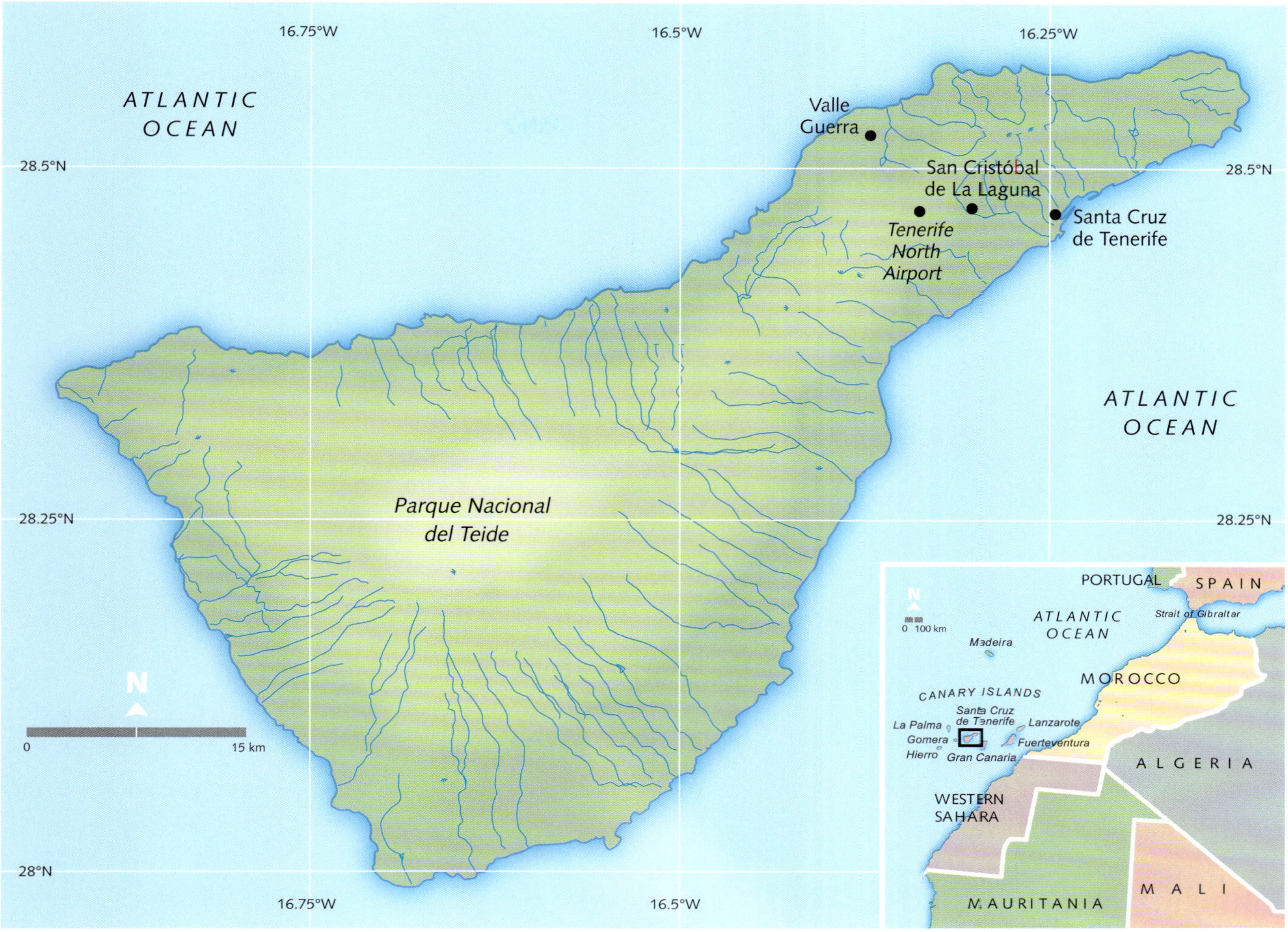

Figure 11.2

Map of Tenerife, in the Canary Islands, Spain.

Eastern Tenerife's climate is moderated by northeast trade winds that originate in an anticyclone position over the neighboring Azores Archipelago during most of the year. Moisture from these winds condenses in the northeast part of the island, producing a consistent cloud layer between 600 and 1800 m in altitude. The climate on Tenerife is also buffered by the Canary Current, a cold marine current that affects the local coastal climate, reducing daily temperature fluctuations and rainfall and increasing humidity. Climatic variations result from the arrival of winter storms from northern Europe, producing rainfall from October to April, and from the dry southeast winds carrying particulates from the Sahara Desert, producing dry hazy weather from May to September.

Based on weather data collected at Los Rodeos Airport (28°48' N, 16°34' W; 617 m above sea level) from 1971 to 2000 (AEMet 2012), the monthly mean temperature ranged from 13°C in January to 21°C in August, while the monthly mean dew point temperature ranged from 8.9°C in January and February to 14.8°C in September (fig. 11.3). Calculated from the monthly mean temperature and dew point temperatures, average monthly humidity ranged from 68% in August to 77% in January. The historical annual exterior mean temperature and humidity were 16.5°C and 73% RH, respectively. The mean annual rainfall total

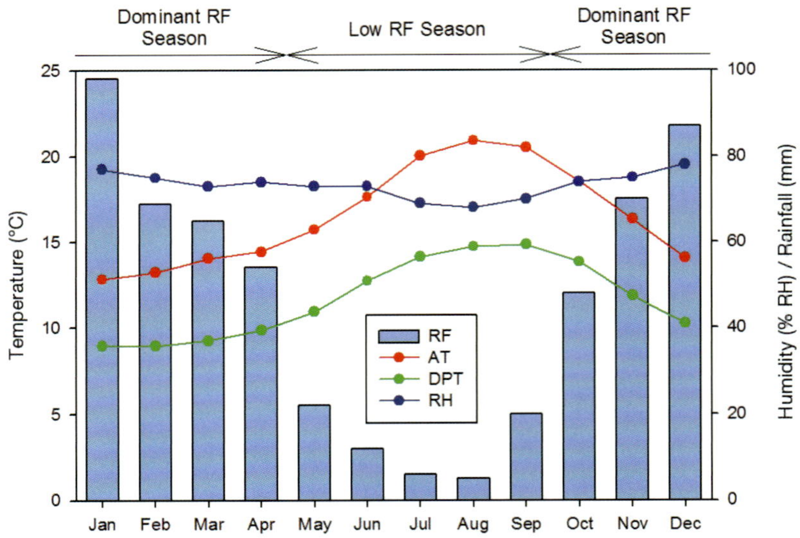

Figure 11.3

Monthly average temperature (AT), humidity (RH), dew point temperature (DPT), and rainfall (RF) at Tenerife North Airport, Tenerife (1971–2000) (data from AEMET).

was 557 mm, and the majority of this rainfall is recorded from October to April, with peak rainfall (98 mm) recorded in January. The dominant rainfall season extends from October to April, and the low-rainfall season lasts from May to September. It should be noted that the climate on Tenerife can be highly localized; for example, southern areas of the island are dry in comparison to the climate in the vicinity of San Cristóbal de La Laguna.

Five years (2004–10) of climate data from Los Rodeos Airport produced a CDD10°C of 2153 cooling degree days (Weather Channel 2014) and HDD18°C of 1108 heating degree days. These indices place San Cristóbal de La Laguna in a Mixed (4) thermal zone. Based on the moisture climate type flowchart presented in chapter 1 (see fig. 1.4), the moisture climate is classified as a Marine (C) type because of the following ordered criteria: the mean temperature of the coldest month (13°C) is between –3°C and 18°C; the mean temperature of the warmest month (21°C) is less than 22°C; at least four months have mean temperature above 10°C (monthly mean temperatures are above 10°C throughout the year); and the cold season month with the heaviest rainfall (98 mm in January) has at least three times more rain than the warm season month with the least rainfall (5 mm in August). Combining the thermal and moisture classification criteria, San Cristóbal de La Laguna is in a Mixed-Marine (4C) climate zone.

Building

The Historic Archive is housed within a massive nineteenth-century masonry building that is representative of civil neoclassical construction in the Canary Islands (fig. 11.4). Originally used as a convent by Dominican nuns, the building was acquired in 1975 by the Municipality of San Cristóbal de La Laguna and adapted to municipal use in 1985. The larger part of this building is taken up by the Municipal Archive, in which the smaller Historic Archive, the subject of this case study, is located on the first floor. The two-story Municipal Archive has an aboveground basement and a central patio. Because of its heavy masonry

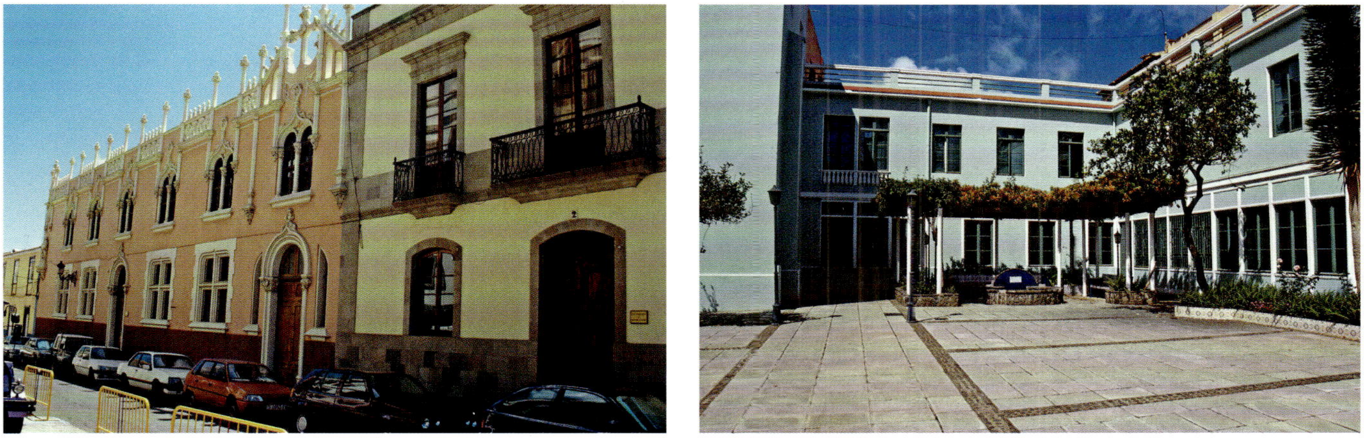

(a) (b)

Figure 11.4
Facade of the municipal building of San Cristóbal de La Laguna (a); the Historic Archive is located on the first floor of the Municipal Archive (painted pink). The Municipal Archive also has a patio (b). Photos: Shin Maekawa. © J. Paul Getty Trust.

walls with plaster and tight construction, this building is categorized as ASHRAE Class IV (ASHRAE 2011).

The Historic Archive is located in two adjacent first-floor rooms at the west corner of the Municipal Archive (fig. 11.5). The building footprint is 825 m^2, and the Historic Archive footprint is 25.4 m^2. Access to the Historic Archive is through a small entry room (1.7 × 2.5 m; height, 4.0 m), which then provides access to a large main room (6.6 × 3.2 m; height, 2.3 m). The walls and ceiling are finished with a cement plaster and painted with a latex paint. The floor is concrete covered with vinyl tiles. A northwest-facing window in the small entry room and a southwest-facing window in the large main room were permanently closed because of security and environmental concerns. The rooms are accessed daily by the archives staff, but public access is limited.

Historic Archive collection

The collection in the Historic Archive consists of historically important documents of the Canary Islands dating from the eighteenth century.

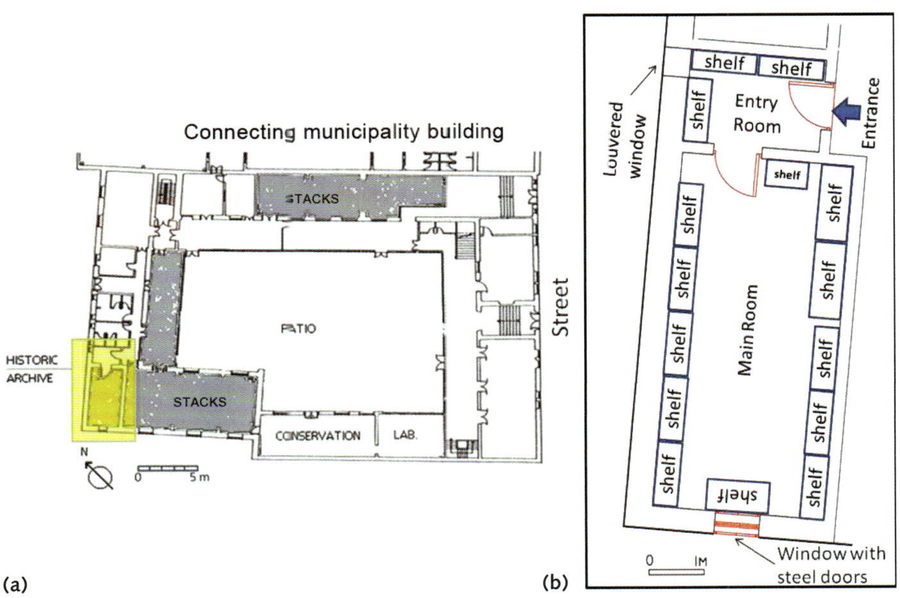

Figure 11.5
The first-floor plan of the Municipal Archive (a), showing the locations of archival stacks (gray areas), the patio, the conservation laboratory, and the Historic Archive (yellow rectangle); and the preinstallation floor plan of the Historic Archive (b).

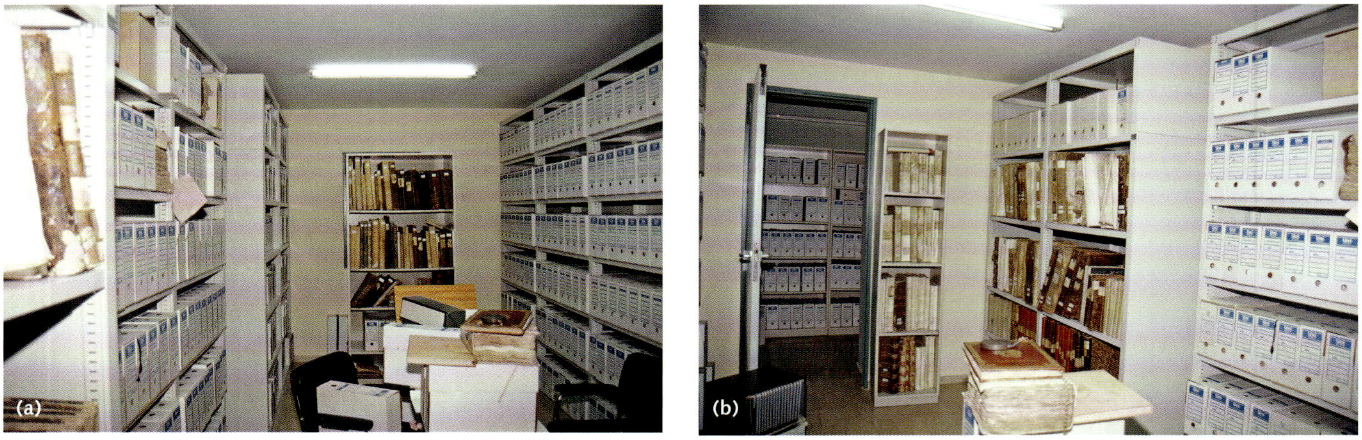

Figure 11.6
The main room of the Historic Archive, where documents were housed in archival boxes and placed on open shelves along interior and exterior walls (a), and the open doorway leading from the Historic Archive's main room to the entry room, visible at the far end of the photo (b). Photos: Shin Maekawa. © J. Paul Getty Trust.

Most of the documents are paper, though some parchment material and leather-bound books are included. The majority of archival documents were unbound and housed in document cases made of acid-free paper or polypropylene. Document cases were stored on open shelving units placed against the walls (fig. 11.6). Bound documents and books with leather spines and covers were also stored on open shelves.

Preinstallation Condition Assessments

Building envelope assessment

The refurbishment of San Cristóbal de La Laguna's municipal building in 1985 retained the facades and most of the exterior walls (approximately 40 cm thick), but a number of internal partitions were removed to allow for more spacious rooms. The majority of exterior doors and windows for the Municipal Archive were permanently closed to improve security and protect against animal and insect intrusion. Although some windows remained operable, the envelope modifications altered the original natural ventilation patterns of the building, resulting in numerous pockets of still air and necessitating the use of dehumidifiers to control elevated humidity levels in underventilated rooms.

The two small rooms forming the Historic Archive are located on the ground floor, and its exterior walls are shaded by surrounding buildings. As a result, the rooms and the interior surfaces remain cool and damp at all times, elevating interior humidity and moisture content of the building materials. While not used to house archival materials, the small entry room has a steel entrance door providing security and segregating the Historic Archive from the remainder of the Municipal Archive. A second steel door separates the entry room from the main room, which contained all Historic Archive materials. Each room in the Historic Archive has one operable window on an exterior wall; however, the windows are kept closed, and bookcases block access (fig. 11.6). The steel door between the two rooms potentially creates different environments. The windows and doors had wide gaps at their frames when closed. With the entrance door to the Historic Archive closed and the shared door open, the air change rate was between 2.1 and 2.2 air changes per hour (ACH); thus the Historic Archive was considered to have a leaky envelope.

The building assessment suggested that the shelving units be rearranged to expose the windows in each room to allow for ventilation within the spaces. Because the Historic Archive is accessed by municipal archivists on a daily basis, its entrance door and its interior door are left open when staff is inside, allowing air exchange with adjacent spaces in the Municipal Archive. Therefore, the building assessment recommended that a vestibule be established at the entrance to the Historic Archive to limit air exchange when the room is accessed.

Exterior environmental assessment

The exterior climate was monitored at the northwest side of the building housing the Historic Archive from May 1999 to January 2002. Although monthly mean air temperatures (14.1°C in February to 21.4°C in August) were similar to those from the historical data collected at Los Rodeos Airport, monthly mean dew point temperature (8.9°C in February to 16.8°C in September) and humidity (73% RH in February to 87% RH in December) during the study period were markedly higher (fig. 11.7). The overall mean rolling 24 hour variations in air temperature, dew point temperature, and humidity were Δ7.1°C, Δ4.2°C, and Δ33% RH, respectively.

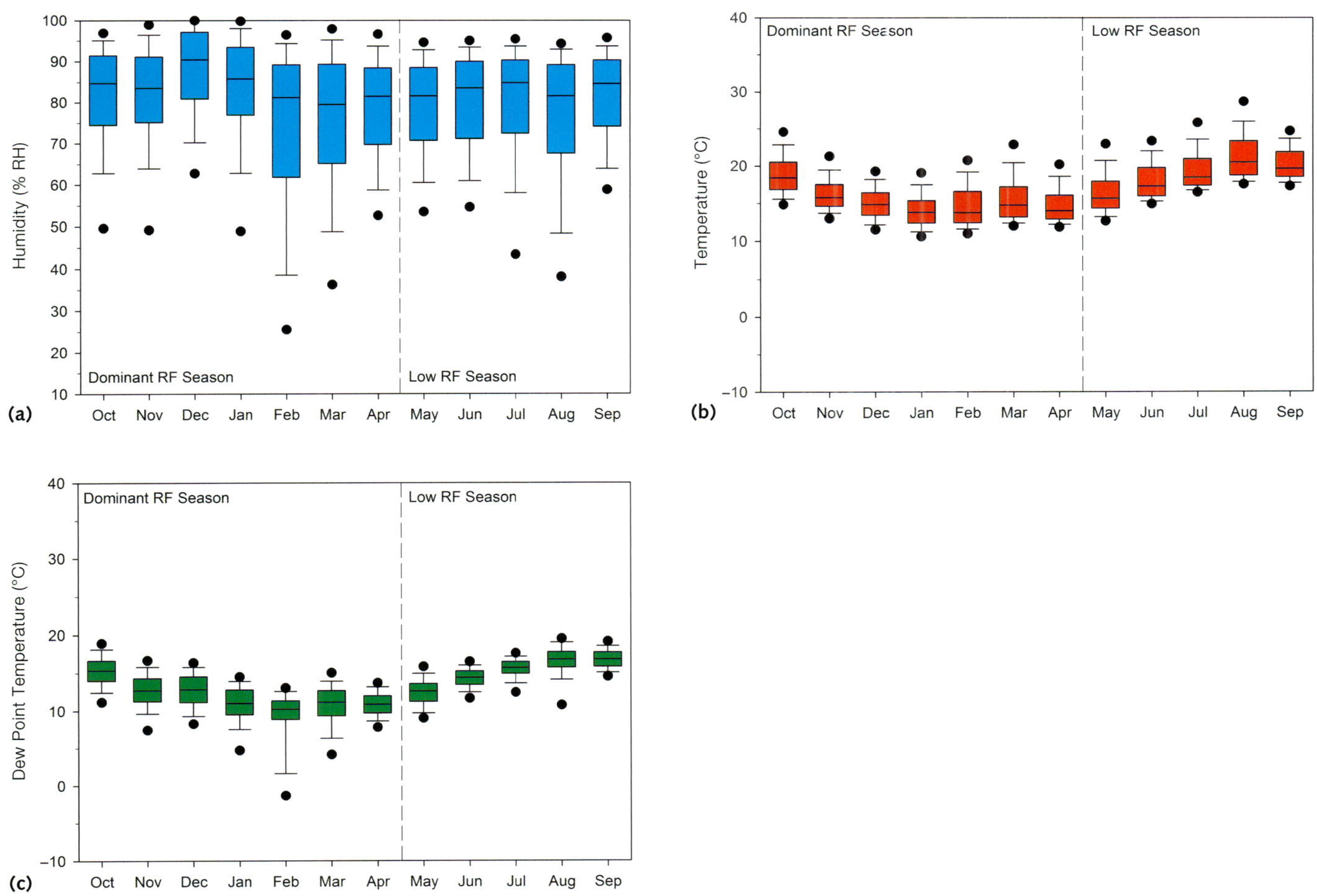

Figure 11.7

Humidity, temperature, and dew point temperature measured outside of the Historic Archive building from May 1999 to January 2002.

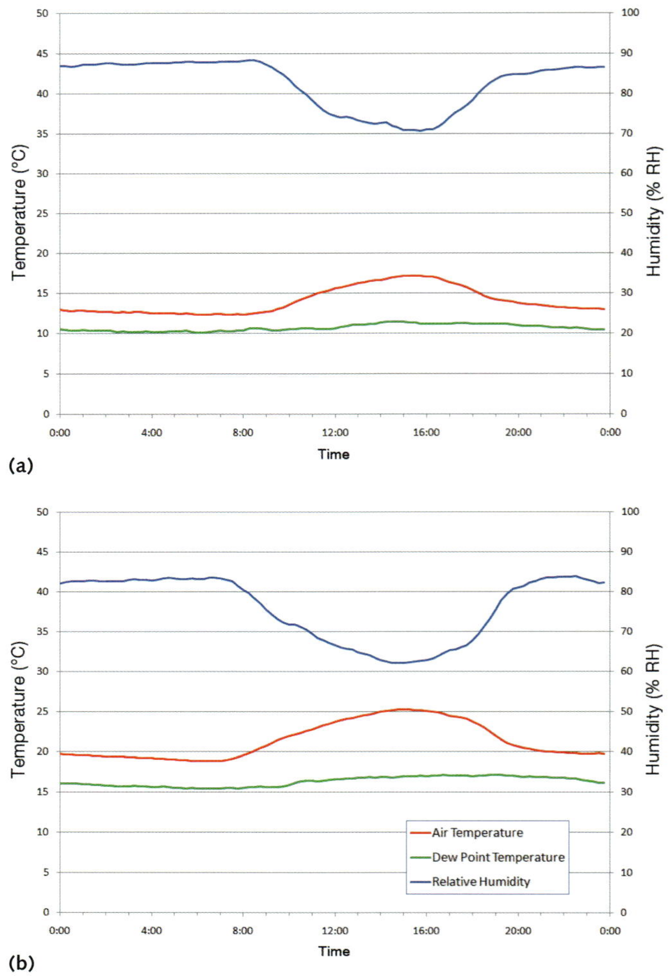

Figure 11.8

Mean daily trends of exterior temperature, dew point temperature, and humidity during (a) the dominant rainfall season (January), and (b) the low-rainfall season (August).

Mean daily trends of exterior temperature, dew point temperature, and humidity during the dominant rainfall and low-rainfall seasons are shown in figure 11.8. The dew point temperature remained constant throughout the day for both seasons. The daily temperature variation during the dominant rainfall season was between Δ3° and Δ4°C and increased to between Δ5°C and Δ6°C during the low-rainfall season. The larger temperature variation during the low-rainfall season resulted in humidity below 70% RH between 11:00 a.m. and 6:00 p.m., providing an opportunity for conservation heating by ventilation in the Historic Archive.

Interior environmental assessment

Humidity

The preinstallation interior environment of the Historic Archive was monitored from May 1999 to April 2000. Monthly mean interior humidity levels ranged from 62% RH in May to 72% RH in August (fig. 11.9). Although the humidity threshold of 75% RH was exceeded during less than 2% (a total of 7 days) of the preinstallation period, humidity levels exceeded 75% RH (with a high of 84% RH) during nearly 20% (a total

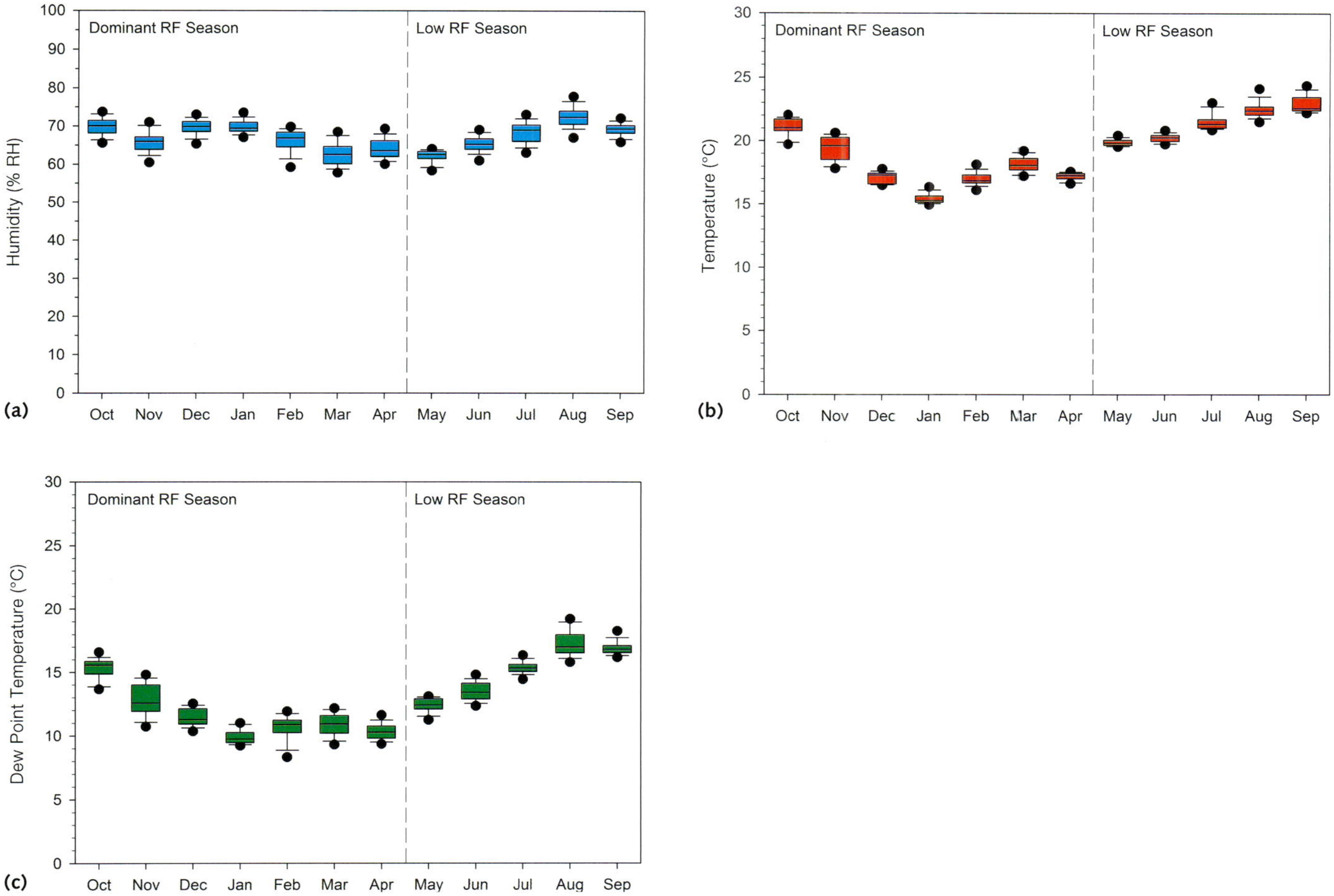

Figure 11.9

Preinstallation humidity (a), temperature (b), and dew point temperature (c) recorded in the Historic Archive from May 1999 to April 2000.

of 6 days) of the month of August, a sufficient continuous duration for microbial germination. Staff archivists generally described the space as cool and damp with a strong and musty mold smell. Monthly mean rolling 24 hour humidity variation ranged from Δ3% to Δ6% RH, but a maximum variation of Δ23% RH was recorded.

Temperature and dew point temperature

Prior to installation of an environmental management system, monthly average temperature and dew point temperature in the Historic Archive ranged from 15.4°C (January) to 22.9°C (September) and from 9.9°C (January) to 17.2°C (August), respectively (fig. 11.9). The maximum recorded temperature was 24.6°C, while maximum dew point temperature was 19.9°C. The monthly mean rolling 24 hour variations of temperature (Δ0.2–Δ0.7°C) and dew point temperature (Δ0.6–Δ1.5°C) were relatively narrow throughout the year.

Preinstallation risk class assessment

The Conservation Environment Classification–HH protocol for the preinstallation interior environment at the Historic Archive is summarized in table 11.1.

Table 11.1

Conservation Environment Classification–HH protocol of the Historic Archive environment before installation of the environmental management system.

Risk		Humidity		Temperature		
Category	Specific Risk (statistic)	Value (% RH)	Class	Value (°C)	Class	
Microbial	Germination threshold (P97.5)	75	C	—	—	
Mechanical	Short-term variation (P95 of rolling 24 hour variation)	Δ10	a	Δ0.8	a	Overall mechanical risk:[1] a
	Seasonal variation (absolute difference of seasonal means, RH means ≦ 70%)	Δ0	a	Δ3.5	a	
	Deviation from historical mean (absolute difference)	Δ6	a	Δ2.9	a	
Chemical	Deviation from historical mean (difference)	—	—	Δ2.9	0	

[1] Overall mechanical risk is based on the highest risk classification in any of the specific humidity- and temperature-based mechanical risks (short-term variation, seasonal variation, and deviation from historical mean).

Environmental Management System

Objectives

The primary environmental management objective is to arrest microbial activity in the Historic Archive's collections while maintaining a stable humidity level that is close to the collection's historical value recorded at Los Rodeos Airport. The specific objectives are:

- Eliminate events/conditions above 75% RH in the space.
- Maintain a stable humidity close to historic average of the site.
- Limit high temperatures to below 30°C and provide outside air to maintain good indoor air quality.
- Install and operate an environmental management system that is low in cost, robust, and low in energy consumption.
- Limit placement of mechanical equipment that can obstruct daily activities of archivists.

Psychrometric strategies

The nonmechanical psychrometric strategies at the Historic Archive focused on source moisture control by allowing air movement throughout the interior space and limiting uncontrolled passage of air through the envelope, which encompassed both exterior and interior walls. Following installation of the selected nonmechanical strategies, the mechanical psychrometric strategies were installed; they were conservation heating by ventilation and conservation heating using a convective heating device to limit interior humidity. This equipment was controlled by the readings of humidity sensors, one outside and another inside the Historic Archive.

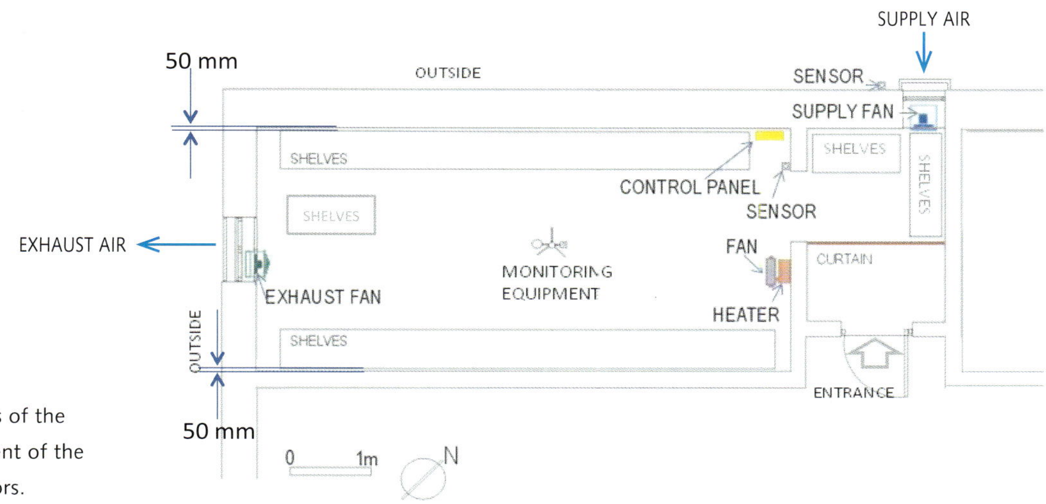

Figure 11.10
Schematic plan of the two rooms of the Historic Archive and the placement of the mechanical equipment and sensors.

Nonmechanical strategies

The nonmechanical strategies were selected to improve ventilation and mixing of air, as well as the performance of the building/room envelope to reduce uncontrolled infiltration/exfiltration (fig. 11.10). They were:

- The entry room was divided in half with a workshop-type curtain consisting of overlapping thick vinyl strips, which created a vestibule, as recommended by the building assessment. This curtain reduced the air change rate when the steel entry door was open.
- The door between the two rooms of the Historic Archive was removed to enable cross ventilation between the two operable exterior windows.
- Wall shelving units were relocated 5 cm away from walls, and the shelf blocking the operable window in the southwest-facing wall was moved toward the center of the main room to improve ventilation.
- Gaps around the windows and the entrance door were sealed with an oil-based putty (an off-the-shelf weather-stripping material) to reduce uncontrolled infiltration/exfiltration.
- Shelving and floor were cleaned with a vacuum cleaner and dry cloths.
- Materials compromised by fungal infestation were removed from the space.

After completion of the above improvements, the air change rate of the two rooms of the Historic Archive was measured using CO_2 as the tracer gas. With the steel entrance door open, the air change rate ranged from 0.7 to 1.2 ACH, a reduction from the prior measurement of 2.1 to 2.2 ACH with the entrance door closed.

Mechanical strategies

As stated above, the mechanical strategies were conservation heating by ventilation and conservation heating with heaters. As shown previously in the daily trends of external temperature and humidity (see fig. 11.8), warm and dry (less than 70% RH) outside air is typically available for ventilation from midday to about 5:00 p.m. during the low-rainfall season. The availability of warm and dry air offers the opportunity to employ conservation heating by ventilation as the primary humidity management strategy during this season. However, during the dominant rainfall season, midday exterior air—generally cooler and more humid than the interior air—is unsuitable for conservation heating by ventilation. Therefore, during the dominant rainfall season, conservation heating via forced-air heating becomes the primary humidity management strategy.

Mechanical system description

Installed in May 2000, the environmental management system at the Historic Archive consists of conservation heating by ventilation using a set of supply and exhaust fans, as well as conservation heating using a mixing fan and convection heater. See table 11.2 for equipment specifications, figure 11.10 for the locations, and figure 11.11 for images of the installed equipment. Supply and exhaust fans were selected to produce 6 to 8 ACH, while the convective heater was sized to produce a small temperature increase (less than Δ5°C) in the space. All equipment installed in the Historic Archive was readily available and appropriate for residential use, and none of it required specialized power sources, such as high-voltage or three-phase power.

Table 11.2

Equipment specifications for conservation heating at the Historic Archive (NA = not available).

Equipment	Type	Fan Capacity	Power	Noise
Supply fan	Propeller	467 m^3/h	150 W	54 dB
Exhaust fan	Propeller	459 m^3/h	200 W	54 dB
Mixing fan	Propeller	85–382 m^3/hr (variable)	80 W	37 dB
Convection heater	Electroresistive	0–170 m^3/hr	1.6 kW	NA

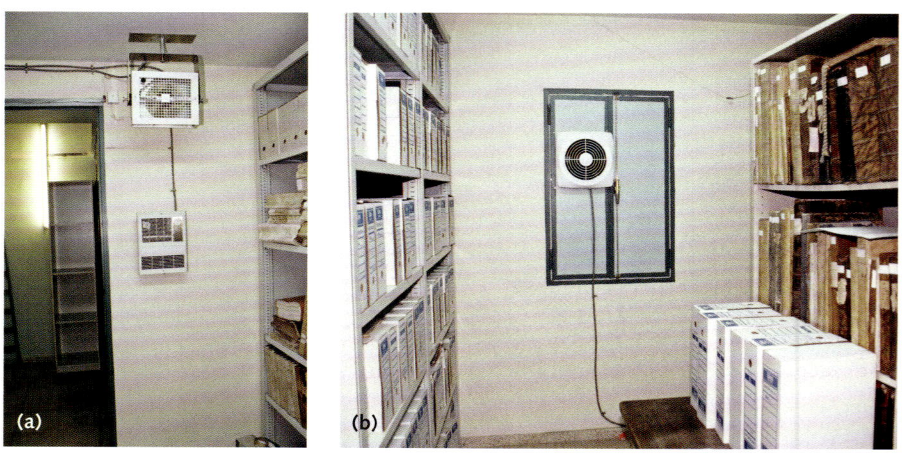

Figure 11.11

Environmental management system installed in the Historic Archive. A convection heater is mounted on the wall, and a recirculating fan is mounted from the ceiling (a); an exhaust fan is mounted on one of the steel window closures (b). Photos: Shin Maekawa. © J. Paul Getty Trust.

Supply and exhaust fans mounted in the windows at opposite ends of the Historic Archive provide forced cross ventilation through the entry room and main room. The supply fan is equipped with a 5 cm thick pleated disposable (MERV 8) filter with antimicrobial treatment, to control the quality of incoming outside air, as well as with a gravity-operated damper to limit infiltration. Spring-loaded dampers were installed at the exhaust fan to limit infiltration. A convection heater and a mixing fan are mounted on the northeast wall and ceiling, respectively, of the main room, near the open doorway to the entry room.

Operational control sequence

A programmable microprocessor-based controller/datalogger controls the environmental management system equipment and collects data. Two sets of temperature and humidity sensors (one in the center of the main room and the other outside the building near the supply fan/window) provide inputs to the controller. Three modes of operation are defined by a comparison of interior and exterior humidity (table 11.3).

Ventilation is enabled when exterior humidity is less than or equal to 70% RH and interior humidity is greater than 75% RH. Ventilation is disabled when interior humidity is less than or equal to 65% RH or when exterior humidity exceeds 70% RH. When the ventilation strategy is applied, the supply fan delivers fresh outside air to the Historic Archive through the entry room window, and interior air is exhausted to the outside through the exhaust fan on the window at the southwest end of the main room (fig. 11.12a).

Conservation heating is enabled if the temperature is less than 30°C and exterior and interior humidity both exceed 75% RH. When the conservation heating strategy is employed, the convection heater and mixing fan operate together to recirculate air through the heater (fig. 11.12b). Conservation heating is disabled when either interior humidity or exterior humidity falls below 70% RH or the interior temperature is less than 30°C. As a safety measure, conservation heating is also disabled if the temperature reaches 30°C. Otherwise, the temperature in the archive is not mechanically controlled.

Table 11.3

Operational control conditions for the environmental management system. Note that heating is disabled when the interior temperature exceeds 30°C (hardware switch).

Psychrometric Strategy	Enable	Disable
Ventilation	$H_{interior} > 75\%$ RH and $H_{exterior} \leqq 70\%$ RH	$H_{interior} \leqq 65\%$ RH or $H_{exterior} > 70\%$ RH
Conservation heating	$H_{interior} > 75\%$ RH and $H_{exterior} > 75\%$ RH and $T_{interior} < 30°C$	$H_{interior} \leqq 70\%$ RH or $T_{interior} \geqq 30°C$
Idle	$H_{interior} \leqq 70\%$ RH or $T_{interior} \geqq 30°C$	$H_{interior} > 70\%$ RH and $T_{interior} < 30°C$

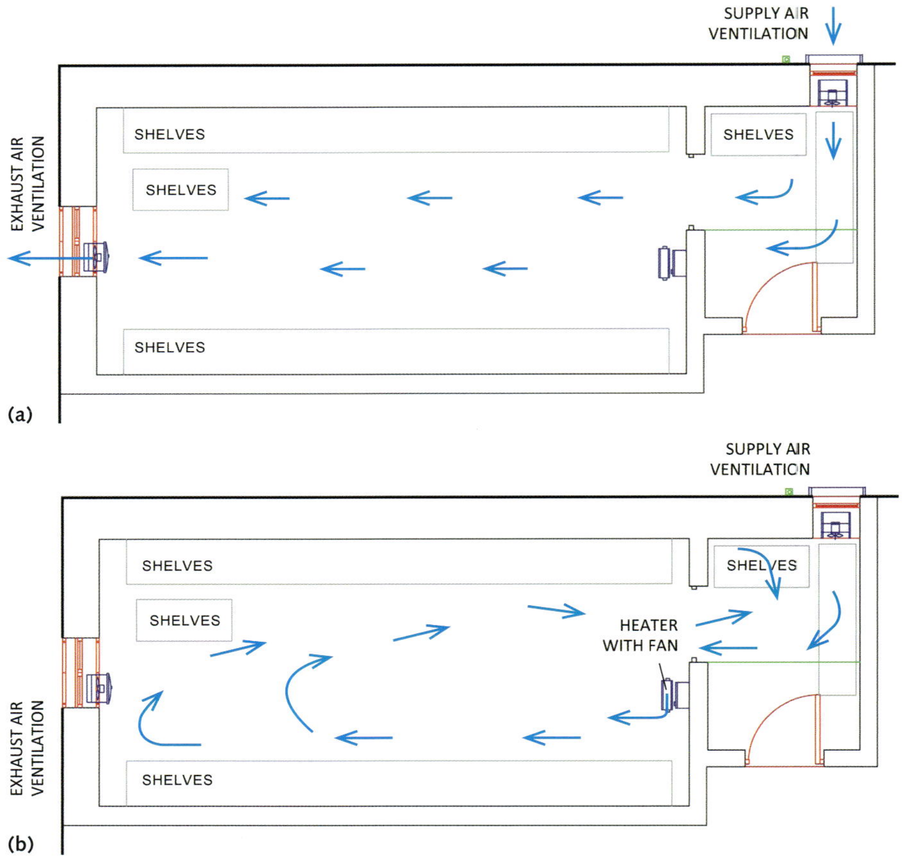

Figure 11.12
Schematics of air movement during ventilation mode (a) and conservation mode (b).

The controller/datalogger can be accessed directly by connecting a PC, or it can be remotely accessed by telephone for real-time monitoring, data download, and diagnostics; both the paper conservation laboratory of the Municipal Archive and the Getty Conservation Institute lab in the United States regularly accessed the controller/datalogger.

Postinstallation System Performance

The environmental management system was installed and operationally commissioned in May 2000, and the resulting postinstallation environment in the Historic Archive was monitored from May 2000 to January 2002.

Humidity

Following installation of the environmental management system in the Historic Archive, monthly mean interior humidity became more stable, ranging from 63% to 66% RH (the preinstallation range was 62%–72% RH) (fig. 11.13a). The system eliminated the frequent high-humidity events recorded in August during the preinstallation period. Furthermore, the 75% RH threshold was exceeded during less than 0.05% of the postinstallation period, and values above 75% RH were recorded only during March (less than 0.7% of the month) and May (less than 0.2% of the

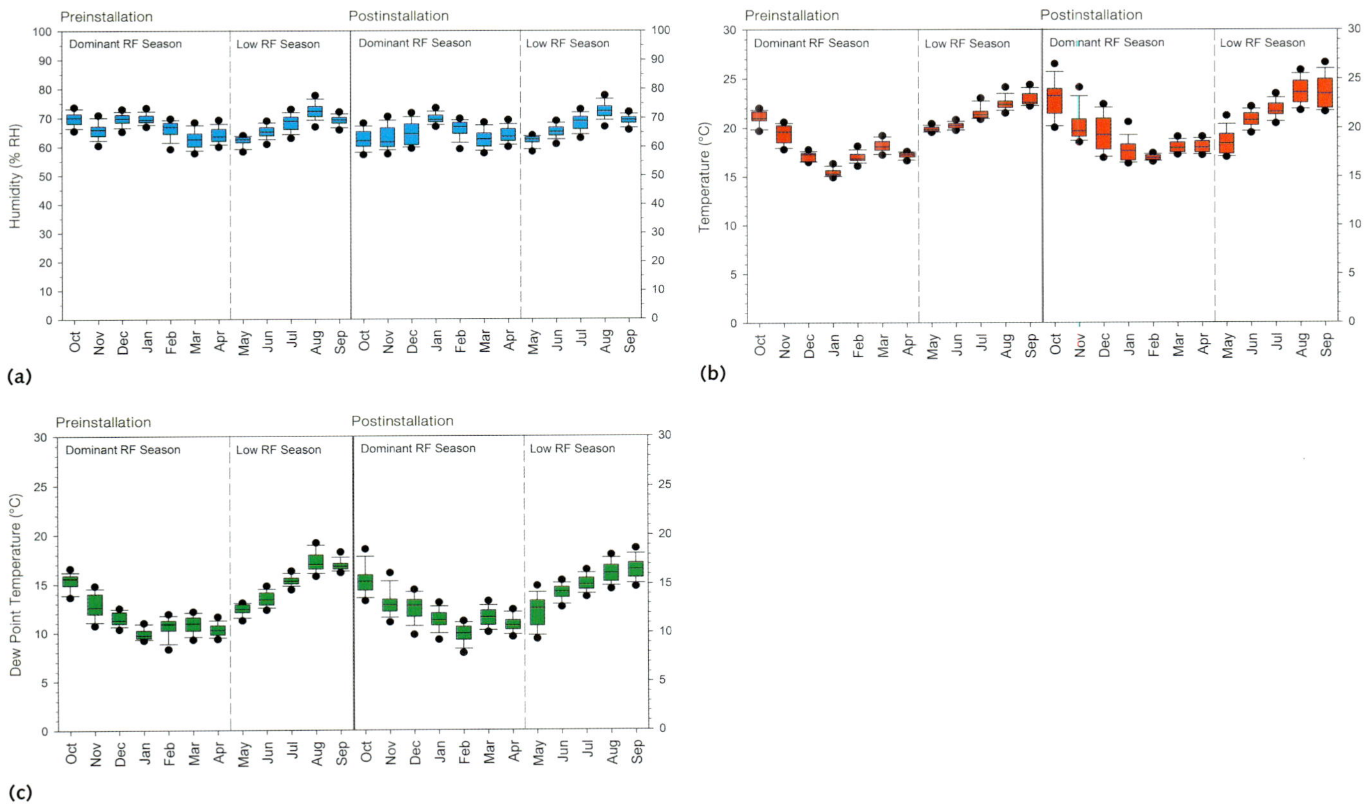

Figure 11.13

Combined plots of humidity (a), temperature (b), and dew point temperature (c) in the Historic Archive during the pre- and postinstallation periods.

month). The monthly average rolling 24 hour variation (short fluctuation) of interior humidity ranged from Δ4% to Δ7% RH, similar to the variation range observed during the preinstallation period.

Temperature

During the postinstallation period, monthly average temperature in the Historic Archive ranged from 17.0°C in February to 23.7°C in August (fig. 11.13b). A comparison of preinstallation and postinstallation mean temperatures by month shows that postinstallation values were elevated by as much as Δ2.3°C over preinstallation values; when examined seasonally, the postinstallation elevation in temperature was more prominent during the low-rainfall season. A maximum postinstallation interior temperature of 27.4°C was recorded in August, Δ2.8°C higher than the maximum value during the preinstallation period. Following system installation, monthly average 24 hour rolling variations of interior temperature ranged from Δ0.6°C to Δ1.1°C, slightly greater than the short-term temperature variation (Δ0.2°C–Δ0.7°C) recorded during the preinstallation period.

Dew point temperature

Monthly mean dew point temperature ranged from 9.8°C (February) to 16.5°C (September)—a range that was similar to values during the preinstallation period (fig. 11.13c). Monthly average 24 hour rolling variations of dew point temperature ranged from Δ1.1°C to Δ2.1°C, slightly wider than the preinstallation range of Δ0.6°C to Δ1.5°C.

Table 11.4

Conservation Environment Classification–HH protocol for the postinstallation Historic Archive environment.

Risk		Humidity		Temperature		
Category	Specific Risk (statistic)	Value (% RH)	Class	Value (°C)	Class	
Microbial	Germination threshold (P97.5)	72	C	—	—	
Mechanical	Short-term variation (P95 of rolling 24 hour variation)	Δ11	b	Δ1.9	a	Overall mechanical risk:[1] b
	Seasonal variation (absolute difference of seasonal means, RH means ≦ 70%)	Δ1	a	Δ2.6	a	
	Deviation from historical mean (absolute difference)	Δ9	a	Δ3.7	a	
Chemical	Deviation from historical mean (difference)	—	—	Δ3.7	+	

[1] Overall mechanical risk is based on the highest risk classification in any of the specific humidity- and temperature-based mechanical risks (short-term variation, seasonal variation, and deviation from historical mean).

Postinstallation risk class assessment

The Conservation Environment Classification–HH protocol for the postinstallation interior environment at the Historic Archive is summarized in table 11.4. Figure 11.14 also shows the pre- and postinstallation environments on psychrometric charts, with values related to classification

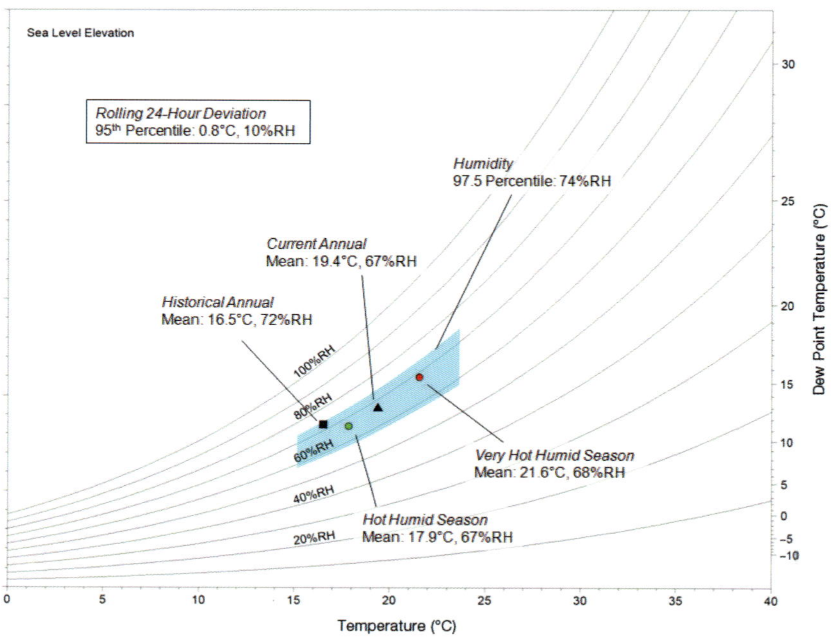

Figure 11.14a

Psychrometric charts showing environmental data collected in the Historic Archive during preinstallation. The boxes are bracketed by the 97.5 and 2.5 percentiles of temperature and humidity.

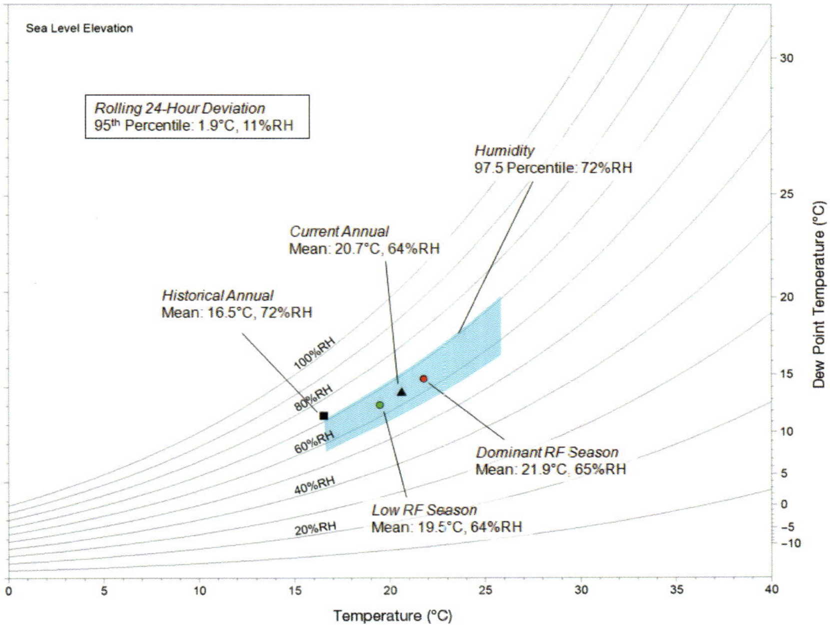

Figure 11.14b
Psychrometric charts showing environmental data collected in the Historic Archive during postinstallation. The boxes are bracketed by the 97.5 and 2.5 percentiles of temperature and humidity.

noted; the boxes represent 95% of the data. Differences between pre- and postinstallation Conservation Environment Classification–HH are discussed below.

The pre- and postinstallation risk classes for the interior environment at the Historic Archive are summarized in table 11.5.

While the pre- and postinstallation microbial risk classification remained the same (Class C, moderate risk), the postinstallation 97.5 humidity percentile was reduced by Δ2% RH over the preinstallation value. The apparent increase in mechanical risk due to short-term humidity variation from Class a (no risk) during preinstallation to Class b (moderate risk) during postinstallation masked the fact that the postinstallation criterion value (95th percentile of rolling 24 hour humidity variation) was only Δ1% RH greater than the preinstallation value. Operation of the environmental management system did increase the chemical risk due to deviation from historical temperature from a preinstallation Class 0 (same risk as historic condition) to a postinstallation Class + (moderately increased risk).

Table 11.5
Comparison of pre- and postinstallation (part B) risk classes at the Historic Archive.

	Microbial Risk		Mechanical Risk														Chemical Risk	
			Humidity						Temperature						Overall[1]			
			Short-Term Variation		Seasonal Variation		Deviation from Historical Mean		Short-Term Variation		Seasonal Variation		Deviation from Historical Mean					
Room	Pre	Post	Pre	Post	Pre	Post	Pre	Post	Pre	Post	Pre	Post	Pre	Post	Pre	Post	Pre	Post
Archive Interior	C	C	a	b	a	a	a	a	a	a	a	a	a	a	a	b	0	+

[1] Overall mechanical risk is based on the highest risk classification in any of the specific humidity- and temperature-based mechanical risks (short-term variation, seasonal variation, and deviation from historical mean).

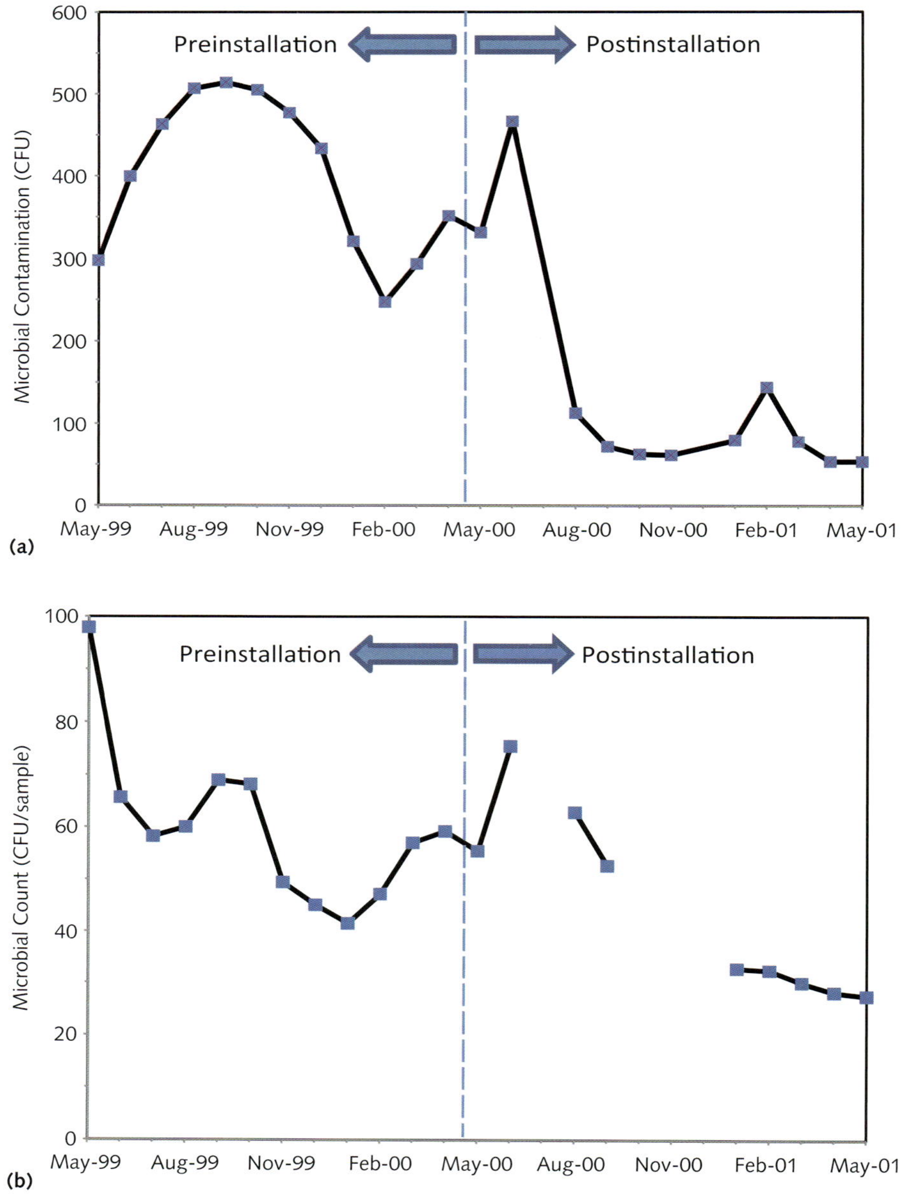

Figure 11.15
Environmental (a) and paper surface (b) contaminations in the Historic Archive of San Cristóbal de La Laguna (Valentín 2000).

Microbes

Microbial activity in both the environment and on documents was monitored monthly before and after installation of the environmental management system (Valentín 2000). Following system installation, microbial monitoring of the air in the Historic Archive demonstrated a drastic reduction in the numbers of colony forming units (CFU), while monitoring on documents stored in boxes showed only small reductions in CFU values (fig. 11.15). The lack of significant change in microbial activity on these documents may indicate difficulties in sampling or a delayed response due to the fact that the stored documents exist in a microenvironment.

Particulate matter

Due to the potential for the introduction of particulates during application of the ventilation strategy, particulate matter was assessed using a quartz crystal microbalance technique. Measurements of dust deposition were taken during the postinstallation period between May 2000 and June 2001. Immediately after system installation, dust deposition was recorded at 52 µg/day, followed by stabilization to 15–25 µg/day. The initial spike in particulate matter was attributed to dust generated by the system installation.

Thermal comfort

The municipal archivists access the Historic Archive on a daily basis, but they remain in the rooms only for brief periods to quickly review documents or take documents to a separate reading room for research. The Historic Archive lacks a working table or reading desk, and reshelving of documents can be done only during a brief stay in the rooms. For short-term visits, the archivists considered the postinstallation interior conditions acceptable. However, the archivists commented that the space was sometimes too cold and drafty or too hot, and it was never comfortable.

Prior to the installation of the environmental management system, an average environment in the Historic Archive during the low-rainfall season was a humidity of 67% RH and a temperature of 21.4°C. Based on the predicted mean vote (PMV) with elevated air speed model, thermal comfort analysis for lightly clothed (0.70 clo) working occupants (1.5 met) and with an air velocity of 0.1 m/s produced PMV and predicted percentage dissatisfied (PPD) values of 0.15 and 5%, respectively, indicating a neutral preinstallation environmental perception (point 1 in fig. 11.16).

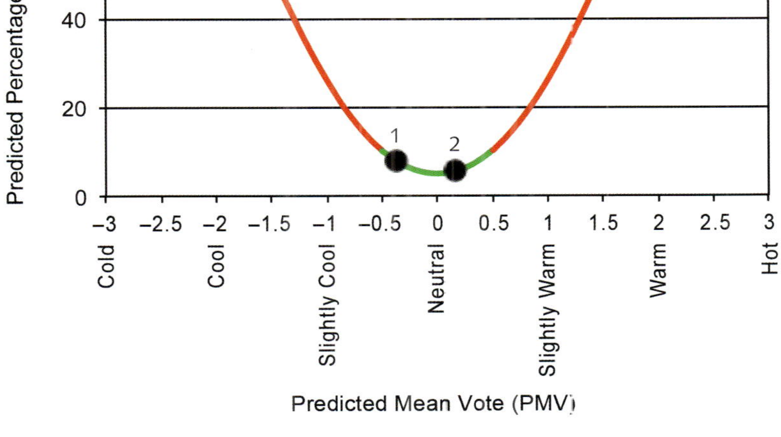

Figure 11.16

Assessment of thermal comfort in the Historic Archive using mean conditions during the low-rainfall season. Point 1 represents the preinstallation environment, while point 2 represents the postinstallation environment.

After the installation of the environmental management system, a typical condition in the Historic Archive during the low-rainfall season was a humidity of 65% RH and a temperature of 21.7°C. Thermal comfort analysis using the same clothing and activity level as for the pre-installation analysis and with air speed increased to 1.0 m/s yielded PMV and PPD values of 0.38 and 8% (neutral), respectively—a condition that complies with ANSI/ASHRAE Standard 55-2013 (point 2 in fig. 11.16) (ANSI and ASHRAE 2013).

Noise

Noise levels at both the supply and exhaust fans were recorded at 54 dbA. Although tolerable, the noise level reflected the presence of hard surfaces on the floor, ceiling, and exposed walls. In contrast, the forced air heater and recirculating fan were considered less noisy (no dB measurement was available) by archivists working in the Historic Archive.

Maintenance

While no equipment failures were observed, the following scheduled maintenance tasks were undertaken during the operational period:

- twice-daily reviews of humidity values via monitoring computer
- weekly review of environmental data in the room via monitoring computer
- monthly visual inspection of individual management system components
- six-month examination of microbial filters installed at the air intake (cleaning or replacement)
- annual calibration of humidity sensor

Energy Metrics

Energy intensity

Based on information provided in table 11.2, the total power of the fans and heaters was 2030 W. Because the floor area of the Historic Archive is 25 m^2, the environmental management system energy intensity was calculated as 81.2 W/m^2. The value was within the range for residential buildings and less than values for commercial facilities.

Energy use intensity

Timed operation of the supply and exhaust fans and the heater was monitored from May 2000 to December 2001 (except for July 2001). The ventilation mode dominated the system operation, as fans were operated for 903 hours, compared to only 35 hours for the heater. The energy use by the system was 375 kWh, and the energy use intensity was estimated to be 9.47 $kWh/m^2/yr$. This value is less than one-tenth of the energy intensity of unrefrigerated warehouses, indicating low energy use.

Conclusions

The concept of an economically sustainable and technologically simple environmental management system for collections conservation was successfully tested in a Mixed-Marine climate zone at the Historic Archive, located in a heritage building of the Municipality of San Cristóbal de La Laguna. Assessments of the Historic Archive's interior environment prior to implementation of the improvement strategy identified periods of elevated humidity that resulted in the microbiological deterioration of the collection. To mitigate this microbial risk, a system consisting of humidistatically controlled fans and a convection heater was installed following improvements to the building envelope, the rearrangement of archive shelves for effective cross ventilation, and the creation of a vestibule area at the entrance. The installation of mechanical components and modifications to the rooms required fewer than twenty man-hours. After an operational period of a year and a half (May 2000–January 2002), the following conclusions were made regarding the system performance and sustainability:

- High-humidity events above 75% RH were reduced from 2% of the preinstallation period to only 0.05% of the operational period.
- Rolling 24 hour humidity variation remained similar before and after the system installation.
- Rolling 24 hour variation of postinstallation temperature and dew point temperature was slightly greater than the preinstallation value.
- Postinstallation dust deposition initially increased, then stabilized to a low level.
- Microbial count of the interior air decreased from 400–500 CFU to less than 100 CFU, while the document surface microbial count pre- and postinstallation remained similar.
- Ventilation was the dominant mode of operation.
- The system showed both low energy intensity and low energy use intensity.
- Installation required minimal building modifications.
- The system is relatively inexpensive, simple, and robust.

It is expected that extended operation of the system will gradually reduce the water activity number[1] in the storage boxes and completely arrest microbial activity on the documents. The environmental management system is also likely to be capable of producing a lower-humidity (55%–65% RH) environment with a slightly elevated ($<\Delta 1°C$) temperature.

The reduction of high-humidity events by the environmental management system did not improve the Conservation Environment Classification–HH for microbial risk, as it remained at Class C (moderate risk). However, reduction in the 97.5 humidity percentile from 74% during preinstallation to 72% during postinstallation, as well as the postinstallation reductions in the microbial counts of interior air, indicate that the system did reduce microbial risk, but to a degree too small to be differentiated by the classification protocol. Adjustment of the humidity set point to a lower value would reduce microbial risk to Class B (low risk) or A (no risk)—but potentially at the expense of increased mechanical and chemical risk to the collection.

The risk for mechanical damage from short-term humidity variation increased slightly during the postinstallation period, while the risk of mechanical damage due to seasonal humidity and temperature variation and deviation from historic mean humidity and temperature remained unchanged. Because of an increase in annual mean interior temperature, the risk of chemical deterioration increased to Class + (moderately increased risk) from the preinstallation Class 0 (same risk as historic condition). Comparison of pre- and postinstallation environmental classifications indicated increased risks of mechanical and chemical deterioration, while the microbial risk remained unchanged in the postinstallation environment.

Additional discussions and information relating to this case study can be found in Instituto Nacional de Meteoroligía 1996–98, Maekawa and Toledo 2002 and 2003, and Valentín 2000.

Lessons Learned from the Historic Archive of San Cristóbal de La Laguna

The installation of a workshop-type curtain in the entry room, creating a vestibule at the entrance, was effective in limiting air exchange. However, archivists disliked the texture and weight of the hanging thick vinyl strips when accessing the Historic Archive. A small revolving door may be a better option for controlling infiltration.

Microbial monitoring indicated that the use of a low-cost humidistatically controlled ventilation and heating environmental system successfully controlled environmental microbial activity even though the Conservation Environment Classification–HH for microbial risk remained the same.

Postscript

The environmental management system was operated until 2007, when a project was developed for a new ducted dehumidifier-based central environmental management system for the entire Municipal Archive, including the Historic Archive. However, shortly after deinstallation of the humidistatically controlled ventilation and heating system, plans for the central system were suspended owing to a lack of funds. Since components of the removed system were damaged during deinstallation, the system could not be reinstalled in the Historic Archive. Personnel had to return to the operation of multiple portable dehumidifiers to control high-humidity conditions.

Project Team and Responsibilities

Municipality of San Cristóbal de La Laguna (Tenerife, Spain)

Rafael Martín Cantos (Paper Conservator)—Local coordination, operation, and conservation assessment

Facilities Department—Installation of nonmechanical and mechanical strategies

Níeves Valentín and Rafael García (Research Biologists, Instituto del Patrimonio Histórico Español, Madrid, Spain)—Microbiological monitoring and analysis

Getty Conservation Institute (Los Angeles, California, United States)

Shin Maekawa (Senior Scientist)—Project direction, strategy design, and engineering design and environmental monitoring

Franciza Toledo (Assistant Scientist)—Architectural design of nonmechanical strategies and supervision of implementation both nonmechanical and mechanical strategies

References

AEMet [Agencia Estatal de Meteorología]
2012 Guía resumida del clima en España (1981–2010), versión 3.0. http://www.aemet.es/es/conocermas/publicaciones/detalles/guia_resumida_2010.

ANSI [American National Standards Institute] and ASHRAE
2013 *Standard 55-2013—Thermal Environmental Conditions for Human Occupancy*. Atlanta: ASHRAE.

ASHRAE
2011 Museums, galleries, archives and libraries. Chapter 23 in *ASHRAE Handbook—HVAC Applications*. Atlanta: ASHRAE.

Instituto Nacional de Meteorología
1996–98 Datos climaticos de Santa Cruz de Tenerife. Estación de Los Rodeos—Aeropuerto Tenerife Norte.

Maekawa, S., and F. Toledo
2002 Controlled ventilation and heating to preserve collections in historic buildings in hot and humid regions. In *ICOM Committee for Conservation 13th Triennial Meeting, Rio de Janeiro, 22–27 September 2002: Preprints*, ed. R. Vontobel, 58–65. London: James & James.

———.
2003 Sustainable climate control for historic buildings in subtropical climates. *Management of Environmental Quality: An International Journal* 14 (3): 369–82. (Also presented at the 18th International Conference on Passive Low Energy Architectures [PLEA], November 2001, Florianópolis, Brazil.)

Valentín, N.
2000 Microbiological analyses in the historical archive in the city of La Laguna in Tenerife Island. Getty Conservation Institute internal report.

Weather Channel
2014 Weather Underground. http://www.wunderground.com/.

Note

1 Water activity, expressed in fractions, is equivalent to humidity (the value for air), but it is used for solids or solid surfaces.

Chapter 12

Museum Storage in Valle Guerra, Tenerife, Spain: Conservation Heating by Ventilation and Conservation Heating

Climate zone 2A (Hot-Humid)

Objectives

- Protect collections from microbial infestation by maintaining a stable environment at less than 70% RH.
- Protect collections from particulate deposition.
- Install low-cost, robust, and low-energy environmental management system.
- Provide a workshop environment for museum staff who occasionally perform the documentation of objects.

Summary

This case study applies environmental management strategies to the Valle Guerra museum storage facility, which houses a collection of nineteenth- and twentieth-century folk artifacts from the Canary Islands, including textiles, pottery, basketry, farm machinery, tools, and furniture. Located in an agricultural zone on the northeastern slope of Tenerife Island, Spain, with a Hot-Humid (2A) climate zone, the storage facility comprises the second and third floors of a contemporary leased six-story building made of reinforced concrete floors and concrete block walls (ASHRAE Class III). Existing environmental management at the storage was limited to infrequent use of open-window ventilation. During the summer of 1999, the storage facility experienced a major insect infestation of an important garment collection on the second floor that had been preceded by a fungal outbreak. Monitoring of the existing interior environment of the second floor highlighted extended periods of high humidity in July and August, with humidity exceeding 75% RH during 20% of each month. The primary conservation goal for the collection was establishing and maintaining a stable and less-humid environment with a low particulate deposition rate.

The alternative environmental management strategy for the Valle Guerra storage facility included nonmechanical and mechanical measures. Nonmechanical strategies included improving the airtightness of the building envelope to reduce infiltration of humid and particulate-laden outside air and thermally insulating single-glazed windows.

Mechanical strategies included installation of a humidistatically controlled ducted supply and exhaust fan and convection heaters. Installed in the second floor of the storage facility in August 2002, the system maintained interior humidity between 55% and 70% RH while allowing temperature to vary between 15°C and 29°C. Monitoring of microbial activity and dust deposition showed reductions in both following system installation. The postinstallation assessment of the collection found no adverse effects from exposure to the new storage environment. However, the conditioned space was considered to be less hospitable for performing work tasks owing to decreased thermal comfort and high noise levels. The environmental management system operated continuously from May 2004 to August 2010, when the building lease expired.

Background

In 1999 the collections storage facility of the Organismo Autónomo de Museos y Centros (OAMC) of Tenerife in Valle Guerra suffered a serious insect and fungal infestation that affected 25% of the collection. The insect infestation was preceded by increased fungal activity in the storage rooms. The biological damage to the collection highlighted the need for humidity reduction and management of the interior environment to reduce the risk of another outbreak. One room of the second-floor storage was periodically used by OAMC staff as a work area to catalogue the contents of the collection, and when that room was occupied, windows and exterior doors were often opened for comfort. This circumstance may have contributed to insect entry into the storage space as well as to a high deposition rate of particulates. Since the storage space was leased by the OAMC, the organization decided against installing a sophisticated environmental control system because of the anticipated high cost of the system and the necessary envelope improvements.

These financial and ownership constraints forced the OAMC to look for a simple, less-expensive, and robust alternative environmental management system that would reduce the risk of future microbial and insect attacks and limit dust deposition in the storage facility. Due to the periodic use of the facility as an active workspace, the OAMC also wanted the system to provide a workshop environment for museum staff. The OAMC decided on adapting the simple and inexpensive environmental management system successfully implemented at the Historic Archive of San Cristóbal de La Laguna, also on Tenerife (see case study, chap. 11) (Maekawa and Toledo 2003), to the Valle Guerra storage facility. This approach was considered a suitable alternative to the use of a conventional air-conditioning system.

Climate

The town of Valle Guerra (28°53' N, 16°39' W) is located on the northwest slope of Tenerife Island (fig. 12.1). A general description of the island's climate is found in the case study for the Historic Archive of San Cristóbal de La Laguna, Tenerife.

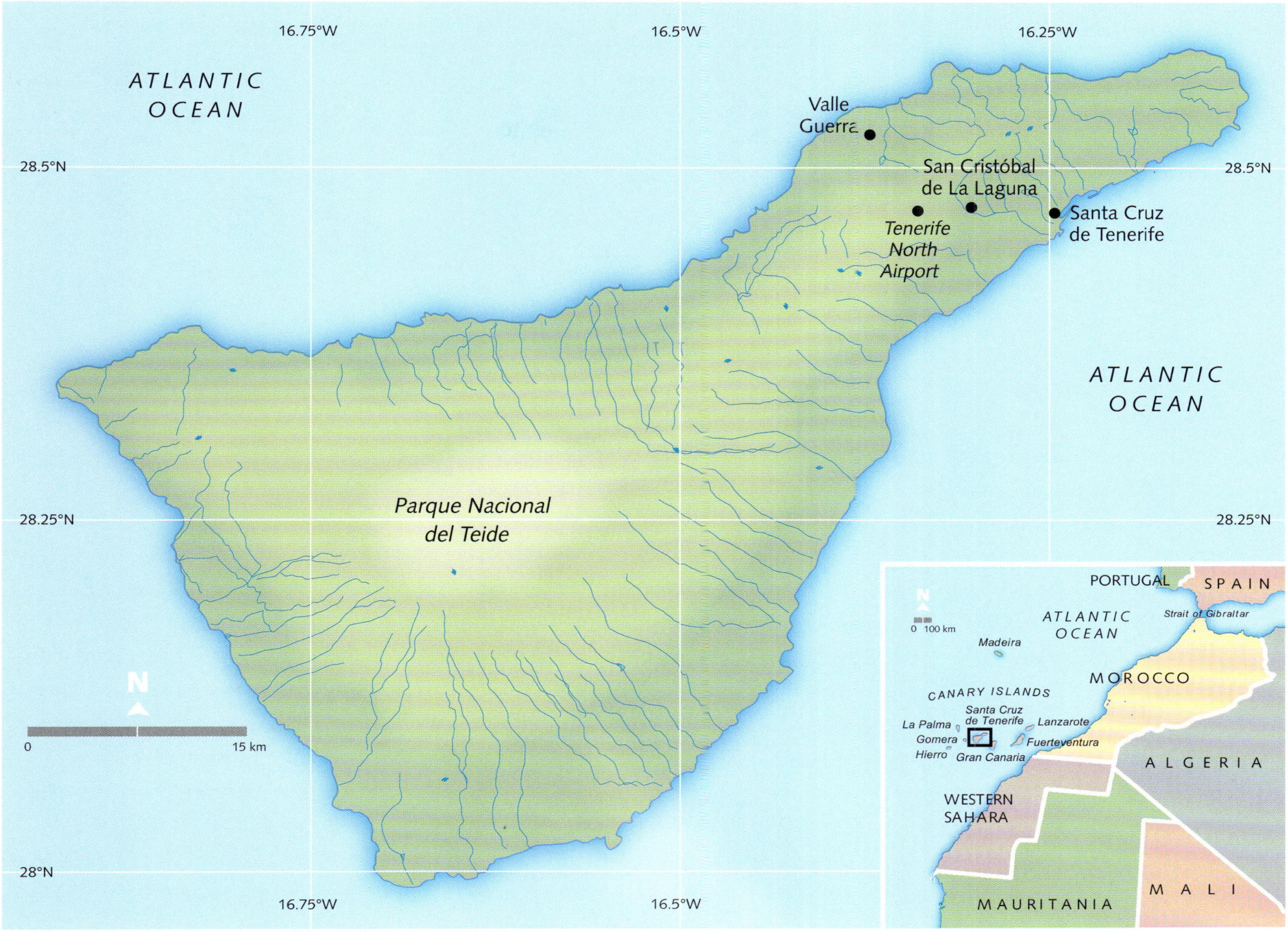

Figure 12.1

Map of Tenerife Island showing the location of Valle Guerra, Santa Cruz de Tenerife, and Tenerife North Airport.

The localized climate at Valle Guerra (196 m above sea level) differs significantly from that of San Cristóbal de La Laguna because of its lower elevation and exposure to trade winds and oceanic moisture sources. Although historical climate data were available from Tenerife North (Los Rodeos) Airport (28°48' N, 16°34' W, 617 m above sea level) for the Historic Archive case study, climate data for Valle Guerra were obtained from the more-distant station at Santa Cruz de Tenerife (28°47' N, 16°25' W, 46 m above sea level)—on the other side of the island and 150 m lower in elevation—because its climate, especially the high temperatures, was considered to be close to that of Valle Guerra.

The temperature at Santa Cruz de Tenerife ranges from the mid-10s in January and February to the upper 20s in July and August. Monthly average temperatures ranged from 17°C in January and February to 25°C in August (weatherbase.com). Monthly averages of humidity remained between 55% and 66% RH throughout the year, and monthly averages of dew point temperature ranged from 9°C in January and February to 15°C in August and September. Historic annual average humidity and temperature were 61% RH and 20°C, respectively. Conforming to data collected at Los Rodeos Airport and typical

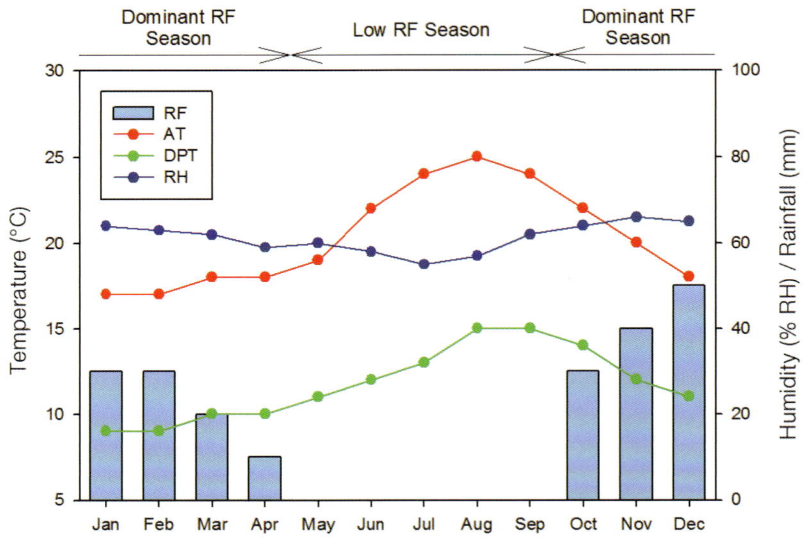

Figure 12.2
Monthly average temperature, humidity, dew point temperature, and rainfall (RF) at Santa Cruz de Tenerife (1971–2000). Source: weatherbase.com.

of northern Tenerife, the bulk of precipitation at the Santa Cruz de Tenerife climate station is recorded from October to April (dominant rainfall season), with no rainfall between May and September (low-rainfall season). Annual total rainfall recorded at Santa Cruz de Tenerife was 250 mm, less than half the annual total collected at Los Rodeos Airport (557 mm) (fig. 12.2).

The CDD10°C value obtained through analysis of the historical temperature data was 4231 cooling degree days (Weather Channel 2014), which placed the climate of Santa Cruz de Tenerife in the Hot (2) climate zone. The climatic moisture assessment excluded it from a Marine designation owing to its warmest monthly mean (25°C) exceeding 22°C. The annual precipitation of 210 mm is less than the Dry-Humid Index of 540, placing it in the Dry-type category (B). Merging the two climate classifications results in a Hot-Dry (2B) climate zone classification for Santa Cruz de Tenerife. This climate classification, however, cannot be fully extended to Valle Guerra.

Though precipitation data at the Valle de Guerra area were not available, the natural vegetative landscape and prominent agricultural activity of northwestern Tenerife Island is an indication of more abundant rainfall than is received on the northeastern side, where Santa Cruz de Tenerife is located. This expectation of increased rainfall coincides with the movement of moist air systems from west to east, resulting in orographic precipitation as the systems pass over the northwestern slope and central highlands of Tenerife. Therefore, the annual rainfall at Valle Guerra is expected to equal or exceed the 550 mm recorded at Los Rodeos Airport, on the central highland of Northern Tenerife (AEMet 2012). Assessment of the moisture classification at Valle Guerra using total rainfall at Los Rodeos Airport and the annual mean temperature (20°C) observed at Santa Cruz de Tenerife results in a Humid-type classification (A). Thus, the climate at Valle Guerra can be classified as Hot-Humid (2A). Seasonal distinctions at Valle Guerra will mimic those of Los Rodeos Airport and Santa Cruz de Tenerife: October to April is the dominant rainfall season, and May to September is the low-rainfall season.

During this case study project, exterior temperature and humidity were recorded at the Valle Guerra storage facility; discussion of the data will be presented in the Environmental Assessment section.

Building

The storage facility at Valle Guerra occupies the second and third floors of a contemporary six-floor building, which is constructed of steel-reinforced concrete frames and concrete masonry walls finished with plaster (fig. 12.3a). Walls are approximately 20 cm thick and contain single-pane windows. Built into a northwest-facing slope, only the four top floors of the building are above ground on the southeast and southwest elevations. On the northeast elevation, the full six-floor height of the building is above ground (fig. 12.3b). Based on its construction and exposure, this structure is classified as ASHRAE Class III (ASHRAE 2011). The southwest elevation of the multipurpose structure fronts a main road leading to a business area, while the northwest and northeast elevations overlook open fields and the ocean, which is at a distance of approximately 2 km.

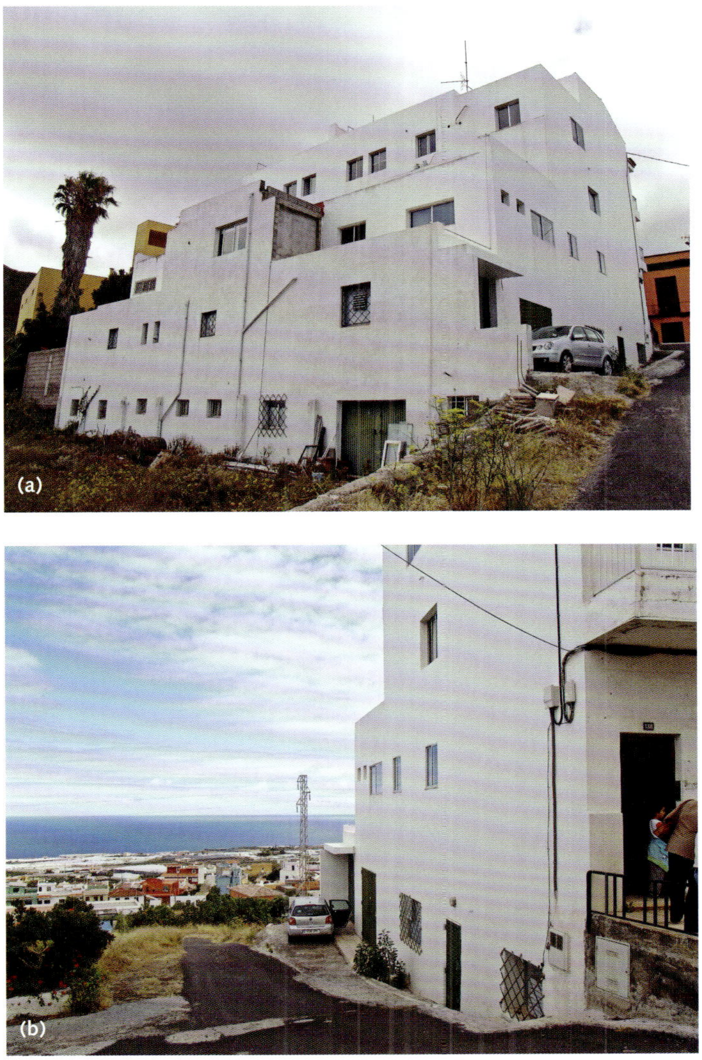

Figure 12.3
Building housing the OAMC museum storage facility at Valle Guerra, seen from the north (a); the lower floor of the storage facility is located at the level of the parked car. The partial view of the building from the west (b) shows the northwestern face, which is partly below ground. The northeast side of the building overlooks the ocean. Photo: Shin Maekawa. © J. Paul Getty Trust.

Environmental management was considered for only the second floor of the building (lower floor of the storage facility), which is isolated from adjacent floors (fig. 12.4). The southeast and southwest walls of this level are below ground, and the northwest and northeast walls contain single-glazed, aluminum-framed windows that can slide open to sea breezes. The northwest side of the building is bordered by a narrow paved road with a depressed driveway. The second-floor storage area can be accessed from the exterior by two entrance doors and a garage door in the northwest wall.

The second-floor storage area encompasses approximately 260 m^2 (18 × 14.5 m); it has a 3 m high ceiling and is divided into five rooms and a restroom (fig. 12.4). The largest spaces are Room B (18 × 8.1 m) and Room A (18 × 3.8 m). These two rooms are connected by a doorway in the dividing wall; each room has a northwest-facing window. Both Rooms A and B have exterior doors on the northwest wall. Room C (2.3 × 5.4 m), Room D (2.3 × 5.4 m), and Room E (2.3 × 4.3 m) are positioned along the northeast wall of the building and open into Room B through interior doors. Rooms C, D, and E also have exterior windows on the northeast wall of each space, while interior windows are located in the partitions between Rooms D and B and Rooms E and B. Room C has an exterior entrance through a garage door on its northwest exterior wall.

The second-floor space was furnished with open steel shelving systems in Rooms A and B and wooden wardrobes in Rooms D and E. Room C, with its large garage door and relatively small size, was considered too easily influenced by the outside climate to serve as storage and was instead designated as a receiving vestibule and quarantine space.

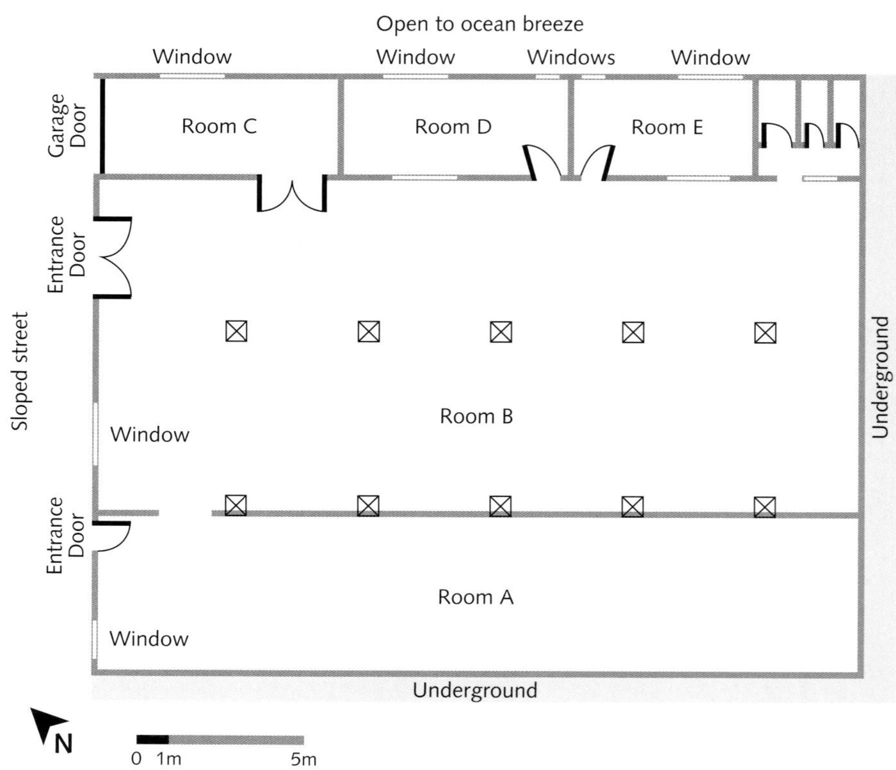

Figure 12.4
Second-floor plan of the building, housing the lower floor of the Valle Guerra storage facility.

Small- and medium-size objects such as pottery, basketry, and household items were placed in Room A, which was the narrower of the two large rooms. Large and bulky objects were housed in Room B, which was larger and had double doors leading to the exterior. The Canary Islands textile collection of traditional dresses and daily clothes was placed in Rooms D and E. Although the northeast wall of these two rooms provided more exposure to the exterior climate through windows, they could be isolated from the other spaces, if necessary, by closing the interior doors shared with Room B.

Collection

The storage facility houses a collection of folk artifacts dating from the late nineteenth through the twentieth century, documenting traditional Canarian activities in agriculture, small industry, and handicraft. Containing important collections of peasant garments, textiles, pottery, basketry, and furniture, the storage facility has approximately three thousand objects made with diverse technologies and a wide range of materials, including many crafted from metal components and organic materials such as sapwood, sticks, leaves, hide, paper, and leather (fig. 12.5).

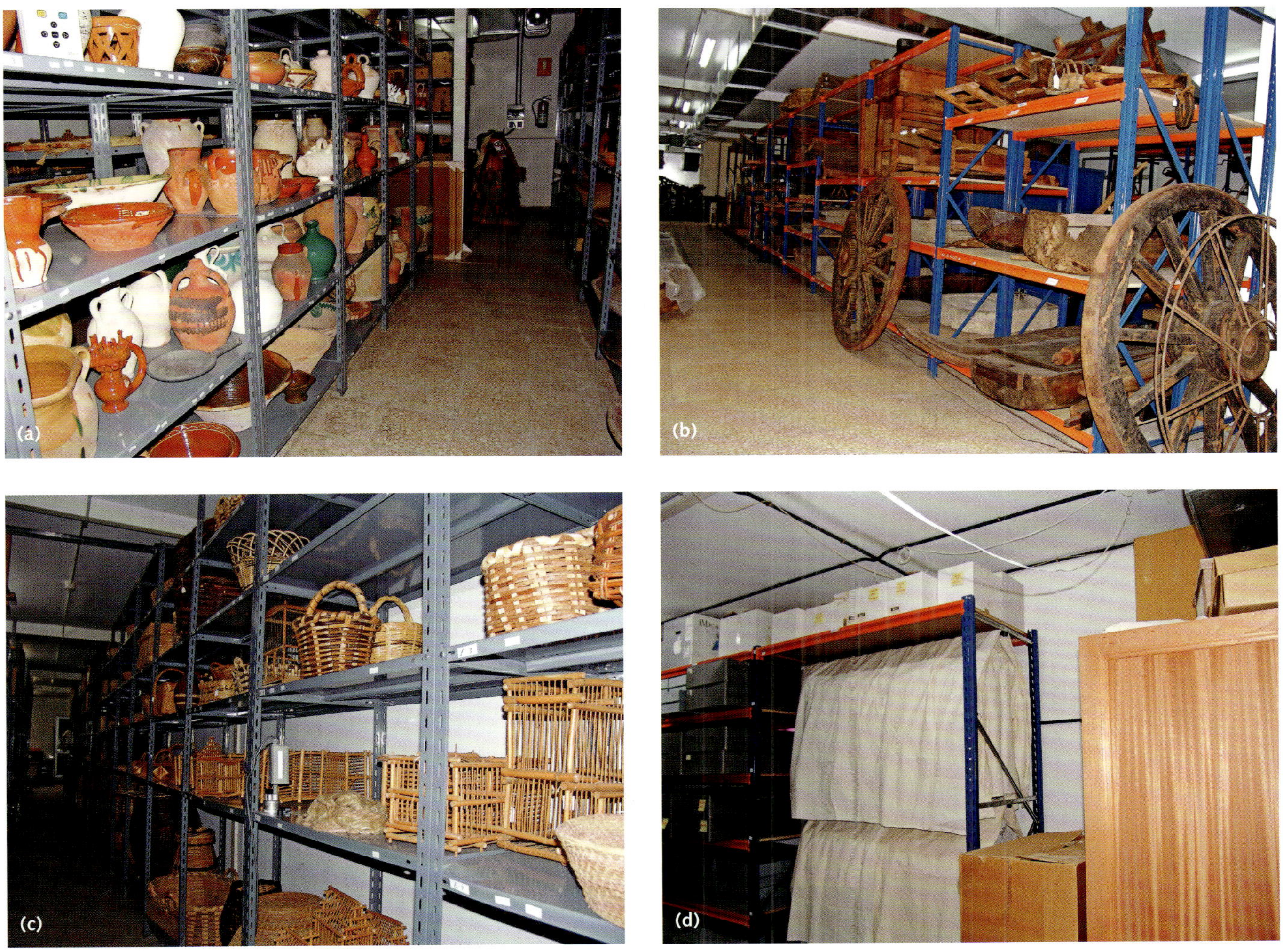

Figure 12.5
Collections housed in the lower floor of the Valle Guerra storage facility: small- and medium-size objects are stored in Room A (a and c), large and bulky objects in Room B (b), and textiles in Room D (d). (Room E is not shown.) Photos: Shin Maekawa. © J. Paul Getty Trust.

Many objects are quite worn and battered from intensive use and subsequent storage and abandonment in cellars, attics, and barns before acquisition by or donation to the OAMC. Generally, these objects have a low sensitivity to accelerated chemical aging due to the historic usage and local storage conditions, as they had already undergone deterioration. However, organic materials in the collection are highly vulnerable to biodeterioration by fungal and bacterial attacks and insect infestation, and the metal components are susceptible to corrosion when exposed to elevated humidity. Textiles and garments are stored in wooden wardrobes, drawers, and acid-free cardboard boxes to shield them from dust deposition. Other objects are placed directly on open metal shelving.

The major conservation goal for the collection was to maintain a stable and less-humid environment with a low particulate deposition rate. Although a secondary priority, thermal comfort of staff members, who periodically perform light work in the storage facility, needed to be addressed.

Preinstallation Condition Assessments

Building envelope assessment

Since the storage space was partially below ground and had no windows on the southwest and southeast walls, air circulation and cross ventilation in Rooms A and B were very poor. Fortunately, no evidence of water infiltration or vapor permeation through the underground walls was observed. The single-glazed, aluminum-frame windows in these two rooms had significant air leakage and a low thermal insulation value. Exterior windows in Rooms D and E had the greatest air leakage, likely due to high wind pressure from the northeast. The fit between exterior doors in Rooms A and B and their door frames was loose, and air leakage was observed when the doors were closed. While the storage space was compartmentalized with doors fitted into all interior passageways, air leakage was observed around a closed interior door between Room B and the restroom, presumably due to high pressure from northwest winds.

Exterior environmental assessment

For the duration of the project (February 2002–July 2005), exterior temperature and humidity were monitored just outside of Room D on the northeast side of the building envelope. Monthly mean temperature at the site ranged from 16.4°C in January to 23.4°C in August, while monthly average dew point temperature ranged from 9.6°C in February to 19.4°C in August (fig. 12.6). (Dew point temperature was calculated from concurrent temperature and humidity values.) Monthly mean humidity was lowest during February (66% RH) and highest during July (79% RH). Further, monthly average humidity exceeded 75% RH during the months of June to October. It should be noted that the seasonal trend of mean humidity at Valle Guerra contrasted with mean humidity seen in the historical data recorded at both the Los Rodeos Airport and Santa Cruz de Tenerife. While the average humidity at Los Rodeos Airport peaked in the

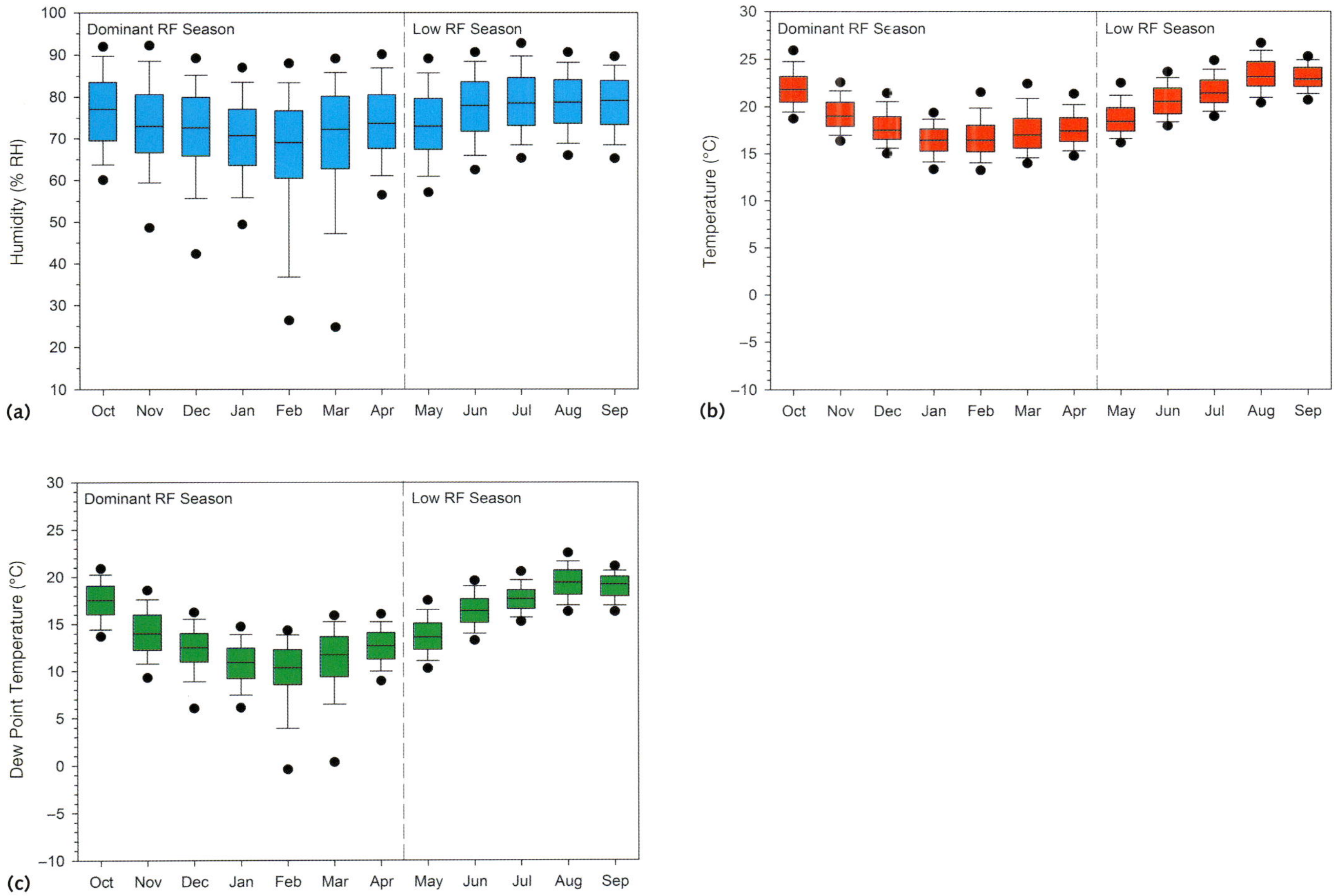

Figure 12.6

Monthly exterior humidity (a), temperature (b), and dew point temperature (c) recorded at Valle Guerra from February 2002 to July 2005.

winter, average values recorded at Valle Guerra peaked during the summer months. This seasonal trend may be attributable to the lower elevation of the site and its exposure to the ocean and trade winds. The annual average exterior temperature and humidity at Valle Guerra were 19.6°C and 73% RH, respectively.

The CDD10°C value obtained through the analysis of the site temperature data was 3521 cooling degree days, which placed the climate of Valle Guerra in the Hot (2) thermal zone. As expected, this classification matched that of Santa Cruz de Tenerife and differed from the Mixed (4) thermal zone of Los Rodeos Airport. The lack of rainfall data at Valle Guerra precluded a direct classification of moisture type, which was then assessed using total rainfall at Los Rodeos Airport (see discussion under "Climate," above).

Examination of mean daily trends also showed diurnal variations in temperature, dew point temperature, and humidity (fig. 12.7). The temperature and dew point temperature rise from the lowest points before sunrise to peak in the afternoon. The mean daily variation in temperature and dew point temperature is roughly Δ3.1°C and Δ1.2°C, respectively, during both January and August, the most extreme months with respect to climate. The daily trend of humidity results in its peak value occurring during the night, followed by a midday depression and a rise again in the

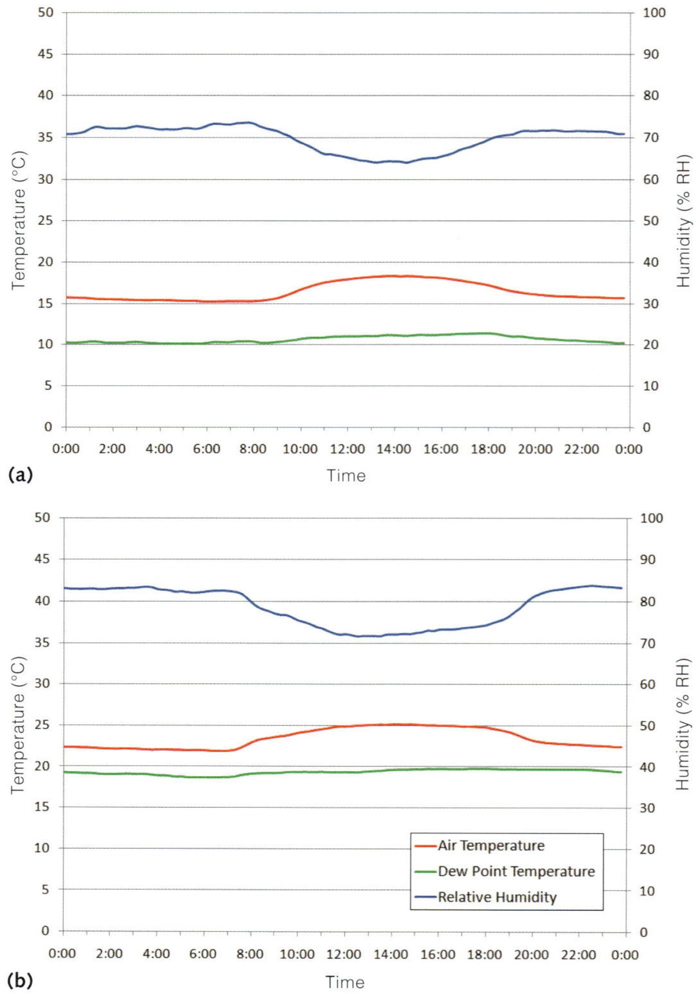

Figure 12.7
Mean daily trends of exterior humidity, temperature, and dew point temperature during (a) January (the dominant rainfall season), and (b) August (the low-rainfall season).

evening. Mean daily variation in humidity is Δ9.7% RH in January and Δ12.4% RH in August. The dry outside air is available for conservation heating by ventilation between 10:00 a.m. and 6:00 p.m. during January (the dominant rainfall season). However, no ventilation opportunity exists in August (the low-rainfall season), since humidity remains above 70% RH throughout the day.

Interior environmental assessment

The preinstallation environment in the storage facility was monitored from February 2002 to May 2003. The opening of large windows and entrance doors to the exterior significantly influenced the interior environment because of the infiltration of humid outside air. In general, the environment in Rooms A and B (the largest rooms) was warmer and more stable than that in the smaller Rooms D and E. The temperature in Rooms D and E was lower and, as a result, they were more humid (by approximately 5% RH) than Rooms A and B. Because of their location against the exposed northeast wall, Rooms D and E had environments that were less stable and more humid, and they presented the most problematic preinstallation environment. Accordingly, the following discussion focuses on the Room D environment, as the Room E environment is similar.

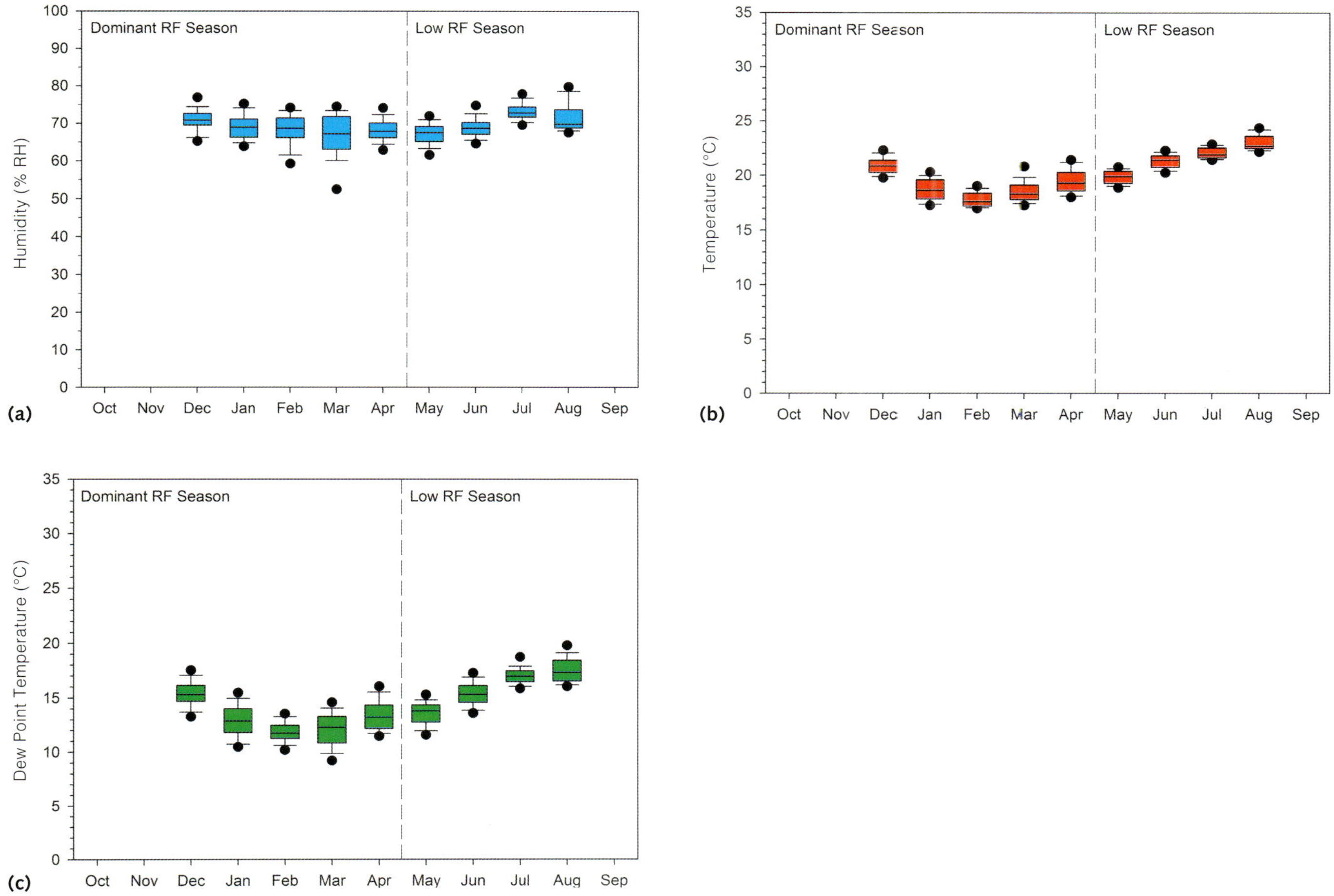

Figure 12.8

Preinstallation comparison of monthly humidity (a), temperature (b), and dew point temperature (c) in Room D. Data were collected from February 2002 to May 2003. Data from September through November 2002 were lost owing to a malfunction of the monitoring equipment.

Monthly mean interior humidity ranged from 66% RH (March) to 73% RH (July), while monthly mean interior temperature ranged from 17.8°C (February) to 23.1°C (August) in room D (fig. 12.8). Monthly average dew point temperature ranged from 11.8°C (February) to 17.5°C (August). The mean rolling 24 hour variations of temperature (Δ0.4°C in February to Δ0.8°C in August), humidity (Δ4% RH in December to Δ6% RH in August), and dew point temperature (Δ1.1°C in July to Δ1.8°C in August) were relatively constant throughout the year.

Although the monthly mean interior humidity was below 75% RH, this threshold for fungal activities was exceeded in Room D, particularly during summer months. During most of the dominant rainfall season and the early months of the low-rainfall season, humidity in Room D exceeded 75% RH during less than 5% of the period. In contrast, when exterior humidity increased during the months of July and August, interior humidity in Room D was above 75% RH during 20% of each month, or approximately 6 days per month. This condition indicated a significant risk of fungal outbreak on organic materials in the collection, though this was not observed on objects in Rooms D and E during the monitored period. The lack of visible biological damage may be due to durations of high humidity that were less than needed for germination of fungal spores. Microbiological analysis showed little microbial contamination within the wardrobes, and contamination outside the wardrobes in Rooms

D and E was similar to that seen in the other rooms when the project was initiated (Valentín 2004).

Preinstallation risk class assessment

The Conservation Environment Classification–HH protocol of the preinstallation interior environment in the Valle de Guerra storage facility is summarized in table 12.1.

Environmental Management System

Objectives

The objectives for this case study were established from the assessments described above. The primary objective of the environmental management strategy was to protect the collection from fungal infestation. This protection can be established and maintained with the storage environment kept at less than 70% RH. Although the historic average exterior humidity at Valle Guerra is presumed to be 73% RH (based on data collected during the case study and used because of lack of local historical climate

Table 12.1

Preinstallation Conservation Environment Classification–HH for Valle Guerra storage facility.

Risk			Humidity		Temperature		
Category	Specific Risk (statistic)	Room	Value (% RH)	Class	Value (°C)	Class	
Microbial	Germination threshold (P97.5)	D	77	F	—	—	
		A	73	C	—	—	
		B	71	C	—	—	
Mechanical	Short-term variation (P95 of rolling 24 hour variation)	D	Δ10	a	Δ1.1	a	Overall mechanical risk:[1] Room D: b Room A: a Room B: b
		A	Δ10	a	Δ1.5	a	
		B	Δ12	b	Δ1.6	a	
	Seasonal variation (absolute difference of seasonal means, RH means ≦ 70%)	D	Δ1	a	Δ2.5	a	
		A	Δ6	a	Δ1.4	a	
		B	Δ4	a	Δ1.8	a	
	Deviation from historical mean (absolute difference)	D	Δ8	a	Δ0.3	a	
		A	Δ3	a	Δ1.5	a	
		B	Δ1	a	Δ0.2	a	
Chemical	Deviation from historical mean (difference)	D	—	—	Δ0.3	0	
		A	—	—	Δ1.5	0	
		B	—	—	Δ0.2	0	

[1] Overall mechanical risk is based on the highest risk classification in any of the specific humidity- and temperature-based mechanical risks (short-term variation, seasonal variation, and deviation from historical mean).

data), the large number of wood objects in the collection requires a lower humidity target. Additional objectives are:

- reducing the particulate deposition rate,
- installing a low-cost, robust, and low-energy environmental management system,
- providing a workshop environment for museum staff who occasionally perform the documentation of objects.

Psychrometric strategies

To achieve the stated objectives of this case study, the nonmechanical psychrometric strategy focused on source moisture controls to limit infiltration of humid air into the storage space, and source thermal control to reduce heat gain from, and loss to, the exterior. Following the implementation of nonmechanical strategies, mechanical psychrometric strategies were applied. These relied on conservation heating by heater and conservation heating by ventilation to maintain a suitable interior conservation environment. Ventilation utilized a particulate filter to reduce particulates.

Nonmechanical strategies

Since the building was a leased property, the environmental management strategy for the Valle Guerra storage facility was limited to minor modifications of the existing building envelope. The creation of new wall openings and permanent closure of existing window and door openings were not permitted, but several reversible modifications to the building envelope improved its thermal and infiltration performance. External windows were either converted for use by supply or exhaust fans, or they were sealed to limit infiltration. Thermal insulation pads were also applied to exterior windows in Rooms D and E to reduce heat gain. All doors were fitted with additional seals and gaskets, and an airtight door was installed in the restroom to limit infiltration into Room B. Access to the storage facility was limited to the garage door in Room C, while other entrance doors on the northwest wall of Rooms A and B were converted to emergency-only exits.

Mechanical strategies

In addition to conservation heating by heater, the environmental management strategy in the Valle Guerra storage facility utilizes conservation heating by ventilation to limit elevated humidity conditions, as was observed in Rooms D and E. Ventilation of warm and dry outside air (less than 70% RH) is used when available, typically during the dominant rainfall season. The use of ventilation was generally restricted during the low-rainfall season, as exterior humidity remained above 70% RH throughout the day and the dew point temperature was consistently near 20°C (see fig. 12.7).

Since the southwest and southeast walls were below ground and without windows, the system design takes filtered supply air from the northwest windows and ducts it to the southeast end of Rooms A and B. Exhaust fans mounted from the ceiling near existing windows on the northeast wall in Rooms D and E result in effective cross ventilation of the entire storage space.

The removal of doors and windows between Rooms B, D, and E established a large single climate zone. Doors between Rooms A and B were removed to allow the ventilation air in Room A to be exhausted from Rooms D and E through Room B. Owing to the small size of Room A and the location of its door opening, the environment of Room A differed from that of Room B. Room A, the most stable and dry environment in the space as a result of its minimal exterior exposure, represents a second smaller climate zone.

After conceptual design of the mechanical system, a local HVAC company was contacted to produce a detailed design and to install the mechanical system. This step posed several obstacles. Although several HVAC companies on Tenerife Island have extensive experience in designing and installing conventional air-conditioning equipment, the conceptual design of the proposed system was a departure from their experience of controlling humidity by condensing moisture from the air through cooling (dehumidification by cooling). Therefore, the start of the project required negotiations with local engineers who were reluctant to embrace an alternative environmental management approach.

A secondary issue was the proposed use of high-speed jet nozzles for turbulent mixing of ventilating air. Although guidelines have not been published on the allowable maximum air velocity on surfaces of collections objects, lower air velocity and less turbulence are recommended to prevent flexing of fragile surfaces and abrasive damage to surfaces. A slow laminar flow, as described in the conceptual design, would provide gentle cross ventilation in the space, but it requires the installation of long ducts to transport fresh air to the opposite side of the rooms. The additional ductwork resulted in modifications to the order during installation, increasing cost and delaying final installation.

After selecting an HVAC company and agreeing to the mechanical layout, a series of meetings between project staff and the contracted engineer were held to develop an installation schedule and to supervise overall progress. Continuous review and supervision were necessary throughout the project to ensure satisfactory execution of the project details.

Mechanical system description

Installed in August 2002, the environmental management system for the Valle Guerra storage facility initially consisted of three supply fans, two exhaust fans, and five convection unit heaters controlled by three humidistats—two interior and the other exterior (fig. 12.9). Equipment specifications are listed in table 12.2. The supply air fans, enclosed within connecting ducts, distributed filtered outside air through ceiling-mounted ducts (fig. 12.10). This supply air was delivered through diffusers in the southeast ends of Rooms A and B. Each supply air intake was equipped with a motorized damper, an insect screen, a washable metal media filter, and a disposable G3 (MERV 5 equivalent) particulate filter. Air from Rooms A and B is exhausted by duct-mounted exhaust fans in short duct

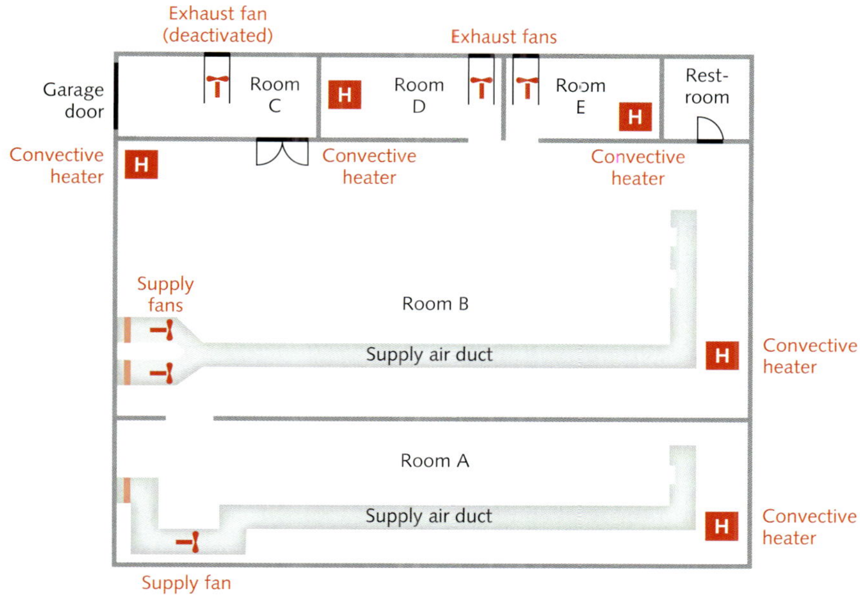

Figure 12.9

Location of equipment for the environmental management system in the Valle Guerra storage facility.

Table 12.2

List of equipment and its specifications for Valle Guerra storage facility (NA = not available).

Number of Units	Equipment	Type of Device	Flow Rate (m^3/hr)	Capacity (watts per unit)	Static Pressure (mm Wg)	Sound (dBA per unit)
3	Supply fan	Centrifugal type	2500	570	25	63
2	Exhaust fan	Centrifugal type	3800	570	NA	63
5	Convection heater	Electroresistive	350	3030	NA	43
3	Particulate filter	G3 (MERV 5)	—	—	—	—

runs near the ceilings of rooms D and E. Gravity dampers were installed on the exhaust fans to limit infiltration.

The exhaust fan in Room C, which served as the reception and quarantine space for incoming objects, was initially part of the environmental management system. However, enabling of the ventilation mode resulted in a negative pressure in the storage facility, making the opening of doors difficult. Consequently, the Room C exhaust fan was taken offline, and the door between Rooms B and C was kept closed. These actions maintained a positive pressure in the storage facility during ventilation mode, preventing infiltration of unfiltered dusty outside air.

Convection heaters were mounted on the ceiling near the south corner of Room A, near the south and north corners of Room B, and in Rooms D and E. Ceiling placement was chosen to allow for adequate projection of heated air and to maximize safety. Operation of the ceiling-mounted heaters in Rooms D and E produced thermal stratification in the two spaces (Δ10°C–Δ12°C differential from floor to ceiling), which was aggravated by the cool floor temperature. The thermal stratification was reduced by the use of freestanding oscillating pedestal fans in Rooms D

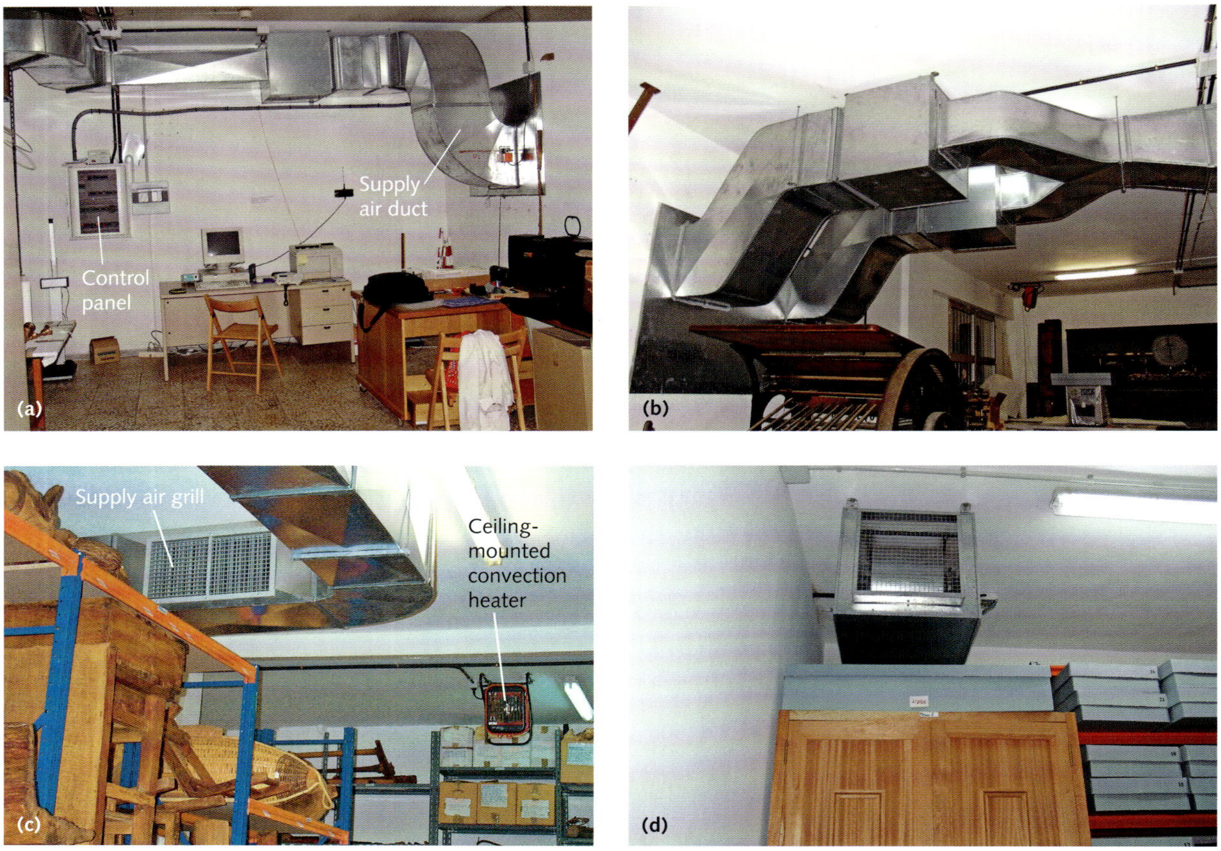

Figure 12.10

Installed equipment at the Valle Guerra storage facility, including (a) supply air duct with a supply fan enclosed within and a control panel mounted on the wall in Room A; (b) two supply-duct-mounted fans in Room B; (c) supply air grille at the end of the duct and ceiling-mounted convection heater in Room A; and (d) an exhaust fan in its ceiling-mounted enclosure in Room E. Photos: Shin Maekawa. © J. Paul Getty Trust.

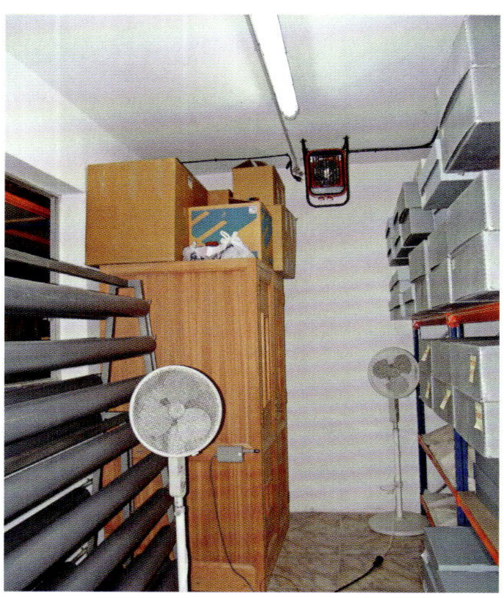

Figure 12.11

Heater and pedestal fans in Room D of the Valle Guerra storage facility. Photo: Shin Maekawa. © J. Paul Getty Trust.

and E that improved mixing of the air mass; these fans were operated in conjunction with the unit heaters (fig. 12.11). Oscillating fans were not necessary in Rooms A and B, as thermal stratification was not observed in those spaces.

Operational control sequence

The environmental management system was designed to operate in three modes: ventilation, heating, or idle mode (table 12.3). Automatic mode selection was based on a comparison of interior and exterior humidity. The operational control sequence for two postinstallation periods will be described. After the successful initial ten-month operation of the system (first postinstallation period), a decision was made to operate the same system with a lower target humidity to evaluate system performance (second postinstallation period). During the first postinstallation period (May 2003–March 2004), the ventilation mode was activated when the interior humidity was greater than 70% RH and exterior humidity was less than or equal to 70% RH (fig. 12.12). In this scenario, interior air with lower temperature and higher humidity is replaced with outside air of higher temperature

Table 12.3

Operational control conditions for the environmental management system for the Valle Guerra storage facility. Note that no temperature enable/disable conditions were included in the psychrometric strategy.

Psychrometric Strategy	First Postinstallation Period (May 2003–March 2004)		Second Postinstallation Period (March 2004–January 2005)	
	Enable	Disable	Enable	Disable
Ventilation	$H_{interior} > 70\%$ RH and $H_{exterior} \leqq 70\%$ RH	$H_{interior} < 60\%$ RH or $H_{exterior} > 70\%$ RH	$H_{interior} > 65\%$ RH and $H_{exterior} \leqq 65\%$ RH	$H_{interior} < 55\%$ RH or $H_{exterior} > 65\%$ RH
Heating	$H_{interior} > 70\%$ RH and $H_{exterior} > 70\%$ RH	$H_{interior} < 60\%$ RH or $H_{exterior} \leqq 70\%$ RH	$H_{interior} > 65\%$ RH and $H_{exterior} > 65\%$ RH	$H_{interior} < 55\%$ RH or $H_{exterior} \leqq 65\%$ RH
Idle	$H_{interior} < 60\%$ RH	$H_{interior} \geqq 70\%$ RH	$H_{interior} < 55\%$ RH	$H_{interior} \geqq 65\%$ RH

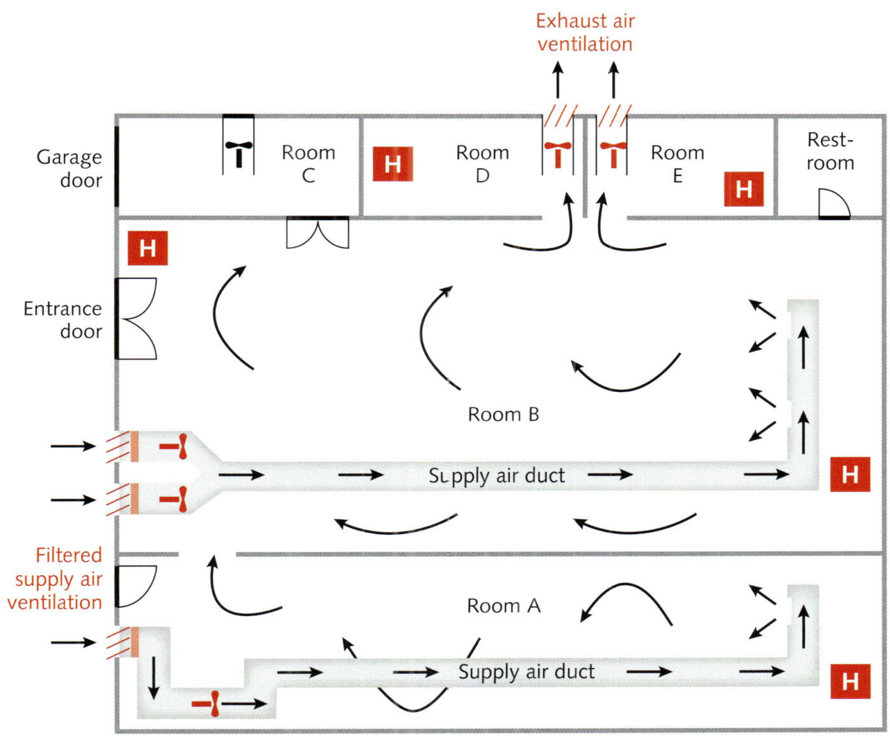

Figure 12.12

Schematic of conservation heating by ventilation mode.

and lower humidity. The heating mode was enabled when both interior and exterior humidity exceeded 70% RH (fig. 12.13). In this mode, the humidity inside the storage is reduced by increasing the interior temperature.

Operation of the ventilation mode ceased when exterior humidity was greater than 70% RH, while operation of the heating mode stopped when exterior humidity was less than or equal to 70% RH. Additionally, both the ventilation and the heating modes ceased when inside humidity fell below 60% RH, resulting in an operational deadband of 10% RH (between 60% and 70% RH).

The second postinstallation period (March 2004–January 2005) maintained a similar control scheme, differing only in the operational range of the environmental management system. During this period,

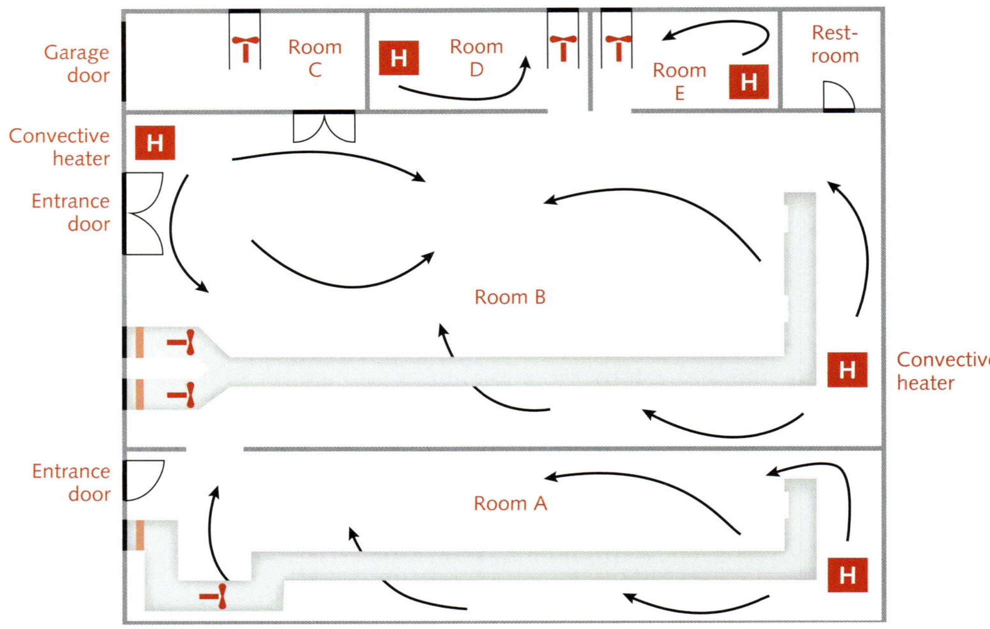

Figure 12.13
Schematic of conservation heating mode.

the ventilation mode was enabled when interior humidity was greater than 65% RH and exterior humidity was less than or equal to 65% RH; the ventilation was disabled when exterior humidity exceeded 65% RH (table 12.3). Heating was enabled when both inside and outside humidity exceeded 65% RH and disabled when exterior humidity was less than or equal to 65% RH. Operation of both modes was disabled when the humidity in the storage facility was less than 55%, maintaining the 10% RH operational deadband between 55% and 65% RH.

Postinstallation System Performance

Following operational commissioning of the mechanical system, the system was operated and monitored during the two postinstallation periods. The first postinstallation period had a target interior humidity range between 60% and 70% RH. The second postinstallation period shifted the target humidity range slightly lower, to between 55% and 65% RH. As noted earlier, discussion of the postinstallation storage facility environments will focus only on Room D. Owing to its position against the exposed northeast wall, this location harbored the most problematic preinstallation environment.

Humidity, temperature, and dew point

During the first postinstallation period, the range of monthly mean humidity in Room D narrowed, while the ranges of monthly mean temperature and dew point temperature expanded. Monthly mean humidity ranged from 61% RH (January) to 64% RH (September) (fig. 12.14a). Further, the maximum interior humidity recorded during the first postinstallation period was 70% RH, consistent with the upper limit of this period's operational range (60%–70% RH). Monthly average interior temperature and dew point temperature ranged from 20.1°C (February) to

27.5°C (August) and from 12.2°C (February) to 20.1°C (August), respectively (figs. 12.14b and 12.14c). In Room D, the maximum temperature was 30.5°C, and the maximum dew point temperature was 23.3°C, both recorded in August. The average rolling 24 hour variations (short fluctuations) of temperature (Δ1.0–Δ2.4°C) and dew point temperature (Δ1.1–Δ2.1°C) were higher than those observed during the preinstallation period, while the average daily variation of humidity (Δ3%–Δ6% RH) was similar to preinstallation levels.

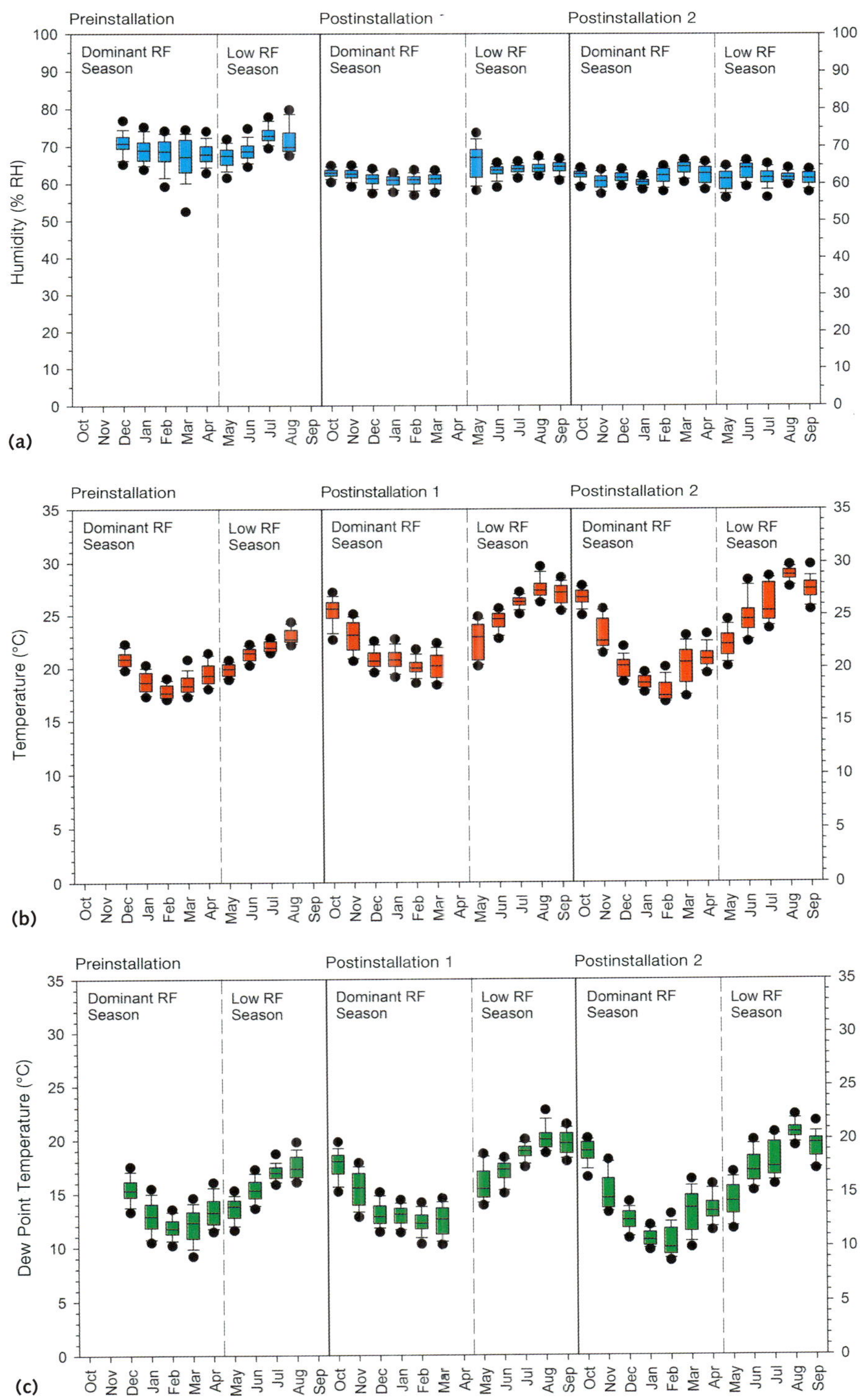

Figure 12.14

Pre- and postinstallation monthly comparisons for Room D: humidity (a), temperature (b), and dew point temperature (c).

During the second postinstallation period, in which the operational humidity range was shifted to between 55% and 65% RH, monthly average humidity in Room D ranged from 60% RH in January to 62% RH in October, with a maximum recorded value of 68% RH in March (fig. 12.14a). Monthly mean interior temperature and dew point temperature ranged from 17.8°C (February) to 28.8°C (August) and from 10.4°C (February) to 20.7° (August), respectively (figs. 12.14b and 12.14c). The highest recorded value for temperature was 30.4°C (August), and the maximum dew point temperature was 22.6°C (August and September). While the mean rolling 24 hour variation of interior temperature (Δ0.4–Δ1.3°C) was higher than preinstallation values, mean daily variations of interior humidity (Δ2% RH–Δ3% RH) and dew point temperature (Δ0.7–Δ1.3°C) were lower than those of preinstallation. In the storage facility, humidity in Room D was maintained between 55% RH and 65% RH (only 0.52% of time below 55% RH and 11.2% of time above 65% RH) throughout the year, and overall mean temperature was less than 25°C and rarely exceeded 30°C (>30°C less than 0.3% of the time).

During the first and second installation periods, monthly average interior temperature exceeded 25°C during the months from July to October, with peak monthly average values (first installation period, 27.5°C; second installation period, 28.8°C) observed in August (see fig. 12.14b).

Postinstallation risk class assessment

The Conservation Environment Classification–HH protocol for both postinstallation environments in Room D of the Valle Guerra storage facility is summarized in table 12.4. For comparison, the preinstallation and second postinstallation conditions are plotted on psychrometric charts in figure 12.15. (The first and second postinstallation conditions occupy similar regions on the psychrometric chart.) Values for determining Hot-Humid Control Classes are also indicated on the plots.

The pre- and postinstallation (first and second) risk classes for the interior environment at the Valle Guerra storage are summarized in table 12.5.

Microbial risk

The first and second postinstallation microbial risk classifications for Room D improved to Class B (low risk) from Class F (high risk) during preinstallation.

Mechanical risk

In Room D, postinstallation mechanical risk due to short-term variation, seasonal variation, and deviation from historical mean remained the same—class a (no risk)—as that observed during preinstallation.

Chemical risk

Operation of the environmental management system increased the chemical risk due to deviation from historical mean temperature from a pre-

Table 12.4

Summary of Conservation Environment Classification–HH for the (a) first and (b) second postinstallation periods.

Risk			Humidity		Temperature	
Category	**Specific Risk (statistic)**	**Room**	**Value (% RH)**	**Class**	**Value (°C)**	**Class**
Microbial	Germination threshold (P97.5)	D	68	B	—	—
		A	71	C	—	—
		B	65	A	—	—
Mechanical	Short-term variation (P95 of rolling 24 hour variation)	D	Δ8	a	Δ2.8	a
		A	Δ12	b	Δ1.7	a
		B	Δ9	a	Δ1.5	a
	Seasonal variation (absolute difference of seasonal means, RH means ≦ 70%)	D	Δ3	a	Δ3.8	a
		A	Δ5	a	Δ3.4	a
		B	Δ3	a	Δ4	a
	Deviation from historical mean (absolute difference)	D	Δ2	a	Δ3.5	a
		A	Δ6	a	Δ4	a
		B	Δ0	a	Δ3.3	a
Chemical	Deviation from historical mean (difference)	D	—	—	Δ3.3	+
		A	—	—	Δ4	+
		B	—	—	Δ3.5	+

Overall mechanical risk:[1]
Room D: a
Room A: b
Room B: a

(a)

Risk			Humidity		Temperature	
Category	**Specific risk (statistic)**	**Room**	**Value (% RH)**	**Class**	**Value (°C)**	**Class**
Microbial	Germination threshold (P97.5)	D	66	B	—	—
		A	67	B	—	—
		B	64	A	—	—
Mechanical	Short-term variation (P95 of rolling 24 hour variation)	D	Δ6	a	Δ2.4	a
		A	Δ4	a	Δ1	a
		B	Δ6	a	Δ1.4	a
	Seasonal variation (absolute difference of seasonal means, RH means ≦ 70%)	D	Δ0	a	Δ4.8	a
		A	Δ3	a	Δ3.3	a
		B	Δ1	a	Δ4	a
	Deviation from historical mean (absolute difference)	D	Δ1	a	Δ3.1	a
		A	Δ2	a	Δ4.5	a
		B	Δ1	a	Δ3.8	a
Chemical	Deviation from historical mean (difference)	D	—	—	Δ3.1	+
		A	—	—	Δ4.5	+
		B	—	—	Δ3.8	+

Overall mechanical risk:[1]
Room D: a
Room A: a
Room B: a

(b)

[1] Overall mechanical risk is based on the highest risk classification in any of the specific humidity- and temperature-based mechanical risks (short-term variation, seasonal variation, and deviation from historical mean).

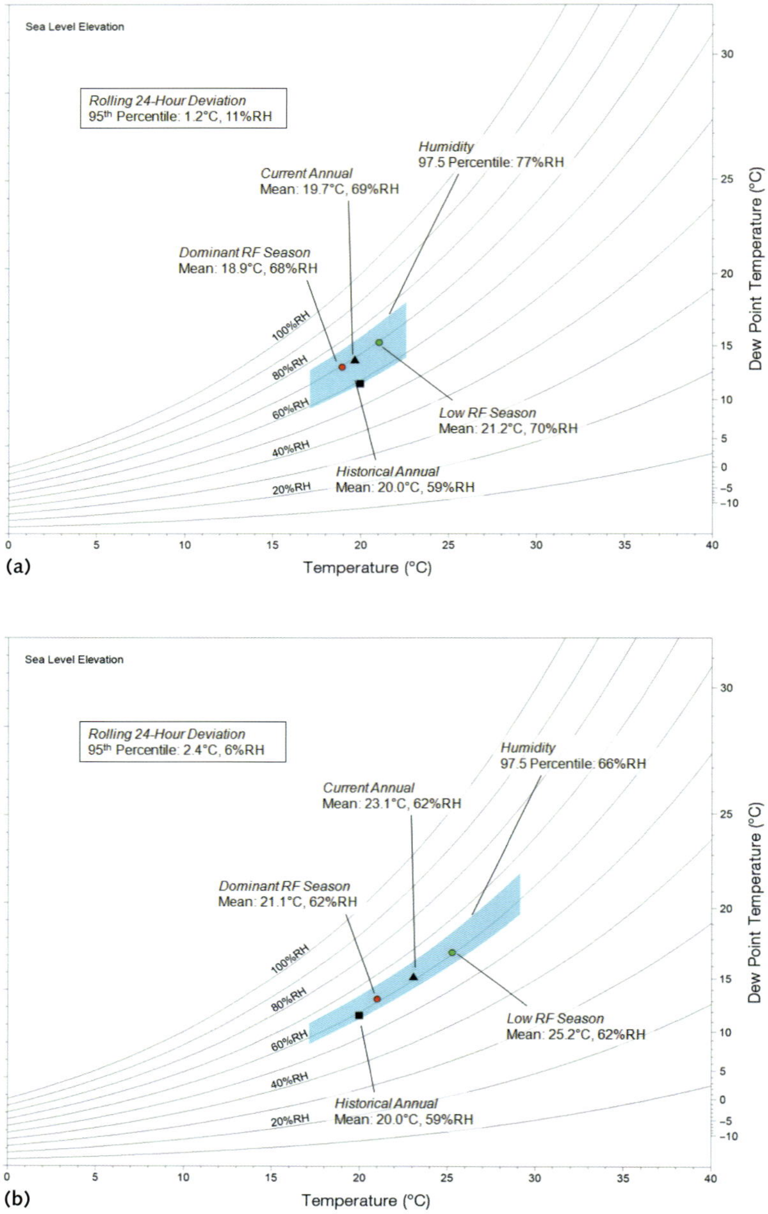

Figure 12.15
Comparison of Conservation Environment Classification–HH for Room D during the preinstallation period (a) and the second postinstallation period (b).

installation Class 0 (same risk as historic condition) to Class + (moderately increased risk) for the first and second postinstallation periods.

Microbes and dust

Air samples to monitor microbial activity were collected in various rooms of the storage facility before installation of the environmental management system and on three occasions during the first year of system operation: at four months, eight months, and twelve months (Valentín 2004). The most common genera of fungi isolated from samples taken inside the storage facility were *Penicillium*, *Rhizopus*, *Cladosporium*, *Mucor*, *Trichoderma*, *Alternaria*, and *Aspergillus*, while less common genera were

Table 12.5

Comparison of pre- and postinstallation (first and second) risk classes at the Historic Archive.

	Microbial Risk			Mechanical Risk																								Chemical Risk		
				Humidity									Temperature									Overall[1]				
				Short-Term Variation			Seasonal Variation			Deviation from Historical Mean			Short-Term Variation			Seasonal Variation			Deviation from Historical Mean								
Room	Pre	Post1	Post2	Pre	Post1	Post2	Pre	Post1	Post2	Pre	Post1	Post2	Pre	Post1	Post2	Pre	Post1	Post2	Pre	Post1	Post2	Pre	Post1	Post2	Pre	Post1	Post2
D	F	B	B	a	a	a	a	a	a	a	a	a	a	a	a	a	a	a	a	a	a	a	a	a	0	+	+
A	C	C	B	a	b	a	a	a	a	a	a	a	a	a	a	a	a	a	a	a	a	a	b	a	0	+	+
B	C	A	A	b	a	a	a	a	a	a	a	a	a	a	a	a	a	a	a	a	a	b	a	a	0	+	+

[1] Overall mechanical risk is based on the highest risk classification in any of the specific humidity- and temperature-based mechanical risks (short-term variation, seasonal variation, and deviation from historical mean).

Phoma and *Fusarium*. Isolated bacteria from the samples were *Bacillus*, *Micrococcus*, *Streptococcus*, *Pseudomonas*, and *Actinomyces*.

Compared to preinstallation levels, microbial activity in the storage facility was reduced by 70% to 80% following installation of the environmental management system. These reduced levels of microbial activity were documented for the first twelve months of operation (fig. 12.16) (Valentín 2004). Though microbial activity was elevated during the eighth month of operation, possibly because of seasonal variation, the recorded level of microbial activity was still much lower than that observed during the preinstallation period. Conservators working in the storage space also reported substantial reductions in dust deposition based

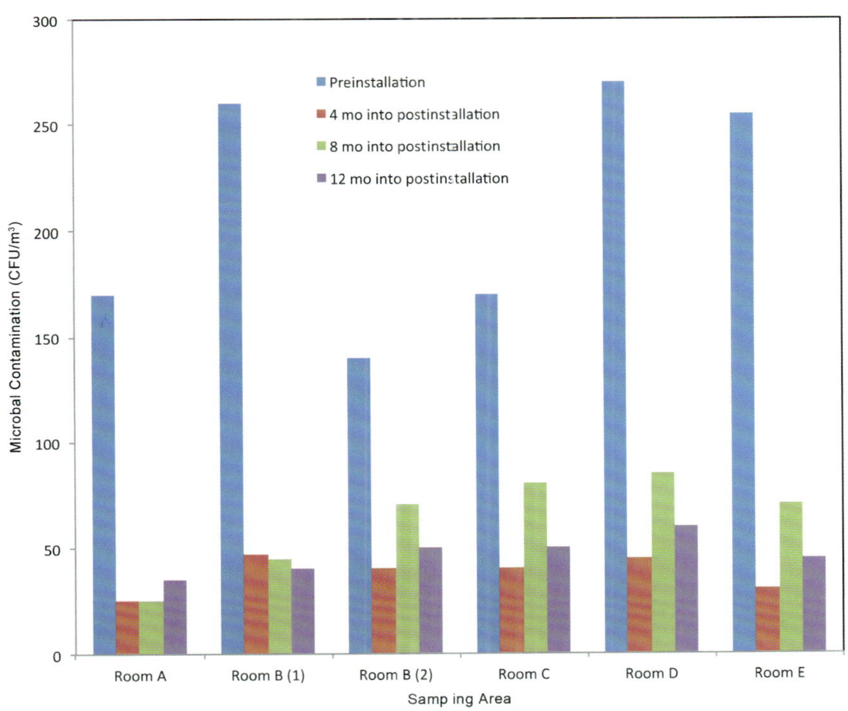

Figure 12.16

Comparison of microbial contamination of air samples collected from various areas in the storage facility during the pre- and postinstallation periods.

on their visual inspections. Following system installation, periodic condition assessments conducted by conservators reported no microbial activity on the collections (Morales 2004).

Thermal comfort

Prior to the installation of the environmental management system, average conditions during the warmest and most humid month, August, were 24.2°C and 71% RH in Room A, which was the hottest location in the storage facility.

Before installation of the environmental management system, museum staff occasionally used Room A for registering and documenting objects. Assuming that mean radiant temperature was equal to the air temperature in Room A, and assuming a metabolic rate of 1.0 met for seated occupants (consistent with the type of work performed in this space), a clothing factor of 0.6 clo (typical summer indoor clothing), and an air speed of 0.1 m/s, the predicted mean vote (PMV) and predicted percentage dissatisfied (PPD) values were 0.29 and 7%, respectively (neutral), complying with ANSI/ASHRAE Standard 55-2013 (point 1 in fig. 12.17) (ANSI and ASHRAE 2013).

During the second postinstallation period, the average conditions in Room A in August shifted to 28.8°C and 63% RH. Using the same assumptions as those in the preinstallation assessment, the PMV and PPD values were calculated as 1.58 and 55%, respectively (slightly warm) (point 2 in fig. 12.17). The operation of fans increased interior air speed up to 1 m/s, resulting in a lowering of PMV and PPD values to 0.33 and 7%, respectively (neutral) (point 3 in fig. 12.17). These indices again comply with ANSI/ASHRAE Standard 55-2013.

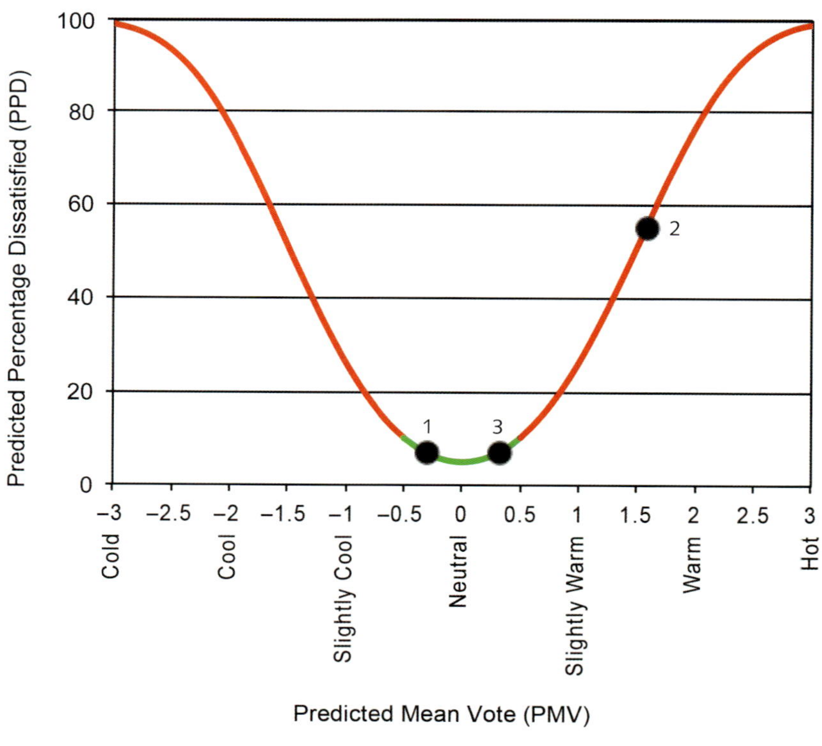

Figure 12.17

PMV-PPD plot showing the thermal comfort of monthly average conditions in August for Room A: (1) before the installation of the environmental management system; (2) after installation, with an air speed of 0.1 m/s; and (3) after installation with an increased air speed of 1 m/s.

Despite verification of acceptable comfort conditions in Room A by the ASHRAE Comfort Calculator, working staff generally felt uncomfortable due to elevated temperatures, particularly during summer months. In addition, complaints were received regarding a lack of fresh air and natural light in this space. Workers would also occasionally open emergency-only doors, allowing uncontrolled outside air to enter into the space and compromising system operation.

Because the environmental management system was designed with the primary objective of collections conservation, occupant comfort in the storage facility may have been less than optimal during the idle mode in which the air becomes stagnant. The freestanding fans can be used to provide air movement for improved thermal comfort so Room A can remain an active work site. The need for fresh air can be resolved by manually operating fans for a limited duration.

Noise

The supply fan noise levels in Room A were 56 dBA at a distance of approximately 1 m, and 54.6 dBA at a distance of 5 m. The presence of two supply fans in Room B added to the noise level in Room A by 0.3 and 0.6 dBA at 1 m and 5 m, respectively. At the diffuser end of the supply air duct in Room A, the noise level was 48 dBA at 1 m and 47 dBA at 5 m. Noise levels at the intake of the exhaust fans in Rooms D and E were 61–62 dB at a distance of 1 m and 55.5–55.9 dB at a distance of 5 m. The higher noise levels recorded at the exhaust fan were due to the exposed installation at the intake ends, and noise may have been exacerbated by the reflection of mechanical noise from the concrete floor and ceiling. However, this fan noise was not audible from outside of the building.

The high fan noise levels of the ventilation system also caused discomfort for workers in the storage facility; sound reflected off concrete walls, floors, and ceilings, compounding the noise problem. As source controls, various noise suppression designs should have been incorporated in the design of the ventilation system. In addition, sound-absorbing materials, such as textile curtains, carpet, and sponge plates, should have been used on walls, floors, and ceilings.

Maintenance

OAMC's facility maintenance personnel were trained by the HVAC contractor to perform the maintenance of the environmental management system, and the collections manager monitored system operation on a weekly basis. Periodic maintenance activities included lubrication of fan bearings, filter replacement as needed, and annual calibration of humidity sensors. The collections manager and the maintenance crew tested system performance using a handheld temperature and humidity meter every six months, or whenever a malfunction was suspected. During the six years of operation (May 2004–August 2010), the system required no replacement of mechanical or electrical parts.

Energy Metrics

Energy intensity

As shown in table 12.2, the total power of fans and heaters used in the Valle Guerra storage facility was 18 kW. Considering the total floor area of the storage (260 m^2), the energy intensity was 69 W/m^2, which is a lower value than that of a typical residential building.

Energy use intensity

Annual energy usage at the Valle Guerra storage facility increased from 21 kWh/m^2/yr before installation to 52 kWh/m^2/yr and 88 kWh/m^2/yr during the first and second postinstallation periods, respectively. These energy use intensities are one-third the energy use of a typical office.

Conclusions

After installation of the environmental management system at the Valle Guerra storage facility, the interior environment in Room D was maintained at a humidity less than 70% RH, as initially proposed. During the second postinstallation period, which maintained an operational range of 55% to 65% RH, mean temperature in Room D of the storage facility ranged from 18°C during the winter to 29°C in the summer.

With respect to microbial risk, the environment in Room D of the Valle Guerra storage facility improved from Class F (high risk) during the preinstallation period to Class B (low risk) during both the first and the second postinstallation periods. The mechanical risk in Room D remained the same, Class a (no risk) during the first and second postinstallation periods. As a result of increased temperature in this space, the chemical risk increased from Class 0 (same risk as historic condition) at preinstallation to Class + (moderately increased risk) during both the first and the second postinstallation periods (table 12.5).

The benefits of consistent humidity levels and protection against fungal and insect damage at the Valle Guerra storage facility will be noticed over the long term, particularly for objects made of wood or vegetable fiber. In the near term, the use of a filtered supply air substantially reduced dust levels in the storage space. These are important achievements considering that the site is on the humid western slope of northwest Tenerife Island, and the storage is surrounded by agricultural lands and located on a heavily traveled road. Since the initial operation of the environmental management system, no signs of deterioration or change have been observed in the stored objects.

Although the target environment for the storage facility focused on collections conservation, the presence of staff members working in the space highlighted human comfort issues. Elevated temperatures during the summer and noise levels from the running of fans were cited by occupants as degradations to the working environment. The thermal comfort issue of the elevated temperatures can be alleviated by the use of freestanding fans for increased local air speed without compromising control. However, the noise comfort issue was overlooked during system design, and no noise-reduction feature was included to reduce noise from intermittent fan operation. Noise-reduction features, such as vibration isolation mounts, sound-insulated enclosures for fan housings, noise-suppression devices both upstream and downstream of fans, and insulated ducts, should be considered in the design for reducing noise from the system. At the same time, the interior surfaces of the building could be treated by sound-absorbing materials.

OAMC staff at the facility also expressed discomfort while working without natural daylight throughout the storage facility. Staff members had relied on the natural daylight coming through the windows for various tasks in the facility, but this opportunity was eliminated when supply air ventilation fans were mounted on the windows of Rooms A and B, and thermal insulation pads covered windows in Rooms D and E. A lighting system for work tasks could substitute for the lost natural light. An alternate solution, especially for tasks that require longer time periods, would be to create a suitable work area in Room C, which is isolated from the conditioned storage space and not subject to environmental management.

Additional discussion relating to the case study can be found in Maekawa and Morales 2006.

Lessons Learned from the Museum Storage in Valle Guerra

- A Conservation Environment Classification–HH microbial risk of Class B is achievable in this Zone 2A (Hot-Humid) climate with the use of a simple system of humidistat-controlled conservation heating and conservation heating by ventilation.
- Envelope improvements must be made to control air infiltration around doors and improve the thermal performance of windows for a Class III building.
- HVAC contractors may be initially reluctant to install novel environmental management systems. It is essential to work with the contractor to arrive at an agreeable plan. Otherwise, find a contractor willing to work with your idea.
- If collections storage areas need to be routinely occupied by staff, consider creating a separate work room for staff that is conditioned for comfort rather than for conservation;
- Although no high-temperature excursions were experienced with the installed heating system, it would be prudent to put a high limit control on conservation heaters in case of humidistat failure; it could also potentially improve thermal comfort.

- Often the use of a space changes over the course of time. For example, storage spaces are often converted to work spaces as the museum's operation changes. Therefore, designs of the environmental management system should include flexibility to respond to future changes in environmental requirements, such as space occupancy.

Postscript

The environmental management system in the Valle Guerra storage facility operated continuously until the end of the building lease agreement in August 2010. No environmentally related damage of an object was reported during system operation. In response to the austerity measures enacted by the Spanish government to reduce the national debt, the collections were subsequently moved to a new OAMC storage facility in a warehouse at Santa Cruz de Tenerife, an act that consolidated all OAMC stored collections in one place. As of December 2013, the OAMC is evaluating possible environmental management strategies for the new facility.

Project Team and Responsibilities

Organismo Autónomo de Museos y Centros (OAMC) (Santa Cruz de Tenerife, Tenerife Island, Spain)

Maria Garcia Morales (Senior Conservator)—Project leader and local coordination, condition assessment of collection before and after installation of the environmental management system

Ruth Rufino and Gilberto González (Conservators)—Project operations

Facilities Department—Monthly maintenance

Fricair S.L. (Contracting HVAC company, Santa Cruz de Tenerife)—Engineering design and installation of the mechanical system

Níeves Valentín (Research Biologist, Instituto del Patrimonio Cultural de España, Madrid, Spain)—Microbiological monitoring and analysis

Getty Conservation Institute (Los Angeles, United States)

Shin Maekawa (Senior Scientist)—Project management, environmental assessment, concept design of environmental management strategies, commissioning, postinstallation monitoring and improvements

Vincent L. Beltran (Assistant Scientist)—Environmental data collection, analysis, and management

References

AEMet [Agencia Estatal de Meteorología]
2012 Guía resumida del clima en España (1981–2010), versión 3.0. http://www.aemet.es/es/conocermas/publicaciones/detalles/guia_resumida_2010.

ANSI [American National Standards Institute] and ASHRAE
2013 *Standard 55-2013—Thermal Environmental Conditions for Human Occupancy*. Atlanta: ASHRAE.

ASHRAE
2011 Museums, galleries, archives and libraries. Chapter 23 in *ASHRAE Handbook—HVAC Applications*. Atlanta: ASHRAE.

Maekawa, S., and M. Morales
2006 Low-cost climate control system for museum storage facility on Tenerife Island. In *Clever Design, Affordable Comfort: A Challenge for Low Energy Architecture and Urban Planning: 23rd International Conference on Passive and Low Energy Architecture, Geneva, Switzerland, 6–8 September 2006: Conference Proceedings*, ed. R. Compagnon, P. Haefeli, and W. Weber, 411–16. Geneva: PLEA 2006 Organizing Committee.

Maekawa, S., and F. Toledo
2003 Sustainable climate control for historic buildings in subtropical climates. *Management of Environmental Quality: An International Journal* 14 (3): 369–82.

Morales, M.
2004 Conservation assessment of the collection in the Valle Guerra storage. OAMC internal report.

Valentín, N.
2004 Análisis microbiológicos realizados en el Museo de Valle Guerra, Tenerife. Departamento Científico, Instituto del Patrimonio Cultural de España, internal report.

Weather Channel
2014 Weather Underground. http://www.wunderground.com/.

Chapter 13

Ethnographic Storage of the Museu Paraense Emílio Goeldi, Belém, Brazil: Conservation Heating by Ventilation and Mechanical Dehumidification

Climate zone 1A (Very Hot-Humid)

Objectives

- Maintain stable humidity that can arrest fungal activities, while maintaining humidity and temperature close to historic values of the Belém area.
- Maintain good air circulation throughout the storage to reduce microbial risk.
- Use equipment that is low-cost, robust, and simple to operate.
- Ensure system survivability long enough to maintain interior conditions during an incident, such as loss of electrical power or equipment failure.

Summary

This case study applies alternative environmental management strategies to a new storage space for the Amazonian Ethnographic collection at the Museu Paraense Emílio Goeldi (Goeldi Museum), an ASHRAE Category III building located near the equator in a Very Hot-Humid (1A) climate zone in northeastern Brazil. Previous storage locations were equipped with window-mounted packaged terminal air-conditioning (PTAC) units and were plagued with unstable environmental conditions, mold, and high power consumption. The goal of the project was to create new storage spaces that could consistently maintain humidity below 70% RH using relatively low-cost equipment that was both robust and simple to operate, while not departing too far from historic humidity and temperature values in the area.

Nonmechanical strategies for the new storage spaces included grading and paving of an area surrounding the storage, building envelope modifications for reduced air infiltration, the installation of thermal insulation in the attic, creation of a vestibule area outside the entrance, and the use of a storage shelving system with perforated side panels and open space drawers to maximize air circulation. The mechanical strategy for environmental management in the new storage space uses humidistat-controlled conservation heating by ventilation and mechanical dehumidification. This system was successfully installed in 2003 and, to date, continues to maintain the target humidity.

Background

The Goeldi Museum is the oldest scientific institution in the Amazon region and the second-oldest natural history museum in Brazil. Its collections have both historical and scientific significance regarding the flora and fauna, physical environment, and social groups that currently dwell or have dwelled in the northern regions of Brazil. The museum's Amazonian ethnographic collection previously occupied two storage rooms (total floor area of 113 m^2) at the Rocinha campus of the Goeldi Museum near the city center of Belém. These rooms had wooden shelves, cabinets, and drawers that were densely filled and poorly organized.

The majority of the previous storage environments for the Goeldi Museum's collections were maintained by multiple wall or window-mounted PTAC units that were operated during the weekdays from approximately 9:00 a.m. to 3:00 p.m. These units were designed to cool for human comfort rather than maintain a stable and low-humidity environment for collections conservation. Mold infestation was common in these facilities owing to high infiltration through the building envelope, inadequate dehumidification, poor air circulation, equipment failures, and frequent and extended power failures. Mold colonized on the exterior surface of the building walls because of condensation and high exterior surface moisture caused by the uninsulated wall (fig. 13.1). Gaps and openings in the envelope allowed for insect entry, resulting in a large quantity of naphthalene being used to protect against insect infestation.

An example of the humidity variations in the original storage space of the Rocinha campus is illustrated by figure 13.2, which

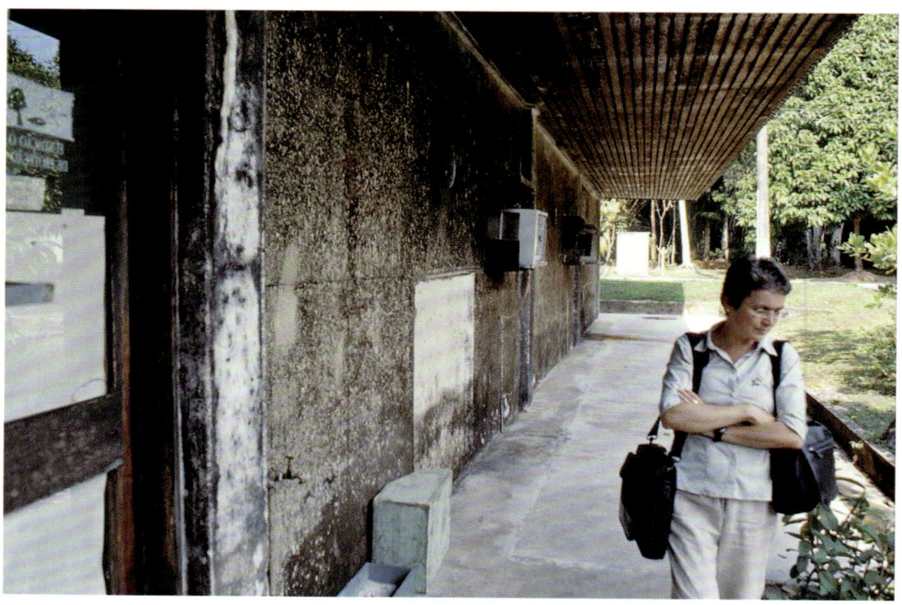

Figure 13.1
Mold on the exterior wall of the Goeldi Museum's old collections storage building. Photo: Shin Maekawa. © J. Paul Getty Trust.

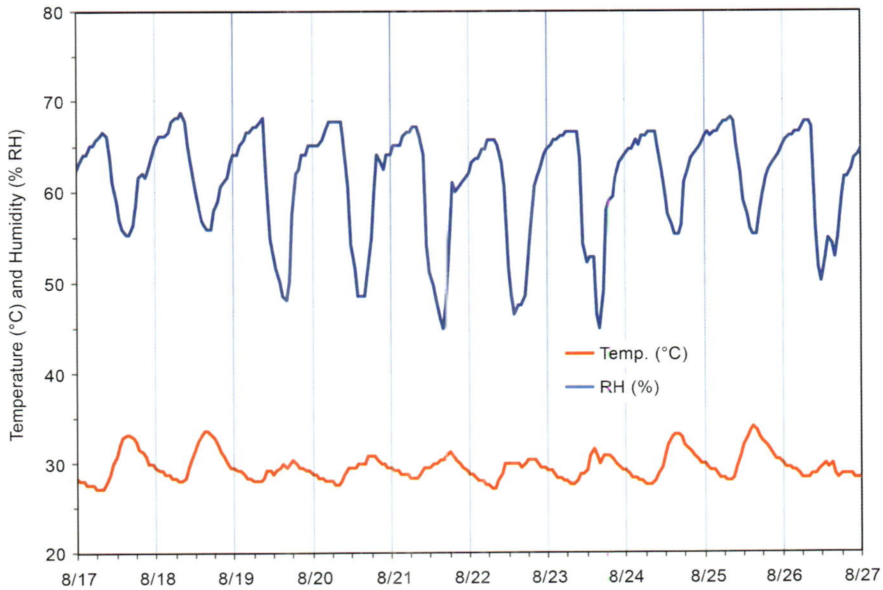

Figure 13.2
Temperature and humidity levels during August 2002 in the original Amazonian ethnographic storage at the Goeldi Museum's Rocinha campus.

shows interior temperature and humidity during a ten-day period in August 2002. Daily humidity levels ranged from approximately 51% RH to 73% RH, with PTACs running only during working hours, from 8:00 a.m. to 12:00–1:00 p.m. on weekdays. The figure shows that humidity daily approached the fungal germination threshold of 75% RH. And it is likely that an extended power or equipment failure will quickly elevate humidity to above the threshold. Operation of PTACs did not reduce peak humidity and resulted in an increase in daily humidity variation of about Δ20% RH—from Δ10% RH without operation of PTACs. The temperature in the storage area was reduced to less than 24°C–30°C when PTACs were operated; otherwise, it ranged daily from 28°C to 31°C and often reached 34°C (not shown in the figure).

PTAC units had been widely used throughout the campus. As a consequence, Goeldi Museum allocated roughly 70% of its operating budget to electrical power for storage facilities and offices. Since energy consumption of the storage facility could not be isolated from the museum's overall energy bill, an estimate of the energy consumption for environmental control of the original Amazonian ethnographic collection storage space—3,780 kWh, or 33 kWh/m^2/month—was based on the power rating of installed equipment (table 13.1). On the heels of the national energy crisis in 2001, the Brazilian government mandated a 70% reduction in energy consumption for all facilities, to be achieved in the ensuing ten years. Therefore, reduction of energy use was a high priority for the museum.

Frequent and extended power outages were common in the region and may have been another factor in fungal and insect outbreaks in the collection, despite heavy and broad use of insecticides and fungicides. The museum was, therefore, searching for an innovative environmental management strategy that would be low in cost for both installation and operation, technologically robust, and capable of

Table 13.1

Estimated monthly energy consumption in the old ethnographic storage space.

Number of Units	Equipment	Watts	Hours	Days	Consumption (kWh)
3	Air-conditioner (daytime operation)	11,700	8	31	2900
2	Air-conditioner (nighttime operation)	4140	16	22	1500
22	Fluorescent lights	880	8	22	150
Total					4550

maintaining a stable conservation environment, even during a period of extended power outage.

In 2003 the museum's Department of Ethnography was allocated a new storage space for its Amazonian ethnographic collection at the museum's research campus (Campus de Pesquisa) near the edge of the city. The new storage space provided the opportunity to employ alternative approaches for conservation environments that would satisfy the needs of the museum storage facilities.

Climate

Located in the Brazilian state of Pará, near the equator, the state capital city of Belém (1°27' S, 48°29' W; elevation, 10 m above sea level) sits at the mouth of the Amazon estuary (fig. 13.3). Historical weather data for

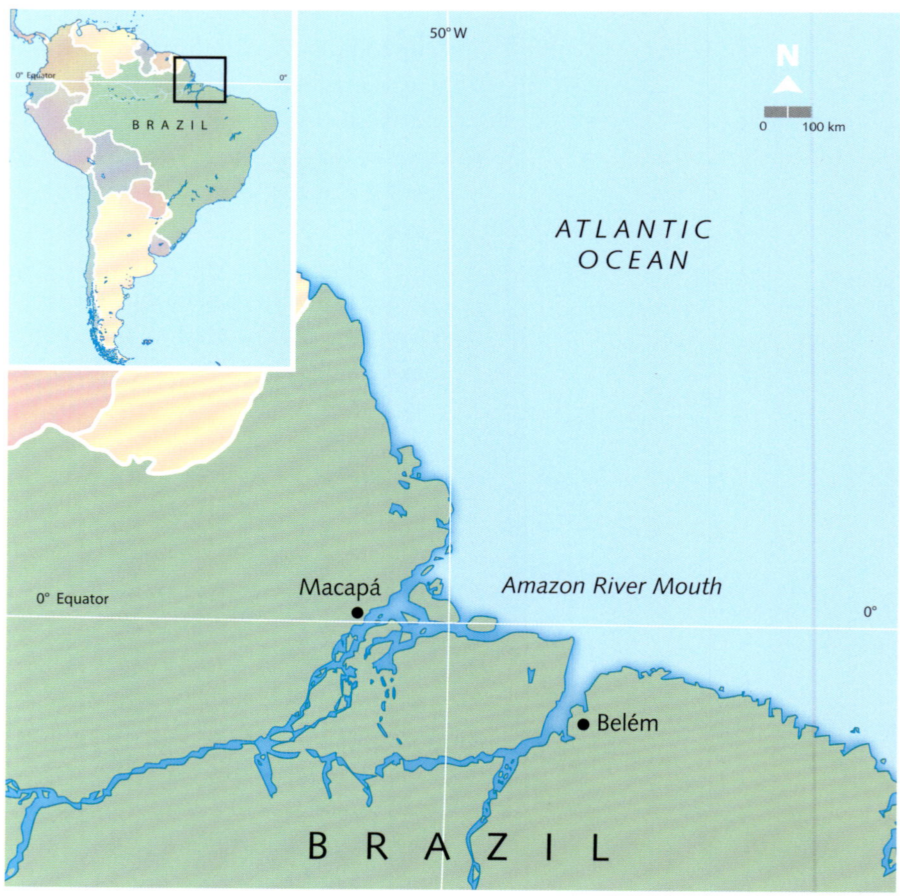

Figure 13.3

Location of Belém on the map of Brazil.

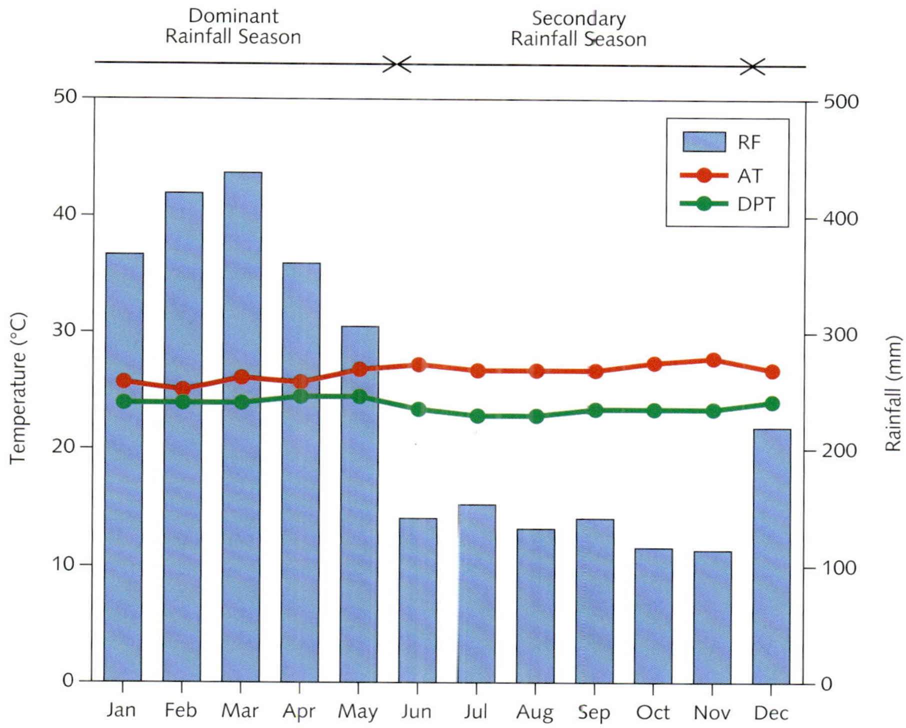

Figure 13.4
Historical weather data recorded at Belém, Brazil. Data for air temperature (AT) and dew point temperature (DPT) were taken from *Engineering Weather Data* (1973–96) (AFCCC and NCDC 1999). Rainfall (RF) data (1961–90) were taken from the Instituto Nacional de Meteorología (INMET 2009).

Belém shows monthly mean air temperatures (25.0°C–27.8°C; annual mean of 26.5°C) and dew point temperatures (22.8°C–24.4°C) to be relatively constant throughout the year (1973–96, AFCCC and NCDC 1999) (fig. 13.4). Maximum air temperature rarely exceeds 32°C, while minimum air temperature is 22°C. Based on historical monthly average air and dew point temperature values, the calculated annual mean humidity is 84% RH. Rainfall is ubiquitous during the year—mean annual total of 2893 mm, with a minimum monthly mean of 112 mm in November—with showers typically occurring in late afternoon and early evening (1961–90, INMET 2009). However, the six-month period from December to May has distinctively higher monthly rainfall totals than the rest of the year; therefore, this period is designated as the dominant rainfall season, and the remainder of the year (June–November) is the secondary rainfall season. With monthly precipitation during the secondary rainfall season exceeding the 60 mm per month threshold, the climate of Belém can be classified as tropical rainforest (*Af*) based on the Köppen-Geiger climate classification (see appendix 2).

The CDD10°C calculation based on temperature collected at the Belém International Airport over a three-year period (2008–10) produced 6460 degree-days per year, which placed Belém in the Very Hot (1) climate zone (Weather Channel 2014). The mean temperature of the coldest month (25°C, February) exceeds 18°C, excluding it from a Marine climate classification. The annual rainfall of 2893 mm is greater than Dry-Humid index of 670, placing Belém's climate in the Humid (A) type category. Therefore, Belém's climate can be classified as Very Hot-Humid (1A).

Building

The new storage space for the Amazonian ethnographic collection is housed in a single-story contemporary urban building (920 m^2 footprint) that is typical of the region and consists of three rooms used for: (1) reception, quarantine, and conservation treatments; (2) inventory and study of collections; and (3) the storage room itself (fig. 13.5a). Measuring 15 by 18 m (270 m^2 footprint), the storage room has a concrete slab-on-grade floor, a concrete slab ceiling at a height of 3 m, and 15 cm thick uninsulated walls made of hollow fired bricks. The walls are finished on both sides with cement plaster and a water-based paint. As found, the building had several openings in the building envelope for PTAC units (not installed), but there were no windows (figs. 13.5b and c). Access to the storage room is through two sets of double doors—one at the center of the room's interior east wall and leading to the corridor, and the other near the east end of the north wall leading to the outside (see fig. 13.5a). Just inside the doors leading to the corridor is a single door that provides access to the reception/quarantine/treatment room. The roof over the storage space consists of corrugated metal with a large ridge vent and eave vents for passive ventilation of the attic space. The roof has a long eave overhang that discharges rainwater to ground-surface drainage away from the walls; it also provides shade. The building is classified as ASHRAE Category III (uninsulated masonry) (ASHRAE 2011).

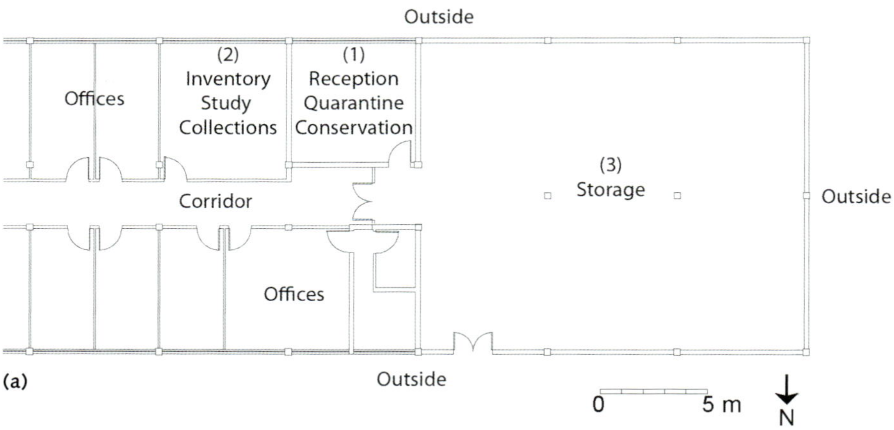

(a)

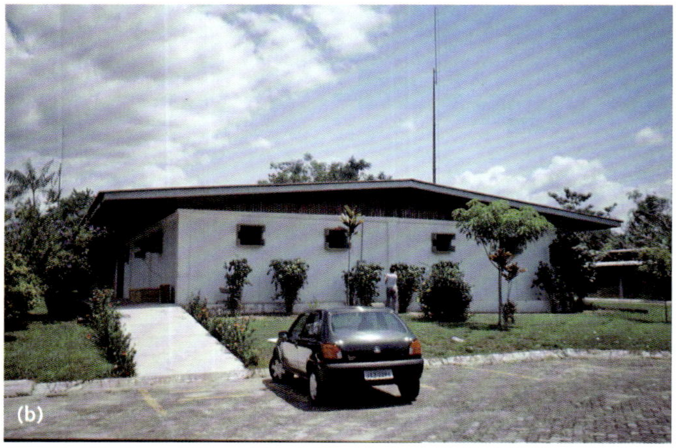

(b)

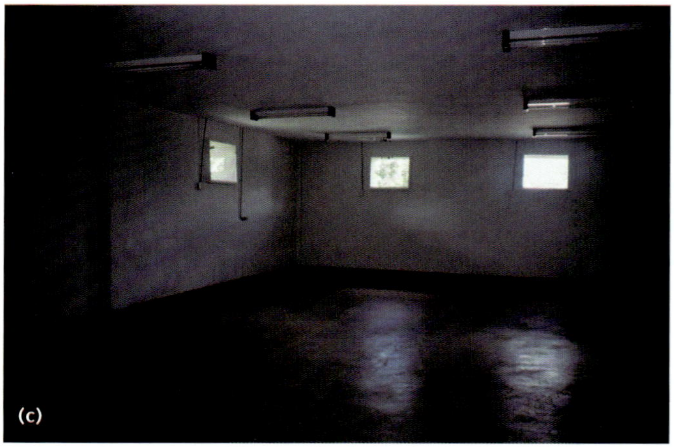

(c)

Figure 13.5
Floor plan (a) of new rooms for the Amazonian ethnographic collection at the Goeldi Museum, plus an exterior view (b) of the structure (seen from the west) and an interior view of the storage room (c) (seen from the east). Note that openings in the building envelope were for PTAC units and not for windows. Photos: Shin Maekawa. © J. Paul Getty Trust.

Collection

The Amazonian ethnographic collection consists of roughly fifteen thousand objects used by different ethnic groups for agriculture, fishing and hunting, food processing, ceremonies, and celebrations (fig. 13.6). Classified according to region and tribe, these items largely consist of basketry, instruments, and ornaments made from plant fibers, wood, seeds, and bird feathers. The ethnographic collection was transferred to the new storage space over a period of several months starting in November 2003, when the environmental management system became operational.

Preinstallation Condition Assessments

Building envelope assessment

The envelope assessment of the new storage facility identified the steel entry door leading from the storage room to the exterior as a potential risk for insect and pest entry, as well as an access point for air and moisture vapor infiltration. Similarly, the unused wall openings intended for

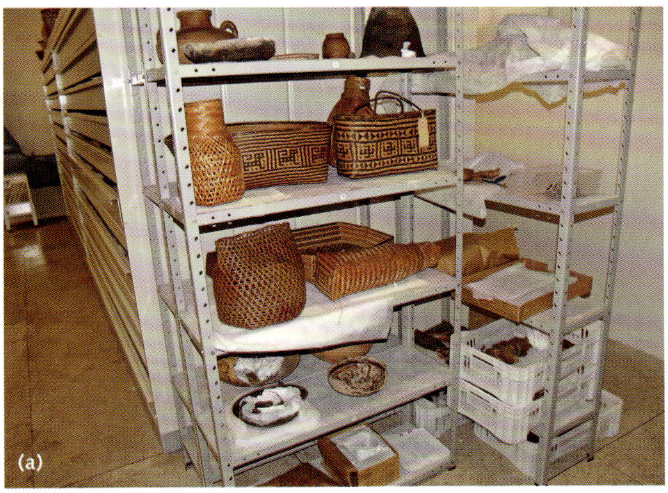

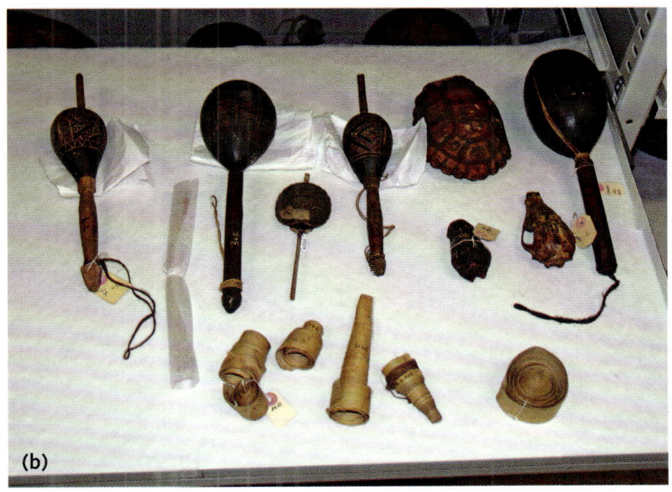

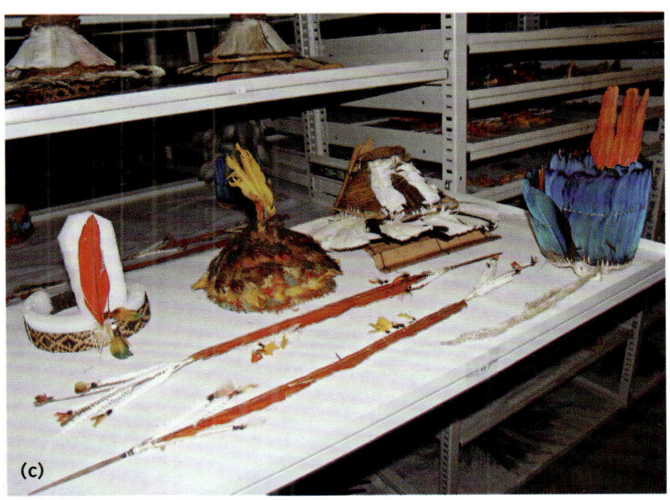

Figure 13.6
Amazonian ethnographic collections: bamboo and plant basketry (a); ceremonial tools and objects made of wood, seeds, and animal bones (b); and headdresses made of bird feathers and plant fibers (c). Photos: Shin Maekawa. © J. Paul Getty Trust.

PTAC units required either permanent closure or conversion to windows with good closure against insects and exterior moisture vapor.

The walls and ceiling of the storage space lacked thermal insulation, and conductive heat transfer through the walls and ceiling was a concern. On the other hand, the envelope walls were painted with a water-based paint on both the exterior and interior, which reduced moisture vapor transmissivity. There was no evidence of bulk moisture entry through the floor slab and ceiling, although vapor transmissivities of the floor slab and ceiling were unknown.

The immediate external surroundings to the north, west, and south of the storage area consisted of exposed soil. The potential for soil erosion from wind and driving rain may affect ground moisture in the vicinity of the storage area, providing a local driving force that could affect building envelope performance.

Exterior environmental assessment

For the duration of the case study (July 2003–March 2006), exterior air temperature and humidity were monitored at 15 minute intervals on the south side of the new storage building, and dew point temperature was calculated from concurrent humidity and temperature values (fig. 13.7). This recent locally collected data indicated higher monthly mean exterior

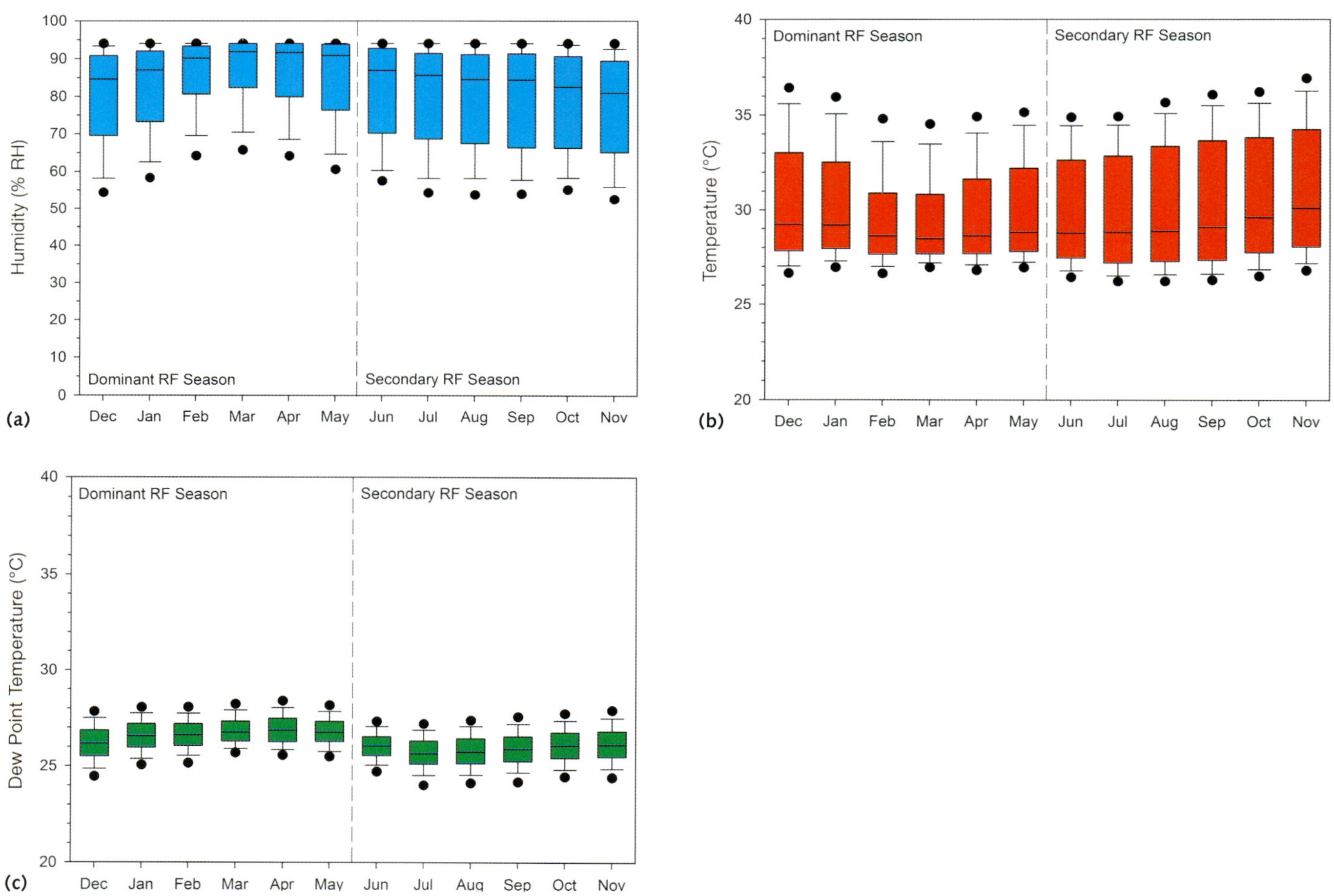

Figure 13.7

Exterior humidity (a), temperature (b), and dew point temperature (c) data recorded onsite at the new Goeldi Museum storage facility from July 2003 to March 2006.

air temperatures (29.4°C–31.1°C; annual mean of 30.1°C) and exterior dew point temperatures (25.7°C–26.9°C) than observed in the historical data. The differences may be due to the geographical orientation/location of the research campus, which was 4–5 km inland from the Belém airport, where the historical climate data was collected, or to localized climatic effects, such as wind blockage or the effect of the thermal mass of the building and pavements. During the case study, the rolling 24 hour variation of exterior air temperature and dew point temperature was roughly Δ9°C and Δ3°C, respectively. Monthly average outside humidity at the site also ranged from a low of 77% RH in November to a high of 87% RH in March (annual mean of 82% RH), although these values may be low owing to the fact that the humidity sensor was unable to record values higher than 94% RH. The monthly mean rolling 24 hour variation of humidity ranged from Δ29% RH in March to Δ41% RH in November.

Average daily trends of temperature, humidity, and dew point temperature in March (during the dominant rainfall season) and November (during the secondary rainfall season) are shown in figure 13.8.

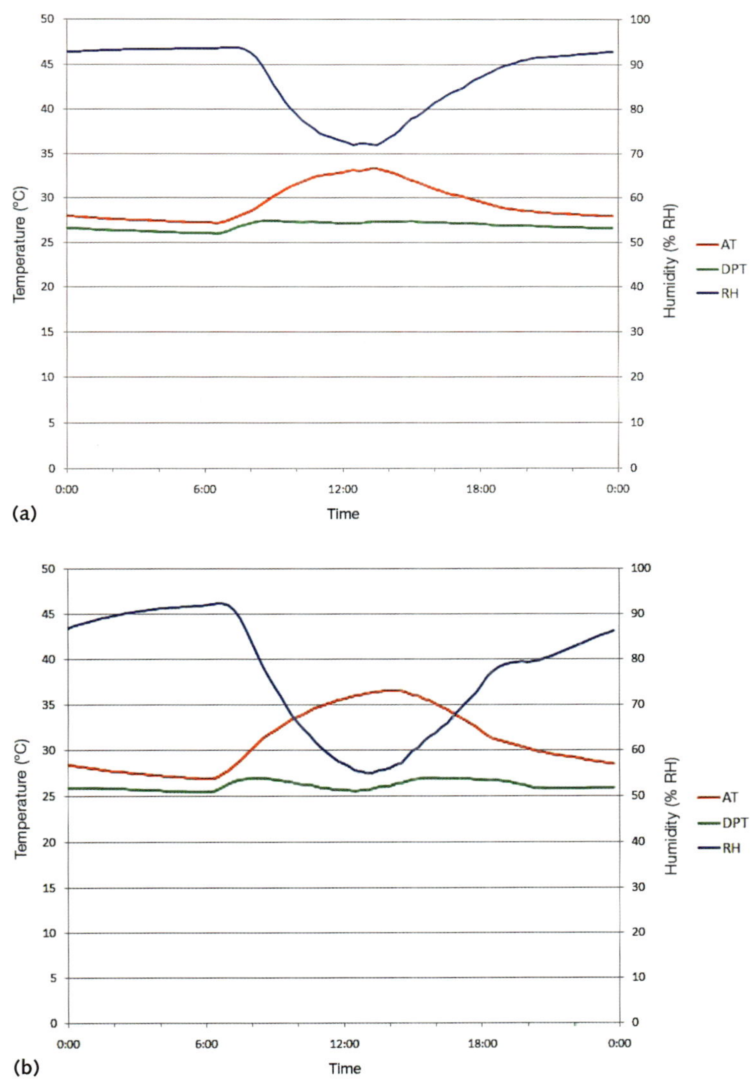

Figure 13.8

Average daily trend of exterior temperature (AT), humidity (RH), and dew point temperature (DPT) near the new Goeldi Museum storage facility in (a) March, during the dominant rainfall season, and (b) November, during the secondary rainfall season.

Dew point temperature remained nearly constant (between 26°C and 28°C) throughout the day in both March and November, while temperature showed a Δ6°C daily variation in March, which widened to approximately Δ10°C in November. This larger temperature variation in November allowed humidity to drop below 70% RH from 10:00 a.m. to 6:00 p.m., while the humidity in March remained above 70% RH throughout the day.

Interior environmental assessment

At the Goeldi Museum's new storage space, interior temperature and humidity were monitored at 15 minute intervals for slightly less than three full months (July–October 2003) before nonmechanical strategies were applied to the building prior to the installation of the mechanical strategies. Dew point temperature was then calculated from paired temperature and humidity values. This limited dataset is representative of a naturally ventilated space.

During this three-month preinstallation period, coinciding with the secondary rainfall season, interior temperature, humidity, and dew point temperature averaged 30.9°C, 74% RH, and 25.8°C, respectively. Interior temperature and dew point temperature ranged from 30.3°C to 32.7°C and 23.3°C to 27.3°C, respectively. The 95th percentile of the rolling 24 hour variation of temperature and humidity was Δ1.5°C. During this period, interior humidity ranged from 61% to 79% RH, and monthly mean humidity ranged from 69% to 77% RH. The 95th percentile of the rolling 24 hour variation of humidity was Δ7% RH (fig. 13.9).

Because of the short duration of the preinstallation interior monitoring period, the dataset represents a naturally ventilated space during only one season—the secondary rainfall season. Table 13.2 summa-

Table 13.2

Conservation Environment Classification–HH for the preinstallation interior environment in the Goeldi Museum's new, naturally ventilated storage space.

Risk		Humidity		Temperature		
Category	**Specific Risk (statistic)**	**Value (% RH)**	**Class**	**Value (°C)**	**Class**	
Microbial	Germination threshold (P97.5)	78	F	—	—	
Mechanical	Short-term variation (P95 of rolling 24 hour variation)	Δ7	a	Δ1.5	a	
	Seasonal variation (absolute difference of seasonal means, RH Means ≦ 70%)	—	f[1]	—	—	Overall mechanical risk:[2] f
	Deviation from historical mean (absolute difference)	Δ9	a	Δ4.4	b	
Chemical	Deviation from historical mean (difference)	—	—	Δ4.4	+	

[1] Seasonal humidity mean is above 70% RH (secondary rainfall: 74% RH, dominant rainfall: n/a), resulting in a class f designation with respect to this specific mechanical risk.
[2] Overall mechanical risk is based on the highest risk classification in any of the specific humidity- and temperature-based mechanical risks (short-term variation, seasonal variation, and deviation from historical mean).

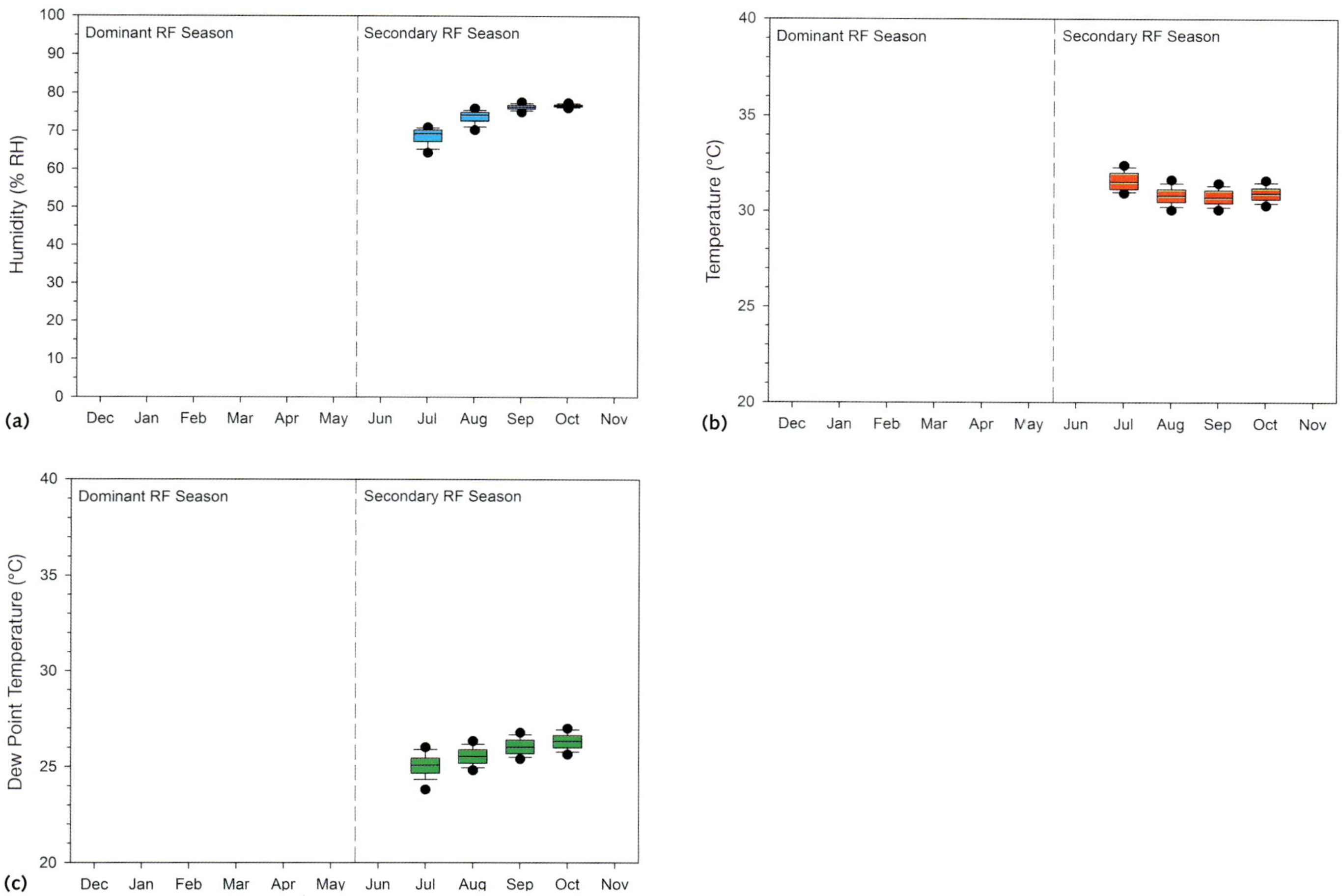

Figure 13.9
Interior humidity (a), air temperature (b), and dew point temperature (c) during the preinstallation period, from July to October 2002.

rizes the Conservation Environment Classification–HH protocol for the storage space's naturally ventilated interior environment during the preinstallation period.

Environmental Management System

Objectives

The primary environmental management objective was to prevent fungal infestation while maintaining a stable humidity that is close to the humidity to which the collections were historically exposed. The specific objectives were:

- Maintain stable humidity below 70% RH and maintain good air circulation throughout the storage to reduce microbial risk.
- Maintain humidity and temperature close to the historic values of the Belém area.
- Use equipment that is low in cost, robust, and simple to operate.
- Ensure system survivability long enough to maintain interior conditions during an incident, such as loss of electrical power or equipment failure.

Psychrometric strategies

The moisture source was controlled by improving the airtightness of the building envelope as well as by reducing the ground soil moisture. The heat source was also controlled by insulating the building envelope for reduced heat gain from the sun.

The environmental management system in the new storage space aimed to maintain humidity below 70% RH by utilizing conservation heating by ventilation when the exterior air was less humid than the interior air; otherwise, dehumidification was used to maintain the interior humidity.

Mechanical dehumidification was chosen over heating to limit the temperature increase in what is already a warm environment, as well as to conserve energy. As described in the "Climate" section above, during the secondary rainfall season, warmer (>31°C) and less-humid (<70% RH) outside air was typically available for ventilation (conservation heating) between 10:00 a.m. and 6:00 p.m. However, during the dominant rainfall season, less-humid air is rarely available, so the use of ventilation is limited; in this case, dehumidifiers were used to mechanically manage the interior environment.

This approach would effectively maintain humidity in the storage space below the threshold for microbial activity, protecting the collection from the primary threat of fungi and bacteria. While the temperature in the storage space would approach that of the exterior midday, chemical aging as a result of this higher temperature was considered a secondary concern, since the objects in the collection were previously exposed to a very hot-humid climate throughout their lifetimes, even before entering the museum collection. Thus, the bulk of chemical damage to the objects had already occurred, and temperature-driven reaction rates have likely slowed.

Nonmechanical strategies

Prior to installation of the mechanical system, two nonmechanical strategies were implemented at the new Goeldi Museum storage space: improvements to the building envelope and selection of furniture to be used within the storage space.

Building envelope

Although the storage room remained the same as in the original plan, adjacent rooms were repurposed to allow for a better object flow from registration to storage. The original inventory study collections room was expanded to serve as a registration site and temporary storage, and the reception/quarantine/conservation room was transitioned into use as a conservation laboratory. In advance of system installation, several architectural modifications were made on the building envelope to improve its airtightness and to limit moisture infiltration. These modifications included (fig. 13.10):

- permanent closure of a steel door leading from the storage room directly to the outside—this action, however,

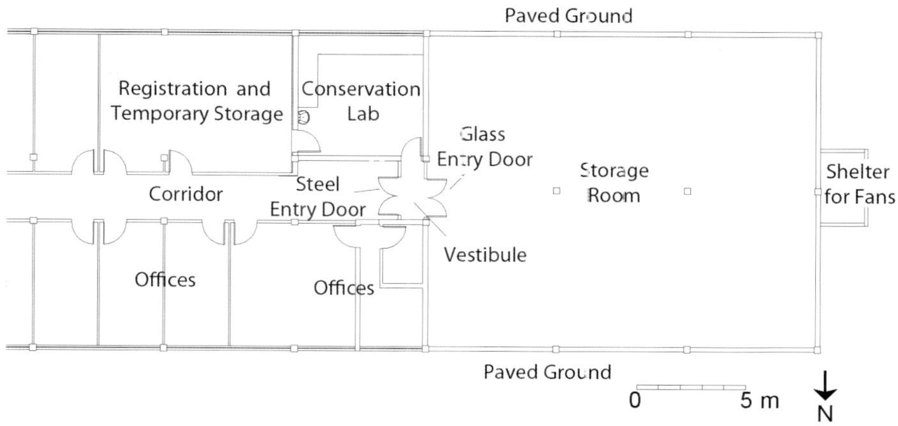

Figure 13.10
Floor plan of the new Amazonian ethnographic storage, showing changes to the building envelope and internal spaces.

limits access to the storage room to only the double doors opening to the corridor and reduces the speed of collection movement in case of evacuation;

- elimination of wall openings intended for PTAC units, with conversion of some into windows;
- installation of thermal insulation in the attic;
- paving of the area surrounding the storage building to improve surface drainage and to limit soil erosion;
- installation of a single fire-resistant metal door to the conservation lab;
- creation of a vestibule to serve as a transition space between the storage and nonstorage areas. The vestibule was created by installation of a double fire-resistant metal door at the storage entrance from the corridor and, just beyond that, installation of a pair of glass entrance doors opening into the storage area.

Storage furniture

Compact shelving systems and steel drawer cabinets with a baked enamel finish (to avoid off-gassing of volatile organic compounds) were required in the storage area to maximize use of the space while allowing for adequate air circulation on all sides. To allow for air circulation even when they are closed, the compact shelving system has perforated side panels, and the steel drawer cabinets have gaps between adjacent drawers (fig. 13.11).

The storage room was divided into four parts with an approximately 3 m wide center (east–west) opening and an approximately 1 m wide open space between perimeter walls and outer ends of the quadrants for good circulation of air as well as people. Two large open desks were placed along the center opening, and shelves and drawers occupied all four quadrants. The compact shelving system occupied the southeastern quadrant, and fixed shelves filled the northeastern quarter of the storage room. Drawers occupied both the southwestern and northwestern quarters.

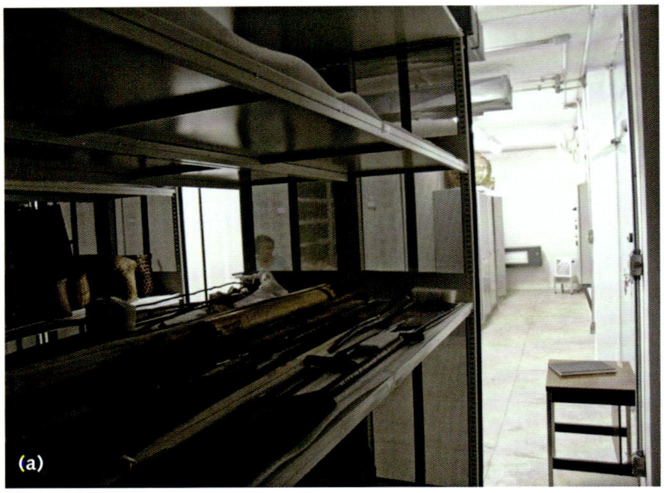

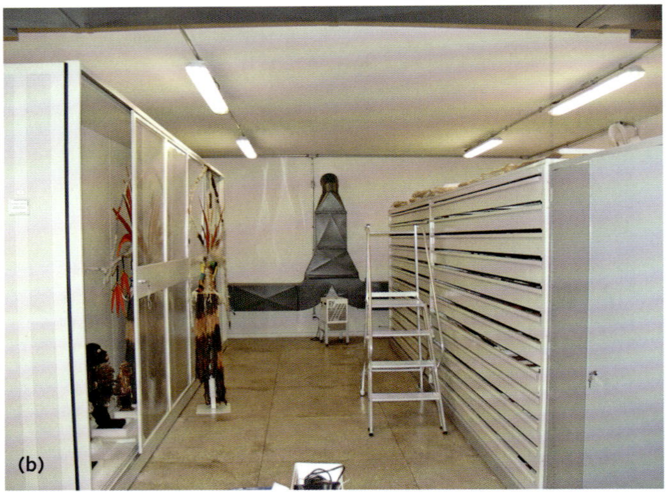

Figure 13.11
Side panels of the compact shelving system are made of perforated sheet metal to allow airflow through the panels even when the shelving units are closed (a); the fronts of the steel drawer cabinets have gaps to permit air circulation even when the drawers are closed (b). Photos: Shin Maekawa. © J. Paul Getty Trust.

Mechanical strategies

Mechanical system description

The conceptual design of the environmental management system was forwarded to a staff architect and a local HVAC company for detailed design, equipment selection, and installation under the supervision of the staff architect. Figure 13.12 shows the location of various components of the environmental management system; images of the installed system are shown in figure 13.13.

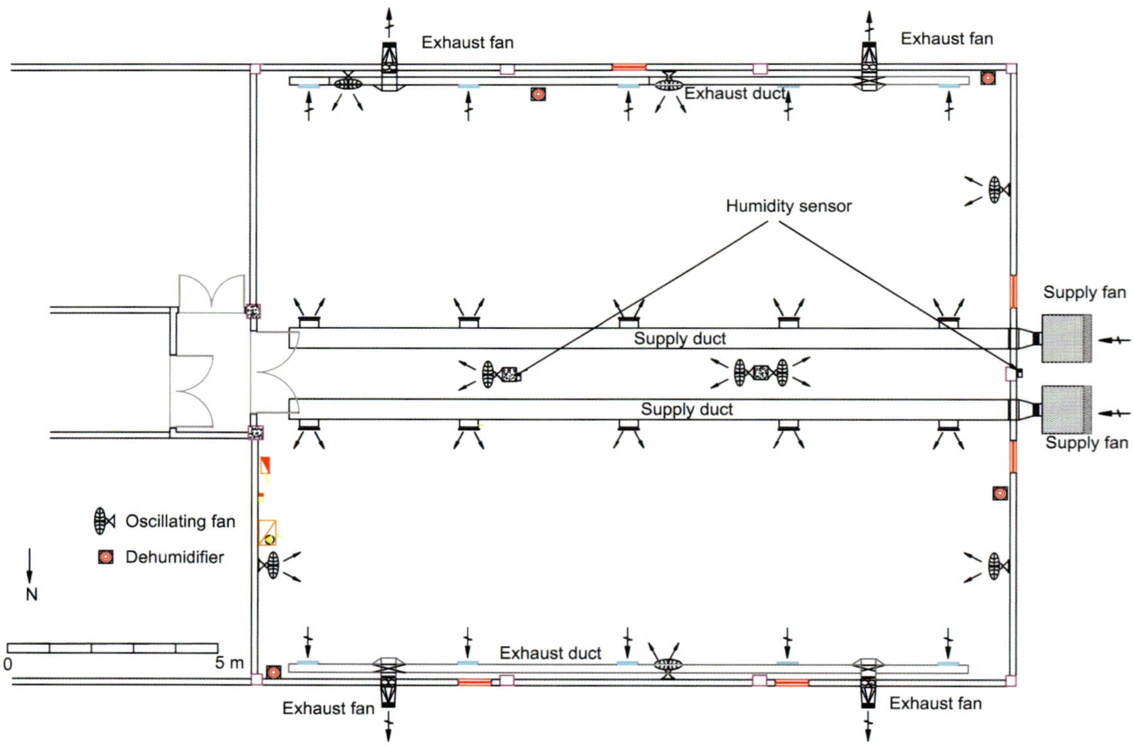

Figure 13.12
Design of the mechanical system for the Amazonian ethnographic collection at the Goeldi Museum's new storage facility.

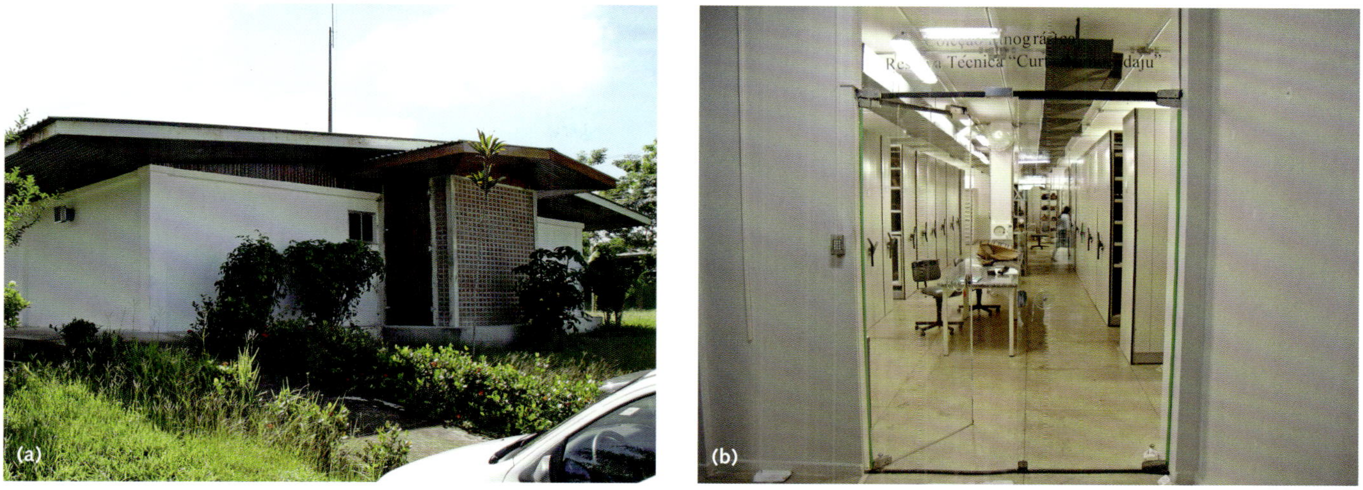

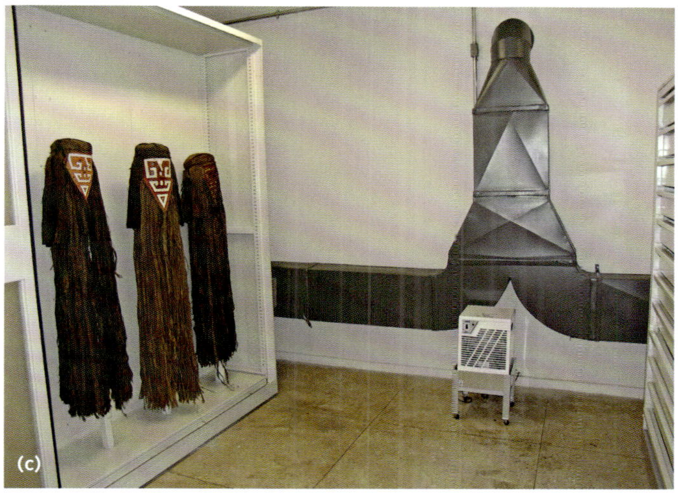

Figure 13.13
Installed environmental management system at the Goeldi Museum's new storage facility: an exterior brick shelter for supply air fans, seen from the west (a); in the storeroom (as seen from the entrance vestibule), two ceiling-mounted central supply air ducts and, underneath them, the compact shelving (b); and an exhaust duct and dehumidifier (c). Photos: Shin Maekawa. © J. Paul Getty Trust.

The environmental management system at the Goeldi Museum's new storage space consists of two centrifugal-type supply fans, four axial-type exhaust fans, six oscillating fans, four portable mechanical dehumidifiers, and two humidistats (see fig. 13.12). The supply air fan units are mounted outside the building along the west wall and provide filtered exterior air at a rate of 6–8 air changes per hour (ACH) into the storage space through two central ducts mounted on the ceiling. The supply ducts discharge air toward the north or south perimeter of the room through five diffusers on each side. Each supply fan is equipped with an insect screen and G3-rated (equivalent to MERV 5) particle filters. An equal amount of exhaust air is collected by two floor-level ducts (five return openings each), positioned along the north and south walls, that are connected to exhaust fans located in the wall openings originally intended for PTAC units. Gravity dampers on the exhaust fans prevent infiltration of outside air and insects. Four portable mechanical dehumidifiers are positioned near the corners of the storage space, and the resulting condensate is discharged to the outside through permanently plumbed drains. Floor dehumidifiers were chosen over a ducted central unit because

Table 13.3

Equipment specifications for the environmental management system at the Goeldi Museum's new storage facility (NA = not available).

Component	Number of Units	Type	Flow Rate	Static Pressure	Voltage/ Phase	Rated Power	Noise Level
Supply fans	2	Centrifugal	1620 m^3/hr	7 mm	220V/3P	0.36 kW	56 dB
Exhaust fans	4	Axial	723 m^3/hr	3.2 mm	220V/3P	0.12 kW	62 dB
Recirculation (oscillating) fans	6	Axial	NA	NA	220V/1P	45 W	NA
Dehumidifiers	4	Mechanical	300 m^3/hr	NA	110V/1P	390 W	NA

of their simplicity of design and maintenance. A backup dehumidifier unit is also available. A total of six oscillating fans mounted near the ceiling on envelope walls operate in conjunction with dehumidification to aid in air mixing. Three more oscillating fans mounted on columns are available for manual operation by staff. Two humidity sensors, one located at the center of the storage space and the other outside the building between the two supply fans, control operation of the ventilators for conservation heating, dehumidifiers, and oscillating fans. Equipment specifications are listed in table 13.3.

Operational control sequence

The environmental management system has three modes of operation (table 13.4): ventilation, dehumidification, and idle. The ventilation mode is enabled when the interior humidity is greater than 70% RH and the exterior humidity is less than or equal to 70% RH (fig. 13.14a). If both the interior humidity and the exterior humidity are greater than 70% RH, the dehumidification mode is enabled, activating both the mechanical dehumidifiers and the oscillating wall fans (fig. 13.14b). Once interior humidity is less than or equal to 60% RH, the system enters into the idle mode, disabling operation of ventilation fans and dehumidifiers. There is no operational condition for temperature in the control logic. Two off-the-shelf humidistats—one for on/off control and the other for selection of either ventilation or dehumidification mode—perform the control logic operations. Each humidistat contains one relay and is connected to one remote humidity sensor. The two relays are then combined to operate the equipment, depending on the comparison of inside and outside humidity.

Table 13.4

Operational control conditions for the environmental management system.

Psychrometric Strategy	Enable	Disable
Ventilation	$H_{inside} > 70\%$ RH and $H_{outside} \leqq 70\%$ RH	$H_{inside} \leqq 60\%$ RH or $H_{outside} > 70\%$ RH
Dehumidification	$H_{inside} > 70\%$ RH and $H_{outside} > 70\%$ RH	$H_{inside} < 60\%$ RH or $H_{outside} \leqq 70\%$ RH
Idle	$H_{inside} \leqq 60\%$ RH	$H_{inside} > 70\%$ RH

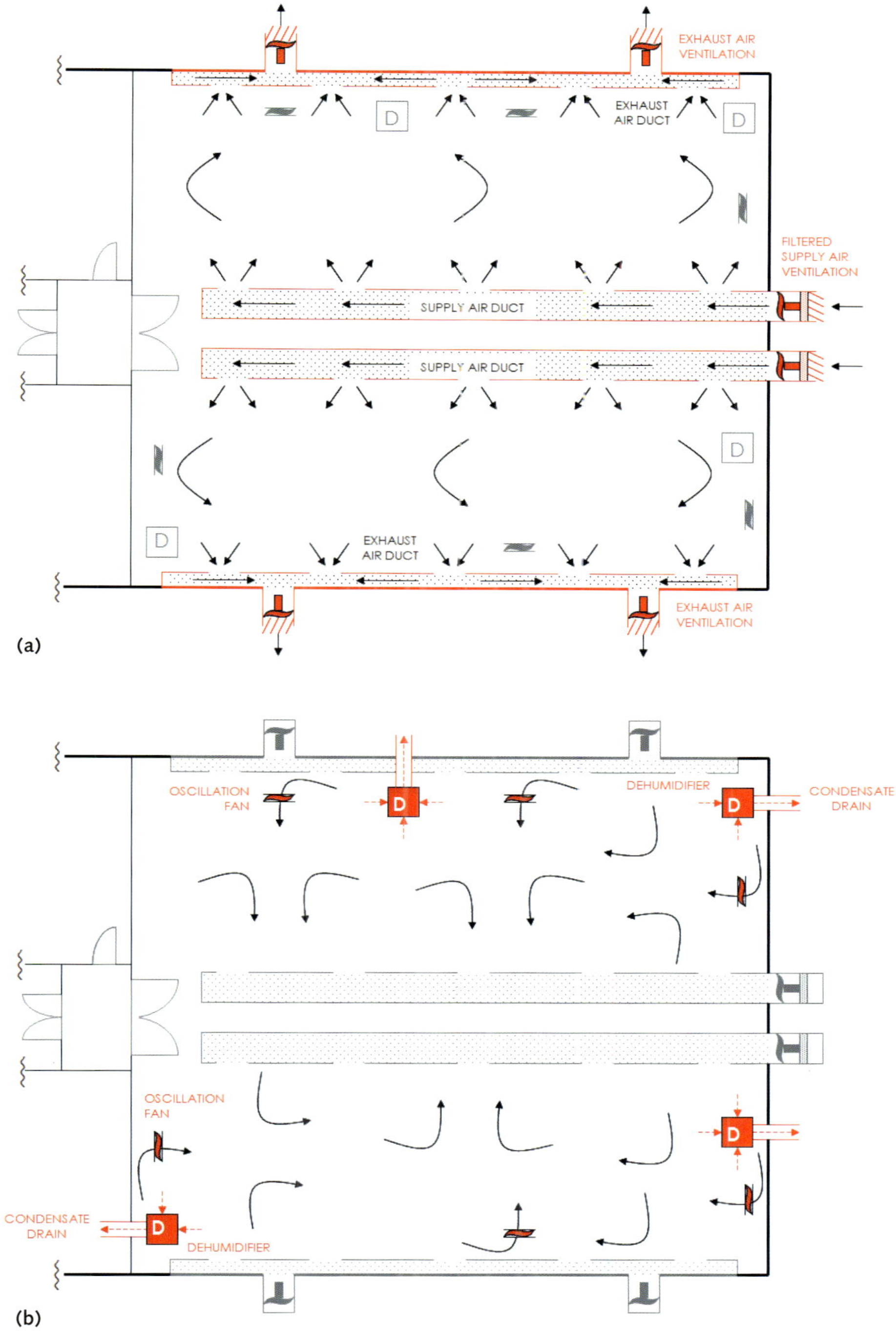

Figure 13.14

Schematics of the psychrometric modes of operation for the environmental management system at the Goeldi Museum's new storage facility: conservation heating by ventilation (a) and dehumidification (b).

Postinstallation System Performance

Following operational commissioning of the environmental management system in October 2003, the performance of the system was monitored from October 2003 to March 2006. As described earlier, although significant differences were found between the published historic climate data collected at the Belém International Airport and the climate data collected at the site, the airport dataset was assumed to be representative of the historic site climate for this analysis.

Several periods of power failure lasting less than one day were recorded during the study. These short-term events had little impact on the storage conditions, which remained near the designed humidity, proving the robustness of the environmental management system.

Humidity

Installation of the environmental management system improved the storage room environment, particularly with respect to humidity. Postinstallation monthly mean humidity ranged from 66% to 69% RH, and no values above 75% RH were recorded (fig. 13.15a). Daily variations of humidity remained approximately Δ10% RH, which matched the deadband range. During the dominant rainfall season, humidity ranged from 60% to 72% RH, with the middle 50% of the dataset between 64% and 69% RH. The humidity range narrowed to 65%–70% RH during the secondary rainfall season, with the middle 50% clustered between 67% and 70% RH.

Temperature

The postinstallation mean monthly interior temperature ranged from 30.7°C to 31.5°C, which was above the seasonal average exterior temperature, slightly higher than the limited preinstallation interior values (typically <31°C), and approximately Δ5°C above the historical seasonal average values (fig. 13.15b). Daily variations of temperature were less than Δ2°C, and the middle 50% of monthly temperature clustered within a narrow Δ1°C range.

Dew point temperature

Dew point temperatures in the storage space were lower than the exterior values during both the dominant rainfall and the secondary rainfall seasons (fig. 13.15c). Though the outside dew point temperatures (see figs. 13.4 and 13.7c) during the dominant rainfall season were higher than those of the secondary rainfall season, the dew point temperatures in the storage room were lower during the dominant rainfall season compared to values during the secondary rainfall season. This trend may be due to mechanical dehumidification being widely used during the dominant rainfall season, whereas conservation heating by ventilation was more common during the secondary rainfall period.

Postinstallation risk class assessment

Table 13.5 summarizes the postinstallation Conservation Environment Classification–HH protocol for the Goeldi Museum's new storage facility. The preinstallation and postinstallation environment are plotted on psychrometric charts in figure 13.16.

The pre- and postinstallation risk classes for the interior environment at the Goeldi Museum storage space are summarized in table 13.6.

Microbial risk

The P97.5 value of humidity improved from 78% RH during preinstallation to 72% RH after installation of the environmental management

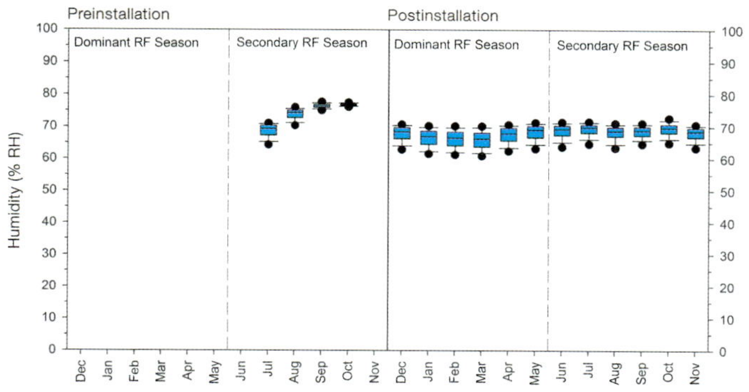

(a)

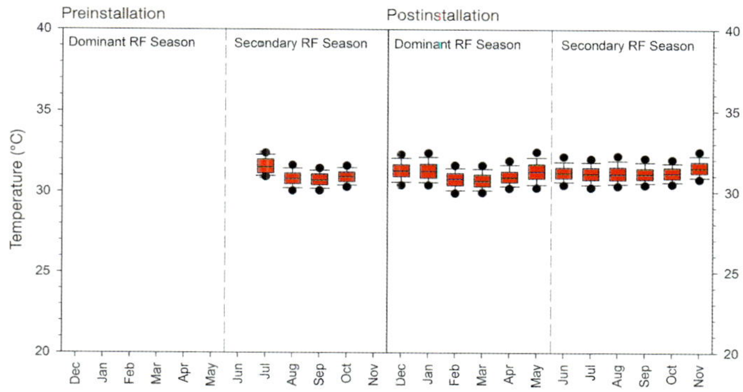

(b)

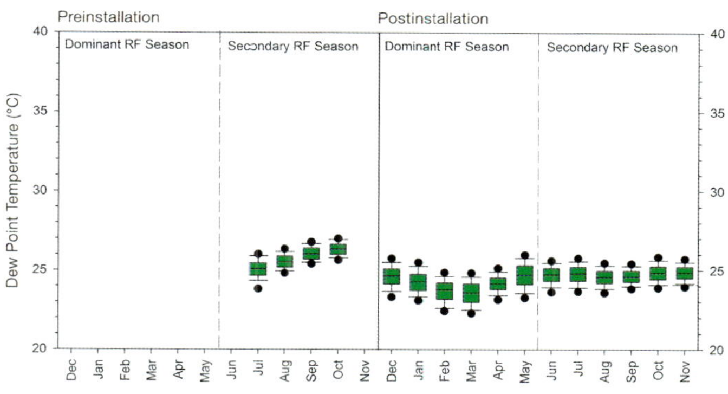

(c)

Figure 13.15

Comparison of pre- and postinstallation interior humidity (a), temperature (b), and dew point temperature (c).

Table 13.5

Postinstallation Conservation Environment Classification–HH results for the Goeldi Museum's new storage facility.

Risk		Humidity		Temperature		
Category	**Specific Risk (statistic)**	**Value (% RH)**	**Class**	**Value (°C)**	**Class**	
Microbial	Germination threshold (P97.5)	72	C	—	—	
Mechanical	Short-term variation (P95 of rolling 24 hour variation)	Δ10	a	Δ1.9	a	Overall mechanical risk:[1] b
	Seasonal variation (absolute difference of seasonal means, RH means ≦ 70%)	Δ2	a	Δ0.1	a	
	Deviation from historical mean (absolute difference)	Δ13	b	Δ4.7	a	
Chemical	Deviation from historical mean (difference)	—	—	Δ4.7	+	

[1] Overall mechanical risk is based on the highest risk classification in any of the specific humidity- and temperature-based mechanical risks (short-term variation, seasonal variation, and deviation from historical mean).

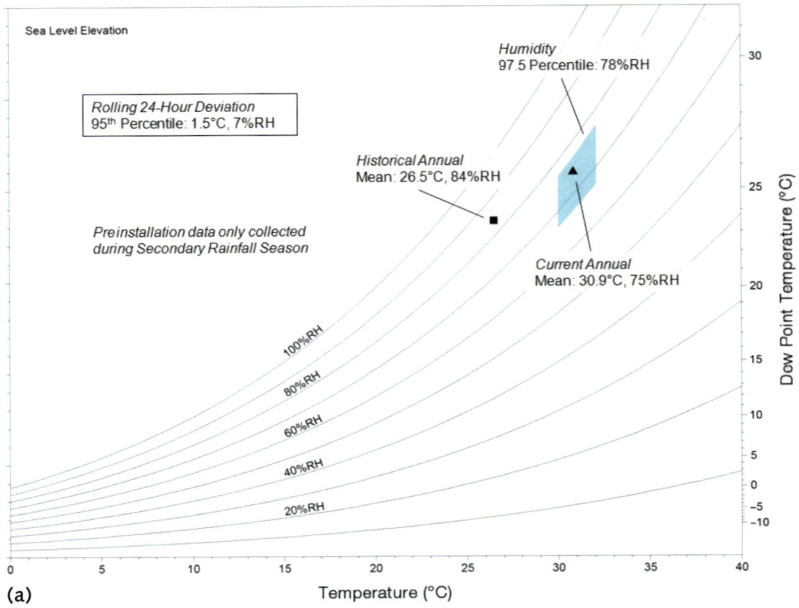

(a)

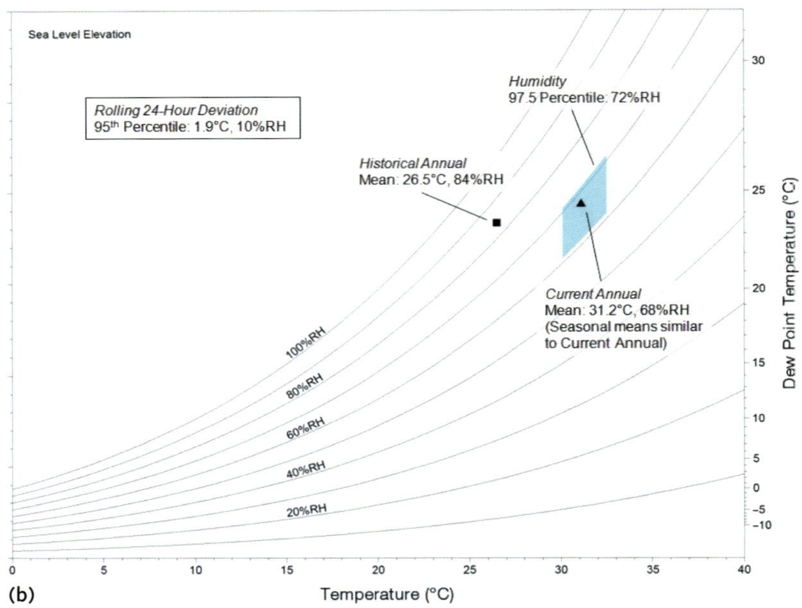

(b)

Figure 13.16

Psychrometric charts showing preinstallation (a) and postinstallation (b) conditions at the Goeldi Museum's new storage facility.

Table 13.6

Comparison of pre- and postinstallation risk classes at the Goeldi Museum storage.

	Microbial Risk		Mechanical Risk														Chemical Risk	
			Humidity						Temperature						Overall[1]			
			Short-Term Variation		Seasonal Variation		Deviation from Historical Mean		Short-Term Variation		Seasonal Variation		Deviation from Historical Mean					
Room	Pre	Post	Pre	Post	Pre	Post	Pre	Post	Pre	Post	Pre	Post	Pre	Post	Pre	Post	Pre	Post
Storage interior	F	C	a	a	f	a	a	b	a	a	—	a	a	a	f	b	+	+

[1] Overall mechanical risk is based on the highest risk classification in any of the specific humidity- and temperature-based mechanical risks (short-term variation, seasonal variation, and deviation from historical mean).

strategies. This change corresponded to the reduction of microbial risk from Class F (high risk) to Class C (moderate risk).

Mechanical risk

Mechanical risk due to seasonal humidity variation was improved from preinstallation Class f (high risk) to postinstallation Class a (no risk). In contrast, mechanical risks due to deviation from historic humidity increased from Class a (no risk) during preinstallation to Class b (moderate risk) during postinstallation—a change due to the overall reduction in interior humidity.

Chemical risk

Following system installation, the chemical risk due to deviation from historical mean temperature remained at Class + (moderately increased risk).

Insects

Insect infestations pose a great threat to organic materials, which comprise the majority of the Amazonian ethnographic collection. The research campus of the Goeldi Museum is located on the fringe of a major rain forest, where insects are abundant.

Insect metabolic rates are known to increase with temperature, and there was concern that the elevated temperature in the storage might promote insect activity and result in substantial infestations. Therefore, in addition to environmental monitoring, insect activity within the storage area was monitored using sticky traps distributed throughout the space. Table 13.7 lists the insect orders and their numbers recorded throughout the storage space over a five-month period between November 2004 and April 2005. The most common species found in the collection was *Psocoptera*. Owing to the small size of this insect (1–2 mm with wings closed, up to 10 mm with wings open), conservators overlooked the infestation in the collection at the original storage location, and *Psocoptera* were transported into the new storage during relocation of the collection. This monitoring established a baseline for future analysis of insect activity.

Insect Order/Class	Common Insects	Total Number Trapped
Araneae	Spiders	69
Coleoptera	Beetles	6
Diplopoda	Millipedes	3
Diptera	Flies	72
Hymenoptera	Wasps, bees, ants	92
Lepidoptera	Moths and butterflies	35
Orthoptera	Grasshoppers, crickets, weta, and locusts	20
Psocoptera	Book lice, bark lice, and barkflies	501
	Total	798

Table 13.7

Total numbers of insects trapped in the Goeldi Museum's new storage facility, November 2004–April 2005.

Pollutants

The level of airborne particulates within the storage space was compared to that in the adjacent corridor and outside the building in August 2005. Table 13.8 shows that postinstallation particulate levels for all size fractions in the storage area were significantly lower than values recorded at the other locations. G3 filters installed at the intake of supply fans are rated for 20%–60% efficiency for average particle sizes of 3.0–10.0 μm. However, significant reductions were recorded for all particle sizes in the storage space.

Thermal Comfort

The storage environment was not designed with occupant thermal comfort in mind, since tasks performed in the storage room were typically of short duration. Longer-duration staff and visitor activity takes place in air-conditioned study areas and in the laboratory adjacent to the storage room. The storage environment typically became uncomfortable in the afternoon when the north wall was heated by the afternoon sun. However, in an effort to improve the thermal comfort in the storage space during these periods, three manually operated oscillating fans were installed in the center of the room to provide air movement. Staff members reported an improved comfort level when the oscillating fans were running.

The predicted mean vote (PMV) and predicted percentage dissatisfied (PPD) values were evaluated for the storage environment using the air speed model in the ASHRAE Thermal Comfort Calculator (fig. 13.7). Seasonal average temperature (31.1°C) and dew point temperature (24.4°C) recorded during the dominant rainfall season were input into the calculation. Additionally, mean radiant temperature was assumed to be equal to the air temperature, the metabolic rate was set at 1.5 met (combination of standing and walking), a clothing factor of 0.6 clo (typical summer clothing) was chosen, and air speed was set at 0.5 m/s. The resulting PMV and PPD values were 2.05 and 79%, respectively. This was too warm, and it did not comply with ANSI/ASHRAE Standard 55-2013 (point 1 on fig. 13.17) (ANSI and ASHRAE 2013).

Assuming manual operation of the three oscillating fans, resulting in an estimated increase in air speed up to 3 m/s, the PMV and PPD values significantly improved to 1.40 and 46% (slightly warm), respectively (see point 2 on fig. 13.17). However, this interior environment remained out of compliance with ANSI/ASHRAE Standard 55-2013.

Table 13.8

Postinstallation number of airborne particles.

Particle Size (μm)	Storage	Outside	Corridor
<0.3	549,331	1,122,695	1,200,616
<0.5	29,961	59,532	80,375
<1.0	4771	5459	8853
<5.0	68	159	408

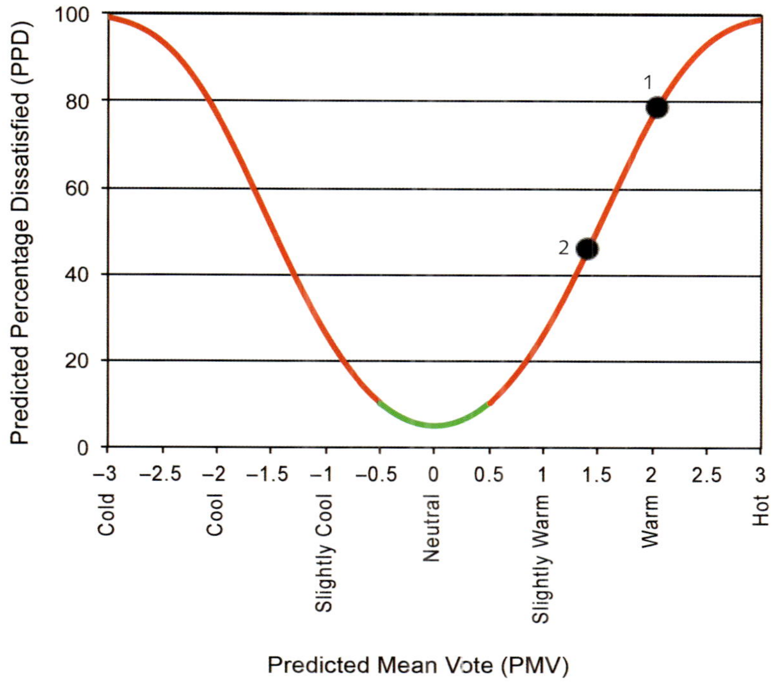

Figure 13.17
PMV-PPD plot showing thermal comfort in the storage room during the dominant rainfall season with an air speed of 0.5 m/s (1), and an air speed of 3 m/s (2).

Noise

Noise levels at the center of the storage area ranged from 52 to 54 dBA during operation of ventilators for conservation heating, and from 60 to 61 dBA with dehumidifiers and oscillating fans running. These values were significantly higher than the typically accepted ambient noise (45 dBA) for offices. However, these noise levels are not hazardous (<85 dBA) and are considered acceptable, particularly since staff were present in the storage space for short durations (less than 15 minutes).

Maintenance

During the first two years of operation, the system required annual calibration of the humidity sensors and filter replacement as needed. In addition to regular maintenance, a failed solenoid switch controlling one of the exhaust fans had to be replaced. This failed component was quickly identified and had minimal impact on the storage environment. Humidity sensors used with this system were found to be inaccurate and difficult to calibrate, affecting system set points. Because of the limited financial resources of the project, replacement of these sensors was not possible.

Energy Metrics

Energy intensity

The original storage space (113 m^2) at the Goeldi Museum's Rocinha campus had an energy density of 384 W/m^2. The floor area of the new storage space at the museum's research campus was 270 m^2, and the total rated power of the equipment listed in table 13.3 was 3.0 kW. Using these values, the energy intensity of the new storage space was 11.2 W/m^2, which represents 3% of the energy intensity recorded at the original storage.

Table 13.9

Monthly energy use intensity, for environmental control and lighting, of the new storage space during the first six months of 2005. June to November are less-humid months; therefore, energy consumption in the ventilation mode is significantly less, as indicated by the value for July (NA = not available).

Month	Monthly Energy Use Intensity (kWh/m^2)
January	1.81
February	2.25
March	2.93
April	2.35
May	1.97
June	NA
July	0.904

Energy use intensity

The environmental management system typically operated in ventilation or dehumidification mode for 4 daytime hours each day. Table 13.9 lists the monthly energy used in the storage space during the first six months of 2005. The energy use intensity for the new storage facility averaged 2.04 kWh/m^2/month (24 kWh/m^2/yr), 6% of the energy cost of the original storage.

It is typical for an annual maintenance contract to cost 10% to 15% of the capital cost of the environmental management system. Since the system's cost was only one-fifth of a conventional air conditioner–based system, such a reasonable cost for a maintenance contract became too small to be accepted by contractors. Therefore, flat daily rates for various scheduled and unscheduled maintenance tasks were agreed upon to maintain the system. Actual maintenance cost over the three-year period following installation was well below 5% of the capital cost.

Conclusions

A conservation-focused environmental management system was designed and implemented at the Goeldi Museum's new Amazonian ethnographic collection storage facility at the museum's research campus on the outskirts of Belém, Brazil, which is in a Very Hot-Humid (1A) climate zone in northeastern Brazil. Following modifications to improve the performance of the building envelope and installation of shelving and cabinetry that allowed for maximum air circulation, a mechanical system consisting of humidistatically controlled ventilators and mechanical dehumidifiers was installed in October 2003. During the performance evaluation, between October 2003 and March 2006, this system successfully maintained interior humidity below 72% RH (P97.5 value) while allowing temperature to drift with the outside climate. Owing to a combination of a high humidity set point, which limited the deviation from the historical mean, and events of local power failure, humidity conditions above the 70% RH set point were recorded during 8% of the monitored period.

Since installation of the environmental management system, conservators have not observed damage to the collection as a result of the modified interior environment. The environmental management project proved its effectiveness in maintaining safe conditions for the Goeldi Museum's ethnographic collection and offered significant savings in equipment, maintenance, and energy costs as well. Based on discussions with local HVAC engineers in Belém, the capital and operational costs of the system are one-fifth and one-twentieth, respectively, of costs associated with a typical air conditioner–based system. The conservation-focused environmental management strategy can be a beneficial alternative to a conventional air-conditioning approach for cultural institutions in Very Hot-Humid (1A) climate zones, especially when resources are limited.

Additional discussion and information relating to this case study can be found in Maekawa and Toledo 2010 and Toledo 2003.

Lessons Learned from the Ethnographic Storage of Museu Paraense Emílio Goeldi

- Use of humidistatically controlled fans and dehumidifiers was successful in maintaining interior humidity below 70% RH in a Very Hot-Humid (1A) climate zone. The interior temperature and dew point temperature ranged from 31°C to 33°C and from 26°C to 27°C, respectively.
- Despite relatively high temperatures within the storage space, staff comfort was addressed through the use of manually operated fans.
- Accurate humidity sensors are essential for good system control.
- In locations with an unreliable electrical infrastructure, consider adding a small emergency generator for operation of the essential climate management system components, such as dehumidifiers, during an extended power outage.
- Historical climate data from nearby locations may not be representative of the site where an environmental management system is being installed.

Postscript

The insect/pest monitoring using sticky traps documented an increase in the population of *Psocoptera* near the end of 2006. This problem coincided with multiple and consecutive power outages that extended for several days, resulting in a significant humidity excursion in the storage space. (This infrastructure problem was not experienced during our monitoring period from November 2004 to April 2005, when power outages typically lasted less than twenty-four hours.) The 2006 power outages triggered not only a *Psocoptera* infestation but also a fungal outbreak in a small part of the collection. In response, when consistent power was restored, the operational range of humidity was permanently reduced from 60%–70% RH to 50%–60% RH, effectively arresting activities of both the *Psocoptera* and the fungi. This lower humidity range provided a larger margin of safety against the risk of microbial damage in the event of long periods of power or equipment failure. In retrospect, the addition of a small emergency generator into the environmental management system design, which could be a future system improvement, would allow for minimum operation during extended power outages.

While reducing the humidity range will increase the differential between the storage RH and mean historical humidity and potentially put the collection at risk of mechanical damage, providing this safety margin against biodeterioration (microbial and insect damages) was thought to be a higher priority. The lower humidity set point also resulted in extended stretches of continuous dehumidifier operation, allowing for more efficient operation. Elevated air temperature, which had been of concern with respect to the increased risk of chemical damage, was not observed in the storage room, and risk of chemical damage remained at the same level, Class +. To date, the climate system at the Amazonian ethnographic storage at Goeldi Museum remains in operation.

Project Team and Responsibilities

Fundação Vitae (São Paulo, Brazil)—funding source for this environmental management project

Gina Machado (Project Manager, Fundação Vitae)—Project management from the funding source

Museu Paraense Emílio Goeldi (Rio de Janeiro, Brazil)

Lucia van Velthem (Goeldi Museum)—Curator of the ethnographic collection and manager of the climate management project

Rosa Arraes (Conservator, Art Museum of Belém, Brazil)—Selection of storage furniture and conservation treatments during the transfer of the collection to the new storage

Alegria Benchimol (Manager, Goeldi Museum)—Manager of the Amazonian ethnographic collection

Thais Toscano (Architect, Goeldi Museum)—Building assessment, designing of building envelope improvements for the new storage facility

Adelina Figueiredo (Contracting HVAC Engineer, Projet AR, Belém)—Engineering design and installation of the environmental management system

Ana Lúcia Nunes (Entomologist, Goeldi Museum)—Insect identification and population monitoring

Staff members of the Amazonian ethnographic collection (Technicians, Goeldi Museum)—Operation of the environmental management system

Getty Conservation Institute (Los Angeles, United States)

Shin Maekawa (Senior Scientist)—Project management and concept design, commissioning, and improvements for the environmental management strategy and environmental monitoring system operational data interpretation

Franciza Toledo (Environmental Consultant, Recife, Brazil)—Project architect (local management of project tasks) and design of nonmechanical strategies

Vincent L. Beltran (Assistant Scientist)—Environmental data collection, analysis, and management

References

AFCCC [Air Force Combat Climatology Center] and NCDC [National Climatic Data Center]
1999 *Engineering Weather Data*. Washington, DC: Air Force Combat Climatology Center. http://purl.access.gpo.gov/GPO/LPS7081.

ANSI [American National Standards Institute] and ASHRAE
2013 *Standard 55-2013—Thermal Environmental Conditions for Human Occupancy*. Atlanta: ASHRAE.

ASHRAE
2011 Museums, galleries, archives and libraries. Chapter 23 in *ASHRAE Handbook—HVAC Applications*. Atlanta: ASHRAE.

INMET [Instituto Nacional de Meteorología]
2009 Normais climatológicas do Brasil, 1961–1990. http://www.inmet.gov.br/portal/index.php?r=clima/normaisClimatologicas.

Maekawa, S., and F. Toledo
2010 A collection climate control system for an ethnographic storage of a museum in north of Brazil. In *ASHRAE Transactions, 2010 Winter Conference, Orlando*, vol. 116, pt. 1, paper OR-10-021.

Toledo, Franciza
2003 Report on the installation of a climate control system in the ethnological storage of the Emílio Goeldi Museum—MPEG, Belém, Pará, Brazil. Getty Conservation Institute, unpublished report.

Weather Channel
2014 Weather Underground. http://www.wunderground.com/.

Chapter 14

Library of Museu Casa de Rui Barbosa, Rio de Janeiro, Brazil: Ventilation (Dilution) and Dehumidification, and Limited Cooling

Climate zone 1A (Very Hot-Humid)

Objectives

- Maintain low and stable humidity to prevent microbial and mechanical deterioration of the collection.
- Reduce both gaseous and particulate pollution.
- Reduce insect population.
- Reduce the exposure of the collection to daylight.
- Maintain thermal comfort for visitors by providing an appropriate combination of temperature, humidity, and air movement.
- Avoid visual intrusions caused by system installation.

Summary

Museu Casa de Rui Barbosa, a nineteenth-century masonry building (ASHRAE Class IV), is the first historic house museum in Brazil; it is located in an urban setting in Rio de Janeiro, in a Very Hot-Humid (1A) climate zone. The former residence of Rui Barbosa de Oliveira, a Brazilian writer, jurist, and politician, the museum houses a collection that includes works of art, furniture, and thirty-seven thousand books. The collection is exhibited within the restored residential structure in a noncontrolled environment that relies on the thermal mass of the stone building to moderate interior conditions and that uses natural ventilation for occupant comfort and daylight illumination via manually operated windows and louvers. The collection had been exposed to fungal and insect attacks, high and fluctuating humidity, and high temperature, as well as high-intensity natural light, particulates, and urban air pollution.

The goals of the environmental management strategy were to (1) create a conservation environment suitable for the building's historic interior and the Library collection, and (2) address thermal comfort for visitors in the suite of rooms encompassing the Library. Prior to installation of the mechanical environmental management system, windows and balcony doors were repaired to reduce uncontrolled infiltration of outside air. Natural ventilation in the basement was restored, and the flow path of an original ceiling ventilation system was also reinstated. Installed in

2006, the system consisted of a ducted dehumidifier in the basement, supply air diffusers and return air grilles positioned along visitor paths, and an exhaust fan in the attic. The project was the first implementation of an alternative environmental management strategy (in contrast to a conventional air-conditioning system) for a historic house museum in Brazil.

Background

Museu Casa de Rui Barbosa (MCRB), a nineteenth-century building in the urban Botafogo district of Rio de Janeiro, is a nationally registered building and Brazil's first historic house museum. The museum is open to visitors on weekdays for supervised tours and now receives approximately ten thousand visitors annually. Daily, groups of students visit the museum as part of their field studies of Brazilian history and politics.

In the museum, the interior environment was moderated by the large thermal mass of the stone building, natural ventilation through existing large, manually operated windows, and natural light passing through manually adjustable louvers. However, this approach to interior environmental management favored human comfort over collections conservation. The collection experienced large temperature and humidity variations, insect attacks, particulate deposition, and color fading. Museum staff were also concerned about increased air pollution levels due to traffic congestion on the adjacent Rua São Clemente. As a result, the institution sought an environmental management strategy to address the dual objectives of collections conservation and visitor comfort. The house's five-room Library suite, which contained the bulk of the book collection within bookcases, was identified as the target area for improved environmental management.

Climate

Located just north of the Tropic of Capricorn in southern Brazil (latitude: 22.95° S, longitude: 43.18° W; elevation: 11 m above sea level), MCRB is situated only 600 m from the shoreline of Guanabarra Bay (fig. 14.1). The annual mean temperature of Rio de Janeiro is 25°C, and monthly mean temperature ranges from 22°C in July to 27°C in February (AFCCC and NCDC 1999). Monthly mean dew point temperature ranges from 17°C in July to 23°C in February (fig. 14.2). The average annual total rainfall is 1096 mm, the majority of which occurs between the months of November and April, when monthly rainfall exceeds 100 mm. During the period from May through October, the average monthly rainfall is as low as 44 mm. The overall climate of Rio de Janeiro is classified as Tropical Savanna (*Aw*) by the Köppen-Geiger climate classification system (see appendix 2). Seasonal climate is distinguished by temperature and rainfall: the period from November through April constitutes the very hot-humid season, while May through October is the hot-humid season.

Based on 2005–9 temperature data available from Weather Underground (Weather Channel 2014), the CDD10°C value of Rio de Janeiro was 5141 cooling degree days, placing the city in the Very Hot (1)

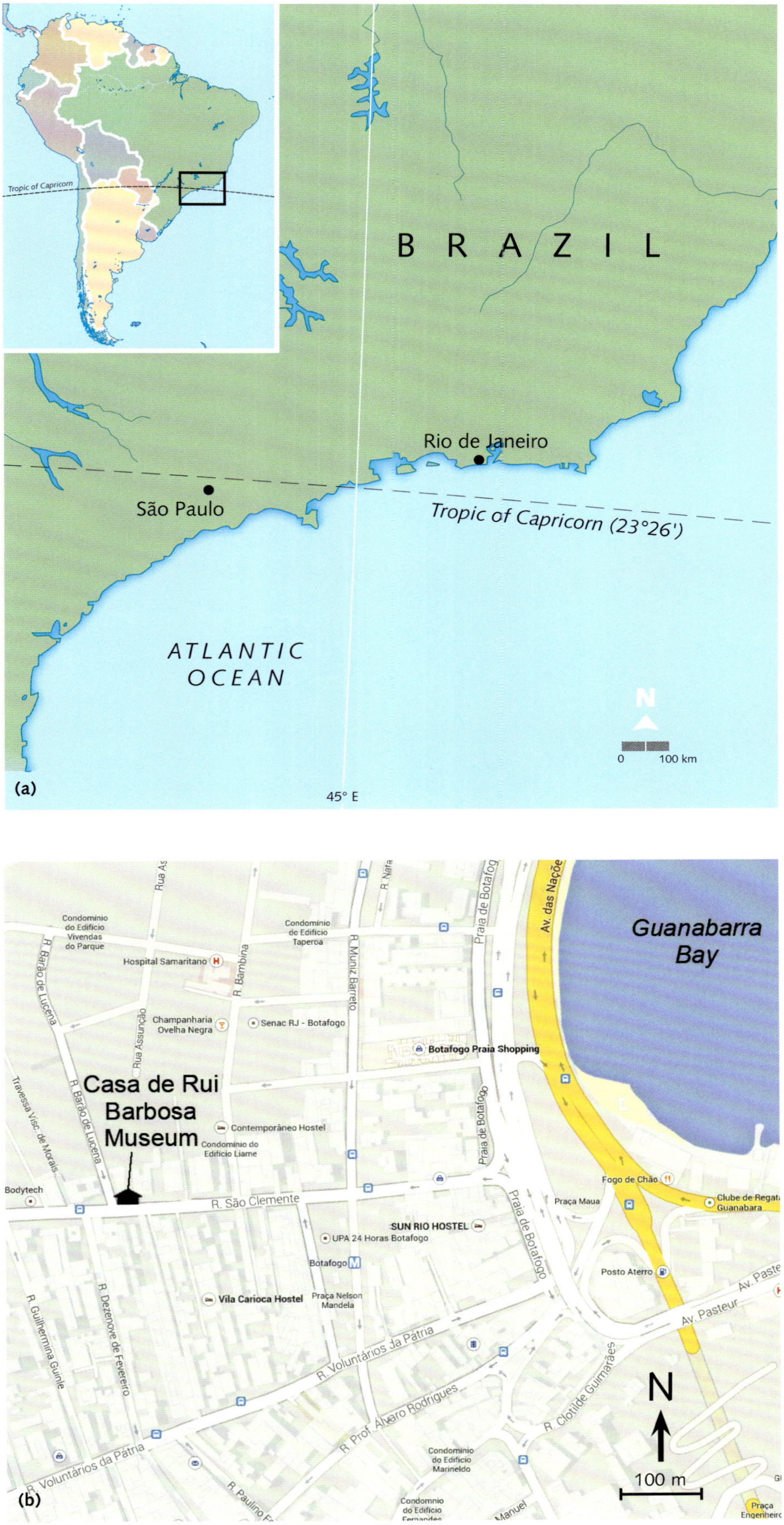

Figure 14.1
Location of Rio de Janeiro, Brazil (a) and the Museu Casa de Rui Barbosa (b).

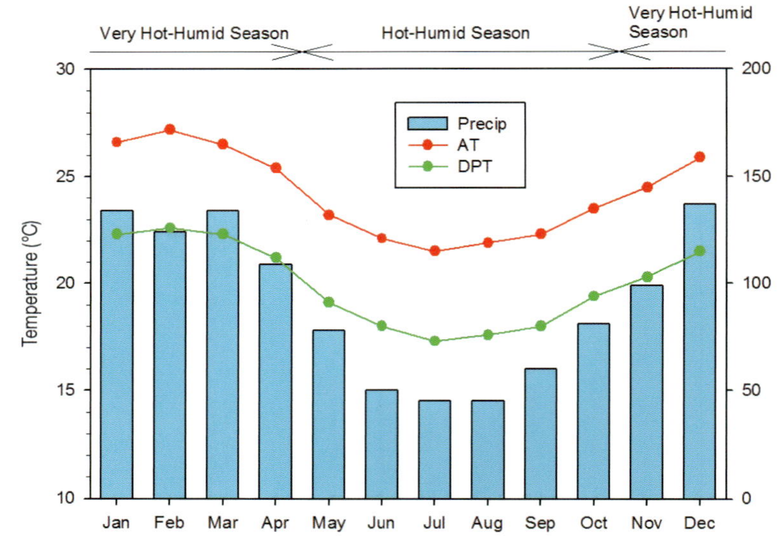

Figure 14.2
Monthly average air temperature (AT), dew point temperature (DPT), and rainfall (precip) at Rio de Janeiro Airport (historical weather data from 1980 to 2009, International Station Meteorological Climate Summary, version 4.0, 1980–2009).

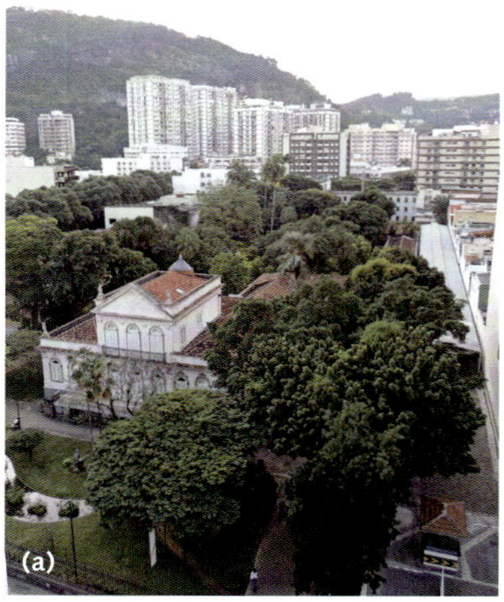

Figure 14.3
Exterior view of the Museu Casa de Rui Barbosa from the southeast (a); a site map shows the location of the museum and surrounding garden (b). Photo and map courtesy of Fundação Casa de Rui Barbosa.

thermal zone in the ANSI/ASHRAE/IES Standard 90.1 climate classification (ANSI, ASHREA, and IES 2013). Rio de Janeiro is excluded from a Marine moisture classification because the mean temperature of its coldest month (22°C in July) exceeds 18°C and the annual rainfall (1096 mm) is greater than the Dry-Humid index of 640—placing the climate of the city in the Humid (A) type. The thermal and moisture classifications combined result in a Very Hot-Humid (1A) climate classification for Rio de Janeiro.

Building

Located in Rio de Janeiro's densely populated urban neighborhood of Botafogo, the MCRB is historically important as the home of Rui Barbosa de Oliveira (1849–1923), a writer and prominent politician who played a major role in drafting the first republican constitution of Brazil. The construction of the original MCRB building was completed in 1850 for Bernardo Casimiro de Freitas, a Portuguese trader and the first baron of the Lagoon District. Two subsequent owners made modifications to the building, giving the house much of its present appearance (fig. 14.3). Rui Barbosa purchased the house in 1893 and further modified the building before occupying it with his family in 1895. Following the death of Rui

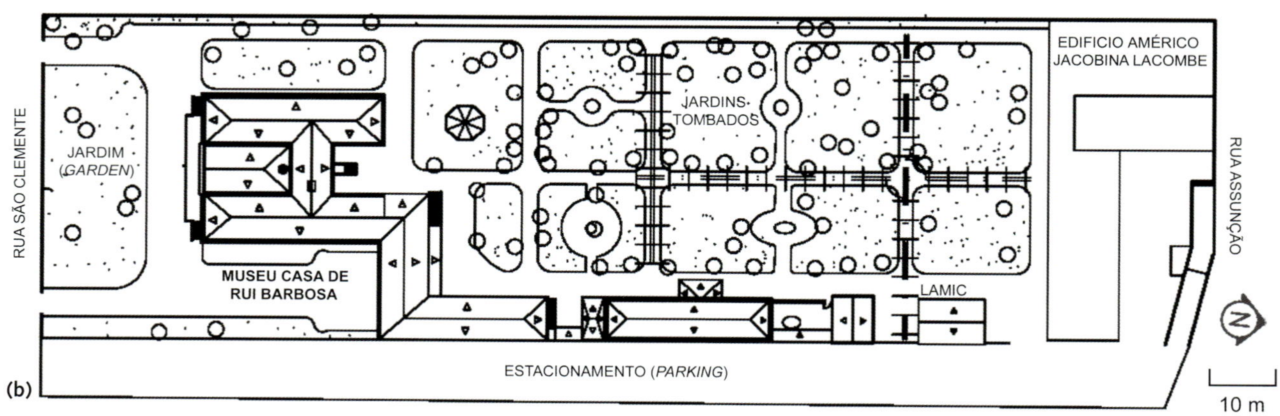

Barbosa in 1923, the Brazilian government purchased the house and his Library collection, opening it to the public on August 13, 1930, as the first historic house museum in Brazil.

In 1938 the MCRB building was one of the first structures to be listed in the national historic registry by the Brazilian Institute of Historic and Artistic Heritage, and as such, it is protected by Brazilian federal law. The building is the oldest surviving residence in the Botafogo district.

The building is located approximately 20 m from the heavily trafficked Rua São Clemente and has a large botanical garden (9000 m^2) to the north. Built in a mid-nineteenth-century neoclassical style, it is constructed of large limestone blocks and mortar and has an arched pediment decorated in low relief, and parapets ornamented by sculptures on the facade of the house. The exterior surface of the building's walls is stucco; the interior surface is plaster. On all four elevations, the building's exterior walls are fitted with numerous wood-framed balcony doors or with windows fitted with louvered blinds for natural ventilation and daylight illumination. The rooms were constructed with wood ceilings, and the ceiling perimeters have recessed openings that vent to the attic, allowing for the natural extraction of buoyant warm air from the rooms (rooms with plaster ceilings lack this feature). The roof is constructed of loosely stacked roof tiles that originally facilitated ventilation and removal of hot air from the attic. The building has a full basement that extends partially above ground; the basement effectively separates the occupied spaces from soil moisture. The ASHRAE classification of this building is Class IV (heavy masonry) (ASHRAE 2011).

Inside the house, the Library suite contains the most significant and environmentally sensitive collections, and thus was targeted as the focus of the environmental management project. The Library suite (fig. 14.4) consists of five interconnected rooms, the names of which were

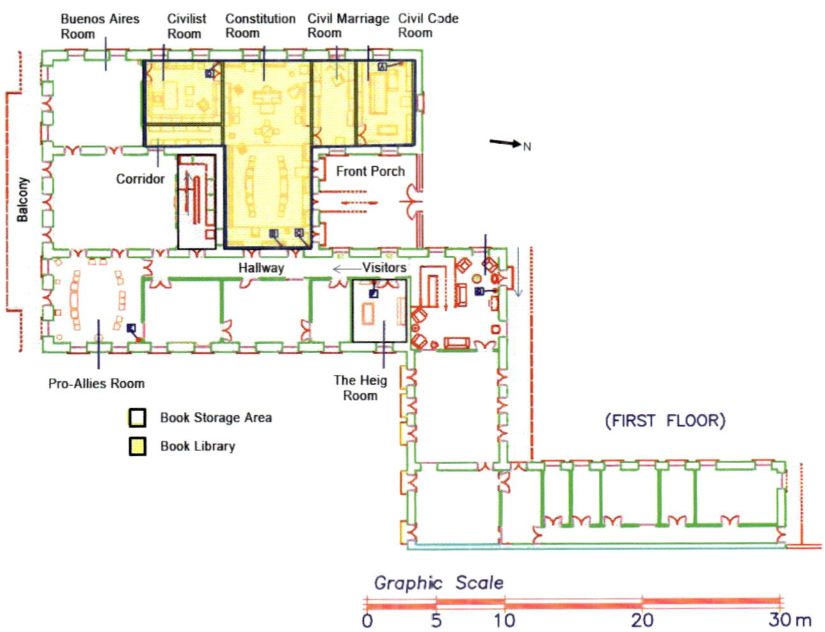

Figure 14.4

Floor plan of the Museu Casa de Rui Barbosa showing the location of the Library rooms (highlighted in yellow). Courtesy of Fundação Casa de Rui Barbosa.

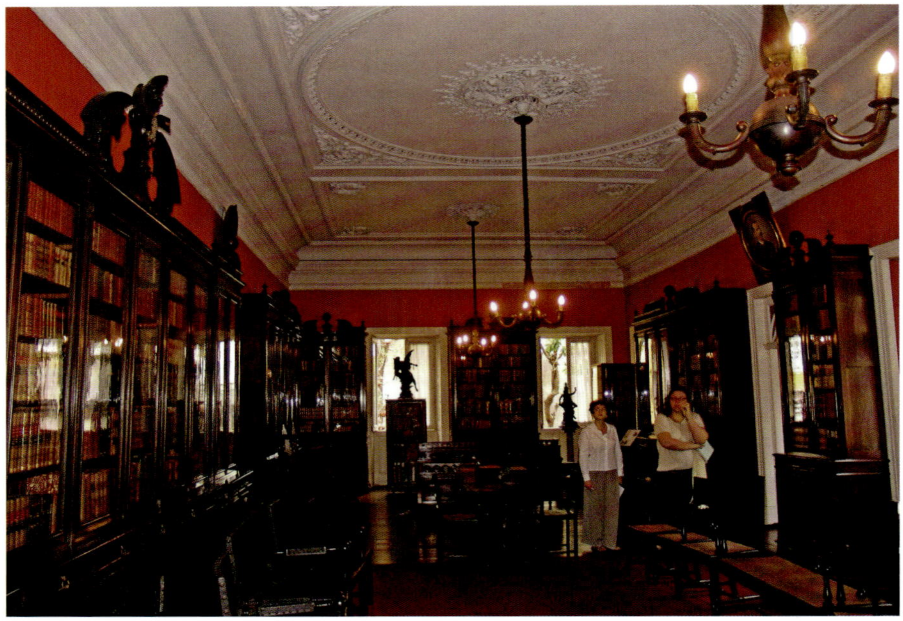

Figure 14.5
Historic interior of the Constitution Room, part of the Library of the Museu Casa de Rui Barbosa. Photo: Shin Maekawa. © J. Paul Getty Trust.

assigned when the building was opened as a museum: the Constitution Room (fig. 14.5), the Civilist Room, the Civil Marriage Room, the Civil Code Room, and the Corridor. The Corridor, which connects the Constitution Room and the Buenos Aires Room (not part of the Library suite), is permanently blocked at the Buenos Aires Room end, and bookcases, which are filled with the book collection, are placed along the east and west walls. These five rooms have a total floor area of 165 m^2 and a ceiling height of 3.8 m, for a total volume of 630 m^3.

Collection

The MCRB collection includes paintings, sculptures, wooden furniture, several automobiles, and the Library collection. The Library collection—consisting of thirty-seven thousand books covering law, humanities, and culture—is considered the heart of the museum. Commissioned by Rui Barbosa, custom-built wooden bookcases with glass doors line two or three walls of each Library room; they house the book collection (fig. 14.5).

The Fundação Casa de Rui Barbosa (FCRB), which manages the Museu Casa de Rui Barbosa, places great importance on the collection in the Library, and so it designated this interior space as the focus of the environmental management project. The high visitation levels in the Library provided an ideal opportunity to apply alternative environmental management strategies that address the dual objectives of collections conservation and occupant comfort.

Preinstallation Condition Assessments

Building envelope assessment

The building envelope assessment was conducted by the FCRB's staff architect; it highlighted the existing nonmechanical environmental man-

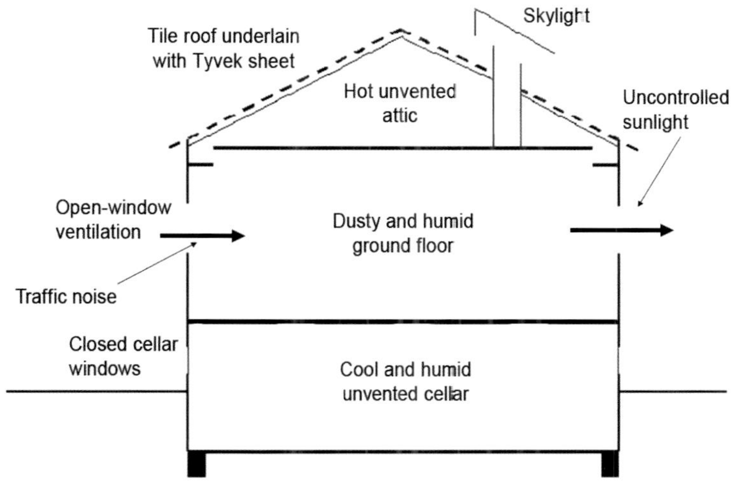

Figure 14.6
Schematic showing compromised features and the resulting interior environmental conditions, as identified by the MCRB building envelope assessment.

agement features of the building, many of which were compromised and not fully utilized (fig. 14.6). For example, in the basement, glass windows had been installed in original ventilation openings in the basement walls to allow the space to be used for storage and for temporary exhibitions. Now closed, these ventilation openings were originally intended as a nonmechanical means of venting moisture from the basement slab floor to the exterior.

The assessment also found that an important original natural ventilation feature of the house, the ventilated ceilings and connecting attic space, had been compromised. As described above, the Library rooms with wood ceilings were perforated at the perimeter to allow buoyant room air to vent to the attic, where the hot air would be exhausted through the loose-fitting roof tiles. However, application of an air infiltration barrier material under the roof tiles, reportedly to reduce particulate infiltration and to catch broken tiles, virtually eliminated natural airflow out of the attic space.

The building assessment also identified numerous doors and windows that did not close properly owing to misaligned or decayed components; these architectural features permitted the uncontrolled infiltration of outside air.

Environmental assessment

The GCI team conducted environmental monitoring at MCRB from 2004 to 2006. The continuous monitoring included temperature and humidity at several locations inside and outside of the Library, as well as inside bookcases. Measurements of gaseous air pollutants and airborne particulates were also performed before and after installation of the environmental management strategies.

Exterior environmental assessment

Figure 14.7a shows 24 hour trends of mean exterior temperature, humidity, and dew point temperature in January (in the very hot-humid season), and figure 14.7b shows the data for July (in the hot-humid season).

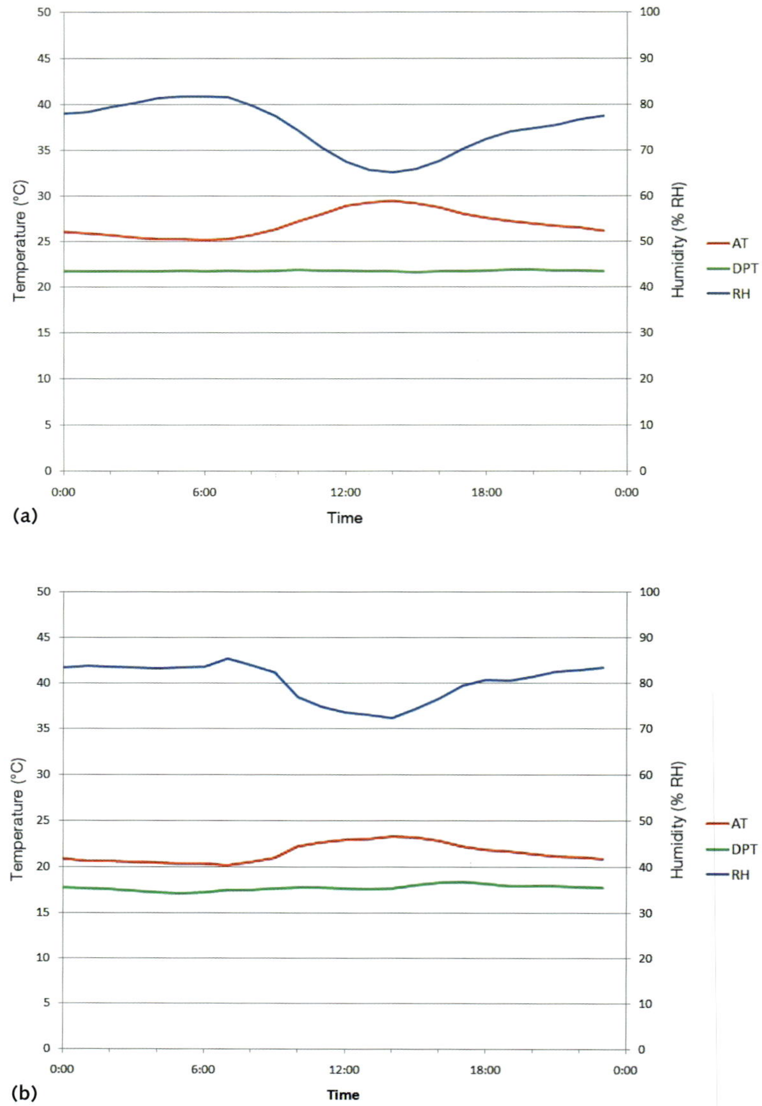

Figure 14.7
Average daily trends of exterior temperature, humidity, and dew point temperature at MCRB in January (a) and July (b).

During both months, temperature was lowest prior to sunrise and peaked at approximately 2:00 p.m. Dew point temperature remained fairly stable, and as a result, humidity varied with temperature; thus, humidity was highest prior to the sunrise and lowest at 2:00 p.m. The mean 24 hour variation in temperature was Δ5°C in January and less than Δ4°C in July. Though mean dew point temperature was elevated in January (22°C) in comparison to July (17°C–18°C), the difference between dew point temperature and temperature was narrower in July than in January. As a result, humidity less than 70% RH was typically observed between 11:00 a.m. and 5:00 p.m. in January, but humidity remained above 70% RH in July.

Interior environmental assessment

Attic and basement

The interior environment assessment documented high-temperature conditions (peaking daily at 30°C–40°C) in the attic above the Library, and humid conditions (annual average of 78% RH in 2005) in the basement below the Library. These extreme conditions in the attic and basement likely influenced the environment in the adjacent Library space (see fig. 14.6). Thus, improvements to the basement and attic environments were planned by the environmental management project.

Library

Figure 14.8 shows monthly variations of interior humidity, temperature, and dew point temperature in the Constitution Room, which is assumed to be representative of the overall environment of the Library suite. Monthly mean humidity (fig. 14.8a) ranged from 65% RH (December) to 78% RH (November), and monthly mean rolling 24 hour variations were between Δ8% RH (February) and Δ11% RH (May). Daily open-window ventilation in the Library likely contributed to large annual fluctuations in humidity (a minimum of 42% RH to a maximum of 93% RH). During the monitoring period, the 75% RH threshold was exceeded during 44% of the overall study period, and five months (April, May, and September to November) had average humidity values of 75% RH or above.

The Constitution Room's interior temperature followed exterior seasonal changes and remained above 25°C for the majority of the monitoring period—and occasionally rose above 30°C during the very hot-humid season. Monthly mean interior temperature (fig. 14.8b) ranged from 22.8°C (July) to 28.2°C (January), and monthly mean dew point

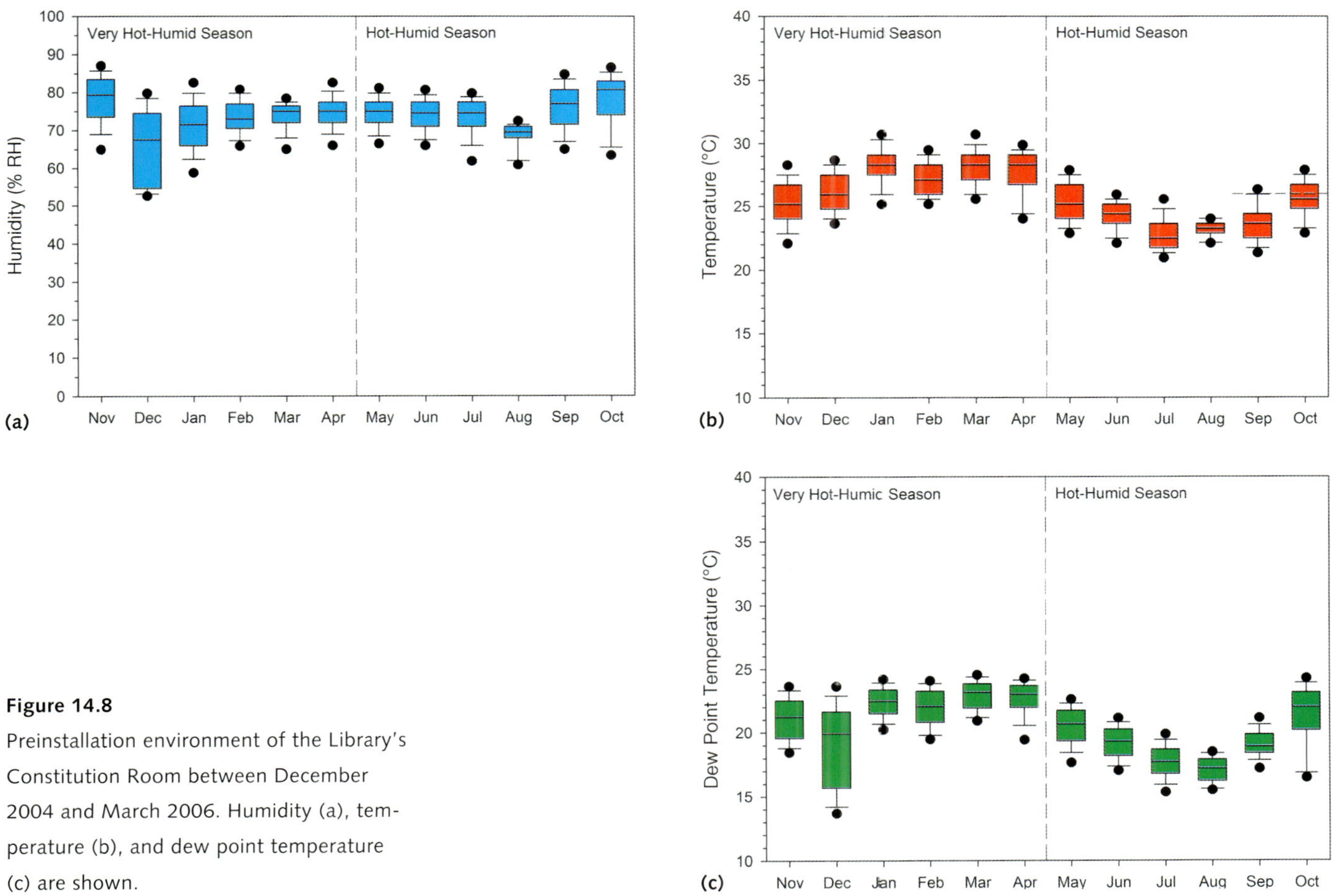

Figure 14.8

Preinstallation environment of the Library's Constitution Room between December 2004 and March 2006. Humidity (a), temperature (b), and dew point temperature (c) are shown.

temperature ranged from 17.0°C (August) to 22.9°C (March) (fig. 14c). Reduced dew point temperatures are clearly seen during the hot-humid season. Monthly average rolling 24 hr. variation for temperature and dew point temperature ranged from Δ1.3°C (February) to Δ2.2°C (August) and from Δ1.6°C (February) to Δ2.9°C (August), respectively. The use of daily open-window ventilation also contributed to large annual fluctuations in Library temperature (a minimum of 19.8°C to a maximum of 32.8°C).

Air pollution monitoring

Gaseous pollutants

Gaseous air pollutants were monitored in the Library in March and September 2005. Passive samplers were used to obtain time-weighted averages of oxidizing pollutants, such as ozone (O_3), sulfur dioxide (SO_2), and nitrogen oxides (NO_X) over a one-month interval. Preinstallation values for ozone and nitrogen dioxide in the Library were 4 ppb and 20 ppb, respectively; sulfur dioxide was near the detection limit. While the ozone level was within the suggested limit of 0.5–5.0 ppb, the nitrogen dioxide concentration was above the suggested limit, 2–10 ppb, for museums—a level that requires immediate action for reduction. The sulfur dioxide level was considered safe for collections.

Particulates

Particulates were monitored in March 2005 with an optical airborne particle counter used to measure two size fractions—0.3–1.0 μm and 1.0–5.0 μm—on a daily interval. Particulate levels in the Library were 19 $\mu g/m^3$ and 0.4 $\mu g/m^3$ for 0.3–1.0 μm and 1.0–5.0 μm particle sizes, respectively. These values were typical of a naturally ventilated urban space. Currently, no guideline exists for concentration limits of airborne particulates for collections conservation. Therefore, the measured particulate values for both size fractions were used as benchmarks for evaluating the filtration efficiency of the environmental management system.

Bookcase environments

The bookcases housing the Library's book collection were of considerable interest because of the potential microenvironments within. Despite the humidity variations observed in the Library rooms, average humidity within the bookcases (with tight door closures) was 70% RH, with a range between 65% RH and 73% RH. Daily humidity variations were less than Δ5% RH. Pollutant measurements within the bookcase microenvironments were 0.2–3.5 ppb for NO_2, while O_3 and SO_2 were at or near detection limits. Pollutant levels in the bookcases were much lower than levels recorded in the Library, a finding that indicated that, with properly fitting and operating doors, the bookcases were effective in isolating their contents from the less-stable and more-polluted Library environment.

Collections condition assessment

FCRB conservators conducted the condition assessment of the furniture and books in the Library. While furniture other than the bookcases was deemed to be in good condition, the bookcases themselves often had misaligned or warped doors and covers, which resulted in poor closure for some units. The assessment of the book collection indicated that numerous volumes were affected by acidification, which causes the typical browning and embrittlement of papers seen in nineteenth-century books. Some books showed evidence of damage due to mishandling, prior fungal or insect attacks, or significant dust accumulation, which can lead to chemical deterioration.

Preinstallation risk class assessment

The Conservation Environment Classification–HH in the MCRB Library prior to installation of the environmental management system is summarized in table 14.1.

Environmental Management System

Objectives

The project goal was to produce and maintain the collections conservation environment in the Library while maintaining limited visitor comfort. Based on the above assessments, the following environmental objectives were established for the Library:

- Maintain low and stable humidity to prevent microbial and mechanical deterioration of the collection.
- Reduce both gaseous and particulate pollution.
- Reduce insect population.
- Reduce the exposure of the collection to daylight.

Risk		Humidity		Temperature		
Category	**Specific Risk (statistic)**	**Value (% RH)**	**Class**	**Value (°C)**	**Class**	
Microbial	Germination threshold (P97.5)	85	F	—	—	
Mechanical	Short-term variation (P95 of rolling 24 hour variation)	Δ19	b	Δ3.1	a	
	Seasonal variation (absolute difference of seasonal means, RH Means ≦ 70%)	Δ3	f[1]	Δ2.8	a	Overall mechanical risk:[2] f
	Deviation from historical mean (absolute difference)	Δ4	a	Δ1.8	a	
Chemical	Deviation from historical mean (difference)	—	—	Δ1.8	0	

[1] Seasonal humidity means are above 70% RH (Very Hot-Humid: 72% RH; Hot-Humid: 75% RH), resulting in a Class f designation with respect to mechanical risk due to seasonal humidity variation.
[2] Overall mechanical risk is based on the highest risk classification in any of the specific humidity- and temperature-based mechanical risks (short-term variation, seasonal variation, and deviation from historical mean).

Table 14.1

Conservation Environment Classification–HH in the MCRB Library before installation of environmental management system.

- Maintain thermal comfort for visitors by providing an appropriate combination of temperature, humidity, and air movement,
- Avoid visual intrusions by caused by system installation.

Psychrometric strategies

The preinstallation environmental assessment highlighted the need to limit infiltration of unconditioned outside air into the building to reduce exterior moisture and pollutant sources. This nonmechanical psychrometric strategy required tightening of the building envelope and the internal envelopes of the Library suite and its bookcases. Furthermore, the accumulation of heat in the Library emphasized the need for nonmechanical source control of thermal energy via insolation through windows, and venting of warm interior air. Following implementation of nonmechanical strategies, mechanical psychrometric strategies focused on the use of dehumidification and conservation heating by ventilation to reduce humidity within the Library when the interior humidity exceeds 65% RH. Ventilation air is filtered to limit the introduction of particulates to the Library space. When temperatures in the Library exceed 28°C, mechanical cooling supersedes humidity control (dehumidification) to provide thermal comfort for visitors. During visiting hours, mechanical ventilation also provides filtered outside air to the Library to maintain good indoor air quality.

Nonmechanical strategies

Building envelope and bookcase improvements

The exterior building assessment highlighted the need to limit the infiltration of unconditioned outside air into the building. Windows and doors were repaired or restored for better fit and for proper operation and closure. Entry and exit doors into the Library space were kept closed to limit infiltration. Misaligned bookcase doors and covers were repaired to reduce air exchange and infiltration of air pollution and particulates into the bookcases. The books were cleaned and returned to the repaired cabinets.

Incorporation of original nonmechanical environmental management features into the new system

The building assessment report emphasized the functional attrition of the original nonmechanical environmental management features, such as basement ventilation and the extraction of warm room air by natural ventilation through the attic. All windows installed in the original basement ventilation openings were removed to once again allow natural ventilation of the cellar. Reinstating the original natural ventilation flow path through the attic with exhaust through loose roof tiles proved difficult owing to changes in the building envelope, but the underlying concept was integrated into the design of the new environmental management system. In the attic, a sheet metal plenum, a chamber for collecting exhaust air, was constructed to collect the warm and buoyant Library air rising through the ceiling perforations. The plenum is connected to a nearby skylight/ventilation shaft with a glass canopy via a large duct, and the air is mechanically exhausted to the outside via the skylight opening.

Visitor management

In the original visitor circulation plan for the museum, visitors entered the Library from the Civilist Room's south door, which was left open during visiting hours; they moved freely through the Library suite, exiting through two east doors of the Constitution Room, which were also left open (fig. 14.9). As part of the environmental improvement plan, the entry and exit doors to the Library remained closed except when in use, in order to limit the infiltration of unconditioned air, although the Library's entry and exit points remained the same. The interior doors between the rooms in the Library suite also remained open to allow air mixing. A new tour route directed visitors through the Library, leading them along a path next to the bookcases on the perimeter of the Constitution Room, rather than allowing them to move about freely. Supply air grilles of the mechanical system were installed along the visitor paths to provide thermal comfort (described below).

Mechanical strategies

The primary focus of the environmental management strategy was to maintain a stable humidity below 65% RH in the Library. This would provide effective protection for the collection from biological and mechanical damage. While interior temperatures would generally be allowed to vary with outside conditions, cooling of the Library air would occur when the inside temperature exceeded 28°C, improving thermal comfort during extremely hot periods. Through restriction of the use of conventional cooling to high-temperature periods, the potential for condensation on the collection and the building surfaces was reduced, especially in the event of power failure during exceptionally hot and humid periods. The design of the environmental management system had to maintain the historic ambience of both the Library interior and the building exterior, without alteration of existing walls and ceilings.

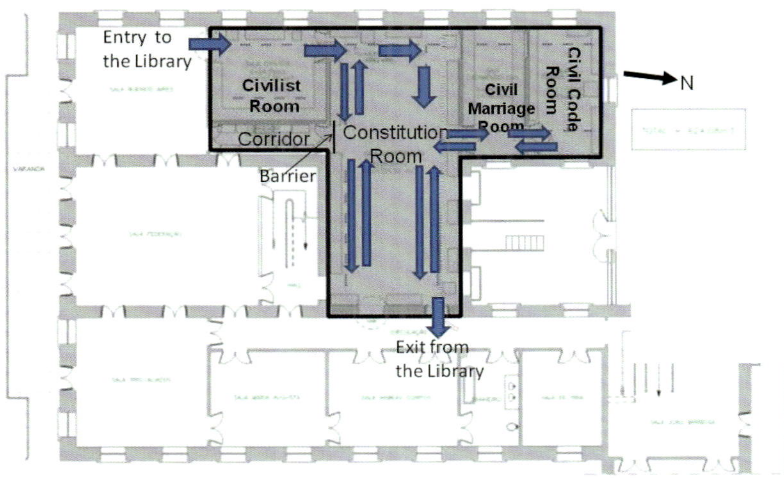

Figure 14.9
New design for visitor access to the Library (shaded area).

A secondary focus of the environmental management strategy was to improve thermal comfort for visitors in the Library and reduce the risk of microbial activity by increasing air movement. Analysis of factors affecting thermal comfort led to the conclusion that increased air movement could effectively improve the sensation of comfort for visitors. This conclusion was arrived at through use of the ASHRAE Thermal Comfort Calculator (see chap. 4) to determine predicted mean vote (PMV) values, assuming casual clothing (0.6 clo) typical of hot-humid climates; a low metabolic activity (1.5 met) for MCRB visitors; and interior conditions with mean radiant temperature of 25°C (due to cool walls and floors), mean air temperature of 27.1°C, and mean humidity of 73% RH, typical of the very hot-humid season. It was determined that an air movement of 1.1 m/s would lead to a PMV value of 0.5 (neutral), compared to a PMV of 1.08 (slightly warm) with 0.1 m/s air movement (acceptable PMV values range from –0.5 to +0.5). Since it was considered not achievable to provide 1.1 m/s of air movement along the visitor paths, the target air movement was set to 0.5 m/s to produce the PMV value of 0.82 (slightly warm) during the very hot-humid season. Although an upper threshold air velocity with respect to collections care has not been established, lower air velocities would obviously pose less risk to collections; the risk to collections from air movement is further mitigated by the protection afforded by the closed bookcases. With visitors directed along a well-defined tour route through the Library, it was possible to provide higher air velocities directly to the visitor pathway, further localizing the air movement risk and avoiding damage to collections.

Adequate amounts of filtered outside air were required to limit the interior concentrations of carbon dioxide and other bioeffluents, in order to maintain acceptable indoor air quality. Assuming an estimated maximum population in the Library of fifty adults, and unconditioned infiltration of 1 to 1.5 air changes per hour (ACH), an additional 2 ACH of fresh air are estimated to be necessary during visiting hours (8:00 a.m.–6:00 p.m.) to maintain a carbon dioxide concentration below 1000 ppm and to meet the recommended fresh air rate of 8 L per person in ANSI/ASHRAE Standard 62.2-2013 (ANSI and ASHRAE 2013b).

The environmental management system in the MCRB Library was therefore designed to provide filtered fresh air and/or conditioned air (55%–65% RH, 22°C–28°C) to the Library rooms using a combination of ventilation, incidental dehumidification, and cooling. The system utilized reheat in combination with cooling to produce supply air at less than 65% RH. To minimize visual intrusion in the Library, most of the system components were installed in the basement, and an exhaust ventilator was installed in the attic (fig. 14.10). The only system components visible to visitors were supply air diffusers and return air grilles in the floor along the tour route in the Library.

Mechanical system description

Installed in October 2006, the mechanical system is shown in figure 14.11. The major mechanical components of the environmental management system were a split direct expansion air-conditioning unit, an elec-

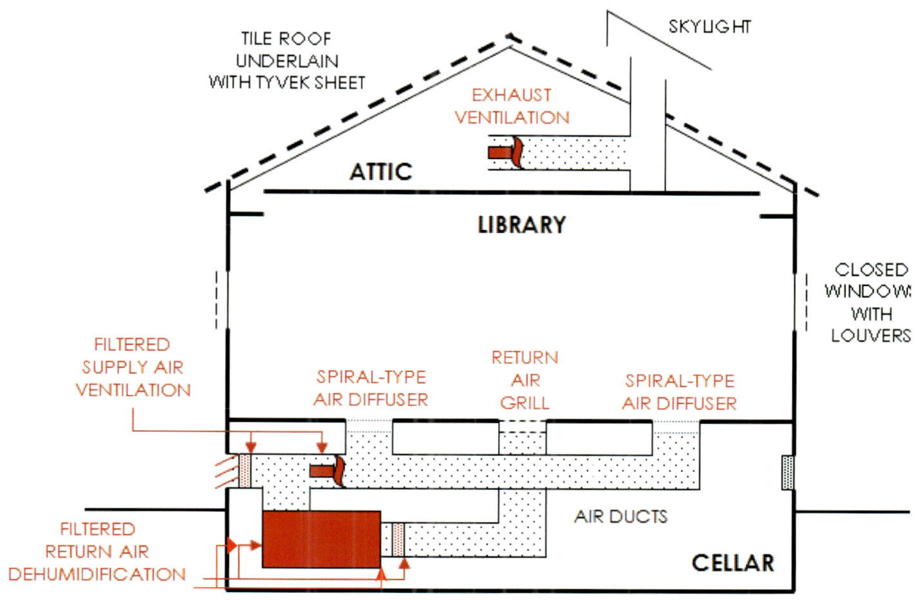

Figure 14.10
Location of the mechanical components of the environmental management system in the MCRB Library.

troresistive reheat coil, and a programmable logic controller (PLC) unit, all located in the basement; supply and return air grilles located in the Library; and an exhaust fan located in the attic (see fig. 14.10). The air-conditioning unit was designed to operate in concert with a downstream reheat coil; together they form an inline dehumidifier. This arrangement was required because, at the time of engineering design, a duct-mounted dehumidifier with heat-recovery condenser was not available in Brazil. Cooling coil condensate was collected and discharged to the exterior via a sump pump.

Figure 14.11
Schematic of the mechanical environmental control system for the MCRB Library. Courtesy of Fundação Casa de Rui Barbosa.

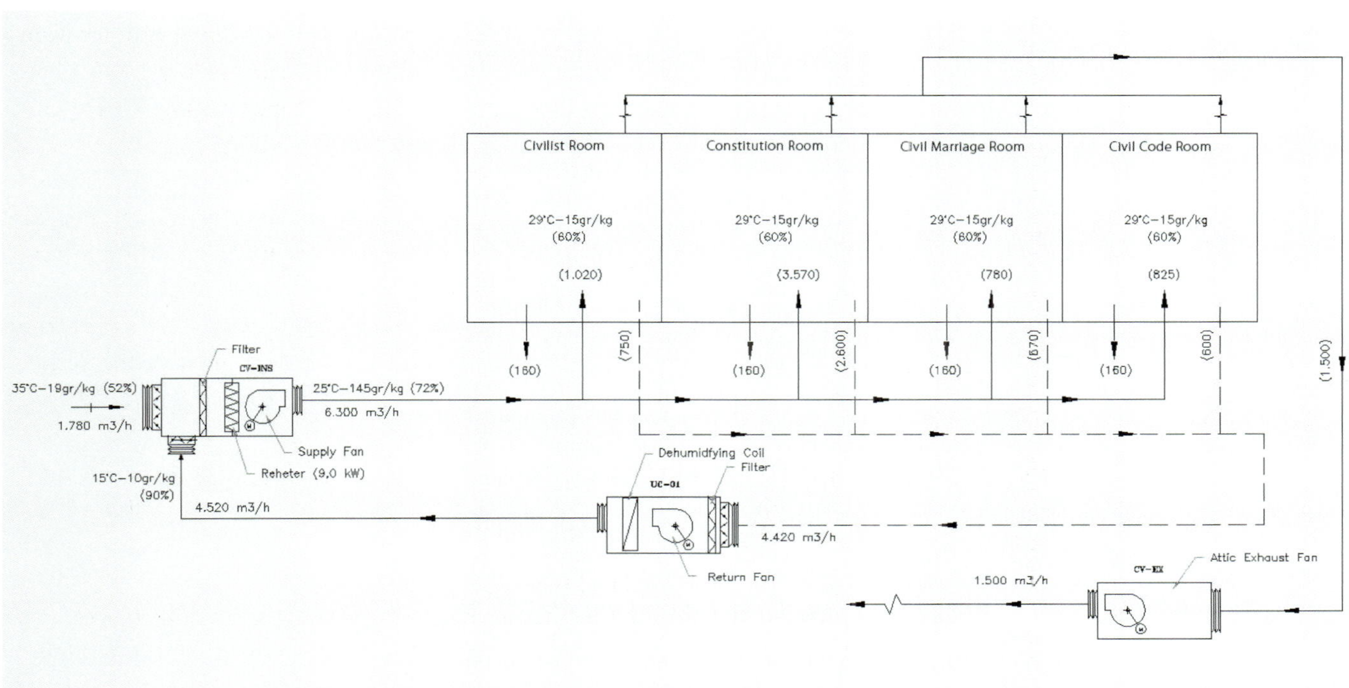

The system was controlled by the PLC and two sensors measuring both temperature and humidity. One sensor monitored the interior environment in the Library, and the second sensor was positioned on an exterior wall on the south side of the building to measure outside conditions. Though not used for system control, a third sensor measuring temperature and humidity was placed in the return air duct to allow for system diagnostics. The specifications of each system component are provided in table 14.2.

Along the visitor tour route in the Library, forty spiral type diffusers were positioned to deliver supply air to the space (fig. 14.12). Return air grilles were positioned on the floor near the east walls of the Constitution Room, the Civil Marriage Room, and the Civil Code Room. Spiral diffusers were selected to provide a large amount of airflow with limited vertical air velocity. Sheet metal ducts were used to transport supply and return air. Located at the supply and return ends of the dehumidifier, MERV 5 (G3) filters were used to filter both outside air and recirculated air.

Air in the Library rooms is either returned to the dehumidification unit via the return air grilles or exhausted through perimeter gaps in the ceiling into the attic, where it is collected by the plenum. An exhaust fan extracts the air from the plenum and discharges it to the exterior via a duct connected to the existing skylight. The exception to this process is the Constitution Room, which has a plaster ceiling that lacks perimeter gaps. The air in the Constitution Room escapes to connected rooms and is exhausted through their ceiling/attic vent and/or is recycled to the dehumidification unit through the return grilles in the room.

Operational control sequence

The environmental management system has five modes of operation: ventilation, dehumidification, hybrid ventilation/dehumidification, cooling, and idle. Modes were selected based on interior and exterior temperature and RH measurements, as summarized in table 14.3.

Table 14.2

Components of MCRB Library environmental management system.

Component	Thermal Capacity	Flow Rate	Power Consumption	Supply Power Requirements
Split air-conditioner	17.58 kW	4800 m³/hr		220V/3 phase
– Compressor	(5TR)		6 kW	
– Condenser			6.5 kW	
– Evaporator			0.7 kW	
– Filter—MERV 5 (G3)				
Reheat coil			6.75 kW	220V/3phase
Fan (basement ventilation) – Filters—MERV 3 and 5 (G1 and G3)		6500 m³/hr	1.5 kW	220V/3phase
Fan (room exhaust located in attic)		1700 m³/hr	0.5 kW	220V/3phase

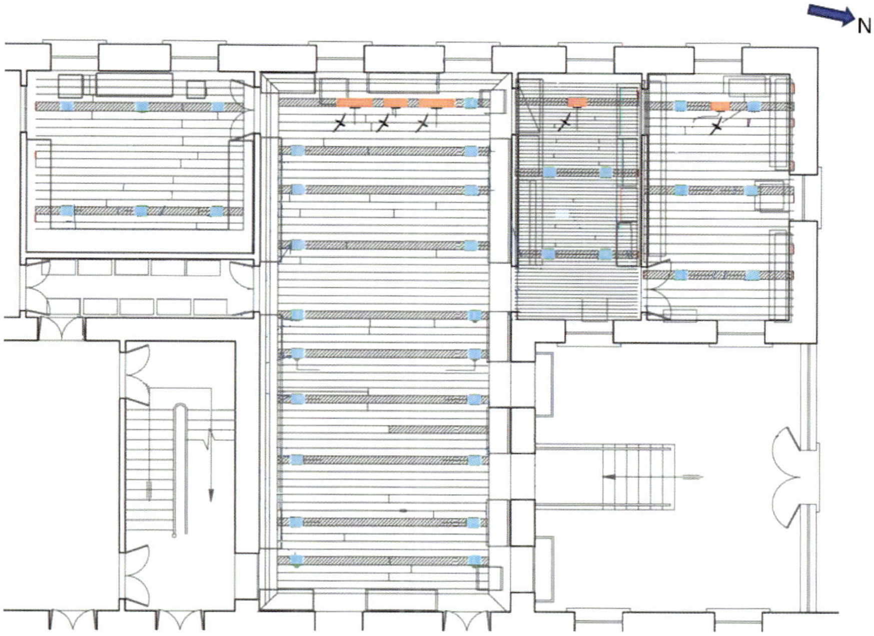

(a)

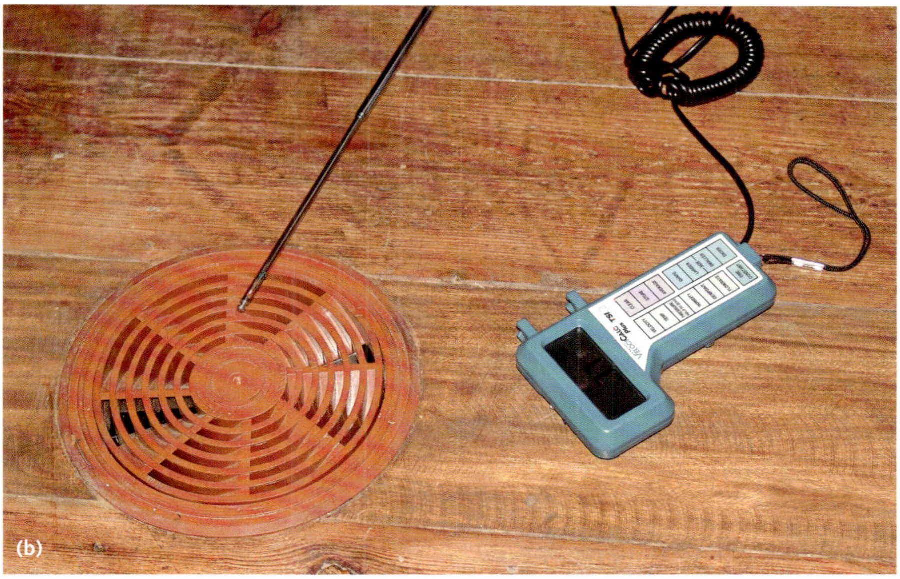

(b)

Figure 14.12
Library floor plan showing floor planks and locations of supply air diffusers (blue) and return grilles (peach) on the floor (a); the original floor planks were removed for installation of the diffusers and grilles and were replaced with new planks (shaded). A spiral-type supply air diffuser on the floor (b) is also shown. Floor plan (a) courtesy of Fundação Casa de Rui Barbosa. Photo (b): Shin Maekawa. © J. Paul Getty Trust.

The ventilation mode opens the fresh-air damper and enables the exhaust fan in the attic (fig. 14.13). In this mode, 2 ACH (1260 m^3/h) of fresh outside air are mixed with 8 ACH (5040 m^3/h) of recirculated air and then supplied to the Library; 2 ACH of air are exhausted through the attic. Used only during visiting hours, the ventilation mode is enabled if the following three criteria are satisfied: interior humidity is above 65% RH, interior temperature is less than or equal to 28°C, and outside humidity is less than or equal to 65% RH. Outside of visiting hours, the ventilation mode is disabled when the humidity in the Library is less than or equal to 55% RH, interior temperature is greater than 28°C, or outside

Table 14.3

Operational control conditions for the MCRB Library environmental management system.

	Psychrometric Strategy	Enable	Disable
Visiting Hours Only	Ventilation	$H_{Interior} > 65\%$ RH and $H_{Exterior} \leqq 65\%$ RH and $T_{Interior} \leqq 28°C$	a) $H_{Interior} \leqq 55\%$ RH or b) $H_{Exterior} > 65\%$ RH or c) $T_{Interior} > 28°C$
Visiting Hours Only	Hybrid (ventilation & dehumidification)	$H_{Interior} > 65\%$ RH and $H_{Exterior} > 65\%$ RH and $T_{Interior} \leqq 28°C$	a) $H_{Interior} \leqq 55\%$ RH or b) $H_{Exterior} \leqq 65\%$ RH or c) $T_{Interior} > 28°C$
Visiting Hours Only	Cooling	$T_{Interior} > 28°C$	$T_{Interior} \leqq 27°C$
Nonvisiting Hours Only	Dehumidification	$H_{Interior} > 65\%$ RH	$H_{Interior} \leqq 55\%$ RH
	Idle	$H_{Interior} \leqq 55\%$ RH	a) $H_{Interior} > 65\%$ RH or b) $T_{Interior} > 28°C$ (T criteria only valid during visiting hours)

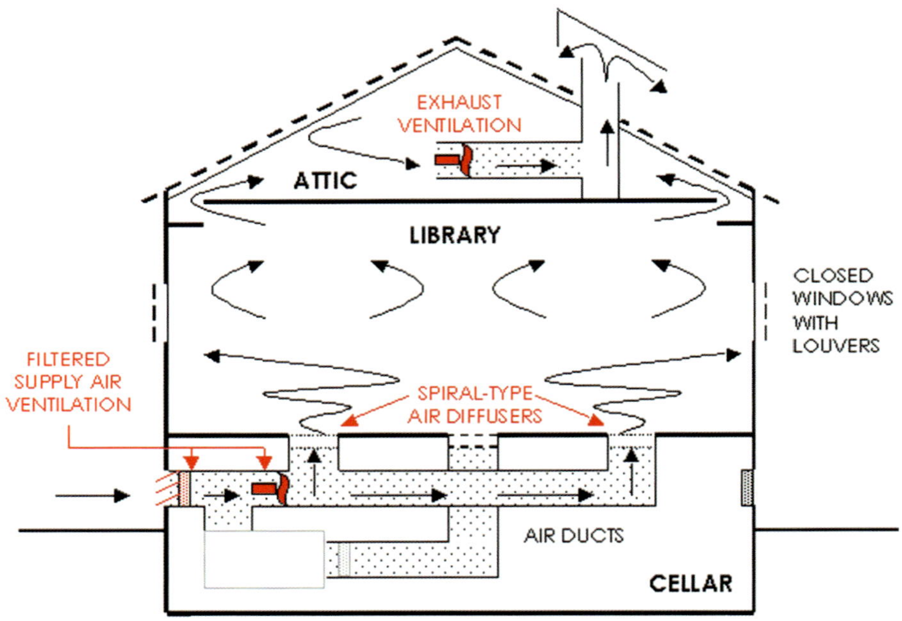

Figure 14.13

Schematic of the ventilation mode for the MCRB Library environmental management system.

humidity is higher than 65% RH. This mode is a low-energy mode, as it does not rely on the more energy-intensive operation of dehumidification.

In the dehumidification mode, both the dehumidifier and supply fans are enabled, but the fresh air damper is closed, and the exhaust ventilator is disabled (fig. 14.14). This mode recycles Library air through the dehumidifier at a rate of 10 ACH (6300 m^3/h); it reduces energy consumption because the environmental management system does not then have to condition outside air. Used only during nonvisiting hours, the dehumidification mode is enabled when interior humidity is higher than 65% RH. This mode is disabled when interior humidity becomes equal to or less than 55% RH.

The hybrid mode is a combination of the ventilation and dehumidification modes (fig. 14.15). During this mode, the dehumidi-

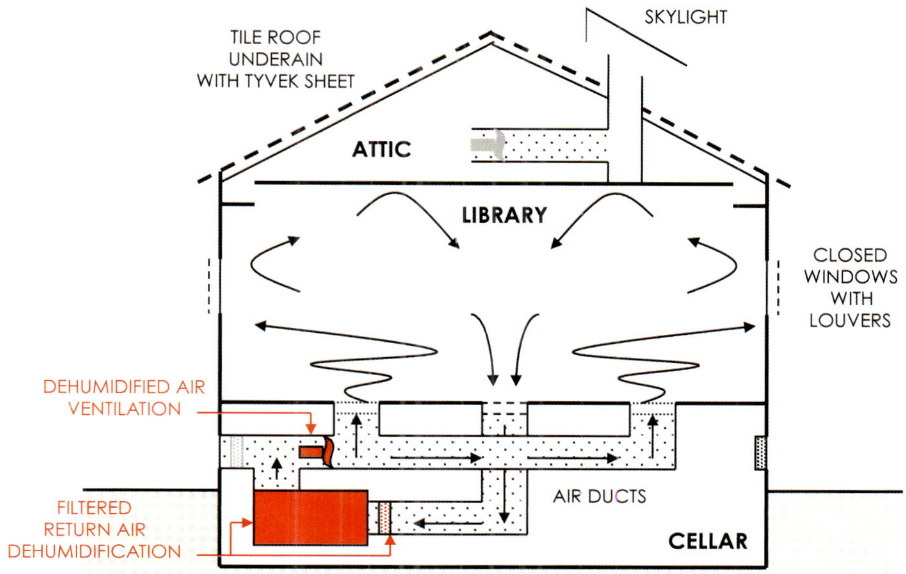

Figure 14.14
Schematic of the dehumidification mode for the MCRB Library environmental management system.

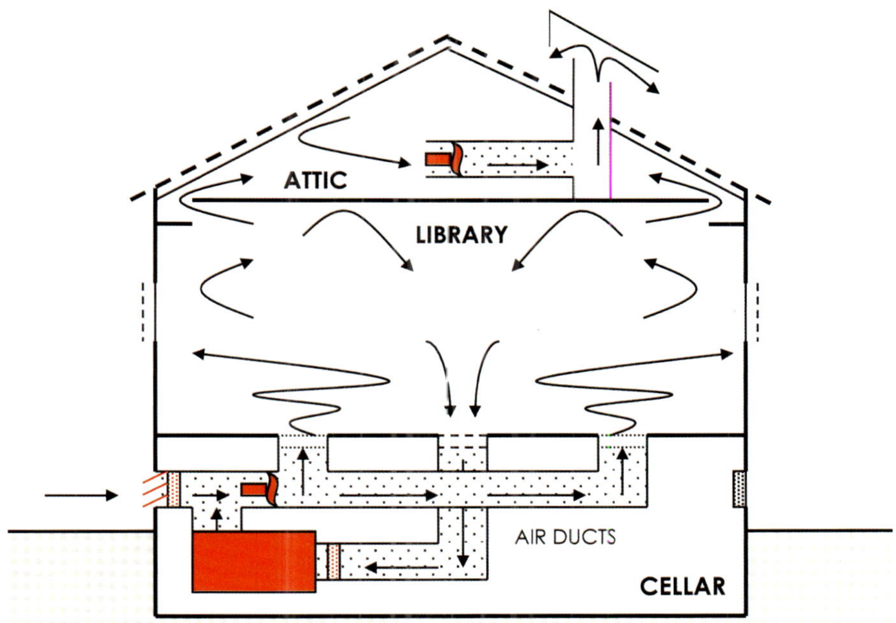

Figure 14.15
Schematic of the hybrid (dehumidification plus ventilation) mode for the MCRB Library environmental management system.

fier and supply and exhaust fans are enabled, and the fresh air damper is opened. The system mixes 2 ACH (1260 m^3/h) of fresh air with 8 ACH (5040 m^3/h) of recirculated and conditioned air, and then supplies the air the Library rooms. It also exhausts 2 ACH of the conditioned Library air to the exterior through the attic. Used only during visiting hours, the hybrid mode is enabled when the following conditions are satisfied: interior and exterior humidity exceeds 65% RH, and interior air temperature is less than or equal to 28°C. The hybrid mode is disabled when the interior humidity becomes equal to or less than 55% RH, exterior humidity is less than or equal to 65% RH, or interior temperature is greater than 28°C.

During the cooling mode, the compressor and air handler are enabled, but the reheat coil is disabled. This mode is similar to the hybrid mode, except the reheat coil is disabled; the result is cooler but humid supply air to provide thermal comfort for visitors. The cooling mode is only used during visiting hours and is enabled when the interior temperature exceeds 28°C. The mode is disabled once the interior temperature in the Library is less than or equal to 27°C.

The environmental management system enables the idle mode when the humidity in the Library is reduced to 55% RH or below, and it disables the idle mode when the interior humidity exceeds 65% RH or interior temperature exceeds 28°C (temperature criteria only valid during visiting hours).

Postinstallation System Performance

As stated above, the environmental management system was installed in October 2006. However, system performance was inconsistent for an extended period and did not meet design specifications. The system's substandard performance was attributable to poor installation quality. During the second year of system operation, the original design/build contractor corrected the problems associated with installation.

Humidity, temperature, and dew point temperature

Following operational commissioning in April 2009, the environmental management system at the MCRB Library performed much closer to design specifications. The following section discusses 15 minute interval data on the postinstallation environment collected by a monitoring system from April 2009 to November 2010 (monitoring sensors were separate from the system's control sensors).

Monthly mean interior humidity ranged from 57% RH in February to 62% RH in October, and the 75% RH threshold was exceeded less than 0.05% of the postinstallation study period (fig. 14.16a). The monthly average rolling 24 hour variation of humidity ranged from Δ4% RH in February to Δ7% RH in July (not shown on figure). With lower humidity and reduced daily variation, the postinstallation humidity conditions were significantly improved in relation to preinstallation values.

The monthly mean interior temperature ranged from 25.4°C in July to 30.9°C in February, and nearly all postinstallation monthly mean temperatures were higher than corresponding preinstallation values. High temperatures in January, February, and November indicated that the cooling capacity of the system was not sufficient (fig. 14.16b). Although not shown in the figure, monthly mean rolling 24 hour variation ranged from Δ1.0°C (January) to Δ1.4°C (July), a slight reduction from preinstallation values.

The annual average dew point temperature decreased from 20.4°C for the preinstallation period to 19.3°C during the postinstallation period (fig. 14.16c). This improvement was confirmed during the very hot-

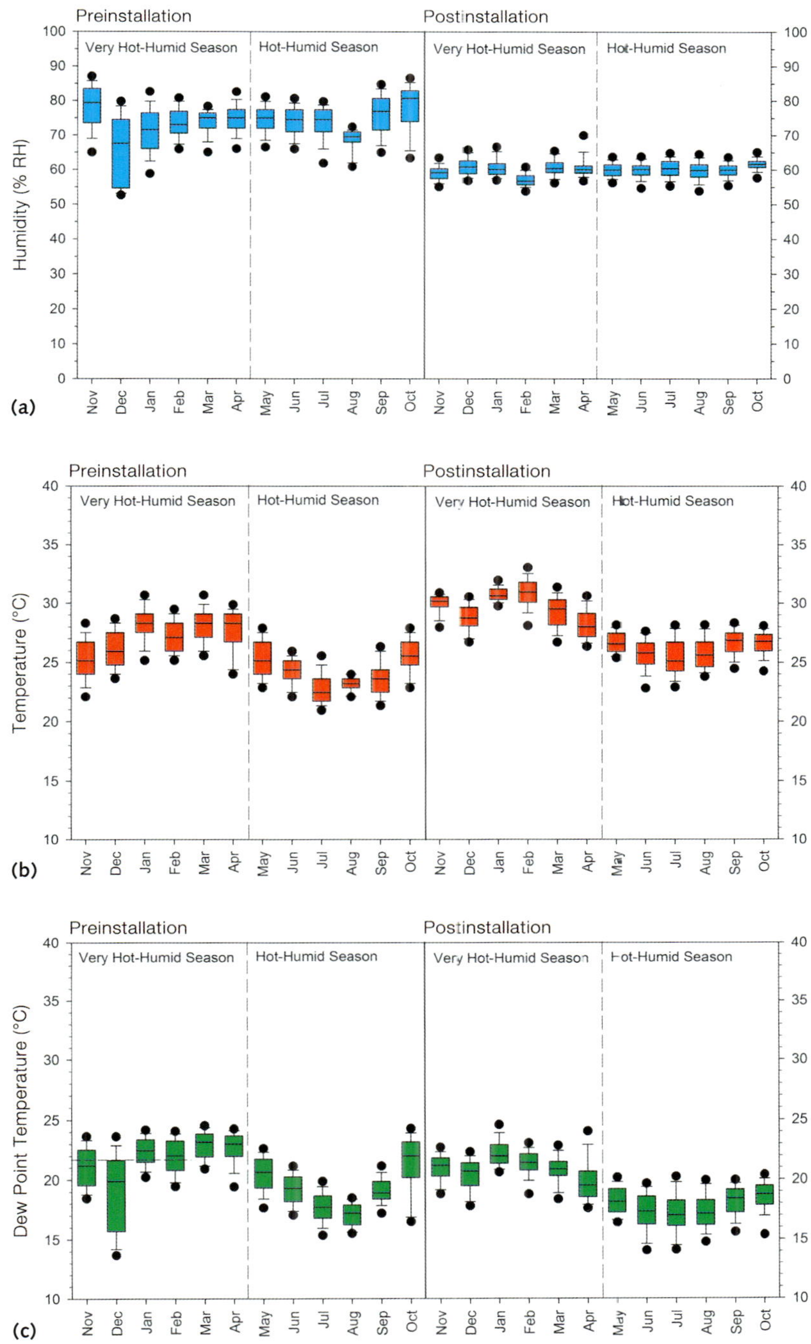

Figure 14.16

Comparisons of humidity (a), temperature (b), and dew point temperature (c) in the MCRB Library during the pre- and post-installation periods.

humid and hot-humid seasons. However, the rolling 24 hour variation in dew point temperature (not shown in the figure) noticeably increased in the postinstallation period as a result of cycling of the dehumidification processor.

Postinstallation risk class assessment

The Conservation Environment Classification–HH protocol for the post-installation Library environment at MCRB is summarized in table 14.4. Figure 14.17 shows on psychrometric charts the preinstallation and post-installation datasets from the Constitution Room, with values pertaining

Table 14.4

Postinstallation Conservation Environment Classification–HH protocol of the MCRB Library.

Risk		Humidity		Temperature		
Category	Specific Risk (statistic)	Value (% RH)	Class	Value (°C)	Class	
Microbial	Germination threshold (P97.5)	66	B	—	—	
Mechanical	Short-term variation (P95 of rolling 24 hour variation)	Δ13	b	Δ 2.2	a	Overall mechanical risk:[1] b
	Seasonal variation (absolute difference of seasonal means, RH means ≦ 70%)	Δ0	a	Δ 3.4	a	
	Deviation from historical mean (absolute difference)	Δ17	b	Δ 3.2	a	
Chemical	Deviation from historical mean (difference)	—	—	Δ 3.2	+	

[1] Overall mechanical risk is based on the highest risk classification in any of the specific humidity- and temperature-based mechanical risks (short-term variation, seasonal variation, and deviation from historical mean).

to the conservation environment classification denoted; the boxes indicate the mid-95% of each dataset. In the figure, the postinstallation humidity range (54%–66% RH) is narrower than the preinstallation humidity range (54%–85% RH). The preinstallation temperature range (21.3°C–29.9°C) is shifted approximately 2°C cooler than the postinstallation range of 23.5°C–31.6°C. Variations of seasonal average temperatures increased from Δ2.8°C during preinstallation to Δ3.4°C during the postinstallation period. However, variations of seasonal average humidity values reduced to Δ0% RH in the postinstallation, from Δ3% RH during the preinstallation period. Postinstallation means also deviate further

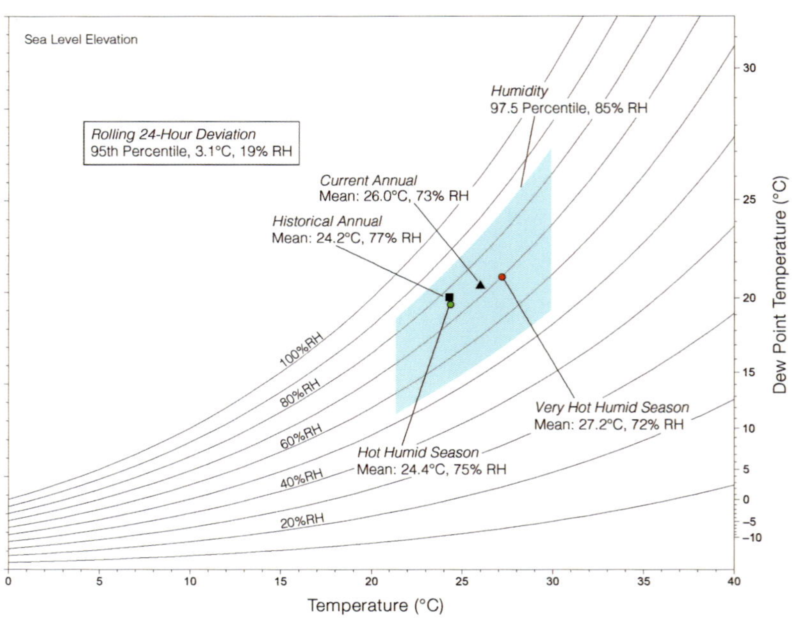

Figure 14.17a

Psychrometric chart showing preinstallation conditions in the Constitution Room at the Casa de Rui Barbosa.

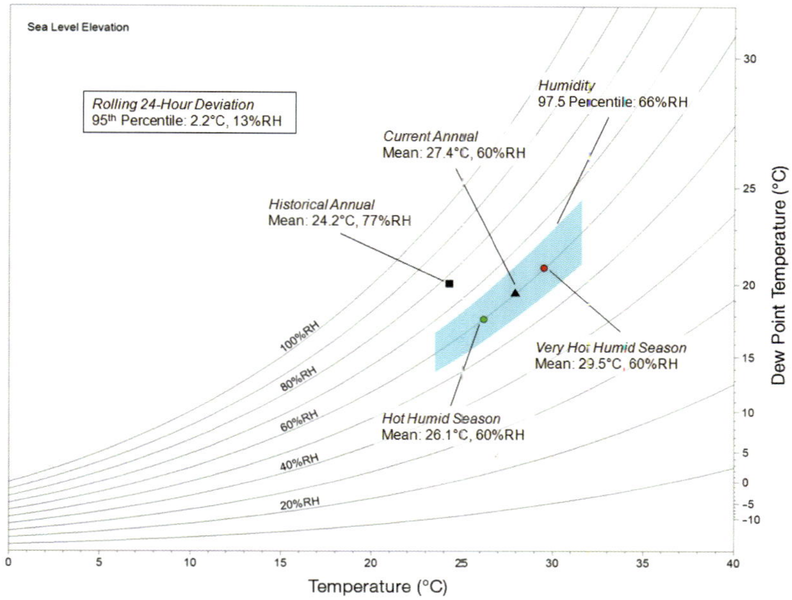

Figure 14.17b

Psychrometric chart showing postinstallation conditions in the Constitution Room at the Casa de Rui Barbosa.

(Δ17% RH and Δ3.2°C) from the historic means than from the preinstallation means (Δ4% RH and Δ1.8°C).

The pre- and postinstallation risk classes for the Constitution Room at MCRB are summarized in table 14.5.

Following an extended period of adjustment due to installation issues, the MCRB Library environmental management system improved the environment in the Library rooms and maintained a safe collections environment with respect to biological deterioration, mechanical damage, and protection against dust and air pollution. Classification of the microbial risk was reduced from Class F (95th humidity percentile of 85% RH) during preinstallation to Class B (95th humidity percentile of 66% RH) following the installation of the environmental management system. Overall mechanical risk was also improved, from Class f during preinstallation (owing to preinstallation seasonal variation risk) to Class b during postinstallation. Mechanical risk due to the deviation from the historical mean humidity did increase from Class a (Δ4°C) during preinstallation to Class b (Δ17°C) during postinstallation. A slightly higher postinstallation interior temperature in the Library space increased the risk of chemical deterioration for chemically active materials from Class 0

Table 14.5

Comparison of pre- and postinstallation risk classes in the Constitution Room at MCRB.

	Microbial Risk		Mechanical Risk														Chemical Risk	
			Humidity						Temperature						Overall[1]			
			Short-Term Variation		Seasonal Variation		Deviation from Historical Mean		Short-Term Variation		Seasonal Variation		Deviation from Historical Mean					
Room	Pre	Post	Pre	Post	Pre	Post	Pre	Post	Pre	Post	Pre	Post	Pre	Post	Pre	Post	Pre	Post
Constitution Room	F	B	b	b	f	a	a	b	a	a	a	a	a	a	f	b	0	+

[1] Overall mechanical risk is based on the highest risk classification in any of the specific humidity- and temperature-based mechanical risks (short-term variation, seasonal variation, and deviation from historical mean).

(Δ1.8°C deviation from historical mean temperature) during preinstallation to Class + (Δ3.2°C deviation from historical mean temperature).

Pollution and particulates

Postinstallation measurements (February 2007) of pollutants and particulates in the Library indicated a general reduction in concentrations. Among air pollutants, the highest decrease relative to preinstallation values was observed in O_3, which was reduced 85% (fig. 14.18) in the Constitution Room, while postinstallation NO_2 concentrations were decreased by 30%. The reduction in these two pollutants may be attributed to surface reactions on the sheet metal ducts prior to air delivery to the Library. (These pollutants can be oxidized as molecules came into contact with duct walls during turbulent mixing in the air ducts.) Reduction may also be attributable to the reduced infiltration of outside air through windows and doors. Results for NO were inconclusive, while SO_2 levels were negligible during both pre- and postinstallation. During both pre- and postinstallation periods, pollutant concentrations measured inside the bookcases were negligible compared to levels recorded in the Library.

The comparison of pre- and postinstallation interior airborne particulate data showed a 75%–85% reduction in the larger size fraction (1.0–5.0 μm). A reduction in this size fraction, however, was also observed in the garden and was unrelated to system operation (fig. 14.19). Reductions in the smaller size fraction (0.3–1.0 μm) were less, ranging from 15% to 60% of preinstallation levels. Reductions in postinstallation interior particulates are attributable to the MERV 5 filters in the supply and return ducts; MERV 5 filters are effective for particles larger than 3 μm (Beltran 2007).

Thermal comfort

The environmental management system was designed to address occupant comfort through the following strategies: a high air change rate,

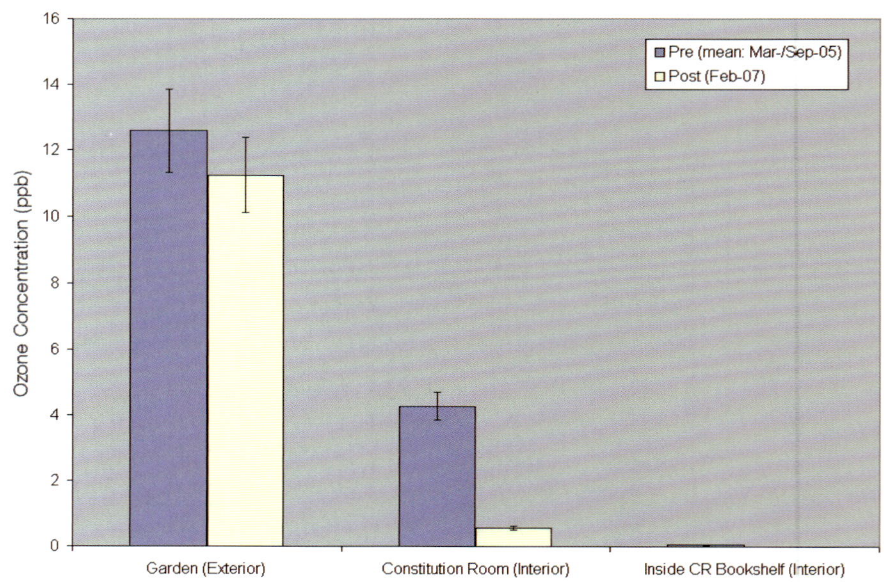

Figure 14.18
Comparison of ozone concentrations before and after installation of the MCRB Library environmental control system.

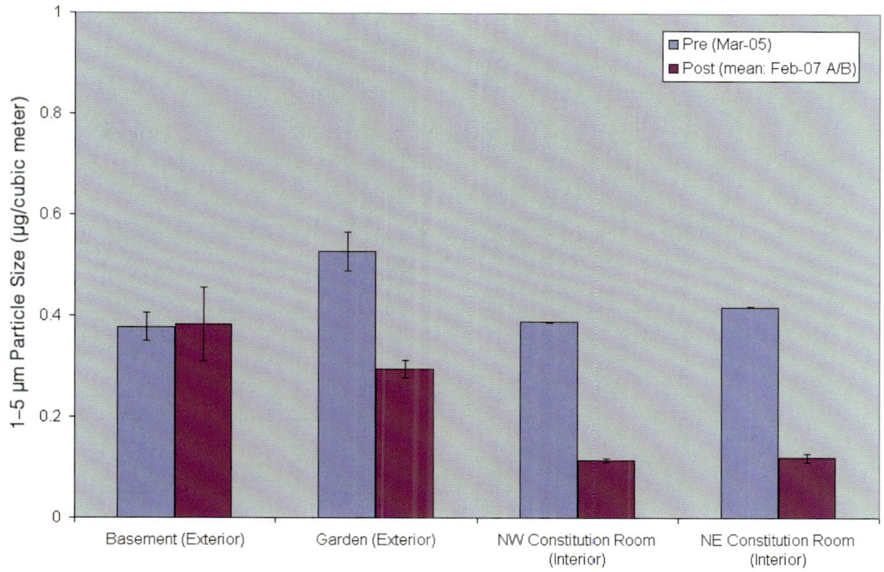

Figure 14.19

Comparison of amounts of 1–5 µm particulate matter before and after installation of the MCRB Library environmental control system.

increased air velocity along visitor pathways, and the use of cooling during extremely warm conditions. Implementation of these thermal comfort strategies was not satisfactory for the following reasons: (1) the designed ACH rate was not achieved because of large leakage at supply diffusers and return grilles, (2) the air velocity along visitor pathways (<0.3 m/s) was less than the design specification of 0.5 m/s, and (3) the designed cooling capacity was not enough to limit high temperatures during the very hot-humid season.

The predicted mean vote (PMV) and predicted percentage dissatisfied (PPD) values for the Library environment were evaluated using the air speed model in the ASHRAE Thermal Comfort Calculator. Prior to installation of the environmental management system, the average temperature and humidity in the Library during the very hot-humid season were 27.1°C and 73% RH, respectively. Assuming a mean radiant temperature of 25°C owing to cooler masonry walls, a metabolic rate of 1.5 met (combination of standing and walking), a clothing factor of 0.6 clo (typical casual summer clothing), and a minimum air speed of 0.1 m/s (the minimum set air speed is 0.1 m/s in the program), the PMV was calculated at 1.08 (slightly warm), and the PPD was calculated at 29% (point 1 in fig. 14.20).

After installation of the system, the average very hot-humid season temperature and humidity in the Library changed to 29.6°C and 60% RH, respectively. Assuming the same minimum air speed (0.1 m/s), mean radiant temperature (25°C), metabolic rate (1.5 met), and clothing value (0.6 clo) as the preinstallation condition, the PMV and PPD were calculated at 1.30 (slightly warm) and 40%, respectively (point 2 in fig. 14.20). If air speed is increased to 0.3 m/s, as was measured above the supply air diffusers, the resulting PMV (1.10) and PPD (30%) were similar to preinstallation thermal comfort levels (point 3 in fig. 14.20).

Although the pre- and postinstallation PMV and PPD values did not comply with ANSI/ASHRAE Standard 55-2013 (ANSI and ASHRAE 2013a), visitors and staff expressed a feeling of improved

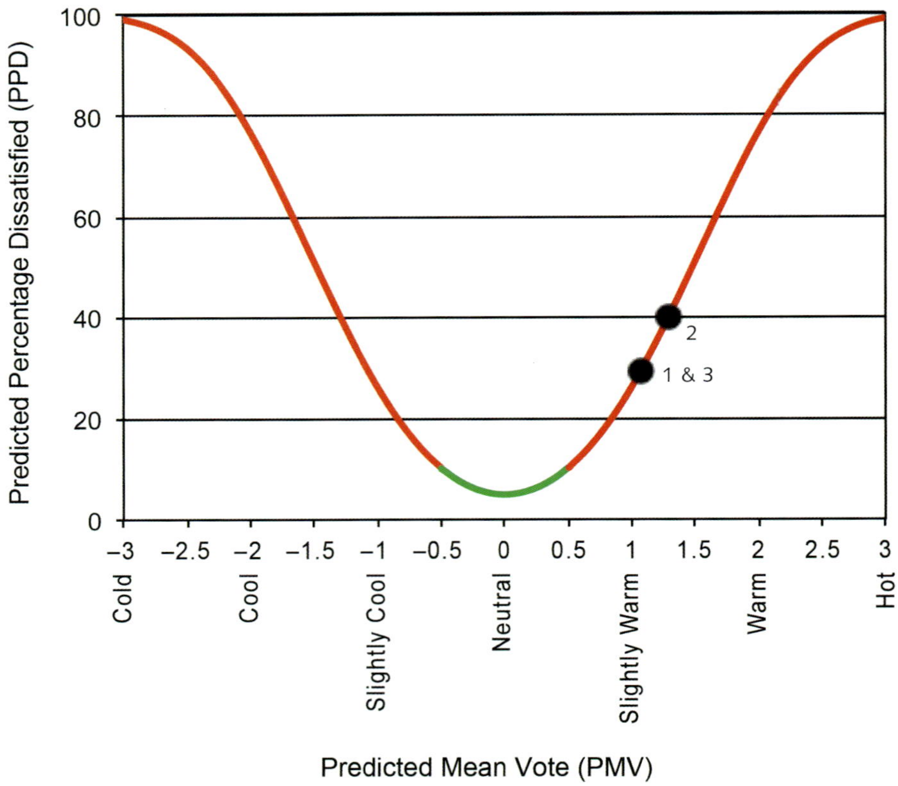

Figure 14.20

PMV-PPD plot showing thermal comfort in the Library during the very hot-humid season: (1) prior to installation with air speed of 0.1 m/s, (2) during the postinstallation period with 0.1 m/s air movement, and (3) during the postinstallation period with the designed air velocity of 0.3 m/s.

thermal comfort in the Library as compared to nonconditioned spaces in the museum. The Library air was described as fresh, dry, and cooler. Closure of entry and exit doors and windows in the Library also reduced noise and limited harsh sunlight, possibly contributing to visitor comfort.

Noise

Noise levels measured at the visitor pathways in the Library were between 45 and 48 dBA at approximately 1.5 m (ear height) above the floor—values close to office ambient levels. While high-frequency "hissing" typical of aeroacoustic noise was not heard in the Library, low-frequency noises, possibly from the operation of basement fans, were faintly audible. Visitors did not express any issues relating to acoustic discomfort.

Insulation was not used in conjunction with system components, including the supply and return fans, a condenser fan, and the refrigerant compressor in the basement. Lack of insulation resulted in in-situ noise levels of 66 to 70 dBA (the hazardous level is considered 85 dBA). Since the equipment was mounted on the basement slab floor, no vibration was transmitted to the Library room above. Outside of the building, noise from the environmental management system was not distinguishable from traffic noise from the adjacent Rua São Clemente.

A centrifugal fan in the attic was housed in an insulated enclosure attached to a pair of ducts, one connected to a manifold over the Library rooms along the western wall and the other connected to the skylight opening. The fan enclosure was bolted to a pair of steel beams, which rested on masonry structural bearing walls of the building using vibration isolation mounts. Noise levels in the attic ranged from 55 to 65 dBA, depending on the distance from the fan, which was located near the skylight that illuminates a hallway leading to the Constitution Room. In the hallway beneath the skylight, a noise level of approximately 50 dBA was registered. However, no negative comments were received from visitors.

Air duct leaks

Air velocities were measured 50 cm above supply air diffusers on the Library floor with a hot-wire anemometer. Measurements indicated air velocities less than 0.3 m/s, below the design specification of 0.5 m/s. Although supply and return duct leaks had reportedly been remedied before the performance monitoring period, leaks were found at several floor-installed supply air diffuser boxes (fig. 14.21) and at the return air grilles; leaks in the grilles led to the infiltration of unconditioned basement air into return air. Although attempts were made to limit the leaks, repairs were difficult because of the close clearances with structural members and the difference in thickness between the original and replacement floorboards in which the grilles were installed.

Diffuser boxes were mounted to the undersides of floor planks at four edges, two on original planks and the other two on replacement planks. The diffuser openings were cut into modern (standard) replacement planks, which were 5 to 10 mm thinner than the original planks; the undersides of the original planks were also irregular and unfinished. The discrepancies between the modern planks and the original ones made it difficult to produce a leak-free mount, and large leaks were often found at the mounting edges of diffuser boxes.

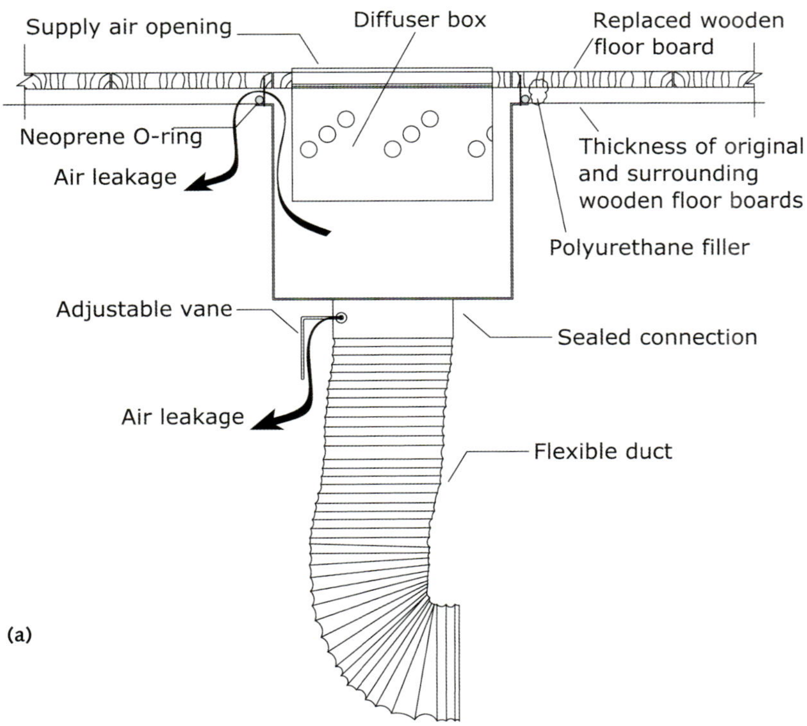

(a)

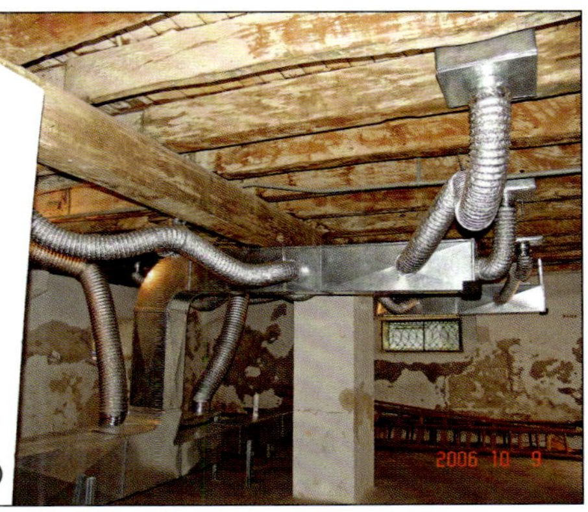

(b)

Figure 14.21
Schematic of supply air diffuser box in the Library floor showing leak locations (a) and supply air ducts and connected diffuser boxes (b). Schematic (a) courtesy of Fundação Casa de Rui Barbosa. Photo (b): Shin Maekawa. © J. Paul Getty Trust.

Maintenance

The facilities maintenance staff of the Fundação Casa de Rui Barbosa (FCRB) were trained by the installation contractor to conduct the day-to-day operation of the environmental management system. Scheduled maintenance, which included the replacement of filters and fan belts, was contracted to a local HVAC maintenance company. The maintenance contract included 24 hour response/repair of the system. A personal computer connected to the local area network (LAN) of the FCRB was linked to the controlling programmable logic controller unit of the environmental management system for Internet Protocol–based remote online monitoring. This configuration allowed for system diagnosis from FCRB offices as well as by the outside contractor. The FCRB architect monitored the operation of the environmental system through the online monitoring system on a daily basis (fig. 14.22).

The environmental management system at the MCRB Library suffered several operational failures or malfunctions during the operational period. As previously stated, air leaks in the supply and return air boxes were problematic, and, though they were identified, satisfactory repairs were not achieved. Malfunctions occurred in the sump pump, which was responsible for draining condensate water beneath the cold plate and from the split unit to the surface drainage system in the garden. Before the sump pump was replaced, there were several flooding events on the basement floor during the first year of operation and within the warranty period.

A circuit breaker protecting the reheat coil and the supply fan occasionally activated, leading to insufficient dehumidification in the Library, which resulted in condensation on the outside of the supply air ducts in the basement. Identified after two years of system operation, the

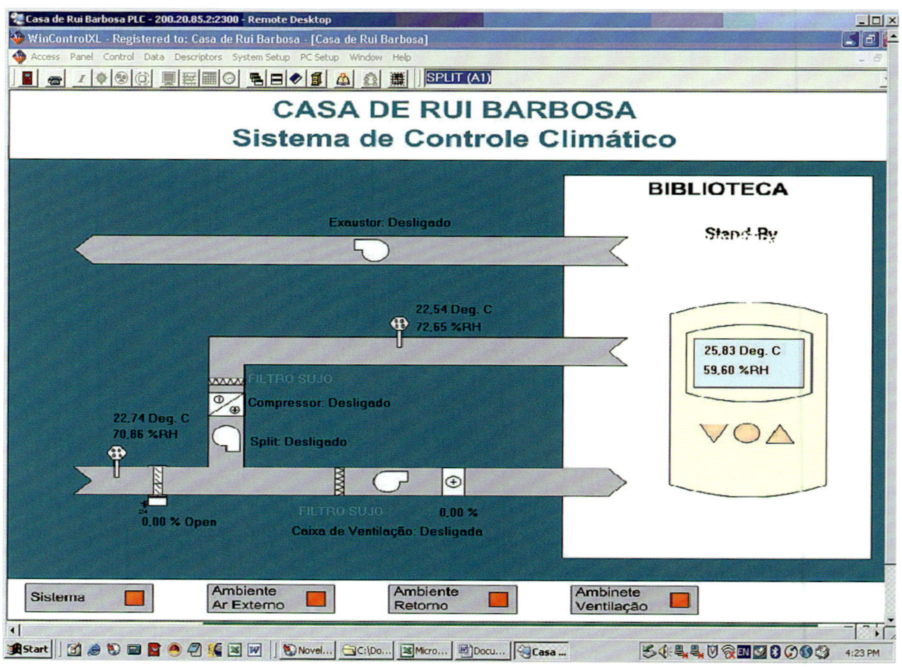

Figure 14.22

Online display of the control system indicating operational status of the environmental management system.

cause of the circuit breaker trip was a combination of the high combined startup current of the circuit and the unstable power supply. When subsequent circuit breaker problems arose, maintenance staff were instructed to shut down and restart the system to correct the problem.

Energy Metrics

Energy intensity

The total floor area of the five Library rooms is 165 m^2, and the total rated power of the environmental management system was 22 kW, resulting in an energy intensity of 148 W/m^2.

Energy use intensity

Since the museum does not have a separate electrical meter for the environmental management system, the energy use intensity was estimated from the power consumption of the environmental management system and the on/off ratio of its compressor. The estimated rate of power consumption of the environmental system was 22 kWh with all components in operation. During 2007 and 2008, the system actively managed the environment during 16% and 75% of the hot-humid and very hot-humid seasons, respectively, and the system typically operated in either the dehumidification or the hybrid mode. The estimated annual energy use intensity was 86,500 kWh, or 524 kWh/m^2. This high consumption was attributed to the use of the electrical heater for supply air reheat.

Conclusions

A ventilation/dehumidification–based environmental management strategy was implemented in the Library at the Museu Casa de Rui Barbosa in Rio de Janeiro. This historic house museum is located in a busy urban district in a Very Hot-Humid (1A) climate zone. After an extended period of adjustment during which installation issues had to be resolved, the strategy improved the environment in the Library rooms as well as maintained a safe collections environment.

Environmental strategies addressing visitor comfort were also included in the environmental management system. Although the thermal comfort defined by ANSI/ASHRAE Standard 55-2013 (ANSI and ASHRAE 2013a) was not achieved, visitors indicated that the Library, with its conditioned environment, was generally more comfortable than the nonconditioned areas of the building.

Lessons Learned from the Library of Museu Casa de Rui Barbosa

- A collections conservation environment can be maintained through ventilation and dehumidification strategies in a 1A climate.

- Special design considerations are needed to minimize leaks when components, such as supply diffusers and return air grilles, penetrate historic fabric, such as floors and walls.
- Devices such as supply air diffusers and return air grille boxes should be positioned where they are easy to install and service, in order to reduce problems such as air leakage.
- The process of meeting design conditions following installation is aided by a rigorous operational commissioning period and active environmental monitoring.
- Collections, building, and environmental condition assessments integrate various staff expertise into the system development process.
- Identification of a building's existing and often disregarded passive environmental management strategies can aid in the conceptual development of an active environmental management system.
- Over the course of a project, any decisions made during project meetings should be documented and confirmed by project team members.

Comments

The working framework of the project followed an outline that may be replicated by other environmental management projects. Initial condition assessments of the building, the collection, and the environment provided essential information for developing improvement strategies, and the assessments also integrated project members' expertise in various areas. Although an extended period of adjustment was necessary before the environmental management system operated near design specifications, continuous environmental monitoring highlighted the remaining problems and provided momentum to complete the needed system adjustments.

This project's position as one of the first in Brazil to implement alternative environmental management for a historic house museum makes it a potential blueprint for the wider application of humidity-based environmental control strategies that use ventilation and dehumidification while allowing the temperature to vary with the outside climate.

Postscript

MCRB continues to operate the environmental management system for the Library. Items previously stored in the basement have been removed, and the basement has been restored to its original function as a buffer zone. Visitor paths have been adjusted to reduce infiltration at entry doors. Following a period of system adjustment in 2011, the system's operation has been stable and has maintained interior humidity at the proposed range of 55% to 65% RH; the interior temperature rarely exceeds 28°C.

Several improvement issues requiring funding separate from the maintenance budget were identified with continued system operation and improved diagnostic experience. These issues include the replacement of the electroresistive reheat element with a hot-gas reheat coil that uses heat from the hot-compressed refrigerant gas generated by the evaporator (dehumidifier), thereby conserving energy; redesign of the supply diffuser and return grille boxes to minimize air leakage; and simplification of the PLC's control program for improved reliability and more effective dehumidification and cooling. Project details were published in papers from an architectural conference (Maekawa et al. 2009).

Project Team and Responsibilities

Fundação Vitae (São Paulo, Brazil)

Gina Machado (Project Manager)—Project manager of the project funding foundation

Franciza Toledo (Consulting Environmental Management Specialist, Recife, Brazil)—Project review, project documentation, and project audit.

Fundação Casa de Rui Barbosa (Rio de Janeiro, Brazil)

Claudia Carvalho (Architect)—Project manager and building assessment, design, and supervision of the nonmechanical strategy implementation

Carla Coelho (Architectural Intern)—Local project support and coordination.

Maria Cristina Jolly (Conservator)—Assessment of the book collection

Jose Manoel Pires (Curator)—Assessment of the museum collection

Ana Pessoa (Director)—Institutional Representative

Integrar Climatização S.A. (Rio de Janeiro, Brazil)

Richardo Barbosa—Engineering design of environmental management system and supervision of the system installation and commissioning

Marcos Barbosa (Technician)—Installation of environmental management system

Flavio Azevedo (Principal, Dia Dia, Rio de Janeiro, Brazil)—Programming and installation of the remote access for the environmental management system

Getty Conservation Institute (Los Angeles, United States)

Shin Maekawa (Senior Scientist)—Project design and management, concept design of environmental management strategies, environmental monitoring, and performance improvements

Vincent L. Beltran (Assistant Scientist)—Air pollution and particulate assessments; environmental data collection, analysis, and management; support of environmental monitoring system

References

AFCCC [Air Force Combat Climatology Center] and NCDC [National Climatic Data Center]
1999 *Engineering Weather Data*. Washington, DC: Air Force Combat Climatology Center. http://purl.access.gpo.gov/GPO/LPS7081.

ANSI [American National Standards Institute] and ASHRAE
2013a *Standard 55-2013—Thermal Environmental Conditions for Human Occupancy*. Atlanta: ASHRAE.

———.
2013b *Standard 62.2-2013—Ventilation and Acceptable Indoor Air Quality in Low-Rise Residential Buildings*. Atlanta: ASHRAE.

ANSI, ASHRAE, and IES [Illuminating Engineering Society]
2013 *Standard 90.1-2013—Energy Standard for Buildings except Low-Rise Residential Buildings* (SI edition). Atlanta: ASHRAE.

ASHRAE
2011 Museums, galleries, archives and libraries. Chapter 23 in *ASHRAE Handbook—HVAC Applications*. Atlanta: ASHRAE.

Beltran, V.
2007 Comparison of pollution and particulate data collected at the MCRB before and after installation of the climate control system. Getty Conservation Institute, unpublished report.

Maekawa, S., C. Carvalho, F. Toledo, and V. Beltran
2009 Collection care and human comfort for a historic house museum in hot and humid climates: An alternative to conventional air-conditioning approach. Paper ID 337, presented at PLEA 2009—26th Conference on Passive and Low Energy Architecture, Quebec City, Canada, 22–24 June, 2009.

Weather Channel
2014 Weather Underground. http://www.wunderground.com/.

Chapter 15

Juanqinzhai, the Qianlong Emperor's Retirement Studio in the Forbidden City, Beijing, China: Seasonal Dehumidification, Limited Cooling, and Recirculation for Filtration

Climate zone 4A (Mixed-Humid)

Objectives

- Prevent low levels of humidity that will pose a risk of mechanical damage due to material desiccation and embrittlement.
- Prohibit mechanical humidification to avoid the risk of condensation (as a result, the building would be unheated in winter).
- Reduce airborne particulates throughout the year.
- Create a conservation environment suitable for the historic collection and interiors, with thermal comfort a secondary consideration.
- Develop an environmental management system that is modest in scale as well as simple and inexpensive to install, monitor, and maintain.

Summary

Located in the Forbidden City in Beijing and in a Mixed-Humid (4A) climate zone, Juanqinzhai is an eighteenth-century building that served as the Qianlong Emperor's retirement studio. The massive timber post-and-beam assembly with masonry infill and tile roof sits on a slight mound at the northwest corner of Qianlong Garden and is an example of traditional, well-constructed Chinese buildings of the period. Although the building exterior is relatively simple and adorned with limited architectural detail, its interior contains fine decorative architectural finishes, such as wall and ceiling paintings on paper and silk, bamboo and jade inlays, and silk embroideries. The building has operable windows and doors on the south facade for ventilation in the summer, and floor heating for thermal comfort in the winter. The building was abandoned for more than a hundred years, and exposure to the elements resulted in damage to the historic interior. In addition, high-humidity conditions during the hot-humid season resulted in fungal infestation on the ground floor.

Under the collaborative restoration project between World Monuments Fund (WMF) in New York and the Palace Museum in

Beijing, Juanqinzhai's building envelope and its interior were restored in 2008, and its building envelope was classified as ASHRAE Class III/IV construction. As part of the restoration, uncontrolled infiltration through the building envelope was reduced by filling gaps between building components and entrance doors, and the thermal performance of the windows was improved. A mechanical environmental management system was installed in June 2008; it utilizes mechanical dehumidification to maintain the humidity below 55% RH, while allowing the temperature to track seasonal changes. The system has a cooling-only mode to limit the maximum interior temperature to 27°C on hot summer days. Winter heating was not installed to prevent the need for concurrent humidification during the comparatively dry winter. The system continuously recirculates interior air through a three-stage filtration system to remove small particulates. As part of the visitor management policy, tours are limited to small, supervised groups of scholars and special guests; all visitors must use disposable shoe covers, and tours are prohibited during the winter months. While designed for continuous operation, the environmental management system was in fact operated fewer than eight hours per day, and only five days per week during the first year, owing to the overnight fire prevention policy of the Palace Museum, which manages the Forbidden City. Despite this operational limitation, implementation of the environmental management strategy reduced the annual range of interior humidity from 10%–80% RH prior to installation to 40%–55% RH following installation. Airborne particulates were also significantly reduced.

Background

Reigning from 1735 to 1796, the Qianlong Emperor, the fifth emperor of the Qing Dynasty (1644–1912), was a well-known patron of the arts and crafts. The Qianlong Garden complex, constructed between 1771 and 1776, was designed for use as his retirement quarters. The complex is situated in the northeast section of the Forbidden City and consists of gardens and twenty-seven structures (fig. 15.1). Among these buildings is Juanqinzhai, or the Studio of Exhaustion from Diligent Service, a building unique for the quality and diversity of arts and crafts used in its construction, decoration, and furnishing. Use of the Qianlong Garden complex declined at the beginning of the twentieth century as the Qing Dynasty came to a close, and the buildings were abandoned during the Republic period (1912–19) and beyond. Over one hundred years of vacancy and neglect resulted in the deterioration of building envelopes and the historic interiors in Qianlong Garden. At the end of the twentieth century, the condition of Juanqinzhai was typical of other vacant buildings in the complex.

Restoration of the Qianlong Garden began in 2002 and was expected to span more than a decade. Juanqinzhai (fig. 15.2) was selected as the pilot restoration project in order to develop methodologies and techniques of conservation/restoration of the buildings' interior and exterior, as well as an environmental management strategy that could

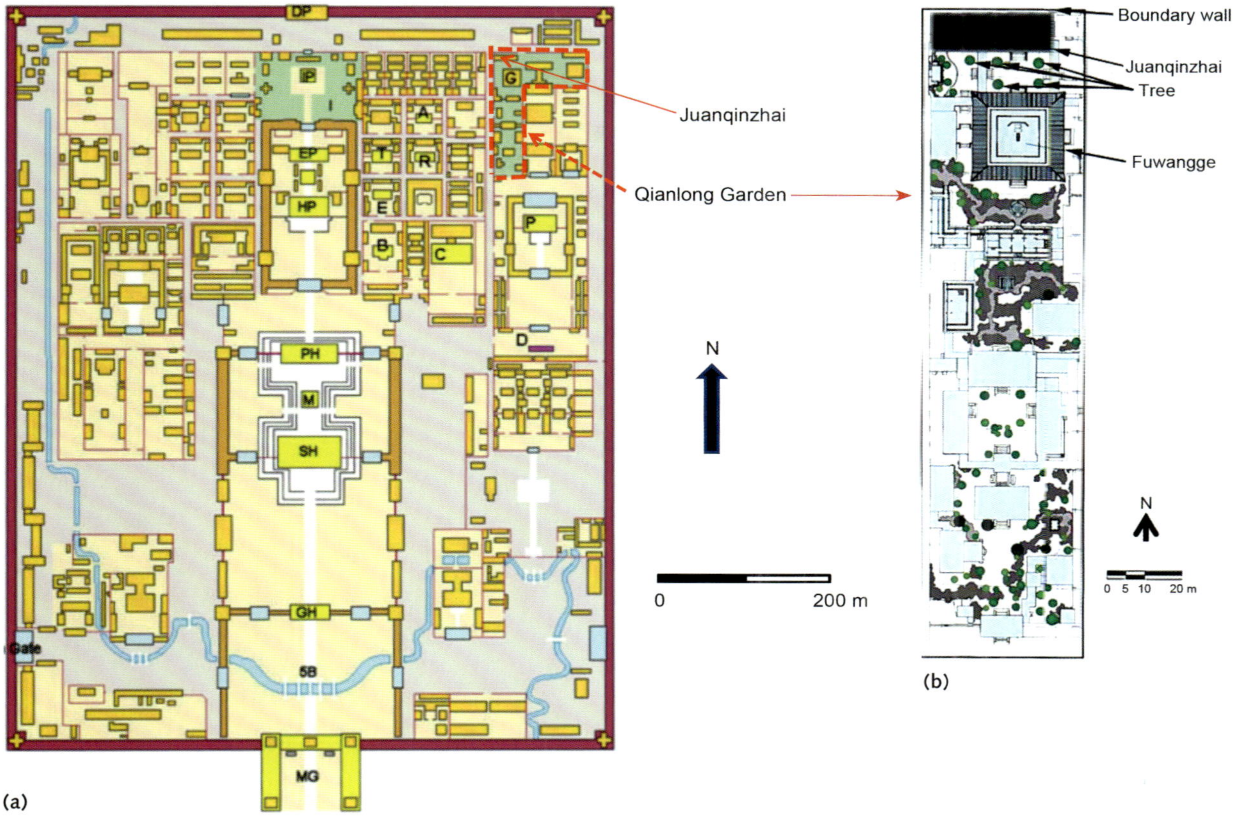

Figure 15.1
Maps of the Forbidden City showing the location of the Qianlong Garden complex (a) and the placement of Juanqinzhai within the complex (b). Images courtesy of Liu Chang.

be applied to the remaining buildings in the complex (Berliner 2008; McClintock 2014).

Climate

Beijing (39°54' N, 116°23' E; 44 m above sea level) is located at the northwestern end of the North China Plain, approximately 150 km northwest of the nearest coastline (fig. 15.3). Mountains border the city from the southwest to the northeast, and the Gobi Desert lies approximately

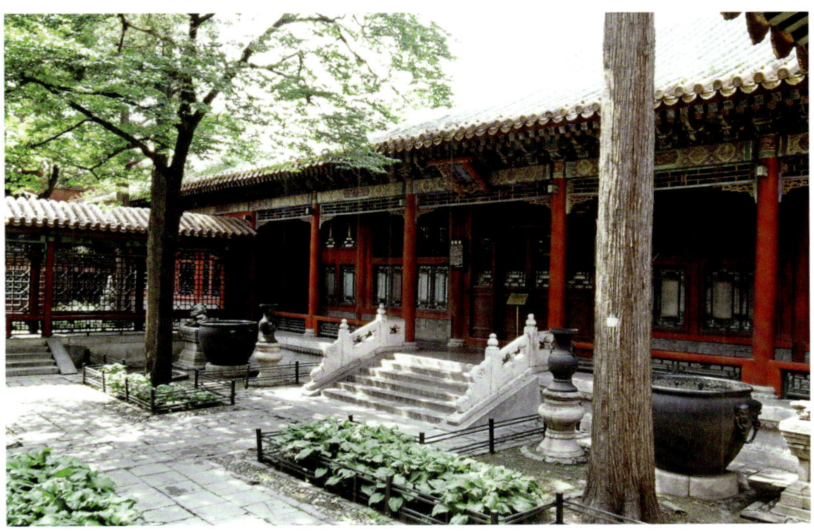

Figure 15.2
View of the south facade of Juanqinzhai. Photo: Shin Maekawa. © J. Paul Getty Trust.

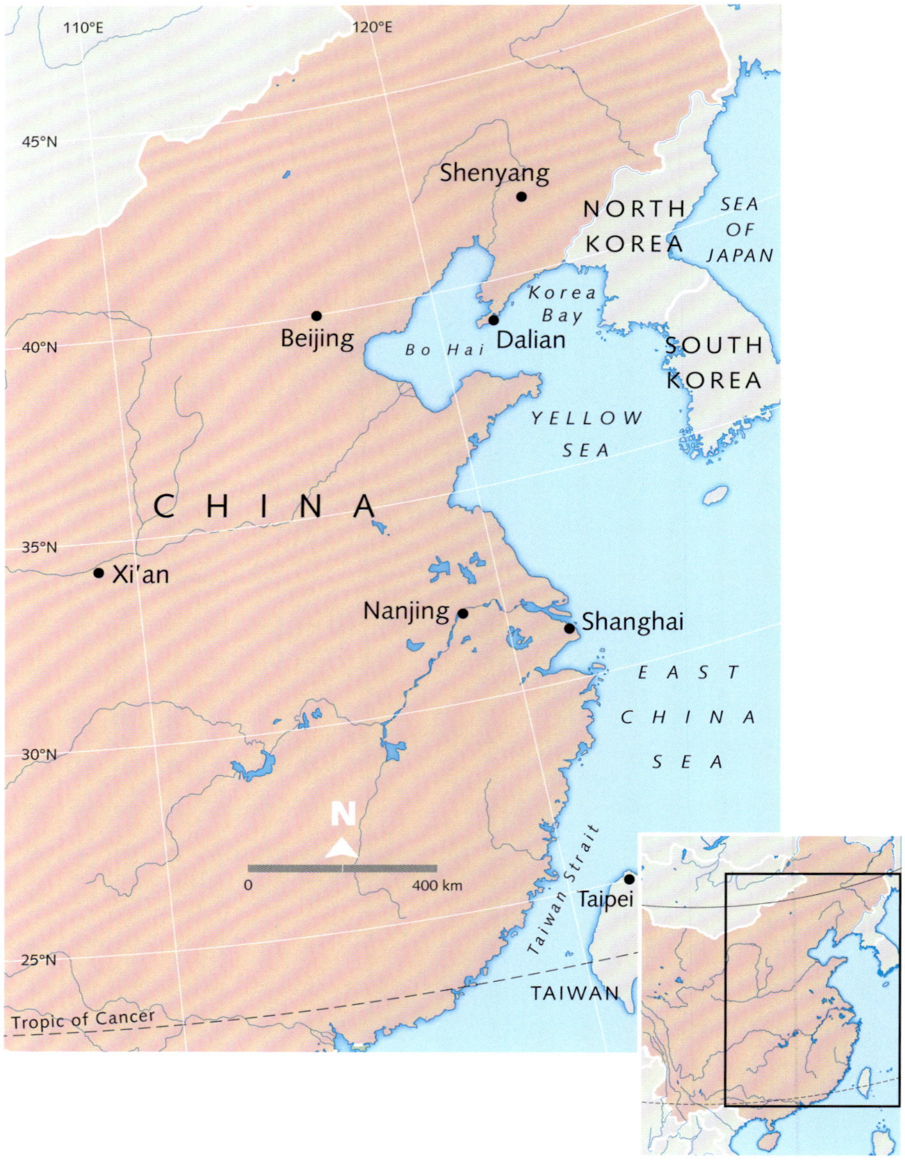

Figure 15.3

Map of eastern Asia showing the location of Beijing with respect to the Tropic of Cancer and the 45°N latitude.

250 km northwest of the city. Large variations in temperature and rainfall define two dominant climatic seasons in Beijing: a hot and humid season during the summer months, and a cold and dry season during the winter. The transitional spring and fall periods are generally dry, with dust storms coming from the deserts to the west and northwest of the city. The climate is classified as a humid continental climate (*Dwa*) by the Köppen-Geiger climate classification (see appendix 2). Monthly mean temperatures range from –2.8°C in January to 26.1°C in July, while monthly mean dew point temperatures range from –16.7°C in January to 20.6°C in July (fig. 15.4) (AFCCC and NCDC 1999). With an annual mean of 574 mm, monthly average rainfall peaks during the months of July (183 mm) and August (178 mm). Except for June (71 mm) and September (48 mm), monthly mean rainfall for the remaining months is less than 25 mm. These seasonal climatic variations provide a significant challenge to preserving both the Juanqinzhai building and its collections.

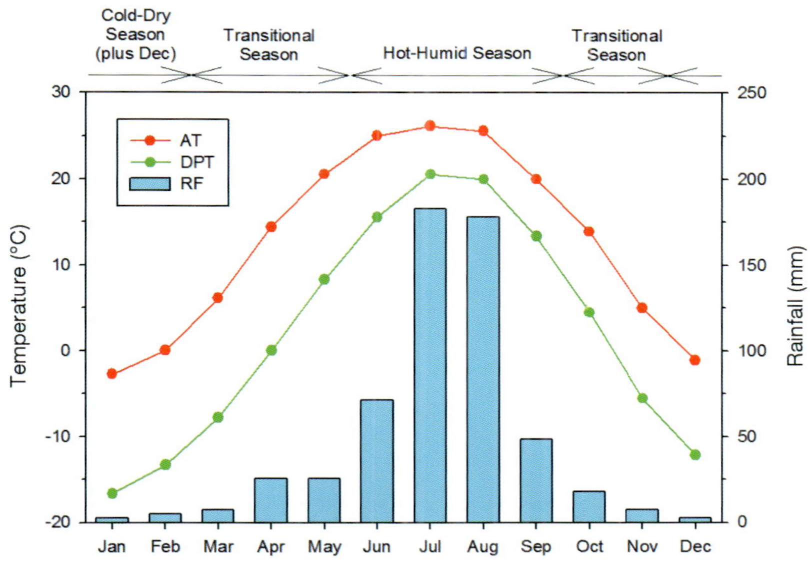

Figure 15.4

Historical weather data for Beijing: air temperature (AT), dew point temperature (DPT), and rainfall (RF). Source: Engineering Weather Data, 1973–96, National Climatic Data Center.

Analysis of a temperature dataset collected from 2006 to 2010 at the Beijing Capital Airport (40°04'48" N, 116°35'47" E; elevation, 30 m), located 22 km northwest of the Forbidden City, resulted in a CDD10°C of 2427 cooling degree days and HDD18°C of 2979 heating degree days (Weather Channel 2014). These values classify Beijing as a Mixed (4) climate zone, based on ANSI/ASHRAE Standard 90.1-2013 (ASHRAE, ANSI, and IES 2013). While a maximum mean monthly temperature (26.1°C in July) above 22°C excludes it from a Marine designation, annual rainfall (574 mm) is greater than the Dry-Humid Index of 394, placing Beijing's climate in the Humid-type category (A). As a result, Beijing is classified as a Mixed-Humid (4A) climate zone. Typical of many locations in a Mixed-Humid zone, this climate is seasonally hot and humid.

High concentrations of airborne particulates and gaseous pollutants are problematic in Beijing, which is often listed alongside Mexico City as one of the most polluted cities in the world. Abrasive sand transported by desert winds is a dominant particulate source, followed by particulates from fossil fuel combustion, domestic cooking, and heating activities. Mean particulate levels during the winter and summer were 296.6 µg m^{-3} and 259.9 µg m^{-3}, respectively (Yu 2005). Gaseous pollutants are also of concern in Beijing, as elevated levels of sulfur dioxide (90 µg/m^3) and nitrogen dioxide (122 µg/m^3) have been observed (WHO 2001).

Building

Juanqinzhai is located in the northwestern corner of Qianlong Garden (39°55'14" N, 116°23'37" E]; elevation, 52 m), which lies within the Forbidden City. Qianlong Garden was intended as the Qianlong Emperor's eventual retirement complex, and its perimeter is enclosed by massive (2 m thick) 8 m high masonry walls.

The north and west walls of Juanqinzhai are situated only 0.8 to 1.0 m from the perimeter walls of the Qianlong Garden. The structure of Juanqinzhai is aligned east–west and was constructed in the traditional method of timber post-and-beam assembly with masonry infill and limited exterior architectural features (fig. 15.5). The ground (first) floor is set 0.8 m above the ground, and the peak of the tiled gabled roof is approximately 8 m from the ground floor. Juanqinzhai is equivalent to an ASHRAE Class IV building type because of the heavy masonry construction of its west and north walls (ASHRAE 2011). However, the south wall houses numerous windows and is equivalent to an ASHRAE Class III building type. The building has a footprint of 224 m^2 and encloses a volume of 1120 m^3. It consists of two finished floors and an unfinished attic. The second floor and attic floor are located 2.3 m and 4.9 m above the ground floor, respectively. The ceiling of the second floor is made of several layers of mulberry paper adhered to wooden lattices.

The building's first floor, with its unique decorative finishes, is divided into two sections: the Reception Hall on the eastern half and the Theater on the western half (fig. 15.6). The Reception Hall is a complex that consists of a two-story-high entrance area on the south side, surrounded by small first- and second-floor rooms interconnected on the same floor level along the north and east walls. The Theater complex consists of a two-story-high open space with a stage at the center of the west end. The stage structure is set under its own inside roof, which reaches to the second floor and has an adjacent musicians' box. A two-level viewing area for the emperor is located at the east end of the

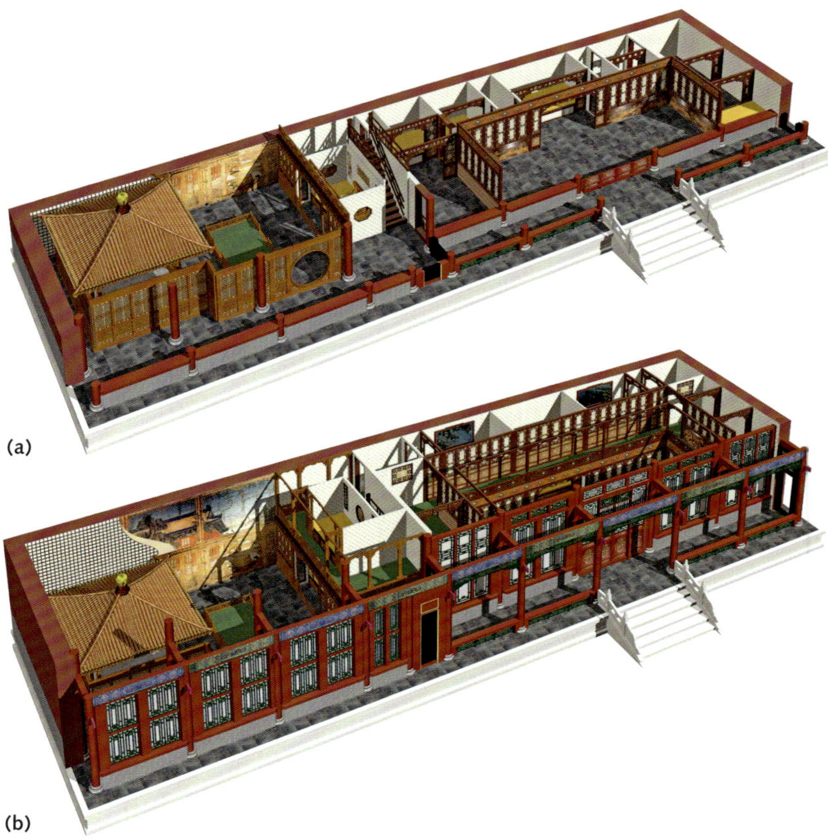

(a)

(b)

Figure 15.5

Computer-rendered images of the first-floor (a) and second-floor (b) interiors of Juanqinzhai. Images courtesy of Liu Chang.

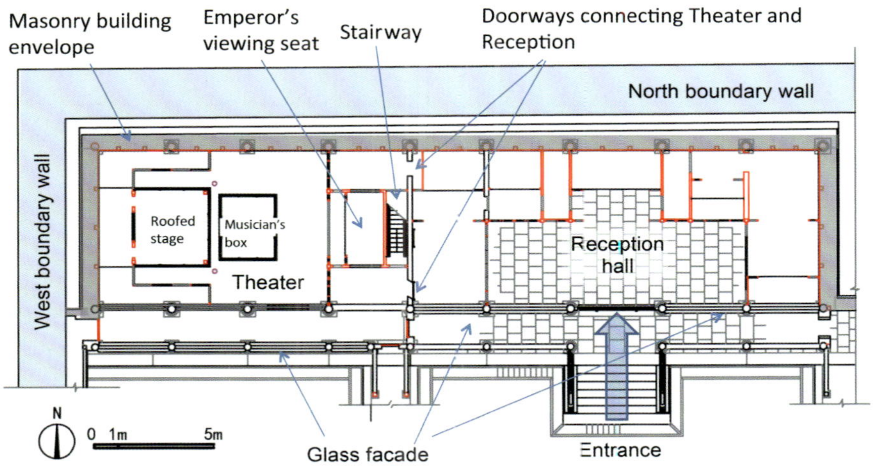

Figure 15.6
Juanqinzhai's first-floor plan showing the Reception Hall (right) and the Theater (left).

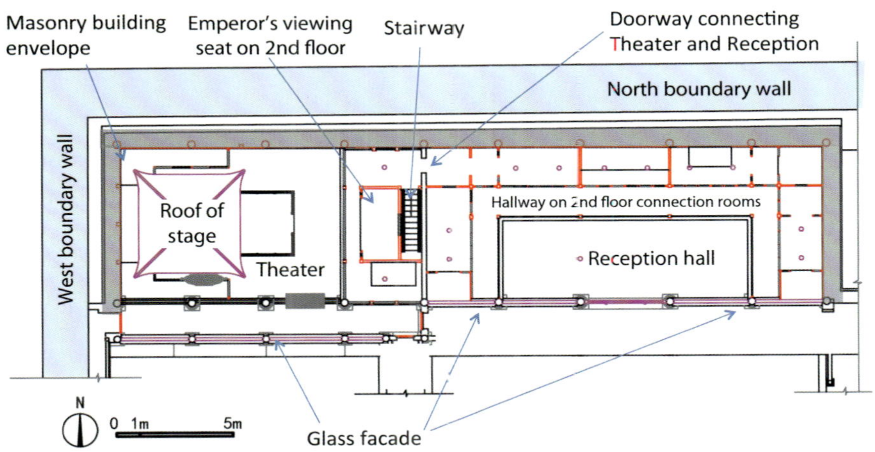

Figure 15.7
Juanqinzhai's second-floor plan.

Theater, set against the wall separating the Theater from the Reception Hall. The first floor on both sides is connected by two doorways approximately 1 m wide and 2 m high.

The second floor is divided in the same manner as the first floor (fig. 15.7). The Theater and the Reception Hall are connected at this level by a 3 m wide opening in the north end of the wall separating these two sections. The Reception Hall side contains small rooms connected by a hallway that surrounds a low barrier around the open entrance area. The rooms can also be accessed by a steep and narrow (approximately 1 m wide) stairway behind the Emperor's second-floor theater viewing box. The attic, accessible through small trap doors in the ceilings above the Theater and Reception Hall, is an unfinished open space without a segregating wall, and the massive timber structural frames are exposed (fig. 15.8).

Originally, coal braziers in subgrade cavities under the Emperor's first-floor seating area, along the north wall, provided underfloor heating in winter. During summer, the large thermal mass of the building's north, east, and west walls and the shading of the south facade provided by overhanging roof eaves, tree canopies, and a large adjacent

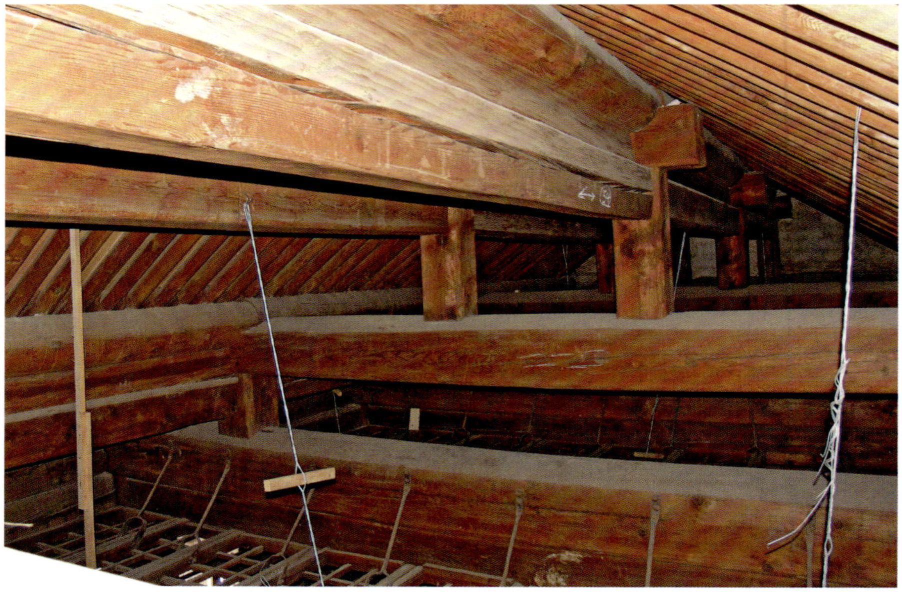

Figure 15.8
Structural timber frames supporting Juanqinzhai's roof, seen in the open and unfinished attic. Photo: Shin Maekawa. © J. Paul Getty Trust.

building (Fuwangge, to the south of Juanqinzhai) reduced solar gain and cooling loads (see fig. 15.1b). Large operable windows and doors in the south facade provided ventilation for additional cooling.

Collection

Though lacking an object-oriented collection, Juanqinzhai contains numerous interior decorative architectural finishes in the Reception Hall and Theater complexes, including the following:

Reception Hall

- polished ironwood used for exposed timber framing
- casement windows with double-sided embroidered silk panels (fig. 15.9a)
- bamboo-thread marquetry over which lacquer laminate was applied and carved and to which jade inserts were added (fig.15.9b)
- wall panels with naturalistic scenes carved in low relief and covered with bamboo veneers
- landscape paintings and works of calligraphy mounted directly on the wallpapered walls

Theater

- Roofed theater stage (fig. 15.10) with a dramatic two-story viewing area (fig. 15.11)
- Wood surfaces painted to resemble a special bamboo wood
- Trompe l'oeil scenes on painted silk attached to the ceiling and walls

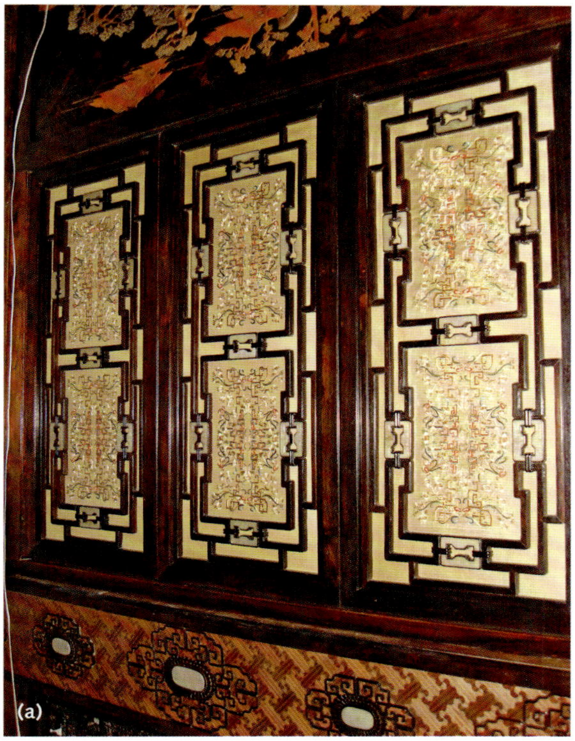

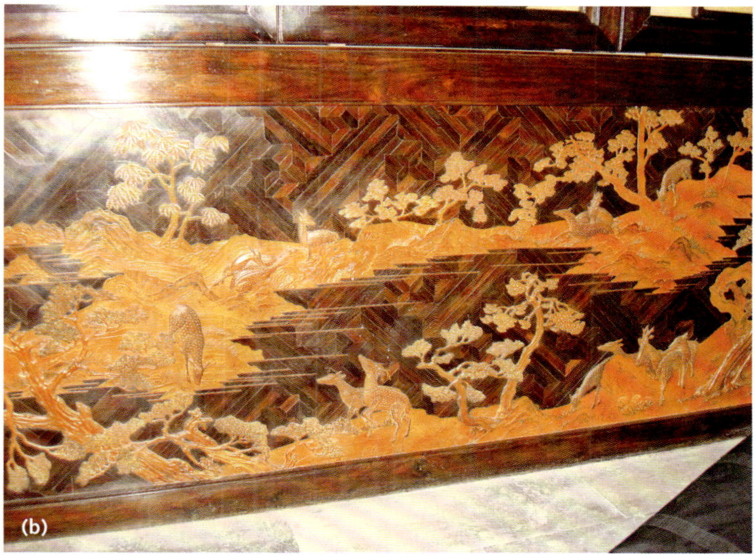

Figure 15.9
Double-sided embroidered silk panels with jade inserts used in casement windows (a), and extruded bamboo-thread marquetry with carved lacquer laminate and jade inserts (b). Photos: Shin Maekawa. © J. Paul Getty Trust.

Preinstallation Condition Assessments

Building envelope and historic interior assessment

Mecklenburg and Liu reported that structural members of Juanqinzhai were generally well preserved (Mecklenburg and Liu 2005). However, the sashes and frames of the operable single-glazed windows on the south facade were warped and had poor closure, allowing high infiltration of outside air. The doors to the entrance area of the Reception Hall showed similar issues with fit. Due to material shrinkage, gaps and loose fits could be observed for many joints between wood members. Joints between roof rafters and wall beams and between base beams of the roof also had large gaps (fig. 15.12). Since these timber joints were in the south facade and the roof plane of the envelope, they presented an air infiltration problem. Walls on the east, north, and west sides of the building were made of approximately 1 m thick masonry and were in good condition with no infiltration.

Exterior and interior environmental assessments

Prior to the design of an environmental management system for Juanqinzhai, exterior and interior temperature and humidity data were collected at 15 minute intervals from March to December 2005 using dataloggers placed by Tsinghua University. Though four seasons—hot-humid, spring transitional, cold-dry, fall transitional—have been recognized, the pre- and postinstallation environmental assessment conducted here will separate the dataset into two seasons: hot-humid, extending

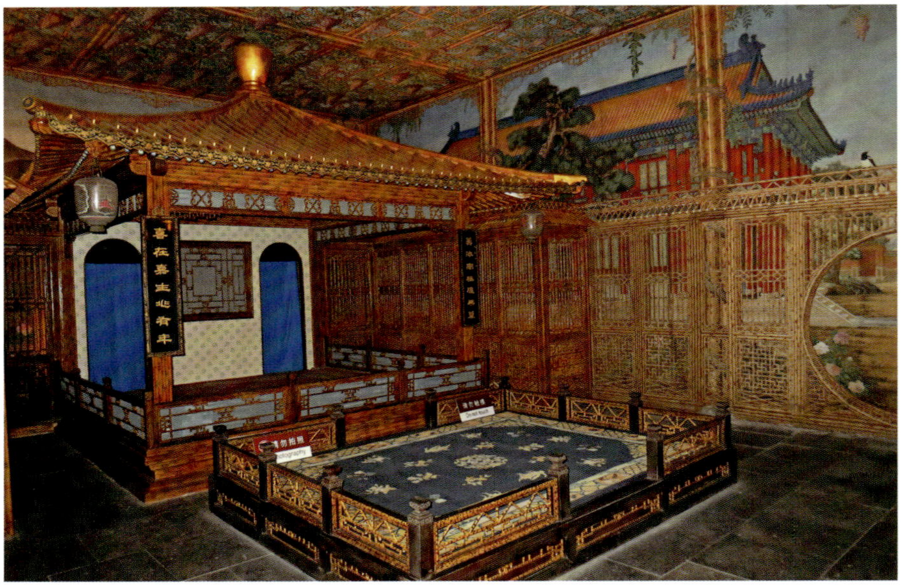

Figure 15.10
Roofed theater stage with wall and ceiling trompe l'oeil paintings in the background. Photo: Shin Maekawa. © J. Paul Getty Trust.

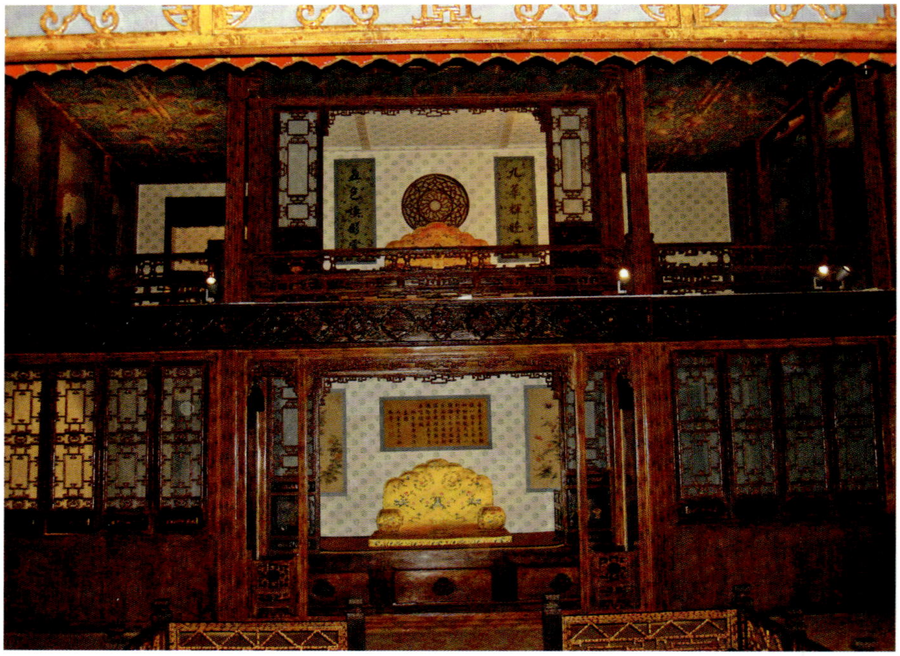

Figure 15.11
The emperor's first- and second-floor theater viewing boxes. Photo: Shin Maekawa. © J. Paul Getty Trust.

from May to September, and cold-dry, extending from October to April. The main purpose of this is to align this case study's data presentation with those from the other case studies and to simplify application of the Conservation Environment Classification–HH protocol, particularly with respect to mechanical risk due to seasonal variation.

Conducted less than a full year, the Tsinghua University study was missing data from January and February, the most extreme months during the cold-dry season. While three locations in the building were monitored, only data collected at the central position was usable since the other datasets were incomplete. Processing and interpretation of the data were performed at the Getty Conservation Institute.

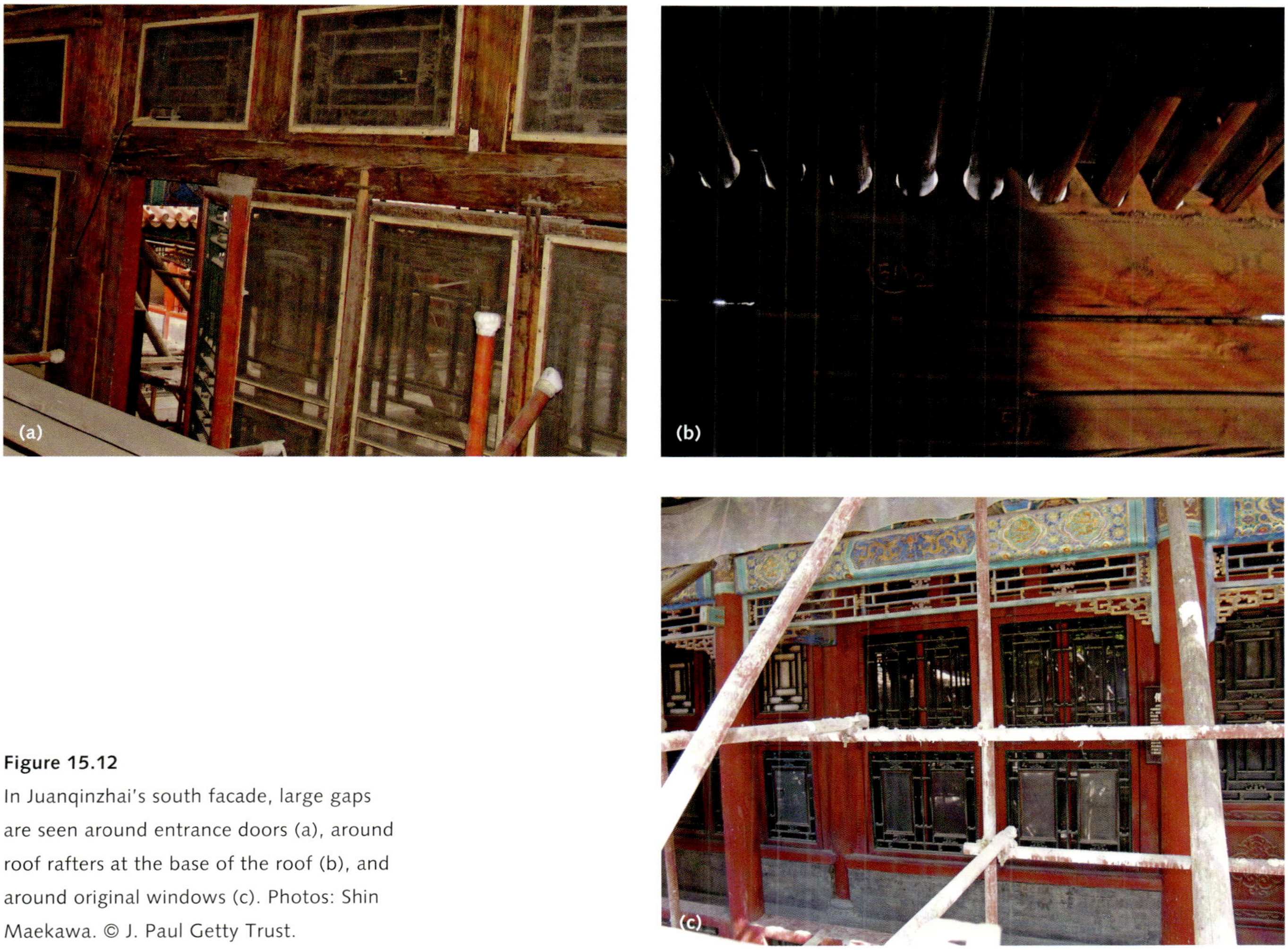

Figure 15.12
In Juanqinzhai's south facade, large gaps are seen around entrance doors (a), around roof rafters at the base of the roof (b), and around original windows (c). Photos: Shin Maekawa. © J. Paul Getty Trust.

During the preinstallation period, monthly mean interior humidity ranged from 17% RH (December) to 75% RH (August), which was similar to the range of monthly mean exterior humidity (26% RH in December to 79% RH in August) (fig. 15.13a). The monthly average rolling 24 hour variation for interior humidity ranged from Δ6% RH (December) to Δ10% RH (July), while that of the exterior ranged from Δ10% RH (May) to Δ27% RH (October). Although interior humidity remained below 75% RH during much of the year, interior humidity exceeded the 75% RH threshold during roughly 40% of both July and August. Conversely, interior humidity frequently dipped below the low threshold of 25% RH during December.

The monthly average interior temperature ranged from 0.2°C (December) to 26.1°C (July), and the monthly average interior dew point temperature ranged from –22.1°C (December) to 22.2°C (July) (fig. 15.13b and c). These monthly mean values were again similar to those recorded for exterior air temperature (–0.7°C in December to 26.0°C in July) and exterior dew point temperature (–18.3°C in December to 20.8°C in July). The monthly mean 24 hour variation ranges for interior

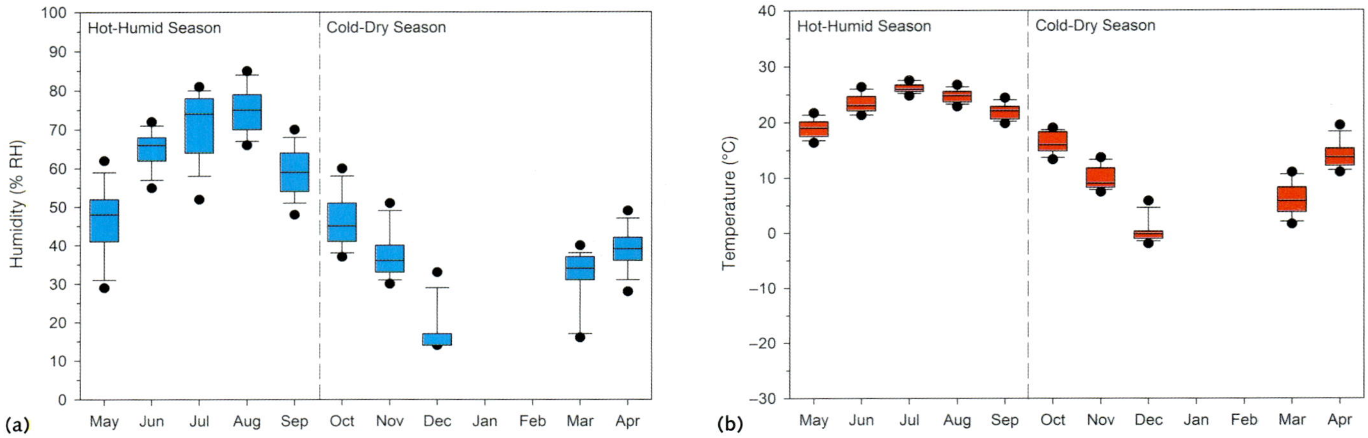

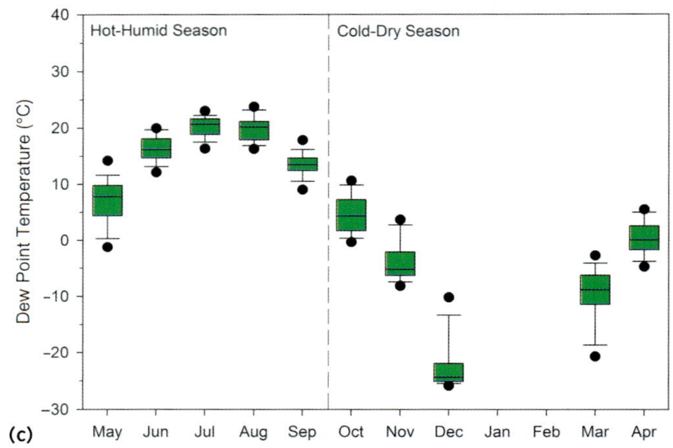

Figure 15.13

Preinstallation humidity (a), temperature (b), and dew point temperature (c) measured from March to December 2005 in Juanqinzhai.

temperature were Δ0.9°C (November) to Δ1.8°C (May), and the monthly mean rolling 24 hour variation ranges of interior dew point temperature were Δ1.9°C (August) to Δ3.8°C (May). Although monthly interior mean values closely tracked with the exterior, the range of monthly mean rolling 24 hour variations in interior temperature and dew point temperature were narrower than those of the exterior (monthly mean rolling 24 hour exterior temperature, Δ1.3°C in July to Δ3.0°C in October; monthly mean rolling 24 hour exterior dew point temperature, Δ2.8°C in July to Δ8.0°C in November).

Collections and decorative interior condition assessments

Assessments of the collections and the decorative interior were jointly performed between the staff members of the Palace Museum and members of the WMF team. Painted silks on ceilings and wall paintings were torn in many places and separated from their lattice supports. Lower portions of wallpapers were water stained. Painted finishes on wood had discolored, cracked, and detached from the substrate. Bamboo veneers were separated from supports. All interior surfaces were heavily soiled, and there were thick deposits of dust throughout the building.

The interior condition assessment indicated that the major cause of deterioration was repeated large variations in interior humidity. This led to the recommendation that humidity be limited to a range between 30% and 65% RH to minimize mechanical stress and damage to the building and decorative materials. The recommendation also included a winter minimum temperature of 10°C to maintain material ductility. Based on the recommended environmental targets, it was determined that implementation of a mechanical environmental management system was necessary to maintain the restored building interior.

Prior to installation of the environmental management system at Juanqinzhai, the deteriorated historic interior materials and finishes were restored or conserved as a part of the restoration project (Berliner 2008).

Preinstallation risk class assessment

The preinstallation Conservation Environment Classification–HH for Juanqinzhai is summarized in table 15.1.

Environmental Management System

Objectives

The aim of the environmental management strategy at Juanqinzhai was to establish and maintain a stable conservation environment for the building's interior decorative finishes, historic decorative art collections, and the structure itself. The thermal comfort of visitors was considered a secondary concern, since future tours of the building would be limited to small groups of scholars and special guests. An integrated environmental

Risk		Humidity		Temperature		
Category	Specific Risk (statistic)	Value (% RH)	Class	Value (°C)	Class	
Microbial	Germination threshold (P97.5)	80	F	—	—	
Mechanical	Short-term variation (P95 of rolling 24 hour variation)	Δ20	b	Δ2.4	a	Overall mechanical risk:[1] f
	Seasonal variation (absolute difference of seasonal means, RH means ≦ 70%)	Δ27	f	Δ13	b	
	Deviation from historical mean (absolute difference)	Δ0	a	Δ4	a	
Chemical	Deviation from historical mean (difference)	—	—	Δ4	+	

[1] Overall mechanical risk is based on the highest risk classification in any of the specific humidity- and temperature-based mechanical risks (short-term variation, seasonal variation, and deviation from historical mean).

Table 15.1

Preinstallation Conservation Environment Classification–HH for Juanqinzhai.

management strategy was proposed that combined minor modifications to the building envelope, the installation of a minimally intrusive mechanical system, and a conservation-focused visitation management policy.

The environmental strategy focused on the management of the interior temperature and humidity by reducing daily and seasonal variations and limiting extreme conditions. The following objectives were developed:

- Prevent high levels of humidity that will pose a risk of fungal damage.
- Prevent low levels of humidity that will pose a risk of mechanical damage due to material desiccation and embrittlement.
- Prohibit mechanical humidification to avoid the risk of condensation (as a result, the building would be unheated in winter).
- Reduce airborne particulates throughout the year.
- Create a conservation environment suitable for the historic collections and interiors, with thermal comfort a secondary consideration.
- Develop an environmental management system that is modest in scale and simple and inexpensive to install, monitor, and maintain.

The target humidity in Juanqinzhai was allowed to change with the season. Humidity would be passively controlled to higher than 30% RH in winter and actively controlled to lower than 60% RH in summer. This summer set point, which was less than the 65% RH target recommended by Mecklenburg and Liu (2005), was chosen to accommodate possible spatial variation of humidity in the building, as well as to provide an additional buffer against prolonged equipment or power failure. The interior temperature will also be allowed to passively follow seasonal changes of the outside. However, it would not exceed 27°C. Though the building would not be pressurized, which would prevent the infiltration of outside unconditioned and unfiltered air, the environmental management system would use filtration to continuously clean the indoor air.

Psychrometric strategies

In implementing the environmental management strategy at Juanqinzhai, the initial effort focused on achieving stable interior moisture levels by reducing temperature variations and limiting the infiltration of unconditioned outside air through the building envelope. Infiltration of the outside air through opening of the entrance doors would be minimized by limiting visitation.

Highly humid conditions typically only occur during the hot-humid season. Dehumidification would be utilized during this season to control humidity while limiting the temperature increase in an already warm interior. When the interior temperature exceeds a set value, dehumidification would be disabled and cooling would be enabled to limit interior temperature.

During the cold-dry season, heating would not be used so as to avoid further reduction of humidity in the interior environment, which would instead be passively conditioned by the moisture contained in the building fabric and interior. Since the visitor policy for Juanqinzhai restricts visitation during the cold-dry season, thermal comfort would not be a concern during the winter months.

Nonmechanical strategies

Envelope improvements

Thermal insulation was added in areas where it did not affect visual aesthetics, such as windows and external doors. Double-glazed windows with internal shades and ultraviolet filtration replaced existed single-glazed windows and were fitted into the original decorative grilles of the windows. Air infiltration was reduced by fitting gaps around entry doors and door frames with thin wood strips wrapped with sponge material and covered with fabric that matched the historic context. Air infiltration was further reduced by injecting polyurethane foam into gaps in the open joints of the historic timber framing, such as the rafters and base beams of the roof (fig. 15.14).

The air change rate of the building was not measured prior to implementation of the envelope improvements. Considering the many openings in the existing building envelope, however, the preimprovement air change rate of Juanqinzhai was estimated to be higher than two air changes per hour (ACH). The improvements to the building envelope successfully reduced this rate to less than 0.5 ACH in both the Reception Hall and Theater complexes (Tsinghua University, April 2012). These improvements upgraded the overall envelope performance from ASHRAE building Class III/IV to Class IV.

Visitor management

Several visitor management strategies were considered to minimize the impact of visitation to Juanqinzhai. Since the historic collections and interior decorative finishes lack physical protection from visitors, the Palace

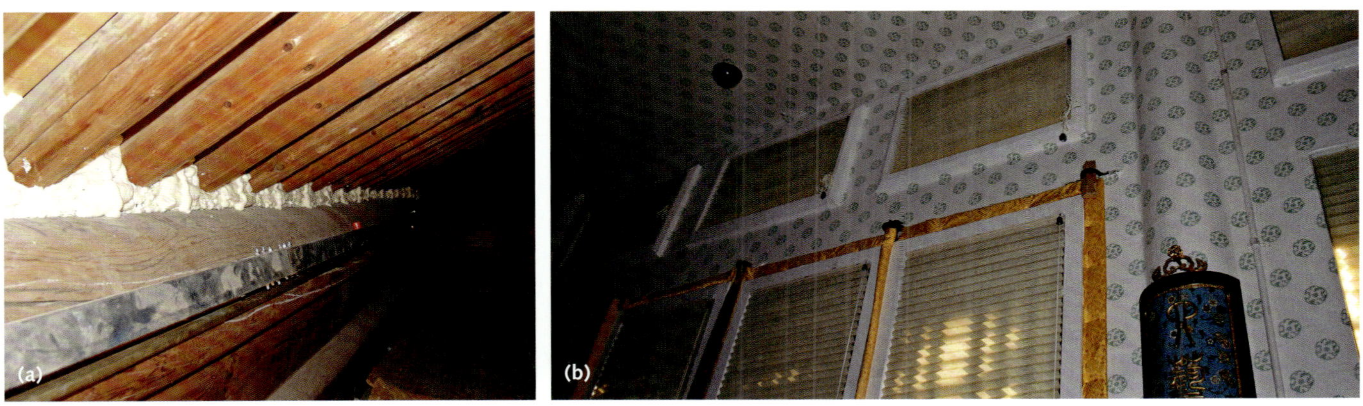

Figure 15.14
Large gaps in timber framing, such as around roof rafters at the base of the roof, were sealed with polyurethane foam to reduce air infiltration (a). To limit infiltration through gaps around entrance doors, synthetic sponge material was stapled to thin wood strips, wrapped in a yellow synthetic cloth with a historic pattern, and mounted along the door frame edges (b). Photos: Shin Maekawa. © J. Paul Getty Trust.

Museum management decided that access to the building's interior will be restricted to a small number of supervised group tours; tours will be prohibited during winter months to avoid vibrations to surfaces that might be embrittled by low humidity or low temperature, and visitors must wear disposable shoe covers to minimize the introduction of particulates and limit abrasion of floors and floor coverings. General visitors to the Qianlong Garden can also view the interior of Juanqinzhai through its windows.

Mechanical strategies

The mechanical strategy addresses the overall environmental management objectives by:

- stabilizing the building's humidity with respect to both short-term and annual fluctuations,
- allowing monthly average humidity to gradually drift from a winter low of 30% RH to a summer high of 60% RH,
- filtering air in the building to reduce deposition of particulates on building interior and collections,
- allowing interior temperature to drop in winter in order to avoid going below the low humidity threshold,
- limiting the interior temperature in summer to a maximum of 27°C.

During winter, operation of the environmental management system filters the air but does not provide heat. During summer, the air handler operates to both filter and dehumidify or cool the air as needed. The system is controlled by a programmable controller, which monitors humidity and temperature in the building.

Central air-handler unit

The mechanical system includes a central air-handler unit (AHU) located in a new mechanical room created by the installation of two partitions in the middle of the attic. This mechanical room divides the attic into two large plenums—one for supply air and one for return air. The AHU circulates conditioned air throughout the whole building by directing supply air into the Theater side of the building while drawing return air from the Reception Hall at the building's opposite side (fig. 15.15). The AHU utilizes short runs of supply and return air ducts within the mechanical room, with the ducting employing sound-silencing attenuators and layouts designed to maximize acoustic absorption.

The environmental management system includes a duct-mounted inline dehumidifier with a direct expansion (DX) cooling coil and a DX hot-gas reheat coil with an auxiliary exterior condenser for heat rejection (see description in chap. 7).

The air-handler unit capacity is rated for 6720 m^3 of air per hour and can recirculate air within the building at a rate up to 6 ACH, allowing particulates to be removed with each pass through the AHU's filters. There is no outside air component in the environmental management

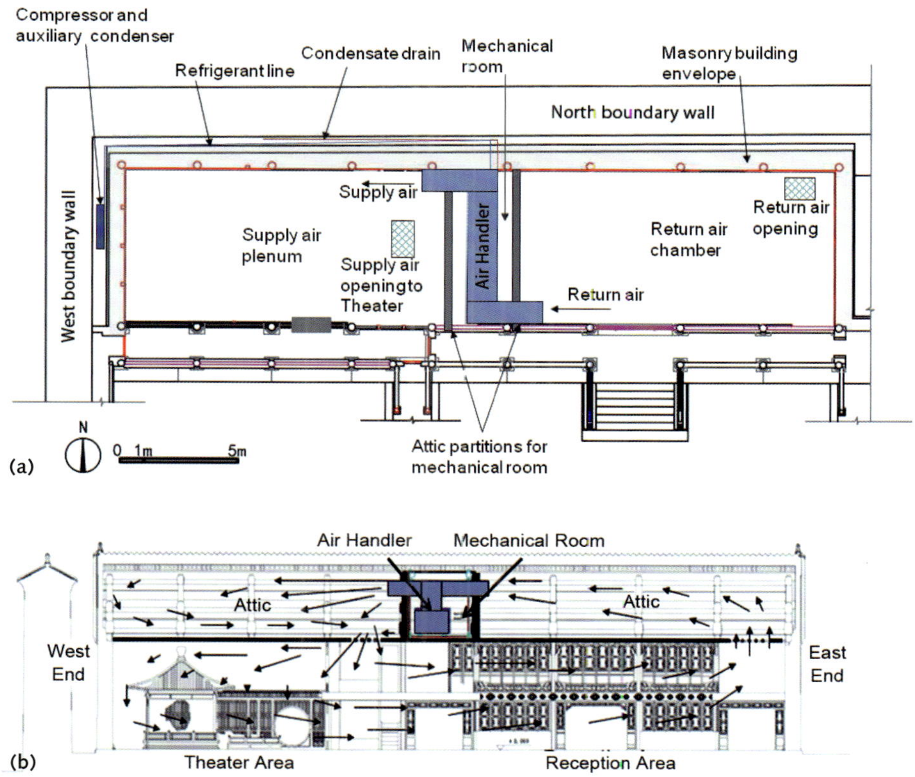

Figure 15.15
Attic floor plan showing the mechanical room and the locations of environmental management system components (a), and the airflow of conditioned air in the building (b). Schematic (b): Shin Maekawa. © J. Paul Getty Trust.

strategy, since the building will not have regular occupants or visitors. Beginning at its return air inlet side, the AHU consists of: two stages of particulate filtration (a disposable pleated MERV 8 filter and a washable MERV 12 bag filter), followed by the dehumidification unit (DX cooling coil and DX hot-gas reheat coil); next, a disposable HEPA box filter (MERV 17), and ending with the fan at the air supply outlet. The AHU also houses spare compartments reserved for future functions if deemed necessary (e.g., additional heating coils and a humidifier) (fig. 15.16). The DX compressor is located outside the building on the west wall, along with the auxiliary condenser for heat rejection to the outside (see fig. 15.15a). The reheat coil utilizes refrigerant hot gas to maintain supply air humidity at its set point of 60% RH, subject to a maximum return air temperature of 27°C. The mechanical system equipment operates with microprocessor-based controls. Table 15.2 lists specifications for the environmental management equipment.

The supply and return air grilles (fig. 15.17) at the duct penetrations through the mechanical room walls are sized for low velocity and are concealed from a visitor's view. Refrigerant lines between the AHU and the exterior condenser/compressor units are routed to the outside of the building along the north wall to avoid placement above the Theater space and to conceal them from view (fig. 15.18a).

Supply air circulates through the west end of the attic and enters the Theater space through a large supply air diffuser located in the ceiling above the emperor's second-floor seat (fig. 15.18b). While

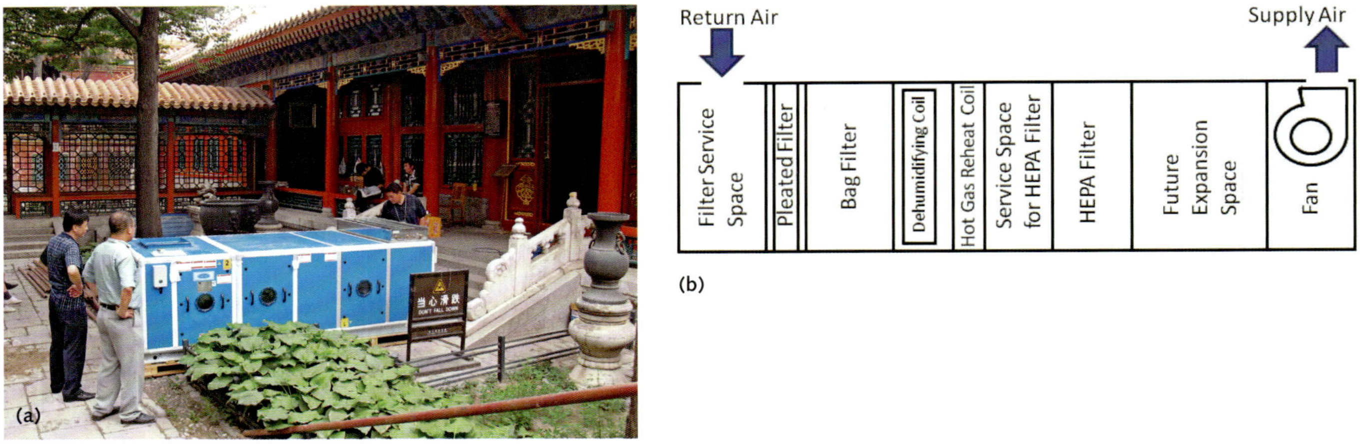

Figure 15.16
Preassembled air-handler unit being prepared for installation at Juanqinzhai (a), and the AHU schematic (b). Photo: Courtesy of Studio TKM.

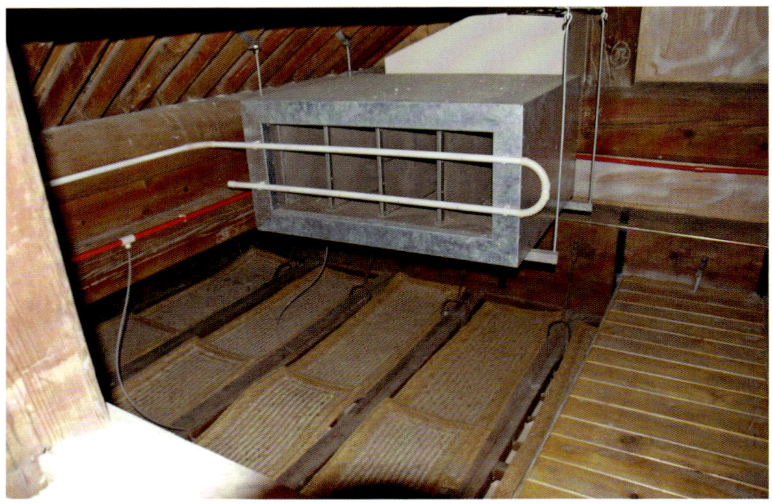

Figure 15.17
Attached to the eastern wall of the machine room in the attic, the return air grille uses perforated airfoils as a noise suppression device. Photo: Shin Maekawa. © J. Paul Getty Trust.

Table 15.2

Specifications for the environmental management system at Juanqinzhai (NA = not available).

Components	Type	Capacity	Static Pressure (Pa)	Power Use	Noise (db)
		Inside Units			
Cooling	Cold refrigerant gas (R22)	21 kW (6 tons)	25 Pa		
Heating	Hot refrigerant gas (R22)	19 kW (5.5 tons)	25 Pa		
Fan	Variable speed centrifugal	6720 m^3/hr	300 Pa	3 kW	72
Filters	Pleated filter	MERV 8			
	Bag filter	MERV 12			
	HEPA filter	MERV 17			
		Outside (Auxiliary) Units			
DX Compressor					65
Fan (cooling) for auxiliary condenser	Propeller	4680 m^3/hr	75 Pa	8.3 kW	NA

placement of the supply air diffuser (see fig. 15.15) at the west end of the Theater ceiling would have been ideal, the presence of the highly significant trompe l'oeil ceiling painting at that position restricted placement of the supply air diffuser at this more central location.

Conditioned air flows from the Theater space to the Reception suite through corridors on the first and second floors. The return air grille is located in the ceiling of a second-floor room at the northeast end of the building, whereby return air enters the attic above the Reception suite and flows to the return air grille in the east wall of the mechanical room.

AHU electrical connections and alarm system

The environmental management system includes an electrical service connection, a service switch, a distribution panel, remote on/off/auto control, local wiring within a conduit, local disconnects, and wiring to the air handler. Sizes of the electrical connections were chosen in accordance with local codes and are suitable for safe operation of the AHU and its controls under all conditions. In case of smoke, a smoke detector located in the AHU will shut down the unit and trigger an alarm in the fire detection and alarm system.

Condensate drain and water leak protection

A small drain pan is located below the cooling coil in the AHU, where it collects condensate. The water is drained by gravity through a pipe to an underground drain behind the north side of the building. Large, flat stainless steel secondary spill pans are also placed beneath the AHU, the condensate drain, and refrigerant lines. These pans are fitted with water sensors that trigger an audible alarm and shut off power to the AHU if water is detected.

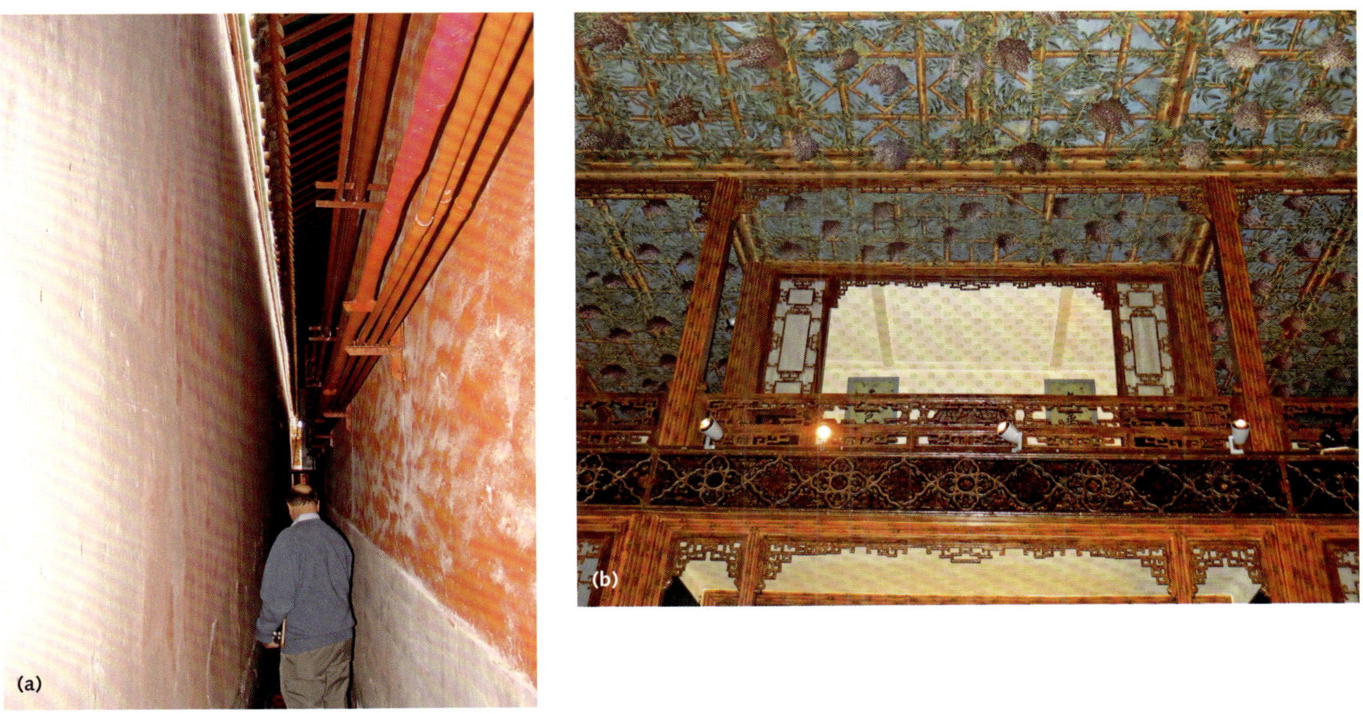

Figure 15.18
Refrigerant lines and electrical conduits are routed along the north wall of Juanqinzhai, which is only 1 m from the wall surrounding the Qianlong Garden (a); air supply ducts are hidden in the ceiling above the emperor's second-floor seat (b). Photo: Shin Maekawa. © J. Paul Getty Trust.

Noise and vibration

The mechanical room, in which the AHU is housed, is lined with sound-absorbing material to reduce the transmission of operational noise to visited spaces. The AHU is mounted on a vibration-isolated frame suspended from beams supporting the roof (a safe level of loading was confirmed by a structural engineer), eliminating vibration transmission through internal walls and the ceiling. The supply and return air openings of the AHU contain four airfoil-shaped wings, made of perforated aluminum sheets, to suppress noise, and the ducts are configured with right angles to reduce sound transmission through the ducts. The ducts are connected to the air handler with flexible boots to reduce transmission of AHU vibrations into the ductwork. The air velocity in the air handler, air ducts, and diffuser grilles is limited to 2.5 m/s to mitigate aerodynamic noise.

Control system

The AHU is controlled by a microprocessor-based system installed in the mechanical room along with temperature and humidity sensors positioned in the return duct. Control components include a central processing unit (CPU), several input and output (I/O) modules, sensors and switches, a data collection/storage module, an Ethernet communication module, and a keypad and display unit (fig. 15.19a). The CPU, I/O modules, and data storage module are placed in a metal cabinet adjacent to the AHU in the mechanical room. Sensor readings are received at the I/O module and forwarded to the CPU module for processing. Commands are then sent to the I/O modules for activating or deactivating various system components. Meanwhile, operational data from the system and sensor readings from the system sensors are stored in the data collection/storage module. The keypad and display unit, mounted on the second-floor wall directly below the attic mechanical room, allow viewing of sensor readings and viewing and modification of all operational parameters. The Ethernet communication module allows a dedicated PC (with proprietary software) in the Palace Museum's network to access the system. Remote PCs, such as one at a maintenance contractor's office, can access the system by logging into the dedicated PC in the Palace Museum via remote access software.

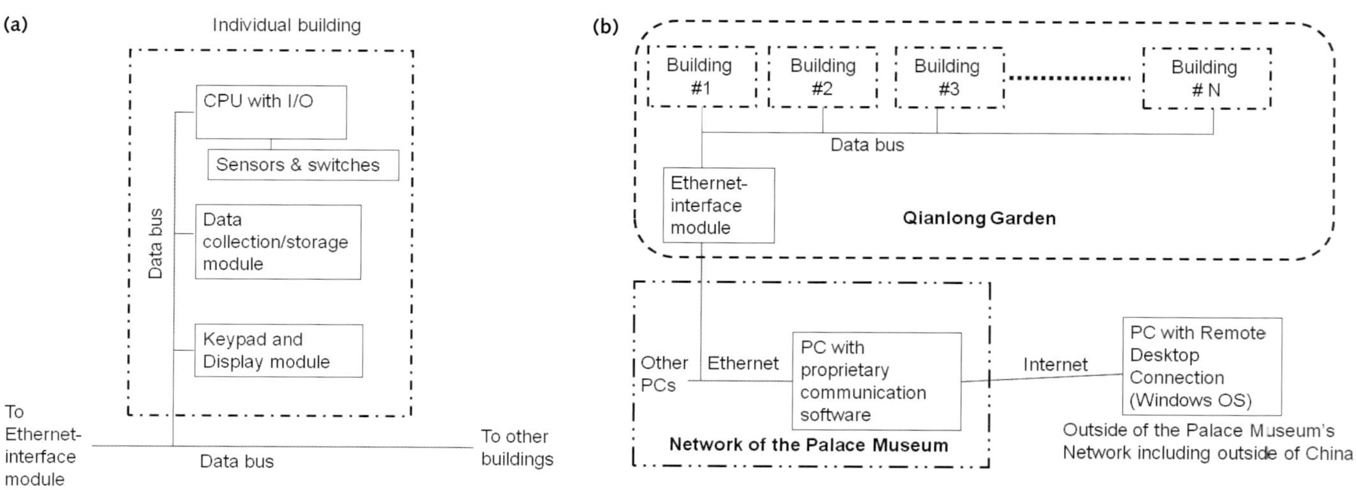

Figure 15.19

Juanqinzhai's control system (a), and the proposed network of control systems for several buildings in the Qianlong Garden complex (b).

In the future, the Juanqinzhai control/communication configuration will be duplicated in other buildings in Qianlong Garden (fig. 15.19b). The intent is to include all the buildings in the local area network (LAN) of the Palace Museum, where one dedicated PC will be able to access and control the operation of all environmental management systems in Qianlong Garden.

Operational control sequence

As summarized in table 15.3, the Juanqinzhai environmental management system has three operational modes: the recirculation-only mode, the dehumidification and recirculation mode, and the cooling and recirculation mode. The recirculation-only mode is enabled when return air temperature is below 15°C (winter conditions) and is disabled when the return air temperature equals or exceeds 15°C. When the recirculation-only mode is enabled, the fan runs continuously but dehumidification or cooling is disabled. When the return air temperature is equal to 15°C or between 15°C and 27°C (spring and fall), the dehumidification and recirculation mode is enabled to maintain humidity at 60% RH or less by running both the fan and the dehumidifier (DX cooling and hot-gas reheat). The dehumidification and recirculation mode produces supply air at a lower humidity while maintaining approximately the same temperature as the return air. The dehumidification and recirculation mode is disabled when any of the following three conditions is met: return air temperature is less than 15°C, return air temperature is equal to or greater than 27°C, or return air humidity is equal to or less than 50% RH. The cooling and recirculation mode is enabled when the return air temperature exceeds 27°C (summer). In this mode, the environmental management system simply cools the air without reheat; the fan and the DX system run in conjunction with the exterior condenser. This results in cooler supply air, but it is at a higher humidity owing to the lack of reheat. This mode is disabled when the return air temperature is equal to or less than 27°C.

A minimum waiting time of 10 minutes was set before any temperature or humidity-based mode changes could enable. This delay was to avoid the complexity of overlapping deadbands—one for temperature and the other for humidity. There is no idle mode, since the system was designed to continuously operate the fan to filter particulates, even when humidity in the building is less than the set point.

In addition to the sensors located in the return duct used for system control, the environmental management system monitors

Table 15.3

Operational control conditions for the Juanqinzhai environmental management system (minimum of 10 minutes before changing modes).

Psychrometric Strategy	Enable	Disable
Recirculation only	$T_{return} < 15°C$	$T_{return} \geqq 15°C$
Dehumidification, filtration, and recirculation	$T_{return} \leqq 27°C$ and $T_{return} \geqq 15°C$ and $H_{return} > 60\%$ RH	$T_{return} < 15°C$ or $T_{return} > 27°C$ or $H_{return} \leqq 50\%$ RH
Cooling, filtration, and recirculation	$T_{return} > 27°C$	$T_{return} \leqq 27°C$

temperature and humidity at the following locations: between the cooling and reheat coils, in the supply air duct, outside, and at eight interior locations throughout the first and second floors. Stored in the system for periodic download, this additional data provides a diagnostic tool when operational problems are encountered.

Postinstallation System Performance

Installed in November 2007, the environmental management system underwent two subsequent inspections, after which minor installation problems were corrected. The system was commissioned in June 2008, and its performance has been monitored since then.

Postinstallation environmental data for the Juanqinzhai interior has been collected by the system at 15 minute intervals since June 2008. The performance analysis presented here focuses on data collected in 2009, since the system was being adjusted during the remainder of 2008.

While the design of the environmental management system at Juanqinzhai assumed that the system would be operated twenty-four hours a day and seven days a week, the system was in fact operated only during visiting hours on weekdays. This reduction in operational hours was due mainly to the Palace Museum's policy for overnight fire prevention, which specifies that all equipment in the Imperial zone, with the exception of the security system, must be turned off during nonvisiting hours, between 5:00 p.m. and 9:00 a.m. During system operation, a staff member in charge of the system must also be present. Consequently, data were only recorded during visiting hours, and between six and nineteen days of data were recorded each month because of the limited operation of the system; no data were recorded in March because of system servicing. Despite these restrictions, this limited dataset provides insight into the efficacy of the environmental management strategy at Juanqinzhai.

The postinstallation analysis utilizes 2009 data from the sensor in the system's return air duct since it is considered to represent interior environmental conditions.

Humidity, temperature, and dew point temperature

Following commissioning of the environmental management system at Juanqinzhai in June 2008 and a subsequent adjustment period for the system (including sensor calibration, revision of the program, and system balancing), monthly mean interior humidity ranged from 39% RH (January) to 57% RH (July), while monthly mean temperature and dew point temperature ranged from –0.7°C (January) to 27.6°C (July) and from –12.8°C (January) to 17.2°C (July), respectively (fig. 15.20). Although these results suggest a significant improvement in humidity (there was no instance of humidity above 65% RH) compared to preinstallation conditions (monthly averages from 17% to 75% RH), the limited dataset represents only daytime hours, when loads from exterior humidity sources are at their lowest, and loads from exterior cooling sources are at their peak.

The monthly average of 24 hour variations for the interior humidity and temperature ranged from Δ2% RH (February) to Δ8% RH (June) and Δ0.9°C (January, February, and December) to Δ2.0°C (July), respectively. The largest 24 hour variations are typically due to the difference between the value recorded at the end of the operation cycle (system shutdown between 4:00 and 5:00 p.m.) and at the end of the nighttime rebound cycle (system startup between 9:00 and 10:00 a.m.)—not during operational hours. These variations are expected to be substantially reduced if the system is operated 24/7.[1]

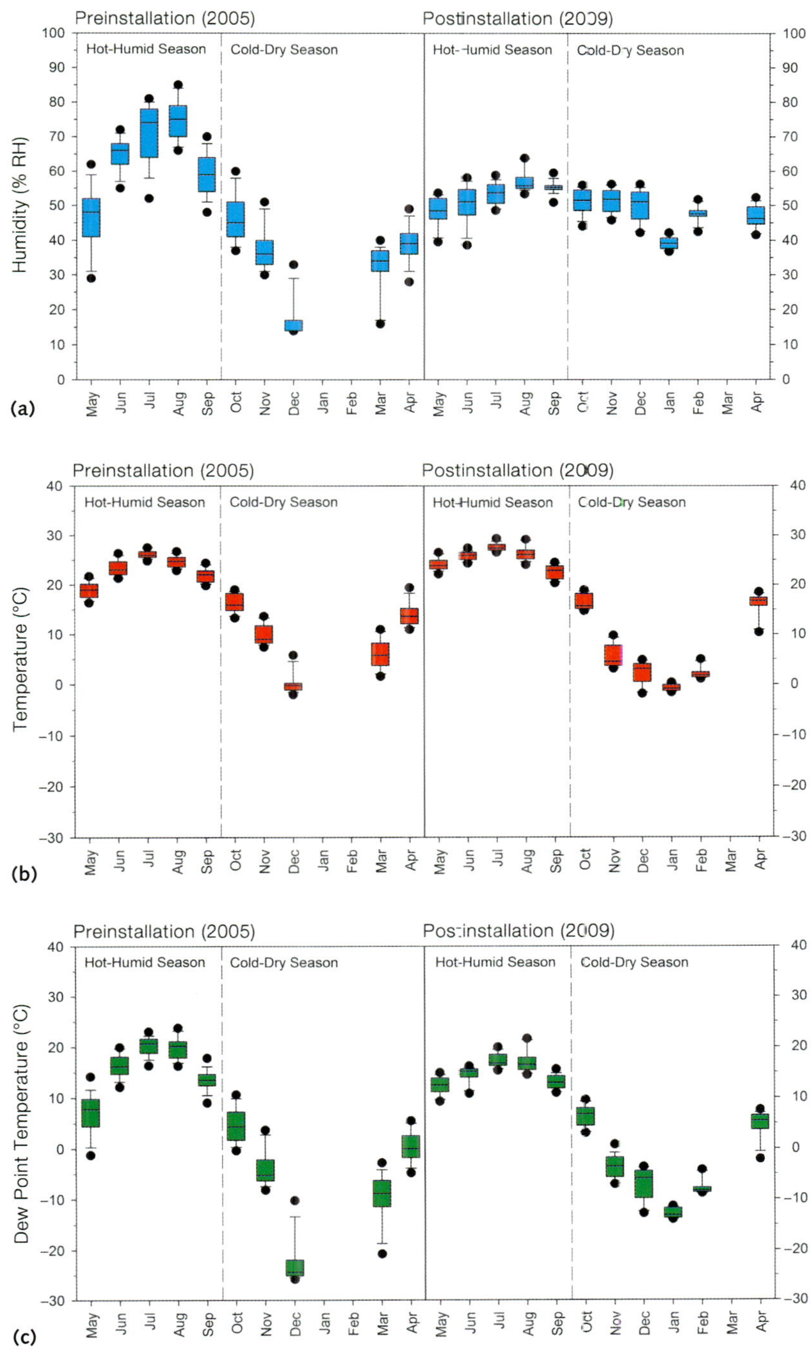

Figure 15.20

Pre- and postinstallation humidity (a), temperature (b), and dew point temperature (c) in Juanqinzhai.

Postinstallation risk class assessment

The postinstallation Conservation Environment Classification–HH for Juanqinzhai is summarized in table 15.4. Figure 15.21 also shows pre- and postinstallation environmental data plotted on psychrometric charts, with values pertaining to the conservation environment classification denoted; the boxes indicate the mid-95% of each dataset. Improved humidity control in the building can be seen in the narrower humidity range achieved in the postinstallation environment. The average postinstallation temperature was also approximately Δ2°C less than that of preinstallation and closer to the historic mean value of 12.7°C.

The pre- and postinstallation risk classes for Juanqinzhai are summarized in table 15.5.

The 97.5 percentile of humidity improved significantly, from 80% RH during preinstallation to 59%–63% RH after the installation of the environmental management strategies. This reduction corresponded to the reduction of microbial risk from Class F (high risk) to Class A (low risk). With respect to mechanical risk, the greatest improvement was observed for seasonal humidity variation. The environment was classified prior to system installation as Class f (high risk), owing to a wide variation between seasonal humidity means; the postinstallation environment was determined to be Class a (no risk) for the return air and the Theater, and Class b (moderate risk) for the Reception Room. While the pre- and postinstallation classes for short-term humidity variation remained largely

Table 15.4

Postinstallation Conservation Environment Classification–HH in Juanqinzhai.

Risk			Humidity		Temperature		
Category	Specific Risk (statistic)	Room	Value (% RH)	Class	Value (°C)	Class	
Microbial	Germination threshold (P97.5)	Return Air	59	A	—	—	
		Reception (East)	63	A	—	—	
		Theater (West)	63	A	—	—	
Mechanical	Short-term variation (P95 of daily variation)	Return Air	Δ12	b	Δ2.3	a	Overall mechanical risk:[1] Return air: b Reception: b Theater: b
		Reception (East)	Δ5	a	Δ1.9	a	
		Theater (West)	Δ11	b	Δ4.3	b	
	Seasonal variation (absolute difference of seasonal means, RH means ≦ 70%)	Return Air	Δ4	a	Δ18	b	
		Reception (East)	Δ11	b	Δ14.6	b	
		Theater (West)	Δ5	a	Δ16.7	b	
	Deviation from historical mean (absolute difference)	Return Air	Δ0	a	Δ1.7	a	
		Reception (East)	Δ4	a	Δ1.7	a	
		Theater (West)	Δ4	a	Δ3.1	a	
Chemical	Deviation from historical mean (difference)	Return Air	—	—	Δ1.7	0	
		Reception (East)	—	—	Δ1.7	0	
		Theater (West)	—	—	Δ3.1	+	

[1] Overall mechanical risk is based on the highest risk classification in any of the specific humidity- and temperature-based mechanical risks (short-term variation, seasonal variation, and deviation from historical mean).

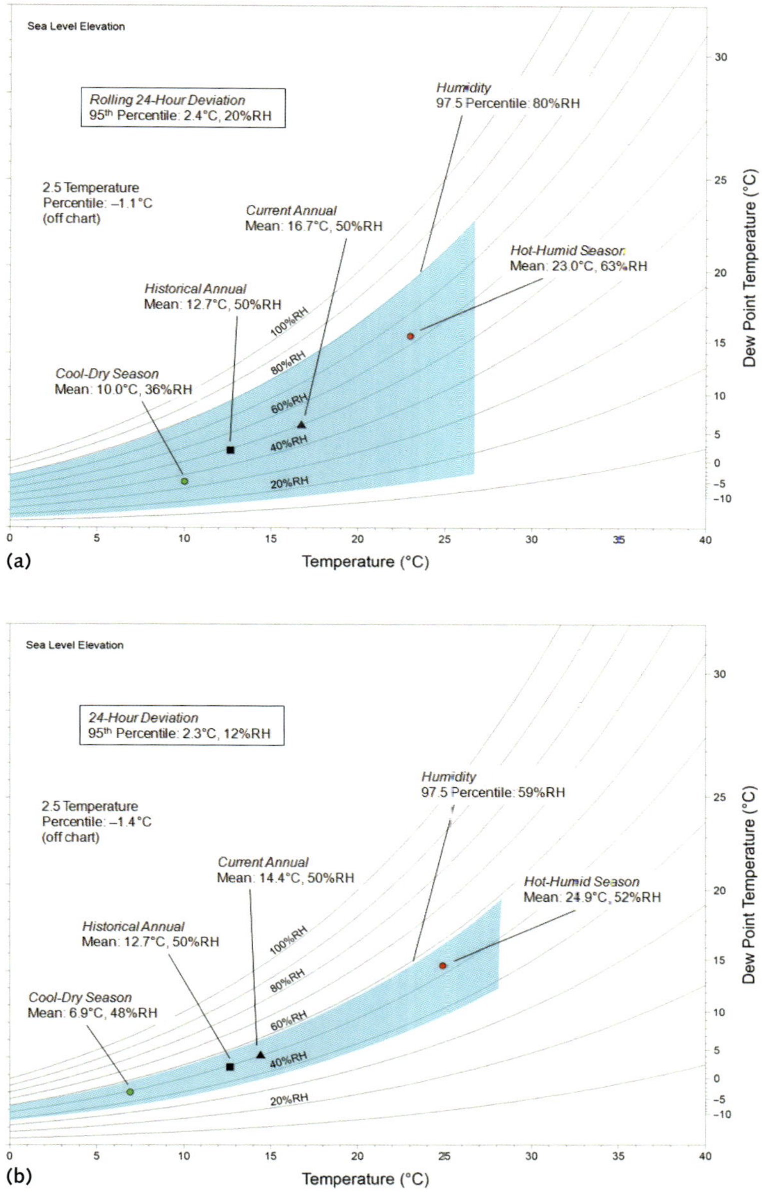

Figure 15.21

Psychrometric charts comparing Juanqinzhai's interior environments before installation (a) and after installation (b) of the environmental management system.

Table 15.5

The pre- and postinstallation risk classes for Juanqinzhai.

	Microbial Risk		Mechanical Risk														Chemical Risk	
			Humidity						Temperature						Overall[2]			
			Short-Term Variation		Seasonal Variation		Deviation from Historical Mean		Short-Term Variation		Seasonal Variation		Deviation from Historical Mean					
Room[1]	Pre	Post	Pre	Post	Pre	Post	Pre	Post	Pre	Post	Pre	Post	Pre	Post	Pre	Post	Pre	Post
Return air		A		b		a		a		a		b		b		b		0
Reception (East)	F	A	b	a	f	a	a	a	a	a	b	b	b	b	f	b	+	0
Theater (East)		A		b		a		a		b		b		b		b		+

[1] Preinstallation monitoring location is in the middle area of the building.
[2] Overall mechanical risk is based on the highest risk classification in any of the specific humidity- and temperature-based mechanical risks (short-term variation, seasonal variation, and deviation from historical mean).

the same (only the Reception Room showed improvement from a preinstallation Class b to a postinstallation Class a), all postinstallation locations reduced their short-term humidity variation by Δ8% RH to Δ16% RH. With respect to mechanical risk due to temperature, when compared to preinstallation conditions, all postinstallation locations showed similar risk, with the exception of short-term temperature in the theater. Chemical risk was also decreased in the return air and Reception Room (both were Class 0, the same risk as historic temperature condition), while the Theater space had the same risk (Class +, moderately increased risk) as preinstallation conditions.

Particulates

The AHU was operated without the high-performance particle air (HEPA) filters during the first six months of operation to protect this costly filter from exposure to a potentially high-dust initial environment. The HVAC installation contractor measured airborne particulates before and after installation of the HEPA filters, which took place six months after installation of the system. Measurements were made at four locations on the first floor and four locations on the second floor. The particulate measurements, shown in figure 15.22, indicate that the combination of reduced air infiltration through the building envelope and the use of HEPA filtration by the mechanical system generally lowered the number of interior particulates for the 10 µm (PM_{10}) and 2.5 µm ($PM_{2.5}$) size fractions. However, no improvement was observed in the Reception Hall's small rooms in the southeast end of the first floor (see fig. 15.6). This may be attributable to insufficient mixing due to the room partition.

Thermal comfort

Prior to the installation of the environmental management system, the average interior temperature and humidity in Juanqinzhai during the Hot-Humid season were 23.0°C and 63% RH, respectively. (Thermal

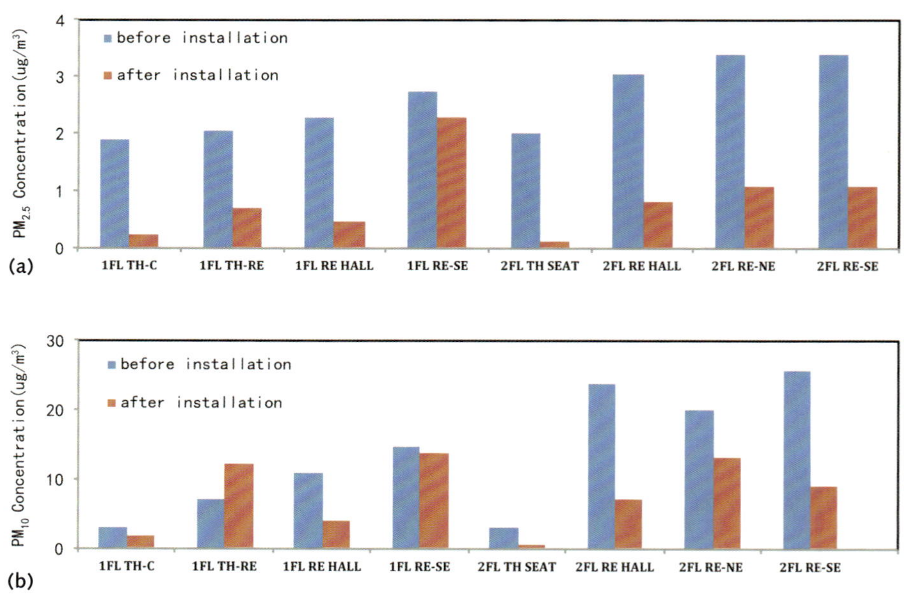

Figure 15.22

Measurements of airborne particulates for (a) small particles ($PM_{2.5}$) and (b) large particles (PM_{10}), before and after the installation of the HEPA filter (1FL = first floor; 2FL = second floor; TH = Theater side; RE = Reception Hall side; TH-RE = area between Theater and Reception; NE = northeastern portion; SE = southeastern portion; C = center; SEAT = viewing seat; and HALL = hall).

comfort during the winter was not assessed, since tours are prohibited then.) Assuming mean radiant temperature equal to the air temperature, a metabolic rate of 1.5 met (combination of standing and walking), a clothing factor of 0.6 clo (typical casual summer clothing), and a minimum air speed of 0.1 m/s, the preinstallation predicted mean vote (PMV) and predicted percentage dissatisfied (PPD) were calculated at 0.27 and 7% (neutral environment), respectively, during the Hot-Humid season (point 1 in fig. 15.23).

Following installation of the environmental management system at Juanqinzhai, the average interior temperature and humidity during the Hot-Humid season shifted to 24.9°C and 52% RH, respectively. With an estimated air speed of 0.6 m/s, mean radiant temperature equal to air temperature, a metabolic rate of 1.5 met, and a clothing value of 0.6 clo, the PMV and PPD were calculated at 0.3 (neutral) and 7%, respectively (point 2 in fig. 15.23). Thus, mean postinstallation conditions at Juanqinzhai during the Hot-Humid season complied with ANSI/ASHRAE Standard 55-2013 (ANSI and ASHRAE 2013), and the interior environment was satisfactory with respect to thermal comfort.

Despite this positive result from the thermal comfort analysis, temperatures above 27°C were recorded on the first-floor rooms of the Reception suite and the Theater in August 2009. This circumstance suggests that either the cooling capacity of the environmental management system was inadequate or the air circulation to these rooms was inadequate—the latter theory is supported by the lack of improvement in particulate concentrations observed in the small first-floor rooms of the Reception Hall before and after the use of HEPA filters, as previously discussed. Although no thermal comfort complaints were registered from visitors, workers in Juanqinzhai did complain of thermal discomfort

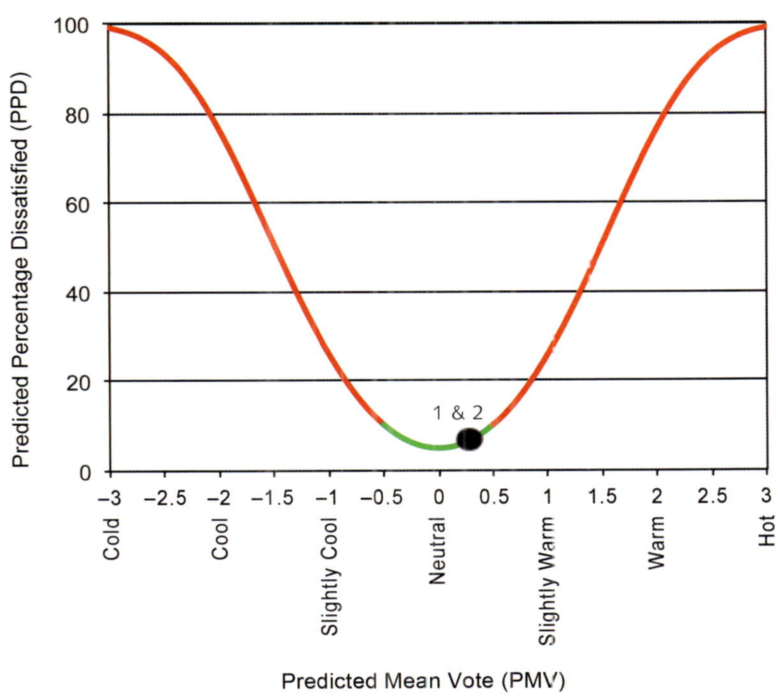

Figure 15.23

Thermal comfort assessment of mean interior environmental conditions in Juanqinzhai during the hot-humid season. Point (1) indicates preinstallation conditions, and point (2) designates postinstallation conditions.

during these warm periods. The apparent thermal comfort satisfaction of visitors may be due to their acclimation to the relatively high outside temperatures and to the short duration of their visit, whereas the worker responses might be attributable to higher levels of physical activity during extended periods inside the building, as discussed in chapter 4.

Noise

The interior noise level adjacent to the AHU in the attic mechanical room is approximately 63 dB; it is less than 50 dB on the second floor directly below the mechanical room, and it is 47 dB in the rest of Juanqinzhai. Since the refrigerant gas compressor, recirculation pump, condenser, and cooling fan are placed in exterior locations and separated from the interior by thick timber and masonry walls, the sources of noise from the environmental management system are limited to AHU operation, including the movement of air (and vibration) through filters, coils, and ducts. Therefore, noise levels were independent of operational modes, as expected.

Outside in front of Juanqinzhai's south facade, noise from the exterior system components was not audible, even during the cooling-only mode, when the auxiliary condenser fan was in operation. The lack of noise may be due to the fact that the components are located nearly 30 m from the entrance and hidden from visitors' view. In addition, Juanqinzhai is located next to the north entrance/exit of Qianlong Garden, where visitors generate noise that may mask noise from the environmental management system.

Maintenance

Daily, weekly, monthly, semiannual, and annual maintenance checklists were developed for Palace Museum staff members in order to maintain normal operation of the environmental management system in Juanqinzhai. The responsible staff member(s) can monitor system operation by remotely logging into the system and reviewing weekly downloads of environmental data (fig. 15.24). A staff member is also responsible for visual inspections of filters and fan belts, reporting of unusual noises and vibrations, testing of overflow sensors, and sensor calibration. Additionally, the Palace Museum contracts out the semiannual and annual maintenance, as well as emergency repairs of the environmental management system, to a local HVAC maintenance contractor at a cost of $22,600 per year in 2010, which was 13% of the installed system cost.

Energy Metrics

Energy intensity

The reheat applied in the dehumidification mode at Juanqinzhai was supplied entirely by recovered heat from the compressed and, therefore, hot

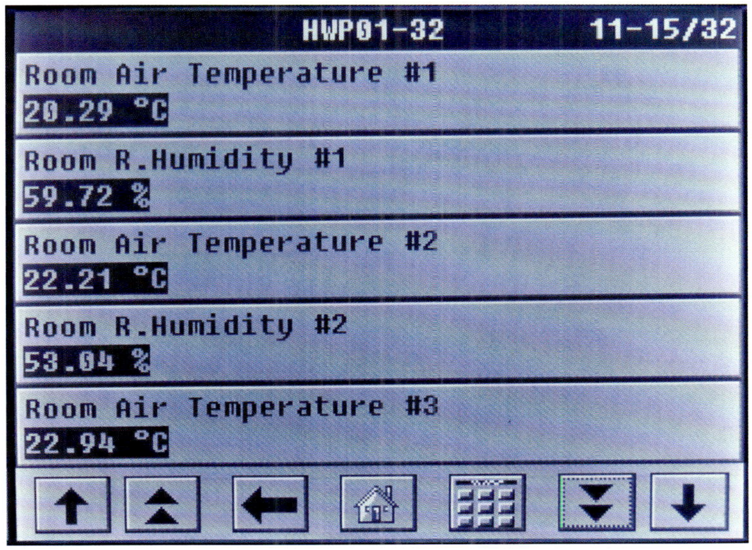

(a)

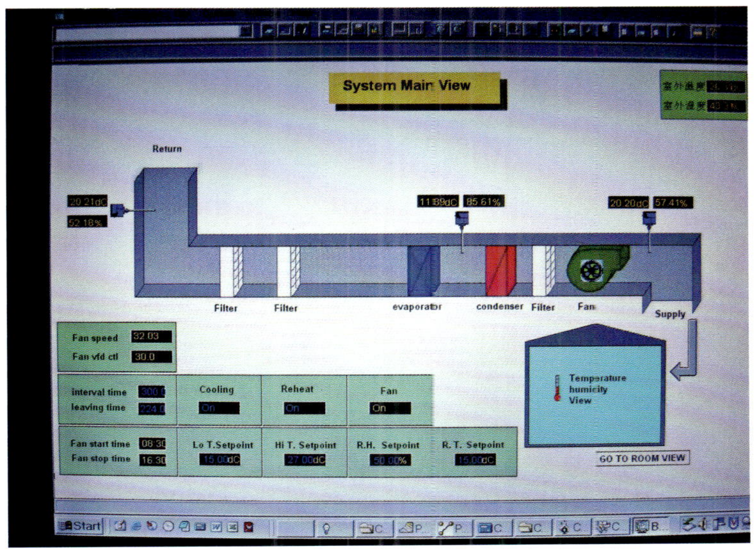

(b)

Figure 15.24
Operation of the environmental management system can be monitored with a keypad and display unit installed in Juanqinzhai (a) or via an office PC that connects remotely to the system through the Palace Museum's LAN (b).

refrigerant gas, and it did not require additional energy input. The cooling capacity of the dehumidification system is 21 kW. Based on the building's total floor area of 447 m^2 (first floor, 224 m^2; second floor, 97 m^2; half of attic floor area, m^2), the energy density of the environmental management system is 47 W/m^2.

Energy use intensity

The energy use of the environmental management system in Juanqinzhai was estimated from the power use rating of its components and hours of operation. The dehumidification and cooling modes were used only during the hot-humid season, from late May through early September; air recirculation through the filtration system was used for the remainder of the year. As described earlier, system operation was limited to approximately eight hours a day throughout the year, although the system was designed to operate twenty-four hours a day and seven days a week. Therefore, the estimated annual energy consumption, 62,000 kWh, for 24/7 operation was estimated by multiplying the recorded value by three. Based on the building's total floor area (447 m^2), the estimated energy consumption is 139 kWh/m^2/yr, assuming system operation twenty-four hours a day and seven days a week.

Conclusions

A conservation-focused environmental management strategy was developed for and implemented at Juanqinzhai. This strategy consisted of nonintrusive building envelope improvements, installation of a minimally intrusive mechanical system for controlling humidity, and a compatible visitor management policy.

Building envelope improvements included the use of thermally insulated windows with internal shades, as well as measures that reduced air infiltration. An air-handler unit—which contains a dehumidification component with hot-gas reheat, a three-stage air filtration system including a HEPA filter, and a fan—was installed in a new mechanical room that divided the attic into a supply air plenum and a return air chamber. The fan-only mode of the environmental management system was used to recirculate conditioned air through openings in the ceiling of the second floor. The system also operated as a dehumidifier to limit interior humidity below 60% RH throughout the occupied spaces and the attic, and as an air conditioner to limit interior temperature to 27°C for the purpose of improving thermal comfort for visitors. The dehumidification and cooling modes primarily operated during the hot-humid season, with the remainder of the year utilizing the fan-only mode. The visitor management policy limits the number and size of tours for scholars and special guests, prohibits visitation during winter to prevent cold-related damage, and reduces particulate loads through the use of disposable shoe covers.

Owing to the Palace Museum's overnight fire prevention policy, the environmental management system was operated fewer than eight hours per day and five days per week through June 2011 (beginning in July 2011, under special permission, the system has been operated approximately eleven hours per day and seven days per week), even though the system design was based on 24/7 operation. Despite this limitation on operational periods and, as a consequence, restriction in the collection of environmental monitoring data, the system stabilized both temperature and humidity, maintained the proposed conservation environment, and significantly reduced microbial and mechanical risks while reducing chemical risk (see table 15.5). Operation of the environmental management system also resulted in an overall reduction in airborne particulates and an interior environment that was in compliance with ANSI/ASHRAE Standard 55-2013 for thermal comfort (ANSI and ASHRAE 2013). Although system operation was generally successful, the lack of air circulation in isolated areas of the building was identified for potential future improvements.

In light of the limited operation of the mechanical system, it was fortunate that the monitored year lacked instances of equipment failure or extended rainy periods. The occurrence of such an event (as will be described below, in the "Postscript") may have resulted in a higher microbial and mechanical risk to the collection as the interior condition worsened during extended periods of nonoperation.

Lessons Learned from Juanqinzhai

- It is essential to operate the environmental management system continuously, twenty-four hours a day and seven days a week (see "Postscript," below).
- Political negotiations and outside forces need to be taken into account when proposing system concept and design.
- Environmental management goals must be reconciled with other risk reduction strategies, such as for fire. In this case, a fire alarm and automatic shutdown system were installed with the environmental management system at Juanqinzhai, although the Palace Museum has not yet allowed 24/7 system operation.
- Localized microenvironments created by bypassing airflow should be avoided.
- It is essential to assign responsibility of mechanical system management to a designated onsite staff member or group, in addition to a maintenance contractor.
- Online access to the climate control system must be available to allow for remote monitoring and diagnostics, leading to more consistent system operation and timely maintenance.

Postscript

Initially, both the staff members of Palace Museum and contracting HVAC engineers did not agree with our expectation to passively maintain humidity above 30% RH in Juanqinzhai during the cold-dry season. However, during January and February 2010, the interior humidity in Juanqinzhai remained above 40% RH, which was higher than the humidity level observed during the first winter of system operation. The achieved increase in humidity during the cold-dry season is a contributing factor in the reduction of mechanical risk due to seasonal humidity variation.

The Palace Museum continued to limit operation of the environmental management system to daytime visiting hours. At the end of June 2010, a system malfunction—no cooling at the cooling coil—was discovered, and return air humidity at Juanqinzhai reached 70% RH. Inspection found fungal infestation on the undersides of floor carpets covering the northeast side of the first floor Reception Hall rooms, where the circulation of the conditioned air is poor and a microclimate with higher than 75% RH had been recorded. The investigation concluded that the high-humidity interior condition, and the resultant fungal infestation, were due to the accumulation of moisture in the building materials during the period of system malfunction during the hot and humid summer. Mold infestation was also linked to cooler areas of the building, such as in the northeast area of the Reception Hall near and on floor tiles and at the base of masonry walls on the first floor. Although the damaged carpets were removed from the first-floor rooms, a similar system malfunction in combination with inadequate calibration of the system sensor resulted in high humidity in the building during June 2011, and a subsequent fungal infestation occurred in the building. In July 2011, after the dehumidifier was repaired, the Palace Museum increased system operational hours to eleven to twelve hours per day seven days per week. This extended operation was the maximum permissible under the existing overnight fire prevention rule, and the effort successfully brought humidity below 60% RH throughout the remainder of the summer of 2011 and the entire hot-humid seasons of 2012 and 2013. Meanwhile, permission from the Palace Museum authority is still being sought for 24/7 use of the environmental management system at Juanqinzhai.

In 2013 the Palace Museum and the World Monuments Fund started restoration of three Qianlong Garden buildings adjacent to Juanqinzhai, as the next phase in the Qianlong Garden restoration project. Environmental management strategies similar to those used in Juanqinzhai, such as airtightening of building envelopes, installation of centralized dehumidification systems, and limited visitation, are proposed for Fuwangge, a large three-story building and one of the three being restored.

Project Team and Responsibilities

Palace Museum (Beijing, China)

Jin Hongkui (Deputy Director, Palace Museum)—Project management

Wang Shiwei (Deputy Director of Historic Architecture Department, Palace Museum)— Project management

Li Yue (Architect, Historic Architecture Department, Palace Museum)—Postinstallation management of environmental management system and project coordination

Wang Fang (Conservation Scientist, Palace Museum)—Postinstallation environmental data collection and processing

Ma Yue (Conservation Scientist, Palace Museum)—Postinstallation environmental data collection and processing

Shanghai Tonghui-Carrier Ltd. (Beijing)—Engineering design and installation of the environmental management system

World Monuments Fund (New York, United States)

Henry Ng (Vice President, New York, United States)—World Monuments Fund Project direction

Liu Chang (China Representative, Beijing)—WMF local project management and building envelope and building interior assessment

T. K. McClintock (Studio TKM, Boston, United States)—Project manager and paper conservation

Marion Mecklenburg (Smithsonian Institution, Washington, DC, United States)—Building envelope and building interior assessment

Ernest Conrad (Conrad Engineers, Fairfield, Connecticut, United States)—Environmental assessment and conceptual design and commissioning of environmental management system

Shin Maekawa (GCI, Los Angeles, United States)—Environmental assessment, conceptual design and commissioning of environmental management system, and performance analysis

Vincent L. Beltran (GCI, Los Angeles, United States)—Environmental data analysis

References

AFCCC [Air Force Combat Climatology Center] and NCDC [National Climatic Data Center]
1999 *Engineering Weather Data*. Washington, DC: Air Force Combat Climatology Center. http://purl.access.gpo.gov/GPO/LPS7081.

ANSI [American National Standards Institute] and ASHRAE
2013 *Standard 55-2013—Thermal Environmental Conditions for Human Occupancy*. Atlanta: ASHRAE.

ASHRAE
2011 Museums, galleries, archives and libraries. Chapter 23 in *ASHRAE Handbook—HVAC Applications*. Atlanta: ASHRAE.

ASHRAE, ANSI, and IES [Illuminating Engineering Society]
2013 *Standard 90.1-2013—Energy Standard for Buildings except Low-Rise Residential Buildings*. Atlanta: ASHRAE.

Berliner, N., ed.
2008 Juanqinzhai in the Qianlong Garden, the Forbidden City, Beijing, World Monuments Fund. London: Scala Publications.

McClintock, T. K.
2014 The conservation of *Juanqinzhai*: A prototype for architectural interiors in the Forbidden City. *Journal of the American Institute for Conservation* 53 (2): 116–39.

Mecklenburg, M., and C. Liu
2005 Trip report: Lodge of Retirement, Forbidden City, Beijing, China, June 11–18, 2005. Smithsonian Center for Materials Research and Education, Smithsonian Institution, Washington, DC.

Weather Channel
2014 Weather Underground. http://www.wunderground.com/.

WHO [World Health Organization]
2001 Environment and people's health in China. World Health Organization, United Nations Development Programme. http://www.wpro.who.int/environmental_health/documents/docs/CHNEnvironmentalHealth.pdf.

Yu, J.
2005 Seasonal variations of number size distributions and mass concentrations of atmospheric particles in Beijing. *Advances in Atmospheric Science* 22 (3): 401–7.

Note

1 Since January 2010, 24/7 monitoring has been conducted using several battery-powered dataloggers to record temperature and humidity, in addition to the limited monitoring of the environmental management system. Temperature and humidity loggers have been placed in the supply and return grille openings, behind the Theater stage, and on the ground at the northeast corner. This dataset included the overnight period between the daily system shutdown and startup. The monitoring data indicated that humidity gradually rebounded between Δ3% and Δ5% RH from the time of shutdown (4:00–5:00 p.m.) to the startup (9:00–10:00 a.m.) of the following day.

Appendix 1

Environmental Monitoring for Alternative Climate Management

Introduction

This appendix provides an overview of environmental monitoring, which supports the design and implementation of environmental management approaches for preservation environments. Although discussion on energy monitoring is not provided, the same principles can be used with appropriate measures of energy usage. The following topics are addressed:

- design of an environmental monitoring program
- devices and software for environmental monitoring
- deployment and implementation of an environmental monitoring program
- analysis and interpretation of environmental data

In practice, environmental monitoring includes both quantitative and qualitative indicators; however, this appendix focuses on quantitative monitoring. The omission of a detailed discussion of qualitative indicators does not imply that observation of qualitative indicators of conditions in the building and collections is unnecessary; on the contrary, qualitative indicators are critical for establishing correlation and causality with the results of quantitative environmental monitoring.

Design of the Environmental Monitoring Program

The purpose of environmental monitoring is to gather representative data on the conditions within a space in the vicinity of collections, or data on the climatic conditions at the site of interest. Monitoring for curatorial purposes documents conditions to which collections have been exposed. Diagnostic monitoring provides insights into the effect of the exterior climate on the interior environment, as well as assists in understanding the performance of the building envelope and environmental management systems. Both monitoring types require temperature and humidity data, but the placement of sensors or dataloggers and their data collection intervals may differ. It is also possible to establish weather stations to monitor meteorological parameters, including temperature, humidity, rainfall, soil moisture, wind speed and direction, barometric pressure, and solar radiation.

Although environmental monitoring is an essential stewardship tool, unnecessary monitoring can divert valuable financial and human resources from vital activities such as maintenance and preventive conservation. Therefore, monitoring programs should be designed with the goal of acquiring only the necessary data and information needed to provide the desired preservation environment. An environmental monitoring program should be jointly designed and carried out by a conservator and a building professional with experience in environmental measurements and the environmental behavior of buildings and their systems.

Devices and Software

Monitoring systems

Monitoring systems consist of environmental sensors; modules for measurement, control, and data communications; memory to store the program and data; and a power supply, such as a battery, to run the system. Integrated systems are available that combine these elements in a low-cost and easy-to-use package, but the flexibility of such system designs is limited with respect to the number and quality of environmental sensors, the method of data communication, memory capacity, and battery life. In contrast, custom environmental monitoring systems provide the flexibility to meet the specific needs of the user or project—but the cost will be higher, and expert assistance will be required.

Sensor placement

Temperature and humidity will vary within an enclosed space. Spatial variations result from forced air movement, natural convection, solar gain on interior surfaces, heat from equipment and lighting, and heat and moisture from occupants. The effects of these influences and the spatial variations they create should be considered when sensors are located. Large spatial variations in temperature and humidity (>20% RH, >10°C) can be problematic and may require diagnostic monitoring so that the cause can be understood, but small variations (<10% RH, <4°C) rarely warrant additional monitoring. For highly sensitive objects or for objects in vitrines, cases, or closed storage cabinets, sensors may need to be placed near the object or within the case, housing, or cabinet.

Data recording intervals

For buildings with mechanical environmental management systems, monitoring or recording intervals of 10–30 minutes typically represent a reasonable balance between the interest in capturing environmental fluctuations and the memory capacity of the datalogger. For buildings that are not mechanically conditioned or for the collection of meteorological data, monitoring intervals of 30–60 minutes are adequate. Occasionally, rapid phenomenological fluctuations in conditions, such as visitor entries or activation and deactivation of environmental management systems, are of diagnostic interest and require shorter datalogging intervals, but these are the exception.

Several manufacturers produce devices that measure and record values of common environmental variables such as temperature, humidity, light intensity, precipitation, and wind speed and direction. Dew point temperature is normally derived from temperature and humidity measurements, because of the high cost and care required for instruments that directly measure this parameter.

Device memory

Memory capacity is a critical factor in device and system selection; device memory must be sufficient to store the data until the next data download. Since the memory capacity of the device is fixed, increasing the number of sensors will reduce either the number of readings that can be stored in the memory or the time interval between data downloads. If the device memory capacity is exceeded before the data can be downloaded, devices can be programmed to either cease recording data or to overwrite earlier data—the latter action, of course, resulting in loss of data. Therefore, device memory should include a reserve capacity in case the download schedule is interrupted by unforeseen events such as staff illness, limited access to the instruments due to weather, or technical issues.

Data retrieval

Sensor data can be retrieved from a datalogger by:

- *Periodic data transfer to a personal computer.* This is accomplished via a proprietary interface device/cable or by transferring data to the computer via a portable memory device, such as a USB flash (thumb) drive or a secure digital (SD) card. These methods require access to the dataloggers, which may limit their placement. If a large number of devices are deployed or if data transfer intervals are frequent, periodic data transfer to a PC can be labor intensive for a long-term monitoring program. However, installation cost is low, making it useful for short-term monitoring (two years or less).
- *Direct hardwired connection of all devices to a dedicated on-site PC.* This method allows continuous remote access to a device's data. This configuration is subject to limitations on available routes for cables and on the cable length over which signals can be transmitted. This approach is useful for sensor locations that are difficult to access, for long-term or permanent monitoring, and, in some instances, for integration with control systems for environmental management equipment.
- *Wireless (radio wave) data transmission between dataloggers and a local PC.* This approach provides the benefits of

hardwired connections without requiring cable installation. A wireless receiver is installed at the computer, and repeaters may be needed when the distance from the datalogger to the computer is long, or when there are physical obstacles to signal transmission. The wireless transmitter increases the power requirement of the monitoring device; therefore, devices must be accessible for periodic battery replacement. Though wireless devices are more expensive than hardwired connections because of embedded transmitters and greater complexity, wireless units are much easier to relocate than hardwired devices. This approach is suitable when staff time must be minimized or when devices will be placed in locations where physical access for downloads is problematic or where installation of a cable connection is not possible.

- *Internet protocol (IP) data communications.* This approach can be used through a local cellular network via a cellular phone (GPRS modem) or a landline-based local area network (LAN) via a Wi-Fi router. GPRS modems are especially useful for remote weather stations or for interior monitoring in heritage buildings with no landline availability. Network access via a GPRS modem must be provided by a wireless telecommunications company with coverage of the site.

Datalogger software

The datalogger manufacturer generally provides proprietary data communications software for programming the datalogger and downloading and managing data from the datalogger. Data management, analysis, and graphical presentation are time-consuming aspects of environmental monitoring, and the user-friendliness, complexity, and flexibility of software will directly affect the time and effort required for these tasks. Therefore, the proprietary software associated with a datalogger is an important aspect of device evaluation and selection. However, if the proprietary software is limited in terms of presentation quality, number of statistical functions, or data processing capability, one can export the monitoring data to spreadsheet or data analysis programs for further processing and presentation.

Deployment and Implementation of an Environmental Monitoring Program

Predeployment check

All monitoring devices should be tested for sensor and clock accuracy before initial deployment. With new, factory-calibrated devices, a simple differential bench check can be accomplished by placing several devices in a small housing for 24 hours and comparing the resultant data; a device showing inconsistent data indicates a potential problem and must be recalibrated or replaced.

Placement

Visibility of the devices is a consideration in selecting location. On the one hand, readily visible devices, proximate to visitors, may invite theft. On the other hand, some institutions view monitoring devices as an educational opportunity and take steps to inform the public of the purpose of the devices. Sensors should also not be exposed to direct sunlight, nor should they be placed near visitors or supply air outlets for mechanical systems. Radio wave communication devices require a line of sight for quality data transmission, and radio frequency interference by building materials or electrical equipment should be avoided. Devices that must be directly connected to a communication cable/interface device should be in easily accessible locations.

Initial data download

After initial deployment, the first dataset should be downloaded at an early stage of the monitoring program and examined in case it points to problems with the device, such as a faulty battery, incorrect battery installation, premature sensor or hardware failure, or unexpected influences on sensor readings related to device location. The initial dataset should be assessed to confirm that the sensors have been placed where conditions are representative of what needs to be known.

As the monitoring program continues, the periodic downloads should include verification of battery reserve power as well as a visual check of the device for indications of damage or contamination. Once a year, all monitoring devices should be calibrated and adjusted or at least be given a differential bench check, as described above, to identify any devices that are no longer accurate.

Analysis and Interpretation of Data

Collected data are analyzed and interpreted in order to satisfy one or more informational objectives: (1) to classify the environmental performance of the space according to a set of criteria, and (2) to determine the interior and exterior factors causing the interior environmental conditions. Biological, mechanical, and chemical risks to the collection and the building interior, as well as thermal comfort for occupants, should be investigated by analysis of the data.

Data interpretation requires expertise and experience in the interaction of buildings and mechanical systems with the exterior environment, as well as knowledge of the response of collections to interior environmental conditions. Skilled interpretation generally requires the collaboration of professionals who can distinguish between statistics that are significant and those that are inconsequential to the collections.

Statistical analysis

Trend plots are useful for visualizing the general behavior of the environment, including sudden shifts, variability, and extremes of data, but they must be supplemented with statistical analyses in order to characterize the environmental performance of the space.

Statistical analysis might include:

- separation of the dataset into monthly or seasonal subsets, each of which might present different combinations of thermal energy and moisture loads on the interior environment;
- breakdown of the data into seasonal average, seasonal extremes, seasonal range, and the maximum short-term (e.g., 24 hour) variation; percentile analysis may be used to exclude outlier data from a dataset;
- dew point temperature or other moisture metrics, such as humidity ratio, absolute humidity, or vapor pressure, which are indications of water vapor control; differences among interior spaces, and between interior spaces and the exterior, can reveal the direction of water vapor transport through the building envelope and within the building itself.

Analysis of the data and interpretation of the results may be enhanced by presenting the information in time-based formats (such as trends or monthly or seasonal boxplots), and/or spatial formats (such as building plans or axonometric diagrams). Annual or seasonal data can also be presented on psychrometric charts that relate temperature, humidity, and dew point temperature conditions, to identify the psychrometric processes required of a mechanical system or to assess the efficacy of an environmental management system.

Visual display of quantitative information can make large sets of data comprehensible and can facilitate analysis and the identification of patterns and trends. However, the visual presentation of data is a highly sophisticated endeavor. A viewer's response to graphical information can be influenced by scale, proportion, color, aesthetics, symbols, and labeling. A valuable resource is Edward Tufte's *Visual Display of Quantitative Information*, which offers guidelines for graphical excellence.

Appendix 2

Köppen-Geiger Climate Classification System

This overview of the Köppen-Geiger Climate Classification System, introduced in chapter 1, will help readers understand the connection between this important climate classification system and the ANSI/ASHRAE/IES Standard 90.1-2013 climate zone classifications used in this publication (ASHRAE, ANSI, and IES 2013).

Standard 90.1-2013 climate zones have not been mapped for countries other than the United States. For locations in other countries, application of Standard 90.1-2013 requires linking the location's climate zone classification to existing Köppen-Geiger climate classes. This overview is intended to help readers select the correct Standard 90.1-2013 class for their location.

Development of the Köppen-Geiger Climate Classification System

The most widely used empirically based method for climate classification was largely developed by Wladimir Köppen, a German botanist and climatologist. It was first published in 1884, and Köppen later modified it in collaboration with German climatologist Rudolf Geiger. The Köppen-Geiger climate classification system was based on the concept that the distribution of vegetation was a reflection of the specific climate to which it was exposed. This system classified the world into five major climate regions, which were then further divided into the subzones listed below.

Notation for the Köppen-Geiger climate classification system uses capital letters (A–E) to denote the major climate groups—Tropical (A), Dry (B), Temperate (C), Continental (D), and Polar (E). Subdivisions within these groups are designated by letters that identify specific climate types. The following is a list of Köppen-Geiger climate classes; the criteria for each are shown in table A2.1.

Group A: Tropical climates

- Tropical Rainforest climate (*Af*)
- Tropical Monsoon climate (*Am*)
- Tropical Savanna climate (*Aw*)

Group B: Dry climates

- Desert climates (*BWh*, *BWk*, *BWn*)
- Semi-arid or Steppe climate (*BSh*, *BSk*)

Group C: Temperate climates

- Mediterranean climates (*Csa*, *Csb*)
- Humid Subtropical climates (*Cfa*, *Cwa*)
- Oceanic climates (*Cfb*, *Cwb*, *Cfc*, *Cwc*)

Group D: Continental/microthermal climates

- Humid Continental climates (*Dfa*, *Dwa*, *Dsa*, *Dfb*, *Dwb*, *Dsb*)
- Continental Subarctic climates (*Dfc*, *Dwc*, *Dsc*, *Dfd*, *Dwd*)

Group E: Polar and Highland climates

- Tundra climate (*ET*)
- Ice Cap climate (*EF*)
- Highland climate (*H*)

Hot and Humid Climates in the Köppen-Geiger Climate Classification System

In the Köppen-Geiger system, the Tropical (A) and Temperate (C) climate zones contain hot and humid climates that coincide with the Standard 90.1 climate zones of 1A, 2A, 3A, 4A, 3C, and 4C that are the focus of this book. Figure A2.1 and table A2.2 show the global distribution of the Köppen-Geiger climate classifications. The Tropical and Temperate climate zones encompass 19% and 13.4% of the overall landmass, respectively. While South America is the continent with the most landmass within a Tropical climate zone (60%), Asia and Africa also contain significant areas classified as Tropical. Temperate climates zones are more evenly distributed across the continents. A recent study that remapped Köppen-Geiger climate classifications showed that the predominant climate subdivision was Dry-Hot Desert (*BWh*), followed by Tropical Savannah (*Aw*) (Peel, Finlayson, and McMahon 2007).

Hot and humid climates can be found in all subdivisions of the Tropical climate zone, which includes Tropical Rainforest (*Af*), Tropical Monsoon (*Am*), and Tropical Savanna (*Aw*). In each of these subdivisions, monthly mean temperature remains above 18°C. With respect to precipitation, Tropical Rainforests (*Af*) show consistent rainfall throughout the year (the driest month exceeds 60 mm precipitation), while a dominant rainfall season is observed for Tropical Monsoon (*Am*) and Tropical Savanna (*Aw*) climate zones (the driest month has <60 mm precipitation) (see table A2.1).

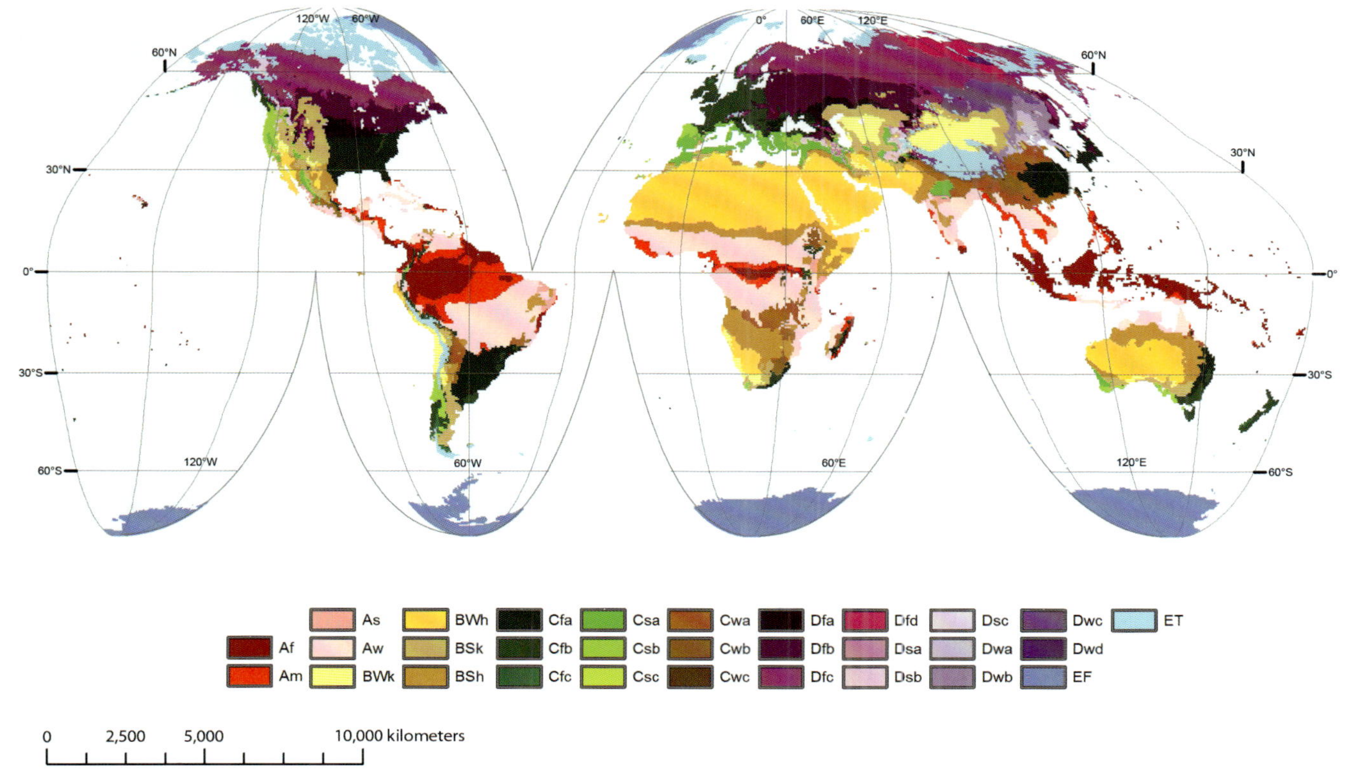

Figure A2.1

World map of the Köppen-Geiger climate classifications. Source: Kottek et al. 2006.

In Temperate climate zones, only the Humid Subtropical climate zones (*Cfa* and *Cwa*) contain hot and humid climates. These climate zones experience hot and humid conditions during the summer months and relatively cool winter conditions. In the Subtropical climate, the warmest mean monthly temperature exceeds 22°C, while the coldest month is between 0°C and 18°C. Climate zone *Cwa* (Temperate–Dry Winter) exhibits seasonal precipitation, with its criteria requiring that the driest winter month receive less than 10% of the precipitation of the wettest summer month (see table A2.1). In contrast, climate zone *Cfa* (Temperate–No Dry Season) shows much less seasonal variation, with its criteria requiring that (a) the driest winter month receive more than 10% of the precipitation of the wettest summer month, (b) the driest summer month exceed 40 mm precipitation, and (c) the driest summer month receive more than 33% of the precipitation of the wettest winter month.

Table A2.1

Köppen-Geiger climate designations and their defining criteria.

1st	2nd	3rd	Description	Criteria[1]
A			Tropical	$T_{cold} \geqq 18°C$
	f		–Rainforest	$P_{dry} \geqq 60$ mm
	m		–Monsoon	Not (*Af*) & $P_{dry} \geqq 100 - P_{am}/25$
	w		–Savannah	Not (*Af*) & $P_{dry} < 100 - P_{am}/25$
B			Dry	$P_{am} < 10 \times P_{threshold}$
	W		–Desert	$P_{am} < 5 \times P_{threshold}$
	S		–Steppe	$P_{am} \geqq 5 \times P_{threshold}$
		h	–Hot	$T_{am} \geqq 18°C$
		k	–Cold	$T_{am} < 18°C$
		n	–Desert	
C			Temperate	$T_{hot} > 10°C$ & $0 < T_{cold} < 18°C$
	s		–Dry Summer	$P_{sdry} < 40$ & $P_{sdry} < P_{wwet}/3$
	w		–Dry Winter	$P_{wdry} < P_{swet}/10$
	f		–Without Dry Season	Not (*Cs*) or (*Cw*)
		a	–Hot Summer	$T_{hot} \geqq 22°C$
		b	–Warm Summer	Not (a) & $T_{mon10} \geqq 4°C$
		c	–Cold Summer	Not (a or b) & $1 \leqq T_{mon10} < 4$
D			Continental	$T_{hot} > 10°C$ & $T_{cold} \leqq 0°C$
	s		–Dry Summer	$P_{sdry} < 40$ & $P_{sdry} < P_{wwet}/3$
	w		–Dry Winter	$P_{wdry} < P_{swet}/10$
	f		–Without Dry Season	Not (*Ds*) or (*Dw*)
		a	–Hot Summer	$T_{hot} \geqq 22$
		b	–Warm Summer	Not (a) & $T_{mon10} \geqq 4°C$
		c	–Cold Summer	Not (a, b or d)
		d	–Very Cold Winter	Not (a or b) & $T_{cold} < -38°$
E			Polar	$T_{hot} < 10°C$
	T		–Tundra	$T_{hot} > 0°C$
	F		–Frost	$T_{hot} \leqq 0°C$
	H		–Highland	

[1] P_{am} = mean annual precipitation in millimeters
T_{am} = mean annual temperature in degrees Celsius
T_{hot} = temperature of the hottest month in degrees Celsius
T_{cold} = temperature of the coldest month in degrees Celsius
T_{mon10} = number of months where the temperature is above 10°C
P_{dry} = precipitation of the driest month in millimeters
P_{sdry} = precipitation of the driest month in summer in millimeters
P_{wdry} = precipitation of the driest month in winter in millimeters
P_{swet} = precipitation of the wettest month in summer in millimeters
P_{wwet} = precipitation of the wettest month in winter in millimeters
$P_{threshold}$ varies according to the following rules:
If 70% of P_{am} occurs in winter, then $P_{threshold} = 2 \times T_{am}$;
If 70% of P_{am} occurs in summer, then $P_{threshold} = 2 \times T_{am} + 28$;
Otherwise, $P_{threshold} = 2 \times T_{am} + 14$.

Table A2.2

Landmass percentages of various climate classes under the Köppen-Geiger climate classification system. Source: Peel, Finlayson, and McMahon 2007.

	A (Tropical)	B (Dry)	C (Temperate)	D (Continental)	E (Polar)
Asia	16.3	23.9	12.3	43.8	3.8
Australia	8.3	77.8	13.9	0.0	0.0
Africa	31.0	57.2	11.8	0.0	0.0
Europe	0.0	36.3	17.0	44.4	2.3
South America	60.1	15.0	24.1	0.0	0.8
North America	5.9	15.3	13.4	54.5	11.0
World	19.0	30.2	13.4	24.6	12.8

References

ANSI [American National Standards Institute], ASHRAE, and IES [Illuminating Engineering Society]
2013 *Standard 90.1-2013—Energy Standard for Buildings except Low-Rise Residential Buildings* (SI edition). Atlanta: ASHRAE.

Kottek, M., J. Grieser, C. Beck, B. Rudolf, and F. Rubel
2006 World map of the Köppen-Geiger climate classification updated. *Meteorologische Zeitschrift* 15 (3): 259–63.

Peel, M. C., B. L. Finlayson, and T. A. McMahon
2007 Updated world map of the Köppen-Geiger climate classification. *Hydrology and Earth System Sciences Discussions* 4 (2): 439–73.

Appendix 3
Climate Calculations for Bangkok and Istanbul

Climate calculations derived from the Standard 90.1-2013 method are illustrated here for Bangkok, Thailand, and Istanbul, Turkey.

Bangkok

Bangkok (100.98° E, 13.37° N; elevation, 4 m), located in Southeast Asia, is classified as a Tropical Wet and Dry climate (*Aw*) under the Köppen-Geiger climate classification system. Climatic data collected at the Bangkok International Airport (table A3.1), show that monthly average temperatures range from 26°C to 30°C during the year, and comparatively wet and dry seasons can be observed in the precipitation data. Because Bangkok has an annual CDD10°C value of 6788 cooling degree days, its climate's thermal classification is Very Hot (1), based on the thermal criterion (5000 degree days < CDD10°C) (see chap. 1, table 1.1).

Application of the moisture criteria shown in chapter 1, figure 1.3, defines the moisture climate zone of Bangkok. A Marine moisture classification is immediately excluded owing to the fact that the temperature during the coldest month (25.6°C in December) exceeds 18°C. Determination of a Dry or Humid moisture classification is based on table A3.1, which shows the city's annual average temperature (T_c, 28°C) and annual average precipitation (P_{mm}, 1497 mm). Since the P_{mm} value exceeds the Dry-Humid Index of 700 (using the comparison of P_{mm} and $20 \times [T_c + 7]$, shown in chap. 1, fig. 1.3), the climate of Bangkok fails the criteria for a Dry-type climate and therefore is classified as a Humid-type climate (A).

The combination of the above thermal and moisture analyses from Standard 90.1-2013 classifies Bangkok's location as a Very Hot-Humid (1A) climate zone.

Istanbul

Istanbul (41°01' N 28°28' E; elevation, 50 m) is positioned at the western end of the Asian continent and surrounded by the Black Sea and the Sea of Marmara. Istanbul's climate is classified as Mediterranean (*Csa*) in the

Table A3.1

Summary of climate data for Bangkok.[1]

	Jan.	Feb.	Mar.	Apr.	May	Jun.	Jul.	Aug.	Sep.	Oct.	Nov.	Dec.	Annual
Average temperature (°C)	26	27	29	30	29	29	28	28	28	28	27	26	28
Average total precipitation (mm)	9	30	29	68	220	149	155	197	344	242	48	10	1497
Average total CDD10°C (degree days)	496	526	613	621	600	575	583	598	570	576	517	513	6788

[1] Climate data source: Hong Kong Observatory site, http://www.hko.gov.hk/wxinfo/climat/world/eng/asia/se_asia/bangkok_e.htm, for years 2008–10.

Köppen-Geiger classification. Based on data shown in table A3.2, the climate of Istanbul shows seasonal variation, with a cold and damp winter followed by a hot and dry summer. Though rarely below freezing, winter temperatures remain cold, and snowfall may occur. The majority of rainfall is recorded between fall and spring, and maximum precipitation is observed in winter. The annual CDD10°C value for Istanbul was calculated to be 2504 cooling degree days, placing it in the Warm (3) thermal zone based on thermal criteria (2500 degree days < CDD10°C ≦ 3500 degree days) from chapter 1, table 1.1.

Application of the moisture criteria shown in chapter 1, figure 1.3, rules out classification of the Istanbul climate as a Marine type, because the mean monthly temperature of its warmest months (23°C in July and August) exceeds the threshold of 22°C. Comparison of the annual average precipitation (P_{mm} = 640 mm) in Istanbul with its Dry-Humid Index of 420 (20 × [T_c + 7], with T_c = annual average temperature of 14°C) satisfies the criteria for a Humid-type climate (A).

The combination of the above thermal and moisture analyses from Standard 90.1-2013 classifies Istanbul's location as a Warm-Humid (3A) climate zone.

Table A3.2

Summary of climate data for Istanbul.[1]

	Jan.	Feb.	Mar.	Apr.	May	Jun.	Jul.	Aug.	Sep.	Oct.	Nov.	Dec.	Annual
Average temperature (°C)	5	5	7	12	16	21	23	23	20	16	11	7	14
Average total Precipitation (mm)	83	66	60	53	29	26	21	25	36	68	74	99	640
Average total CDD10°C (degree days)	14	15	35	100	257	399	477	515	339	219	93	41	2504

[1] Data source for Istanbul CDD calculation (2006–10): www.wunderground.com. Balance of Istanbul data from: http://www.weatherbase.com.

Appendix 4

Structured Decision-Making Strategies

The following are formal decision-making methodologies that can be applied to the analysis of data (Van Domelen 2009).

Multiple Attribute Decision Analysis (MADA)

MADA utilizes either *weighting methods* or *sequential elimination methods*. Weighting methods assign multiplicative weights to each attribute (or objective) and scaled numerical values for each attribute, yielding a total value for each option. The weighting method presumes that the highest total value indicates the best choice. Sequential elimination methods order the attributes by importance and compare the options sequentially for each attribute—the best overall option being found by elimination. MADA is sensitive to the assigned weighting and importance factors, which can be manipulated for a biased outcome.

Analytical Hierarchy Process (AHP)

AHP, which is recommended by the Whole Building Design Guide (WBDG), an Internet resource managed by the National Institute of Building Sciences, requires the use of software for the evaluation process. ASTM E1765—*Standard Practice for Applying Analytical Hierarchy Process (AHP) to Multi-attribute Decision Analysis of Investments Related to Buildings and Building Systems* provides a complete description of the method.

Choosing by Advantages (CBA)

The US National Park Service (NPS) defines CBA as "a system of concepts and methods to structure decision-making. CBA quantifies the relative importance of nonmonetary advantages or benefits for a set of alternatives and allows subsequent benefit and cost consideration during decision-making. CBA may be used as an evaluation method during the evaluation phase of the value analysis job plan, in lieu of the more traditional weighted-factor analysis. CBA is the preferred evaluation method where critical non-monetary benefits need to be evaluated"[1] (NPS 2002).

Economic Decision Making, Based on Life-Cycle Costing (LCC)

LCC considers initial capital cost, replacement cost, and operating and maintenance costs, resulting in an estimate of the total cost of a material or system over its estimated service life. LCC analysis requires data on service life expectancy, which are limited and highly variable and which can strongly influence the results. LCC analysis is useful for evaluating overall cost and, indirectly, energy consumption, but it does not account for noneconomic values, such as risk mitigation for collections.

Life-Cycle Assessment (LCA)

LCA addresses the environmental burdens associated with a product, process, or activity by quantifying energy and material usage and environmental releases. "The assessment includes the entire life cycle of the product, process, or activity, from the extraction of raw materials through manufacturing, use/reuse/maintenance, recycling, and final disposal" (SETAC 1993). LCA is data driven, but current databases are incomplete and inconsistent, and implementation of LCA could require substantial investigation and cost.

References

NPS [National Park Service]
2002 Director's order #90: Value analysis. http://www.nps.gov/policy/DOrders/DO90.htm.

SETAC [Society for Environmental Toxicology and Chemistry]
1993 *Guidelines for Life-Cycle Assessment: A Code of Practice*. Brussels: SETAC.

Van Domelen, S. K.
2009 The choice is yours: Considerations and methods for the evaluation and selection of substitute materials for historic preservation. Master's thesis, University of Pennsylvania. repository.upenn.edu/hp_theses/119/.

Glossary

Absolute humidity. Amount of water vapor contained in a unit volume of air. It can be expressed as either the partial pressure of water vapor or the mass ratio of the water to dry air.

Accuracy. Degree of agreement between the instrument reading and the true value of the variable being measured. This is important if it is necessary to compare subtle differences in readings from devices at different locations.

Barometric pressure. Barometric pressure is the force per unit area exerted by weight of a column of air above a specific location. Weight is the force exerted by gravity on unit mass (in this case, the air mass). SI: pascal (Pa); I-P: pound per square inch (psi).

Boolean. A data type that records a reading of either true or false. In contrast, arithmetic data consist of numerical values.

Calibration. Establishing the accuracy of a device through comparison to a known standard and documenting the deviation or adjusting the device to within a range of acceptable accuracy. Calibration is done prior to, and after, use of a device; under long-term programs it is checked at specified frequencies during use.

Coefficient of moisture (hygroscopic) expansion. The degree of expansion of a material divided by the change in moisture.

Coefficient of thermal expansion. The degree of expansion of a material divided by the change in temperature.

Conservation heating (also known as humidistatic heating). The use of heating to reduce interior humidity. While humidity is reduced, the dew point temperature or water content of the air is unchanged.

Datalogger. Device that records the measurement data for future transmission or retrieval.

Deadband. An interval of a signal domain or band where no action occurs, with the purpose of preventing oscillation or repeated activation/deactivation cycles.

Dew point temperature (T_{dp}). The surface temperature at which the air volume is saturated with water vapor, and below which dew (liquid condensation) or frost (solid deposition) first forms. SI: degree Celsius (°C) or Kelvin (°K); I-P: degree Fahrenheit (°F) or Rankine (°R).

Drift. Change over time in the reading that corresponds to a specific quantity; drift may be a simple shift of all readings over the entire range, or it may be skewed or nonlinear. Drift is important in long-term monitoring programs and in establishing calibration frequencies.

Dry bulb temperature (T_{db}). The temperature indicated by an ordinary thermometer when the sensing element is dry. SI: degree Celsius (°C) or Kelvin (°K); I-P: degree Fahrenheit (°F) or Rankine (°R).

Efflorescence. The loss of water of crystallization from a hydrated salt to the atmosphere on exposure to air.

Energy intensity (EI). The installed total capacity of a mechanical system per unit floor area in watts per square meter (W/m^2).

Energy use intensity (EUI). The energy use of the climate management system over a year in kilowatt-hour per square meter per year ($kWh/m^2/yr$).

Equilibrium moisture content. The moisture content at which a material is neither gaining nor losing moisture; this however, is a dynamic equilibrium and changes with relative humidity and temperature.

Exfiltration. Uncontrolled passage of indoor air out of a building through unintended leaks in the building envelope (e.g., cracks between wall sections, wall-floor connections, corners, the roof-wall interface, or around windows and doors).

Germinate. The emergence of cells from resting spores and the growth of hyphae from spores in fungi.

Humidistatic control. Mechanical equipment operation controlled by humidity set points.

Humidistatic heating. See *conservation heating.*

Humidity ratio (W). The ratio of the mass of water vapor to the mass of dry air in a volume of moist air. It is defined by the equation: $W = m_{WV}/m_{DA}$, where W is the humidity ratio, m_{WV} is the mass of water vapor in the space of a sample of moist air, and m_{DA} is the mass of dry air. SI: gram of water per kilogram of dry air (g/kg); I-P: grain per pound (gr/lb).

Hyphae. Each of the branching filaments that make up the mycelium of a fungus.

I/O module. A general-purpose circuit module that enables a microprocessor to interface with relays, switches, and transducers (*I*, or input, is data received by the system; *O*, or output, is data sent from the system).

Integrated Pest Management (IPM). A broad-based approach that integrates a range of practices for economic control of pests.

Isohumes. A line indicating constant relative humidity.

Isotherms. A line indicating constant temperature.

MERV rating. MERV, or minimum efficiency reporting value, is a measurement scale designed in 1987 by the American Society of Heating, Refrigerating and Air-Conditioning Engineers (ASHRAE) to rate the effectiveness of air filters.

Measurement. Quantifying the value of a variable using an instrument or device.

Measurement range. Range of values of a measured variable over which the instrument is useful. The useful range of an instrument may be limited by its accuracy or precision, or by environmental or physical factors (such as pressures or temperatures that may render the device inoperable or inaccurate).

Microclimate. The climate of a very small or restricted area, especially when it differs from the climate of the surrounding area.

Moist air. Air that is a mixture of dry air and any amount of water vapor.

Monitoring. Collecting and recording measured data in a systematic, consistent, and repetitive manner.

Mycelium. The vegetative part of a fungus, consisting of a mass of branching, threadlike hyphae.

Observation. May be visual or by measurements taken by instruments and devices.

Precision. Degree of agreement between successive readings for a fixed variable. This is important if it is necessary to draw conclusions from differences in values from the same instrument over time.

Preventive conservation. Actions taken to prevent or delay the deterioration of cultural heritage. The primary goal is to identify and reduce potential hazards to heritage items with thoughtful control of their surroundings.

Process paths. The series of states that a system passes through as it moves from an initial state to a final state.

Psychrometry. The field of engineering concerned with the determination of the physical and thermodynamic properties of gas-vapor mixtures.

Relative humidity (RH). The percent ratio of actual water vapor pressure to the saturation water vapor pressure at the same dry bulb temperature. RH is defined by the equation: $RH\ (\%) = 100 \times P_{wv}/P_{wvs}|t_{db}$, where P_{wv} is the actual water vapor pressure, and $P_{wvs}|t_{db}$ is saturation vapor pressure at the same dry bulb temperature.

Resolution. Smallest change increment in a measured variable that will result in a response from the instrument. This is important if it is necessary to measure small, discrete changes.

Sensitivity. Ratio of the instrument output signal to a change in the measured variable. Sensitivity is important in assembling an instrument system, since the sensor, transmitting device, and indicating/recording device should be well matched with respect to sensitivity.

Sensor. Element that responds directly to a measured variable, producing a related signal, usually a small, variable electrical current or voltage.

Service life. The expected lifetime, or the acceptable period of use, in service. It is the time that any manufactured item can be expected to be "serviceable" or supported by its manufacturer.

Set point. A reference condition (typically temperature or humidity) set by a control device that activates mechanical air-conditioning equipment.

Sorption isotherm. The equilibrium of the sorption of a material at a surface boundary at constant temperature. It represents the amount of material bound at the surface (the sorbate) as a function of the material present in the gas phase and/or in the solution.

Specific volume. The volume per unit of the dry air component. SI: cubic meter per kilogram of dry air (m^3/kg_{DA}); I-P: cubic foot per pound (ft^3/lb).

State points. The psychrometric properties of air at a given set of conditions—e.g., temperature, dew point temperature, and humidity.

Subflorescence. Salt crystallization within a material, which can cause damage to the internal structure.

Transmitter. Converts the signal to a usable form, either digital or analog, and may include the wires or circuitry between the sensor and the indicating/recording element.

Water activity (a_w). The partial vapor pressure of water in a substance divided by the standard state partial vapor pressure of water. Higher water activity values are required to support microorganism activity.

Water vapor pressure. The pressure exerted by water vapor molecules, whether in a container by themselves or sharing a container (or the atmosphere) with other gaseous molecules, such as oxygen and nitrogen. SI: pascal (Pa); I-P: pound per square inch (psi).

Wet bulb temperature (T_{wb}). The temperature indicated by an ordinary thermometer with its sensing device covered with a wetted sleeve. This is also the lowest temperature that can be attained by evaporatively cooling air at 100% saturation efficiency (temperature of evaporation). SI: degree Celsius (°C) or Kelvin (°K); I-P: degree Fahrenheit (°F) or Rankine (°R).

Supplemental Bibliography

American Conference of Governmental Industrial Hygienists [ACGIH]
1999 *Bioaerosols: Assessment and Control,* ed. Janet Mather. Cincinnati: ACGIH.

———.
2001 *Industrial Ventilation: A Manual of Recommended Practice,* 24th ed. Cincinnati: ACGIH.

American National Standards Institute [ANSI] and ASHRAE
2001 *Standard 140-2011—Standard Method of Test for the Evaluation of Building Energy Analysis Computer Programs.* Atlanta: ASHRAE.

American Society for Testing and Materials [ASTM], ed.
1995 *Test Method for Shrinkage Factors of Soils by the Wax Method.* D 4943-95. West Conshohocken, PA: ASTM.

Andreu, I. C., and T. Oreszczyn
2004 Architects need environmental feedback. *Building Research and Information* 32 (4): 313–28.

Arnold, A., and K. Zehnder
1991 Monitoring wall paintings affected by soluble salts. In *The Conservation of Wall Paintings: Proceedings of a Symposium Organized by the Courtauld Institute of Art and the Getty Conservation Institute, London, July 13–16, 1987,* ed. S. Cather, 103–35. Los Angeles: Getty Conservation Institute.

Ashley-Smith, J., N. Umney, and D. Ford
1994 Let's be honest—realistic environmental parameters for loaned objects. In *Preventive Conservation: Practice, Theory and Research: Preprints of the Contributions to the Ottawa Congress, 12–16 September 1994,* ed. A. Roy and P. Smith, 28–31. London: International Institute for Conservation of Historic and Artistic Works.

ASHRAE
2013 Air contaminants. Chapter 11 in *ASHRAE Handbook—Fundamentals.* Atlanta: ASHRAE.

———.
2013 Thermal comfort. Chapter 9 in *ASHRAE Handbook—Fundamentals.* Atlanta: ASHRAE.

Association for Preservation Technology International and American Institute for Conservation of Historic and Artistic Works
1992 *New Orleans Charter for the Joint Preservation of Historic Structures and Artifacts.* http://cool.conservation-us.org/bytopic/ethics/neworlea.html.

Baker, J., and F. Lugano
1999 A passive approach to practical climate control. In *Postprints of the Wooden Artifacts Group,* 34–45. Washington, DC: Wooden Artifacts Group, American Institute for Conservation.

Berglund, L., and W. Cain
1989 Perceived air quality and the thermal environment. In *Proceedings of IAQ '89: The Human Equation—Health and Comfort*, 93–99. Atlanta: ASHRAE.

Blades, N., T. Oreszczyn, B. Bordass, and M. Cassar
2000 *Guidelines on Pollution Control in Museum Buildings*. London: Museums Association.

Bläuer-Böhm, C., K. Zehnder, H. Domeisen, and A. Arnold
2001 Climate control for the passive conservation of the Romanesque painted wooden ceiling in the church of Zillis (Switzerland). *Studies in Conservation* 46 (4): 251–68.

Bordass, W., and C. Bemrose, eds.
1996 *Heating Your Church*. London: Church House Publishing.

Bordass, W., and A. Leaman
1997 Design for manageability: Unmanageable complexity is a major source of chronic problems in building performance. *Building Research and Information* 25 (3): 148–57.

Bratasz, Ł., S. Jakieła, and R. Kozłowski
2005 Allowable thresholds in dynamic changes of microclimate for wooden cultural objects: Monitoring in situ and modelling. In ICOM Committee for Conservation, *14th Triennial Meeting, The Hague, 12–16 September 2005: Preprints*, ed. I. Verger, 582–89. London: James & James, Earthscan.

Bratasz, L., and R. Kozłowski
2005 Laser sensors for continuous in-situ monitoring of the dimensional response of wooden objects. *Studies in Conservation* 50 (4): 307–15.

Bratasz, L., R. Kozłowski, D. Camuffo, and E. Pagan
2007 Impact of indoor heating on painted wood: Monitoring the altarpiece in the church of Santa Maria Maddalena in Rocca Pietore, Italy. *Studies in Conservation* 52 (3): 199–210.

Brimblecombe, P., N. Blades, D. Camuffo, G. Sturaro, A. Valentino, K. Gysels, R. Van Grieken et al.
1999 The indoor environment of a modern museum building, the Sainsbury Centre for Visual Arts, Norwich, UK. *Indoor Air* 9 (3): 146–64.

Brimblecombe, P., and D. Camuffo
2003 Long term damage to the built environment. In *The Effects of Air Pollution on the Built Environment*, ed. P. Brimblecombe, 1–30. London: Imperial College Press.

British Standards Institution
2010 *BS EN 15757:2010—Conservation of Cultural Property. Specifications for Temperature and Relative Humidity to Limit Climate-Induced Mechanical Damage in Organic Hygroscopic Materials*. London: British Standards Institution.

———.
2012 *PAS 198:2012—Specification for Managing Environmental Conditions for Cultural Collections*. London: British Standards Institution.

Bullock, L., B. Hayes, S. Staniforth, and A. Singleton
1997 Conservation heating by cable and radio: National Trust houses. *Museum Practice* 2 (1): 71–74.

California Office of Environmental Health Hazard Assessment
2012 All OEHHA acute, 8-hour and chronic reference exposure levels (chRELs) as of February 2012. http://www.oehha.org/air/allrels.html.

———.
2014 All OEHHA acute, 8-hour and chronic reference exposure levels (chRELs) as of June 2014. http://www.oehha.org/air/allrels.html.

Camuffo, D.
1998 *Microclimate for Cultural Heritage: Developments in Atmospheric Science.* Amsterdam: Elsevier BV.

Camuffo, D., and A. Bernardi
1995 The microclimate of the Sistine Chapel. *European Cultural Heritage Newsletter on Research* 9 (2): 7–32.

Camuffo, D., P. Brimblecombe, R. Van Grieken, H.-J. Busse, G. Sturaro, A. Valentino, A. Bernardi et al.
1999 Indoor air quality at the Correr Museum, Venice, Italy. *The Science of the Total Environment* 236 (1–3): 135–52.

Camuffo, D., G. Sturaro, and A. Valentino
1999 Thermodynamic exchanges between the external boundary layer and the indoor microclimate at the Basilica of Santa Maria Maggiore, Rome, Italy: The problem of conservation of ancient works of art. *Boundary-Layer Meteorology* 92 (2): 243–62.

Clarke, T.
2004 Music and chemistry: Organ failure. *Nature* 427:8–9.

Conrad, E. A.
1995 Energy conservation issues for modern buildings. In *Preserving the Recent Past,* ed. D. Slaton and R. Shiffer, 4, 137–40. Washington, DC: Historic Preservation Education Foundation.

———.
1996 Balancing environmental needs of the building, the collection, and the user. In *Abstracts of Papers Presented at the Twenty-Fourth Annual Meeting, Norfolk, Virginia, June 10–16, 1996, by the American Institute for Conservation,* 15–18. Washington, DC: American Institute for Conservation.

———.
1996 The dews and don'ts of insulating. *Old House Journal* 24 (3): 36–41.

———.
1999 The realistic preservation environment. Paper presented at the National Archives and Records Administration (NARA) 14th Annual Preservation Conference, "Alternative Archival Facilities." http://www.archives.gov/preservation/storage/realistic-preservation-environment.html.

Cook, J.
2001 Openness as a factor of comfort: Architectural choices from experience, not experiments. Paper presented at the Cooling Frontiers Symposium, Arizona State University, Tempe, 4–7 October 2001.

Crispino, L. C. B., ed.
2006 *As origens do Museu Paraense Emílio Goeldi: Aspectos históricos e iconográficos (1860–1921).* Belém, Brazil: Paka-Tatu.

Crowley, J. E.
2001 *The Invention of Comfort: Sensibilities and Design in Early Modern Britain and Early America.* Baltimore: Johns Hopkins University Press.

Daniel, V., S. King, C. Pearson, and I. Cole
2005 Stories from historic buildings. In ICOM Committee for Conservation, *14th Triennial Meeting, The Hague, 12–16 September 2005: Preprints,* ed. I. Verger, 625–31. London: James & James.

Daniel, V., C. Pearson, I. Cole, W. Ganther, and S. King
2000 Behaviour of museum buildings in tropical climates. In *Tradition and Innovation: Advances in Conservation: Contributions to the Melbourne Congress, 10–14 October 2000*, ed. A. Roy and P. Smith, 45–50. London: International Institute for Conservation of Historic and Artistic Works.

Dardes, K., and A. Rothe, eds.
1998 *The Structural Conservation of Panel Paintings: Proceedings of a Symposium at the J. Paul Getty Museum, 24–28 April, 1995*. Los Angeles: Getty Conservation Institute.

della Torre, S., and V. Pracchi, eds.
2003 *Le chiese come beni culturali: Suggerimenti per la conservazione*. Milan: Electra.

Dionisi Vici, P., P. Mazzanti, and L. Uzielli
2006 Mechanical response of wooden boards subjected to humidity step variations: Climatic chamber measurements and fitted mathematical models. *Journal of Cultural Heritage* 7 (1): 37–48.

Dupont, A.-L., and J. Tétreault
2000 Cellulose degradation in an acetic acid environment. *Studies in Conservation* 45 (3): 201–10.

Eames, R. M.
1980 The historic house in climate extremes: Problems and proposals. In *Conservation within Historic Buildings: Preprints of the Contributions to the Vienna Congress, 7–13 September 1980*, ed. N. S. Brommelle, G. Thomson, and P. Smith, 32–33. London: International Institute for Conservation of Historic and Artistic Works.

Environmental Protection Agency [EPA]
1992 EPA's approach for assessing the risks associated with chronic exposure to carcinogens. Background Document 2. Washington, DC: EPA. http://www.epa.gov/iris/carcino.htm.

———.
2004 *Air Quality Criteria for Particulate Matter*. 2 vols. EPA 600/P-99/002aF–bF. Washington, DC: EPA. http://cfpub.epa.gov/ncea/cfm/recordisplay.cfm?deid=87903.

———.
2005 Guidelines for carcinogen risk assessment. EPA 630/P-03/001F. Washington, DC: EPA. http://www.epa.gov/cancerguidelines/.

———.
2008 National ambient air quality standards (NAAQS). Washington, DC: EPA. http://www.epa.gov/ttn/naaqs/.

Eshøj, B., and T. Padfield
1993 The use of porous building materials to provide a stable relative humidity. ICOM Committee for Conservation, *10th Triennial Meeting: Washington, DC, 22–27 August 1993: Preprints*, ed. J. Bridgland, 605–9. Paris: ICOM Committee for Conservation.

European Commission Environment Leather Project
1996 *Deterioration and Conservation of Vegetable Tanned Leather*. Ed. R. Larsen. Hvidovre, Denmark: L. P. Nielsen.

Forest Products Laboratory [US]
2010 *Wood Handbook: Wood as an Engineering Material*. Centennial ed. Madison, WI: US Dept. of Agriculture, Forest Service, Forest Products Laboratory. http://purl.fdlp.gov/GPO/gpo11821.

Foster, M., and T. Oreszczyn
2001 Occupant control of passive systems: The use of venetian blinds. *Building and Environment* 36 (2): 149–55.

Gatley, D. P.
2005 *Understanding Psychrometrics*. 2nd ed. Atlanta: ASHRAE.

Givoni, B.
1994 Effectiveness of mass and night ventilation in lowering the indoor daytime temperatures. In *Architecture of the Extremes: Proceedings of the Eleventh International PLEA Conference, Dead Sea, Israel, July 3–8, 1994*, ed. Y. Etzion, 151–56. Sede-Boker, Israel: Published on behalf of the International PLEA Organization by Desert Architecture Unit, Jacob Blaustein Institute for Desert Research, Ben-Gurion University of the Negev.

Grzywacz, C. M.
2006 *Monitoring for Gaseous Pollutants in Museum Environments*. Los Angeles: Getty Conservation Institute.

Gut, P., and D. Ackerknecht
1993 *Climate Responsive Building*. Saint Gallen, Switzerland: SKAT.

Gysels, K., F. Delalieux, F. Deutsch, R. Van Grieken, D. Camuffo, A. Bernardi, G. Sturaro, H.-J. Busse, and M. Wieser
2004 Indoor environment and conservation in the Royal Museum of Fine Arts, Antwerp, Belgium. *Journal of Cultural Heritage* 5 (2): 221–30.

Harriman, L., and J. Lstiburek
2009 *The ASHRAE Guide for Buildings in Hot and Humid Climates*. 2nd ed. Atlanta: ASHRAE.

Henry, M. C., and K. Switala-Elmhurst
2001 Climate monitoring and management system for Drayton Hall: Project design document. Unpublished report. Watson & Henry Associates, Bridgeton, NJ, for the National Trust for Historic Preservation, Washington, DC.

Hödl, I.
2004 Schimmel und Gesundheit: Das 10-Punkte-Hygienepaket des Steiermärkischen Landesarchiv [Mould and health: The 10-point-hygiene package of the Steiermärkisches Landesarchiv]. In *Schimmel: Gefahr für Mensch und Kulturgut durch Mikroorganismen* [Fungi: A threat for people and cultural heritage through microorganisms], ed. A. Rauch, S. Miklin-Kniefacz, and A. Harmssen, 237–42. Bonn: VDR, Verband der Restauratoren.

Holmberg, J. G., and T. Burt
2000 Preventive conservation methods for historic buildings: Improving the indoor climate by changes to the building envelope. In *Conservation without Limits: IIC Nordic Group XV Congress, August 23–26, 2000, Helsinki, Finland*, ed. R. Koskivirta, 71–88. Loimaa, Finland: Loimaan Kirjapaino.

Huynh, C. K., H. Savolainen, T. Vu-Duc, M. Guillemin, and F. Iselin
1991 Impact of thermal proofing of a church on its indoor air quality: The combustion of candles and incense as a source of pollution. *Science of the Total Environment* 102:241–51.

International Institute for Conservation of Historic and Artistic Works [IIC]
2008 Climate change and museum collections. Roundtable discussion.

Jakieła, S., Ł. Bratasz, and R. Kozłowski
2007 Acoustic emission for tracing the evolution of damage in wooden objects. *Studies in Conservation* 52 (2): 101–9.

———.
2008 Acoustic emission for tracing fracture intensity in lime wood due to climatic variations. *Wood Science and Technology* 42 (4): 269–79.

———.
2008 Numerical modelling of moisture movement and related stress field in lime wood subjected to changing climate conditions. *Wood Science and Technology* 42 (1): 21–37.

Ke-Jian, Y.
1988 Short notice about the climate control of storage room. In *Control of Museum Climate in Asia and the Pacific Area: 18 September 1988, Kyoto, Japan*, ed. N. Kamba and S. Miura, 26–27. Kyoto: Japanese Organizing Committee of IIC Kyoto Congress.

Kerschner, R. L.
1992 A practical approach to environmental requirements for collections in historic buildings. *Journal of the American Institute for Conservation* 31 (1): 65–76.

———.
2000 Implementation of practical climate control strategies at the Shelburne Museum. In *The Conservation of Heritage Interiors* [La conservation des intérieurs patrimoniaux], 161–66. Ottawa: Canadian Conservation Institute.

———.
2006 Providing safe and practical environments for cultural properties in historic buildings . . . and beyond. Paper presented at the 20th Annual National Archives Preservation Conference, "Beyond the Numbers: Specifying and Achieving an Efficient Preservation Environment," March 16, 2006. http://www.archives.gov/preservation/conferences/2006/kerschner.pdf.

Kerschner, R. L., and J. Baker
1998 Practical climate control: A selected, annotated bibliography. http://cool.conservation-us.org/byauth/kerschner/ccbiblio.html.

King, S., V. Daniel, and C. Pearson
2000 Air conditioning small tropical museums: A technical note. *AICCM Bulletin* 25:33–36.

Kozłowski, R.
2000 Mikroklimat w zabytkowym drewnianym kościele w Dębnie Podhalańskim [Microclimate in the historic wooden church in Dębno Podhalańskie]. *Biuletyn informacyjny konserwatorów dzieł sztuki* 11 (4): 16–22, 96–99.

Kozłowski, R., C. Wittenburg, and A. Zeunert
2000 Stucco marble in the Dukes Chapel in Krzeszów, Poland: Microclimate and degradation mechanisms. *International Journal for Restoration of Buildings and Monuments* 6 (4): 463–80.

Künzel, H. M.
1991 *Bauphysikalische Untersuchungen in unbeheizten und beheizten Gebäuden alter Bauart A. Bauphysikalische Grundlagen und generelle Zusammenhänge über die Temperatur- und Feuchteverhältnisse auf Grund von Langzeituntersuchungen.* Stuttgart: Fraunhofer IRB Verlag.

———.
1994 Verfahren zur ein- und zweidimensionalen Berechnung des gekoppelten Wärme- und Feuchtetransports in Bauteilen mit einfachen Kennwerten. PhD diss., Universität Stuttgart.

Lafontaine, R. H.
1982 Humidistatically-controlled heating: A new approach to relative humidity control in museums closed for the winter season. *Journal of the International Institute for Conservation, Canadian Group* 7 (1–2): 35–41.

Landi, S.
1992 *The Textile Conservator's Manual.* 2nd ed. Butterworth-Heinemann Series in Conservation and Museology. Oxford: Butterworth-Heinemann.

Lawrence, M. G.
2005 The relationship between relative humidity and the dewpoint temperature in moist air: A simple conversion and applications. *Bulletin of the American Meteorological Society* 86 (2): 225–33.

Lloyd, H., and S. Staniforth
2000 Preventive conservation and "a madness to gaze at trifles": A sustainable future for historic houses. In *Tradition and Innovation: Advances in Conservation: Contributions to the Melbourne Congress, 10–14 October 2000*, ed. A. Roy and P. Smith, 118–23. London: International Institute for Conservation of Historic and Artistic Works.

Loomans, M. G. L. C.
1998 The measurement and simulation of indoor air flow. PhD diss., Universiteit Eindhoven.

Louw, H., and R. Crayford
1999 A constructional history of the sash-window, c. 1670–c. 1725 (part 2). *Architectural History* 42:173–239.

Lstiburek, J., and J. Carmody
1996 *Moisture Control Handbook: Principles and Practices for Residential and Small Commercial Buildings.* New York: John Wiley & Sons.

Maekawa, S., and V. Beltran
2004 Climate controls for historic buildings: A new strategy. *Conservation: The GCI Newsletter* 19 (1): 21–24.

Maekawa, S., C. Carvalho, and F. Toledo
2006 Um sistema para controle ambiental na biblioteca Rui Barbosa. Paper presented at Preservação do Patrimônio Cultural: Gestão e Desenvolvimento Sustentável: Perspectivas: XII Congresso Nacional da ABRACOR, Fortaleza, Ceará, Brasil, 28 de Agosto a 1 de Setembro de 2006.

Maekawa, S., F. Toledo, and R. Arraes
2006 Um sistema de controle climático para a reserva técnica etnográfica do Museu Paraense Emílio Goeldi, Belém, Pará. Paper presented at Preservação do Patrimônio Cultural: Gestão e Desenvolvimento Sustentável: Perspectivas: XII Congresso Nacional da ABRACOR, Fortaleza, Ceará, Brasil, 28 de Agosto a 1 de Setembro de 2006.

Maekawa, S., and N. Valentín
2003 Ventilation and heating-based microbial controls for cultural institutions. In *Moulds, Health and Heritage*, ed. A. Rauch. Braunschweig, Germany: Braunschweigisches Landesmuseum.

Mecklenburg, M. F., M. McCormick-Goodhart, and C. S. Tumosa
1994 Investigation into the deterioration of paintings and photographs using computerized modeling of stress development. *Journal of the American Institute for Conservation* 33 (2): 153–70.

Mecklenburg, M. F., and C. S. Tumosa
1998 Structural response of painted wood surfaces to changes in ambient relative humidity. In *Painted Wood: History and Conservation*, ed. V. Dorge and F. C. Howlett, 464–83. Los Angeles: Getty Conservation Institute.

Mecklenburg, M. F., C. S. Tumosa, and N. Wyplosz
1995 The effects of relative humidity on the structural response of selected wood samples in the cross-grained direction. In *Materials Issues in Art and Archaeology IV: Symposium Held May 16–21, 1994, Cancún, Mexico*, ed. J. R. Druzik, J. L. Galván

Madrid, I. C. Freestone, and G. S. Wheeler, 305–24. Warrendale, PA: Materials Research Society.

Michalski, S.
1993 Relative humidity in museums, galleries and archives: Specification and control. In *Bugs, Mold and Rot II: A Workshop on Control of Humidity for Health, Artifacts, and Buildings*, ed. W. B. Rose and A. TenWolde, 51–62. Washington, DC: National Institute of Building Sciences.

———.
1994 A systematic approach to preservation: Description and integration with other museum activities. In *Preventive Conservation: Practice, Theory and Research: Preprints of the Contributions to the Ottawa Congress, 12–16 September 1994*, ed. A. Roy and P. Smith, 8–11. London: International Institute for Conservation of Historic and Artistic Works.

———.
1995 Wooden artifacts and humidity fluctuation: Different construction and different history mean different vulnerabilities (chart, version 3.0). Ottawa: Canadian Conservation Institute.

———.
1996 Environmental guidelines: Defining norms for large and varied collections. In *Preservation of Collections: Assessment, Evaluation, and Mitigation Strategies: Papers Presented at the Workshop, Norfolk, Virginia, June 10–11, 1996*, 28–33. Washington, DC: American Institute for Conservation of Historic and Artistic Works.

———.
1996 Quantified risk reduction in the humidity dilemma. *APT Bulletin* 27 (3): 25–29.

———.
1997 Humidity and temperature guidelines for digital media. In *The Preservation of Digital Media: A Workshop Presented by the Canadian Association for Conservation of Cultural Property CAC in Association with the Association of Canadian Archivists: Abstracts*, 59–60. Ottawa: National Archives of Canada.

———.
1998 Climate control priorities and solutions for collections in historic buildings. *Historic Preservation Forum* 12 (4): 8–14.

———.
1999 *Relative Humidity and Temperature Guidelines for Canadian Archives*. Ottawa: Canadian Council of Archives and Canadian Conservation Institute.

———.
1999 Setting standards for conservation: New temperature and relative humidity guidelines are now published [Normes de conservation: Publication de nouvelles directives sur la température et l'humidité relative]. *CCI Newsletter = Bulletin de l'ICC* 24:3–4.

———.
2000 *Guidelines for Humidity and Temperature for Canadian Archives*. CCI Technical Bulletin no. 23. Ottawa: Canadian Conservation Institute.

———.
2003 Archives: Analog to digital: Preservation requirements. *ASHRAE Journal* 45 (10): 25–31.

Mills, M. J., and G. Fore
2000 Environmental monitoring and conservation study of Drayton Hall in Charleston, South Carolina. *APT Bulletin* 31 (2–3): 63–70.

Monowar Jahan, S. A. M.
1988 Hot and humid climate: A challenge for conservators. In *Control of Museum Climate in Asia and the Pacific Area: 18 September 1988, Kyoto, Japan*, ed. N. Kamba and S. Miura, 29–33. Kyoto: Japanese Organizing Committee of IIC Kyoto Congress.

Morey, P. R., J. C. Feeley, and J. A. Otten
1990 *Biological Contaminants in Indoor Environments*. ASTM Special Technical Publication 1071. Philadelphia: American Society for Testing and Materials.

Morey, P. R., M. J. Hodgson, W. G. Sorenson, G. J. Kullman, W. W. Rhodes, and G. S. Visvesvara
1984 Environmental studies in moldy office buildings: Biological agents, sources and preventive measures. *Annals of the American Conference of Governmental Industrial Hygienists* 10:21–35.

Morey, P. R., and B. A. Jenkins
1989 What are typical concentrations of fungi, total volatile organic compounds, and nitrogen dioxide in an office environment? In *The Human Equation: Health and Comfort: Proceedings of the ASHRAE/SOEH Conference, IAQ '89, April 17–20, 1989, San Diego, California*, 67–71. Atlanta: ASHRAE.

National Research Council. Committee on Indoor Pollutants
1981 *Indoor Pollutants*. Washington, DC: National Academies Press.

———.
1983 *Risk Assessment in the Federal Government: Managing the Process*. Washington, DC: National Academies Press.

Neilen, D., H. L. Schellen, and M. A. P. van Aarle
2003 Characterizing and comparing monumental churches and their heating performance. In *Research in Building Physics: Proceedings of the 2nd International Conference on Building Physics, 14–18 September, 2003, Antwerpen, Belgium*, ed. J. Carmeliet, H. S. L. C. Hens, and G. Vermeir, 793–802. Lisse, Netherlands: A. A. Balkema.

Neuhaus, E., and H. L. Schellen
2006 Conservation heating to control relative humidity and create museum indoor conditions in a monumental building. In *EPIC 2006 AIVC Proceedings: Technologies and Sustainable Policies for a Radical Decrease of the Energy Consumption in Buildings. The 4th European Conference on Energy Performance and Indoor Climate in Buildings. Conference of the IEA Programme on Energy Conservation in Buildings and Community Systems. The 27th Conference of the Air Infiltration and Ventilation Centre. Lyon, France, 20–22 November 2006*, ed. G. Guarracino, vol. 1, 45–50. Vaulx-en-Velin, France: École nationale des travaux publics de l'État.

Occupational Safety and Health Administration [US]
2006 Permissible exposure limits (PELs). https://www.osha.gov/dsg/topics/pel/index.html.

Olesen, B. W., and G. S. Brager
2004 A better way to predict comfort. *ASHRAE Journal* 46 (8): 20–28.

Olstad, T. M.
1994 Medieval wooden churches in a cold climate: Parish churches or museums? In *Preventive Conservation: Practice, Theory and Research: Preprints of the Contributions to the Ottawa Congress, 12–16 September 1994*, ed. A. Roy and P. Smith, 99–103. London: International Institute for Conservation of Historic and Artistic Works.

Olstad, T. M., A. Haugen, and T.-N. Nilsen
2001 *Polychrome Wooden Ecclesiastical Art: Climate and Dimensional Changes*. NIKU Publications 110. Oslo: Norwegian Institute for Cultural Heritage Research. http://www.niku.no/filestore/Publikasjoner/NIKUPublikasjoner110.pdf.

Oreszczyn, T.
2004 Our innate ability to think of new ways to use energy. *Energy and Environment* 15 (6): 1011–14.

Oreszczyn, T., and S. E. C. Pretlove
1999 Condensation Targeter II: Modelling surface relative humidity to predict mould growth in dwellings. *Building Services Engineering Research and Technology* 20 (3): 143–53.

Oreszczyn, T., I. Ridley, S. H. Hong, and P. Wilkinson
2006 Mould and winter indoor relative humidity in low income households in England. *Indoor and Built Environment* 15 (2): 125–35.

Padfield, T.
1987 The dangers of installing air conditioning in historic buildings. In *Old Cultures in New Worlds: Washington, District of Columbia, United States of America, October 10–15, 1987: Symposium Papers* [Cultures anciennes dans les mondes nouveaux: Washington, District of Columbia, États-Unis d'Amérique, 10–15 Octobre 1987: Communications du colloque], 432–39. Washington, DC: ICOMOS United States Committee.

———.
1994 The role of standards and guidelines: Are they a substitute for understanding a problem or a protection against the consequences of ignorance? In *Durability and Change: The Science, Responsibility, and Cost of Sustaining Cultural Heritage*, ed. W. E. Krumbein, P. Brimblecombe, D. E. Cosgrove, and S. Staniforth, 191–99. Chichester, UK: John Wiley.

———.
1996 Low-energy climate control in museum stores: A postscript. In *Proceedings of the ICOM-CC Conference, Edinburgh, September 1996*, ed. J. Bridgland, vol. 1, 68–71. London: James & James.

———.
1998 The role of absorbent building materials in moderating changes of relative humidity. PhD diss., Technical University of Denmark.

———.
1999 On the usefulness of water absorbent materials in museum walls. In ICOM Committee for Conservation, *12th Triennial Meeting, Lyon, 29 August–3 September 1999: Preprints*, ed. J. Bridgland, 83–87. London: James & James.

———.
2005 How to keep for a while what you want to keep for ever. http://www.conservationphysics.org/phdk/phdk_tp.php.

———.
2014 Conservation physics: Index. http://www.conservationphysics.org/.

Padfield, T., H. Berg, N. Dahlstrom, and A.-G. Rischel
2002 How to protect glazed pictures from climatic insult. In ICOM Committee for Conservation, *13th Triennial Meeting, Rio de Janeiro, 22–27 September 2002: Preprints*, ed. R. Vontobel, 80–85. London: James & James.

Padfield, T., and P. Jensen
1990 Low energy climate control in museum stores. In ICOM Committee for Conservation, *9th Triennial Meeting, Dresden, German Democratic Republic, 26–31 August 1990: Preprints*, ed. K. Grimstad, 596–601. Paris: ICOM Committee for Conservation.

Padfield, T., and J. S. Johnsen
1996 The breath of Arrhenius: Air conditioning in photographic archives. In *Research Techniques in Photographic Conservation: Proceedings of the Conference*

in Copenhagen, 14–19 May 1995, ed. M. S. Koch, T. Padfield, J. S. Johnsen, and U. B. Kejser, 59–63. Copenhagen: Royal Danish Academy of Fine Arts, School of Conservation.

Padfield, T., and P. K. Larsen
2004 How to design museums with a naturally stable climate. *Studies in Conservation* 49 (2): 131–37.

———.
2005 Low-energy air conditioning of archives. In ICOM Committee for Conservation, *14th Triennial Meeting, The Hague, 12–16 September 2005: Preprints*, ed. I. Verger, 677–80. London: James & James, Earthscan.

Padfield, T., and A. Robinson
1996 The climate behind pictures mounted against the outer walls of the chapel of Ledreborg, Denmark. In ICOM Committee for Conservation, *11th Triennial Meeting, Edinburgh, Scotland, 1–6 September, 1996: Preprints*, ed. J. Bridgland, 72–75. London: James & James.

Park, S. C.
1991 *Heating, Ventilating, and Cooling Historic Buildings: Problems and Recommended Approaches*. Preservation Briefs 24. Washington, DC: US Dept. of the Interior, National Park Service, Cultural Resources, Technical Preservation Services.

Perré, P., R. Rémond, and J. Gril
2006 Simulation of the effects of ambient variations. In *Mona Lisa: Inside the Painting*, 54–55. New York: Abrams.

Podany, J., M. L. Anderson, N. Bell, C. M. Grzywacz, S. Michalski, K. Colby Strothart, and T. Drayman-Weisser
2010 *The Plus/Minus Dilemma: The Way Forward in Environmental Guidelines*. Dialogues for the New Century. London: International Institute for Conservation of Historic and Artistic Works. https://www.iiconservation.org/sites/default/files/dialogues/plus-minus-en.pdf.

Podany, J., S. Staniforth, C. Sabbioni, M. Cassar, J. M. Reilly, M. C. Henry, and N. Serota
2008 *Climate Change and Museum Collections*. Dialogues for the New Century. London: International Institute for Conservation of Historic and Artistic Works. https://www.iiconservation.org/sites/default/files/dialogues/climate-change-en.pdf.

Pretlove, S. E. C., and T. Oreszczyn
1998 Climate change: Impact on the environmental design of buildings. *Building Services Engineering Research and Technology* 19 (1): 55–58.

Randl, C.
2005 *The Use of Awnings on Historic Buildings: Repair, Replacement, and New Design*. Preservation Briefs 44. Washington, DC: US Dept. of the Interior, National Park Service, Cultural Resources, Technical Preservation Services.

Reis Filho, N. G.
1970 *Quadro da arquitetura no Brasil*. Coleção Debates 18. São Paulo: Editôra Perspectiva.

Roaf, S., D. Crichton, and F. Nicol
2005 *Adapting Buildings and Cities for Climate Change: A 21st Century Survival Guide*. Oxford: Elsevier Architectural Press.

Rode, C., A. Holm, and T. Padfield
2004 A review of humidity buffering in the interior spaces. *Journal of Thermal Envelope and Building Science* 27 (3): 221–26.

Rogers, G. D.
1976 The ideal of the ideal environment. *Journal of the International Institute for Conservation, Canadian Group* 2 (1): 34–39.

Roy, A., and P. Smith, eds.
1994 *Preventive Conservation: Practice, Theory and Research: Preprints of the Contributions to the Ottawa Congress, 12–16 September 1994*. London: International Institute for Conservation of Historic and Artistic Works.

Royal Ontario Museum
1979 *In Search of the Black Box: A Report on the Proceedings of a Workshop on Microclimates Held at the Royal Ontario Museum, February, 1978*. Toronto: Royal Ontario Museum.

Ryhl-Svendsen, M., T. Padfield, V. A. Smith, and F. De Santis
2003 The indoor climate in historic buildings without mechanical ventilation systems. In *Healthy Buildings 2003: Energy-Efficient Healthy Buildings: Proceedings of ISIAQ, 7th International Conference, 7th–11th December 2003, Singapore*, ed. T. K. Wai, C. Sekhar, and D. Cheong, 278–83. Singapore: National University of Singapore. http://www.conservationphysics.org/ppubs/indclim.pdf.

Salkinoja-Salonen, M. S., J. Peltola, M. A. Andersson, and C. Saiz-Jimenez
2003 Microbial toxins in moisture damaged indoor environment and cultural assets. In *Molecular Biology and Cultural Heritage: Proceedings of the International Congress on Molecular Biology and Cultural Heritage, 4–7 March 2003, Sevilla, Spain*, ed. C. Saiz-Jimenez, 93–105. Lisse, Netherlands: Balkema.

Sandwith, H., and S. Stainton, eds.
1984 *The National Trust Manual of Housekeeping*. London: A. Lane in association with the National Trust.

Sangiorgi, F.
2004 Recovery and reuse of traditional rural buildings in Italy. In *First International Conference on Science and Technology in Archaeology and Conservation: August 12–17, 2002, Jordan: Actas*, ed. T. S. Akasheh, 401–11. Granada: Fundación el Legado Andalusí.

Santamouris, K., and P. Wouters
2006 *Building Ventilation: The State of the Art*. London: Earthscan.

Schellen, H. L., D. Neilen, A. W. M. van Schijndel, and M. A. P. van Aarle
2004 Indoor air humidity of monuments and hygrothermal surface conditions. Paper presented at the 6th Indoor Air Quality Meeting (IAQ 2004), Padua, Italy, 10–12 November 2004. http://www.isac.cnr.it/iaq2004/.

Schijndel, A. W. M. van, and H. L. Schellen
2006 Application of a combined indoor climate and HVAC model for the indoor climate performance of a museum. *International Journal for Restoration of Buildings and Monuments* 12 (3): 219–28.

Schijndel, A. W. M. van, H. L. Schellen, and W. J. Timmermans
2010 Simulation of the climate system performance of a museum in case of failure events. *Energy and Buildings* 42 (10): 1790–96.

Schoffelen, M. E. A., D. Neilen, and H. L. Schellen
2004 Measurement set-up for the verification of CFD modeling used in a design study for a local bench heating system for churches. In *21st PLEA Conference, September 19–21, 2004, Eindhoven, the Netherlands: Proceedings*, ed. M. H. de Wit, vol. 2, 821–26. Eindhoven: Technische Universiteit Eindhoven.

Schotes, P.
1972 Kirchenheizung: Bericht über die Tagung der Diözesanbaumeister und Baureferenten in Mainz. *Das Münster* 25 (4): 257.

Siau, J. F.
1995 *Wood: Influence of Moisture on Physical Properties*. Blacksburg, VA: Virginia Polytechnic Institute and State University, Dept. of Wood Science and Forest Products.

Spolnik, Z., A. Worobiec, J. Injuk, D. Neilen, H. Schellen, and R. Van Grieken
2004 Chemical characterization of airborne particles in St. Martinus Cathedral in Weert, the Netherlands. *Microchimica Acta* 145 (1–4): 223–27.

Staniforth, S.
1990 Benefit versus costs in environmental control. In *Managing Conservation: Papers Given at a Conference Held Jointly by the United Kingdom Institute for Conservation and the Museum of London, October 1990*, ed. S. Keene, 2–30. London: UK Institute for Conservation.

———.
1995 Environmental control in historic buildings. *Transactions: Association for Studies in the Conservation of Historic Buildings* 20:33–42.

———.
1995 Preventive conservation in National Trust houses. In *International Symposium on the Conservation and Restoration of Cultural Property: Cultural Property and Its Environment, October 11–October 13, 1990*, 145–66. Tokyo: Tokyo National Research Institute of Cultural Properties.

———.
1997 Today's work for tomorrow's enjoyment: The role of preventive conservation. In *The Victoria Memorial Hall, Calcutta: Conception, Collections, Conservation*, ed. P. Vaughan, 12–28. Mumbai: Marg Publications, on behalf of the National Centre for the Performing Arts.

Staniforth, S., W. M. Ballard, E. N. Caner-Saltik, R. Drewello, I.-L. Eckmann, W. E. Krumbein, T. Padfield et al.
1994 Group report: What are appropriate strategies to evaluate change and to sustain cultural heritage? In *Durability and Change: The Science, Responsibility, and Cost of Sustaining Cultural Heritage*, ed. W. E. Krumbein, P. Brimblecombe, D. E. Cosgrove, and S. Staniforth, 217–24. Chichester, UK: John Wiley.

Staniforth, S., and I. Griffin
2000 Damp problems in the chapel at Cliveden. In *The Conservation of Heritage Interiors* [La conservation des intérieurs patrimoniaux], 177–84. Ottawa: Canadian Conservation Institute.

Staniforth, S., and B. Hayes
1987 Temperature and relative humidity measurement and control in National Trust houses. In *8th Triennial Meeting: Sydney, Australia, 6–11 September 1987: Preprints*, ed. K. Grimstad, 915–26. Los Angeles: Getty Conservation Institute.

Staniforth, S., B. Hayes, and L. Bullock
1994 Appropriate technologies for relative humidity control for museum collections housed in historic buildings. In *Preventive Conservation: Practice, Theory and Research: Preprints of the Contributions to the Ottawa Congress, 12–16 September 1994*, ed. A. Roy and P. Smith, 123–28. London: International Institute for Conservation of Historic and Artistic Works.

Staniforth, S., R. Kerschner, J. Ashley-Smith, J. Druzik, and J. Levin
2004 Sustainable access: A discussion about implementing preventive conservation. *Conservation: The GCI Newsletter* 19 (1): 10–16.

Staniforth, S., and K. Lithgow
2003 When conservator meets architect and engineer. In *Conservation of Historic Buildings and Their Contents: Addressing the Conflicts*, ed. D. Watt and B. Colston, 66–79. Shaftesbury, UK: Donhead Publishing.

Stewart, J., S. Julien, and S. Staniforth
2004 An integrated monitoring strategy at Chedworth Roman villa (Gloucestershire). In *Preserving Archaeological Remains in Situ? Proceedings of the 2nd Conference, 12–14 September 2001*, ed. T. J. P. Nixon, 179–87. London: Museum of London Archaeology Service.

Straube, J.
2006 *Moisture and Materials*. Building Science Digest 138. Westford, MA: Building Science Press. http://www.buildingscience.com/documents/digests/bsd-138-moisture-and-materials?topic=doctypes/digests.

Taylor, J., N. Blades, M. Cassar, and I. Ridley
2005 Reviewing past environments in a historic house library using building simulation. In ICOM Committee for Conservation, *14th Triennial Meeting, The Hague, 12–16 September 2005: Preprints*, ed. I. Verger, 708–15. London: James & James.

Tetreault, J., and E. Stamatopoulou
1997 Determination of concentrations of acetic acid emitted from wood coatings in enclosures. *Studies in Conservation* 42 (3): 141–56.

Toledo, F.
1999 Passive environmental control for museums in hot-humid climates: Guidelines for building design and fabric improvements. PhD diss., University of London.

Toledo, F., and C. Price
1998 A note on tropical, hot, and humid museums. *Journal of Conservation and Museum Studies* 4:11–15.

Trechsel, H. R., ed.
1994 *Moisture Control in Buildings*. ASTM Manual Series MNL 18. Philadelphia: ASTM.

Troi, A., and G. Hausladen
2002 Heating effects in South Tyrolean churches. In *Indoor Air 2002: Proceedings, 9th International Conference on Indoor Air Quality and Climate, Monterey, California, June 30–July 5, 2002*, ed. H. Levin, G. Bendy, and J. Cordell, 848–53. Santa Cruz, CA: Indoor Air 2002.

Ughenu, V.
2000 Museums as a source of history: The National Museum of Colonial History, Aba. In *Museums and History in West Africa*, ed. C. D. Ardouin and E. Arinze, 49–52. Washington, DC: Smithsonian Institution Press.

University of Michigan
2006 Threats to global biodiversity. University of Michigan Global Change Program. http://www.globalchange.umich.edu/globalchange2/current/lectures/biodiversity/biodiversity.html.

Valentín, N., R. García, J. L. Ibañez, and S. Maekawa
2001 Tratamientos con ventilación controlada para detener el crecimiento microbiano en materiales de archivo. *Archivamos: Boletín ACAL* 39–40:40–44.

———.
2002 Air ventilation for arresting microbial growth in archives. In *La conservation à l'ère du numérique: Actes des quatrièmes journées internationales d'études de l'ARSAG, Paris, 27–30 mai 2002*, 139–50. Paris: Association pour la recherche scientifique sur les arts graphiques.

Valentín, N., R. García, O. de Luis, and S. Maekawa
1998 Microbial control in archives, libraries and museums by ventilation systems. *Restaurator* 19 (2): 85–107.

Velthem, L. Hussak van, F. L. Toledo, A. Benchimol, R. Arraes, and R. Souza
2004 A coleção etnográfica do Museu Paraense Emílio Goeldi: Memória e conservação. *MUSAS: Revista Brasileira de Museus e Museologia* 1:121–34.

Waller, R.
1994 Conservation risk assessment: A strategy for managing resources for preventive conservation. In *Preventive Conservation: Practice, Theory and Research: Preprints of the Contributions to the Ottawa Congress, 12–16 September 1994*, ed. A. Roy and P. Smith, 12–16. London: International Institute for Conservation of Historic and Artistic Works.

Waller, R., and S. Michalski
2005 A paradigm shift for preventive conservation, and a software tool to facilitate the transition. In ICOM Committee for Conservation, *14th Triennial Meeting, The Hague, 12–16 September 2005: Preprints*, ed. I. Verger, 733–38. London: James & James, Earthscan.

Watkinson, D.
2010 Preservation of metallic cultural heritage. In *Management and Control of Corrosion*. Vol. 4 of *Shreir's Corrosion*, ed. T. J. A. Richardson, R. A. Cottis, and L. L. Shreir, 3307–40. Amsterdam: Elsevier.

Wit, M. H. de
2001 WAVO: A simulation model for the thermal and hygric performance of a building. FAGO report, Technische Universiteit Eindhoven.

———.
2006 *HAMBASE: Heat, Air and Moisture Model for Building and Systems Evaluation*. Bouwstenen 100. Eindhoven: Technische Universiteit Eindhoven, Faculteit Bouwkunde. http://alexandria.tue.nl/repository/books/709317.pdf.

Wit, M. H. de, and G. J. Donze
1990 A second order model for the prediction of indoor air humidity. NOVEM contract FAGO-W/E, 90.08.W, April 1990, Technische Universiteit Eindhoven.

Wit, M. H. de, and H. H. Driessen
1988 ELAN—a computer model for building energy design. *Building and Environment* 23 (4): 285–89.

Zhou, B.
1988 The measures of controlling the environment in museums. In *Control of Museum Climate in Asia and the Pacific Area: 18 September 1988, Kyoto, Japan*, ed. N. Kamba and S. Miura, 21–25. Kyoto: Japanese Organizing Committee of IIC Kyoto Congress.

Index

Page numbers followed by *f* and *t* indicate figures and tables, respectively.

N

O

P

R

S

About the Authors

Shin Maekawa is a senior scientist at the Getty Conservation Institute, conducting research on microenvironmental technologies for the conservation of cultural objects and the built environments of heritage buildings. Previously he conducted research on oxygen-free environments for conservation. He holds an MS in mechanical engineering and a PhD in conservation science. He is a registered engineer with the State of California.

Vincent L. Beltran is an assistant scientist at the Getty Conservation Institute. In addition to his work on environmental management in hot and humid climates, his research has focused on the effect of anoxia on light-induced color change and an assessment of the transportation environment for cultural heritage objects. He has a BS from the University of California, Los Angeles, and an MS from the University of Hawaii at Manoa.

Michael C. Henry, PE, AIA, is principal engineer/architect with Watson & Henry Associates, Bridgeton, New Jersey. He specializes in the preservation of historic buildings and museum environments and in the engineered stabilization of large artifacts. He is adjunct professor of architecture in the Graduate Program in Historic Preservation at the University of Pennsylvania, Philadelphia, and is a guest lecturer at the Winterthur/University of Delaware Program in Art Conservation. He was a 2006 Fulbright Distinguished Scholar at the Centre for Sustainable Heritage at University College London.